Geophysical Monograph Series

Geophysical Monograph Series

Geophysical Monograph 254

Dynamic Magma Evolution

Francesco Vetere
Editor

This Work is a co-publication of the American Geophysical Union and John Wiley and Sons, Inc.

AGU
ADVANCING EARTH
AND SPACE SCIENCE

WILEY

This Work is a co-publication between the American Geophysical Union and John Wiley & Sons, Inc.

This edition first published 2021 by John Wiley & Sons, Inc., 111 River Street, Hoboken, NJ 07030, USA and the American Geophysical Union, 2000 Florida Avenue, N.W., Washington, D.C. 20009

Published under the aegis of the AGU Publications Committee

Brooks Hanson, Executive Vice President, Science
Carol Frost, Chair, Publications Committee
For details about the American Geophysical Union visit us at www.agu.org.

Wiley Global Headquarters
111 River Street, Hoboken, NJ 07030, USA

For details of our global editorial offices, customer services, and more information about Wiley products visit us at www.wiley.com.

Library of Congress Cataloging-in-Publication data has been applied for

ISBN: 9781119521136 [hardback]

Cover Design: Wiley
Cover Images: Magma © Unsplash

Set in 10/12pt Times New Roman by SPi Global, Pondicherry, India

Printed and bound by CPI Group (UK) Ltd, Croydon, CR0 4YY

10 9 8 7 6 5 4 3 2 1

CONTENTS

CONTRIBUTORS

Tamara L. Carley
Department of Geology and Environmental Geosciences, Lafayette College, Easton, PA, USA

Lissie Connors
Department of Geology and Environmental Geosciences, Lafayette College, Easton, PA, USA

Luca Caricchi
Department of Earth Sciences, University of Geneva, Geneva, Switzerland

Adrian Fiege
Department of Earth and Planetary Sciences, American Museum of Natural History, New York, NY, USA

Letizia Giuliani
Department of Engineering and Geology (InGeo), University G. d'Annunzio of Chieti-Pescara, Italy

François Holtz
Institute of Mineralogy, Leibniz University Hannover, Hannover, Germany

Gianluca Iezzi
Department of Engineering and Geology (InGeo), University G. d'Annunzio of Chieti-Pescara, Italy and National Institute of Geophysics and Volcanology (INGV) Rome, Italy

Jürgen Koepke
Institute of Mineralogy Leibniz University Hannover, Hannover, Germany

Marc-Antoine Longpré
School of Earth and Environmental Sciences, Queens College, City University of New York, NY, USA

Silvio Mollo
National Institute of Geophysics and Volcanology (INGV), Rome, Italy and Department of Earth Sciences, Sapienza University of Rome, Rome, Italy

Sabrina Nazzareni
Department of Physics and Geology, University of Perugia, Perugia, Italy

Maurizio Petrelli
Department of Physics and Geology, University of Perugia, Perugia, Italy

Caterina Petrillo
Department of Physics and Geology, University of Perugia, Perugia, Italy and National Institute for Nuclear Physics (INFN), Perugia Section, Perugia, Italy

Stefano Rossi
Department of Physics and Geology, University of Perugia, Perugia, Italy

Francesco Sacchetti
Department of Physics and Geology, University of Perugia, Perugia, Italy

Nobumichi Shimizu
Woods Hole Oceanographic Institution, Woods Hole, MA, USA

John Stix
Earth and Planetary Sciences, McGill University, Montreal, Quebec, Canada

Atsushi Toramaru
Department of Earth and Planetary Sciences, Kyushu University, Fukuoka, Japan

Francesco Vetere
Department of Physics and Geology, University of Perugia, Perugia, Italy

Shunsuke Yamashita
Department of Earth and Planetary Sciences, Kyushu University, Fukuoka, Japan. Now at Idemitsu Kosan Co., Ltd., Tokyo, Japan

Marco Zanatta
Department of Informatics, University of Verona, Verona, Italy

Georg F. Zellmer
Volcanic Risk Solutions, Massey University, Palmerston North, New Zealand

Chao Zhang
Institute of Mineralogy Leibniz University Hannover, Hannover, Germany and State Key Laboratory of Continental Dynamics, Northwest University, Xi'an, China

PREFACE

How and when a volcano will erupt, and the damage and disruption it may cause, is a question of great scientific and public concern. However, up to now we do not have the relevant knowledge to answer this fundamental question. While volcanic events can be catastrophic, they remain mostly unpredictable. More insights are needed into the key mechanisms involved in these dramatic natural events.

One of the areas requiring greater study is understanding the magmatic processes responsible for the chemical and textural signatures of volcanic products and igneous rocks. This includes investigating both equilibrium scenarios (e.g., phase stabilities, element partitioning) and kinetically controlled events such as the function of changing magma chemistry (replenishment), cooling, decompression (e.g., crystal and bubble nucleation and growth), devolatilization, and shear stress.

The coexistence of crystals, bubbles, and liquid phases needs to be considered as a dynamic environment in which temporal evolution depends on the system's characteristic features (e.g., bulk composition) and on its thermodynamic conditions. In this scenario, kinetics play a major role allowing possible liquid to solid phase transitions occurring at different times. Knowing the rate at which such transformations occur will provide fundamental information as it defines the eruption behavior of volcanoes. Kinetic processes are, indeed, responsible for abrupt changes in the rheological behavior of magmas, resulting in rapid and unexpected changes from low- to high-energy eruptions. It is then obvious that a better understanding of magma dynamics will improve our abilities to monitor volcanic activity and mitigate its potentially devastating impact.

The complexity of the physico-chemical processes involved during active volcanism and dynamic magmatism, which cover wide ranges of dimensions (from submicron to intercontinental) and timescales (from seconds during explosive eruptions to thousands of years for volcano-tectonic processes), demands multidisciplinary research approaches. Covering all scientific aspects related to this topic is far beyond the scope of this book, but this compilation of geochemical, petrological, physical, and thermodynamic studies provides a comprehensive overview about recent progress, often achieved using unconventional and novel techniques. The book can be a useful compendium for lecturers and students, as well as a reference for researchers developing new and innovative ideas and projects.

This book is divided into two sections. The first section (chapters 1–4) focuses on geochemical, petrological, and geophysical studies to understand dynamic magma evolution and the timescales involved, including processes such as magma recharge, mixing, and mingling. The second section (chapters 5–9) focuses on physical, thermodynamic, and theoretical approaches to investigating magmas according to the different geological scenarios that occur naturally.

Chapter 1 by Petrelli and Zellmer deals with rates and timescales of crustal magma transfer, storage, emplacement, and eruption. The authors review the most pertinent unresolved questions in this field, and highlight geochemical and geophysical methods that are available to address these questions. Long-storage magmatic timescales are discussed in detail as well as the influence of volatiles in volcanic environments and possible storage depths or ascent rates.

Chapter 2 by Longpré, Stix, and Shimizu, deals with melt inclusions as a unique material to retrieve information on the volatile budgets and compositional diversity of magmatic systems. Their results document a rare case of boundary-layer melt entrapment in natural magmas, indicating that melt inclusions hosted by fast-grown spinel are not a reliable record for melt evolution. The authors highlight the importance of boundary-layer melt entrapments as a tracer for dynamic magmatic processes.

Chapter 3 by Connors, Carley, and Fiege is dedicated to the properties of apatite as a tool for understanding the evolution of volatile and trace elements in magmatic systems. Torfajökull, a historically active Icelandic volcano erupting apatite-bearing magmas, is a key study site since it exhibits a unique compositional history, showing transitioning from more-evolved peralkaline rhyolites (Pleistocene) to less-evolved metaluminous rhyolites (Holocene). This study reveals the potential of apatite to elucidate magmatic and volcanic processes in Iceland.

Chapter 4 by Yamashita and Toramaru investigates the crystal size distribution (CSD) of plagioclase phenocrysts in four historical lavas from Sakurajima volcano, southern Kyushu, Japan. The results reveal a correlation between CSD and geological data (volumes and intervals between eruptions) for large historical eruptions, suggesting that the supply rate from the mantle controls the triggering of eruptions.

Chapter 5 by Giuliani, Iezzi and Mollo investigates the solidification of magmas occurring by cooling- ($\Delta T/\Delta t$)

and/or degassing-induced decompression ($\Delta P/\Delta t$), as a function of solidus temperature, glass transition temperature, and melting temperature, respectively. The study evaluates the influence of bulk chemical composition of a silicate liquid, *P*, *T*, fO_2, and H_2O on magma solidification. The most relevant solidification conditions of magmas leading to possible crystallization paths and relative physical models and reconstruction of magmatic intensive variables using mineral composition variability, are discussed. Chemical attributes of minerals are reviewed in order to discriminate their equilibrium and disequilibrium formations and, finally, using chemical composition of solid phases the authors reconstruct magmatic intensive variables.

Chapter 6 by Nazzareni, Rossi, Petrelli, and Caricchi is dedicated to the Main Ethiopian Rift (MER). The MER represents a young continental rift with an associated large volume of magmatism that forms one of the major large igneous provinces. The authors used clinopyroxene crystals in order to record variations of *P*, *T*, and fO_2 with the aim of reconstructing the geological history of host rocks. Clinopyroxene geobarometry was performed combining X-ray diffraction with mineral chemistry to highlight a complex polybaric plumbing system active since 7.5–3.7 Ma. The continuous polybaric MER clinopyroxene crystallization from the lower to middle crust is explained by a plumbing system composed of a dyke complex where magmas rise, stall, and, finally, crystallize.

In Chapter 7, Vetere and Holtz compare viscosity data from three different compositions: basalt, andesite, and a synthetic pyroxenite. In addition to the determination of melt viscosities, viscosity data were collected at subliquidus conditions in partially crystallized systems, under controlled shear rates of 0.1 and 1 s^{-1}. Experimental data show that changes of shear rates from 0.1 and 1 s^{-1} may cause a viscosity difference of half to one order of magnitude, pointing to the so called "shear-thinning effect." The authors suggest that this effect should be taken into account when considering magmatic processes occurring in volcanic conduits as it could drastically change the dynamics of the magmatic system.

Chapter 8 by Zanatta, Petrillo, and Sacchetti propose the time-resolved neutron diffraction as a tool for the study of crystallization kinetics of glasses and supercooled liquids. In the situation of multiple crystallizing phases, the analysis of the Bragg peaks can provide an unambiguous identification of the crystal structures with proper lattice parameters and crystallization timescales for each phase. Results show that the isotopic sensitivity of neutrons could be exploited to highlight a single species with respect to the surrounding medium, thus facilitating data interpretation for complex systems such as ascending magmas. Neutron-based techniques are particularly suitable to measure bulk samples controlling environmental parameters such as *T* and *P*. This can be pivotal for geological studies aiming at in situ measurements during time-dependent processes such as crystallization in magmas. Moreover, an empirical model for the interpretation of the crystallization kinetics in supercooled liquids is presented.

Chapter 9 by Koepke and Zhang sheds new light on the complexity of magmatic and metamorphic processes ongoing within and at the roof of axial melt lenses (AMLs), with a focus on the petrological and geochemical record provided by fossilized AMLs. The International Ocean Discovery Program (IODP) Hole 1256D in the equatorial Pacific is the location, where for the first time, the transition between sheeted dikes and gabbros in an intact fast spreading crust was penetrated, providing a core with a continuous record of the upper part of an AML. This location can be regarded as Rosetta Stone to answer long-standing questions on the complex magmatic evolution within an AML, as well as on metamorphic and anatectic processes ongoing at the roof of a dynamic AML.

The topics presented in this book cover a wide range of observations and measurements on dynamic magmatic processes, merging insights from fieldwork and experimental results, theoretical approaches, and computational modeling.

The coexistence of crystals and melt must be seen as a dynamic process for which time evolution depends on various intrinsic (e.g., chemical compositions, temperature, pressure) and extrinsic parameters (e.g, tectonic environment). Kinetics play a fundamental role allowing a variety of possible phase transitions: liquid to solid, liquid immiscibility or exsolution in minerals, mineral reactions, and formation of compositional zoning. Knowing the rate at which such transformation occurs is fundamental as it defines the eruption behavior in volcanoes. Kinetic processes are responsible for abrupt changes in the rheological behavior of magmas, resulting in low- to high-energy eruptions. Thus, a better understanding of magma dynamics will improve our ability to monitor and forecast volcanic activity. The results presented in this book represent another step forward in this direction.

Francesco Vetere, *University of Perugia*
January 2020

Part I
Timescales and Time Sensor

1

Rates and Timescales of Magma Transfer, Storage, Emplacement, and Eruption

Maurizio Petrelli[1] and Georg F. Zellmer[2]

ABSTRACT

The rates and timescales of crustal magma transfer, storage, emplacement, and eruption are a key to understanding subvolcanic processes, characterizing volcanic hazards, and developing mitigation strategies. In this chapter, we review the most pertinent open questions in this field, as well as the many geochemical and geophysical methods that are available to address these questions. Results point to long storage timescales, of up to ~10^6 years, in deep (i.e., ~20–30 km), crustal hot zones. Estimated ascent velocities from deep reservoirs to shallower systems span a vast range of ~10 orders of magnitude, and are a function of the thermophysical parameters of the ascending magma (e.g., density, viscosity, and overpressures in the reservoirs) and the host rocks. At mid- to upper crustal levels (i.e., < 15–20 km), we elucidate the cold storage of magma mushes for long periods, which can be unlocked during short-term events to form ephemeral magma chambers. Unlocking timescale estimates range from minutes to thousands of years, indicating a variability of about ~8 to ~10 orders of magnitude. This large variation results from the interplay among many processes, often nonlinearly coupled, occurring before an eruption. For example, exsolved volatile species have a significant role in modulating preeruptive dynamics and relative timescales; they increase the buoyancy of magmas, affect phase equilibria, promote convective dynamics, and may ultimately trigger eruptive events. As a consequence, understanding the role of volatiles in subvolcanic magmatic processes, including ascent rates, storage mechanisms, and relative timescales, will be paramount for future studies.

1.1. INTRODUCTION

There are about 500 active volcanoes on Earth with approximately 500 million people living on or close to them (Siebert et al., 2011, 2015). On average, about 50 volcanoes erupt each year (Siebert et al., 2015), generating a direct hazard for the people living in their surroundings. For example, the 1815 eruption of Tambora volcano, Indonesia, killed more than 90,000 people, including deaths resulting from subsequent crop loss and famine (Stothers, 1984). Taking into account population growth through time, an equivalent eruption today, in populated areas (e.g., Indonesia and the Philippines), would potentially involve more than 20 million people (Newhall et al., 2018). Further, explosive eruptions constitute a significant economic risk on a global scale. The modest ash cloud generated by Eyjafjallajökull volcano (Iceland) in April 2010 paralyzed the air traffic of a large part of Europe for about 1 week (Sigmundsson et al., 2010), causing an economic loss of several billion Euros as millions of travelers were grounded and industrial production was decelerated or halted. Often, explosive eruptions occur with virtually no warning time or with only weak geophysical precursors. Examples are the eruptions of Chaitén (2 May 2008; Castro & Dingwell,

[1] *Department of Physics and Geology, University of Perugia, Perugia, Italy*

[2] *Volcanic Risk Solutions, Massey University, Palmerston North, New Zealand*

Dynamic Magma Evolution, Geophysical Monograph 254, First Edition. Edited by Francesco Vetere.

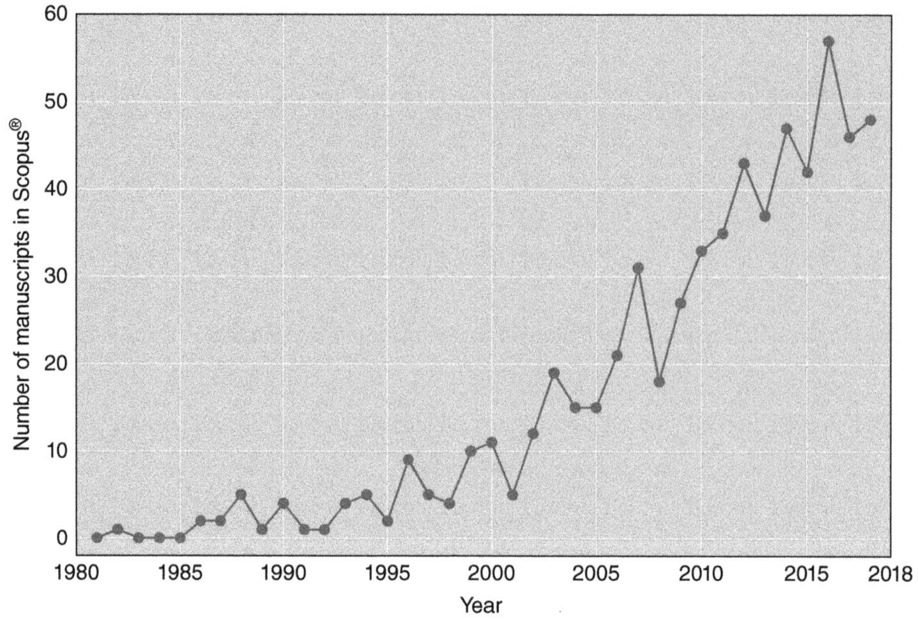

Figure 1.1 Number of manuscripts published from 1980 to 2018 containing the words "magma" and "timescales" in title, abstract, or keyword fields. [*Source*: Scopus®).]

2009; Wicks et al., 2011) and Calbuco volcanoes (22 April 2015; Arzilli et al., 2019; Romero et al., 2016) in Chile.

Understanding the processes occurring in crustal sections below active volcanoes and their time evolution is therefore essential in the provision of prompt and reliable information to decision makers and institutions in order to define appropriate hazard mitigation plans. To achieve this goal, we need to improve our knowledge of rates and timescales of the processes occurring below active volcanoes before an eruption. It is of note that the number of manuscripts involving estimates of timescales for magmatic processes is now approaching more than 50 per year (Figure 1.1); some well-documented review studies are included among these (K.M. Cooper, 2019; Hawkesworth et al., 2004; S. Turner & Costa, 2007). One chapter of the *Treatise on Geochemistry* (Turekian & Holland, 2013) is about the timescales of magma transfer (i.e., the process of magma migration, typically from deeper to shallower crustal regions) and magma storage (i.e., the mechanisms and the processes governing the stagnation of magmas at a fixed depth) in the crust (Reid, 2003). Four chapters of *The Encyclopedia of Volcanoes* (2nd edition; Sigurdsson, 2015) deal with the modeling and timescale estimations of melt migration (Daines & Pec, 2015), magma ascent and storage (Browne & Szramek, 2015), magma transport in dikes (Gonnermann & Taisne, 2015), and magma ascent and degassing at shallow levels (Burgisser & Degruyter, 2015). Also, focused reviews and extensive studies are available on the timescales for specific fields of study and techniques (e.g., Bachmann, 2010; K.M. Cooper, 2019). Examples are the modeling of zoning patterns in crystals (Costa et al., 2008; S. Turner & Costa, 2007), uranium series isotopes (K.M. Cooper & Reid, 2008), magma ascent rates (Burgisser & Degruyter, 2015; Crisp, 1984; Gonnermann & Manga, 2013; Rutherford, 2008; Scandone et al., 2007), magma degassing (M.B. Turner et al., 2013), volatiles accumulation (Petrelli et al., 2018), and pluton assembly (Caricchi et al., 2012). Finally, entire books are focused on the evolution of volcanic plumbing systems (defined as the structural framework of pathways and storage regions through which magma travels on its journey from its source region to the Earth's surface; Burchardt, 2018), timescales of magmatic processes (Dosseto et al., 2010), and petrochronological methods (Kohn et al., 2018).

All of these works contributed to unblurring our vision of preeruptive processes, and their evolution in time and space. In detail, early models of volcanic plumbing systems considered a long-lived, melt-dominated magma chamber, typically emplaced at shallow (i.e., below ~10 km depth) crustal levels (Montagna et al., 2015; Oldenburg et al., 1989). The simplest models assume the magma chamber to behave as a closed system, filled by a Newtonian melt (Oldenburg et al., 1989), where crystals nucleate and progressively grow (Petrelli et al., 2016). Other models, in addition, account for periodic refilling with new magma, thus considering the magma chamber as an open system (Annen, 2009). The most evolved models further account for the non-Newtonian rheology of magmas, as well as volatile saturation and exsolution, occasionally including the effects of overpressure in the system due to, for example, refilling with new magma,

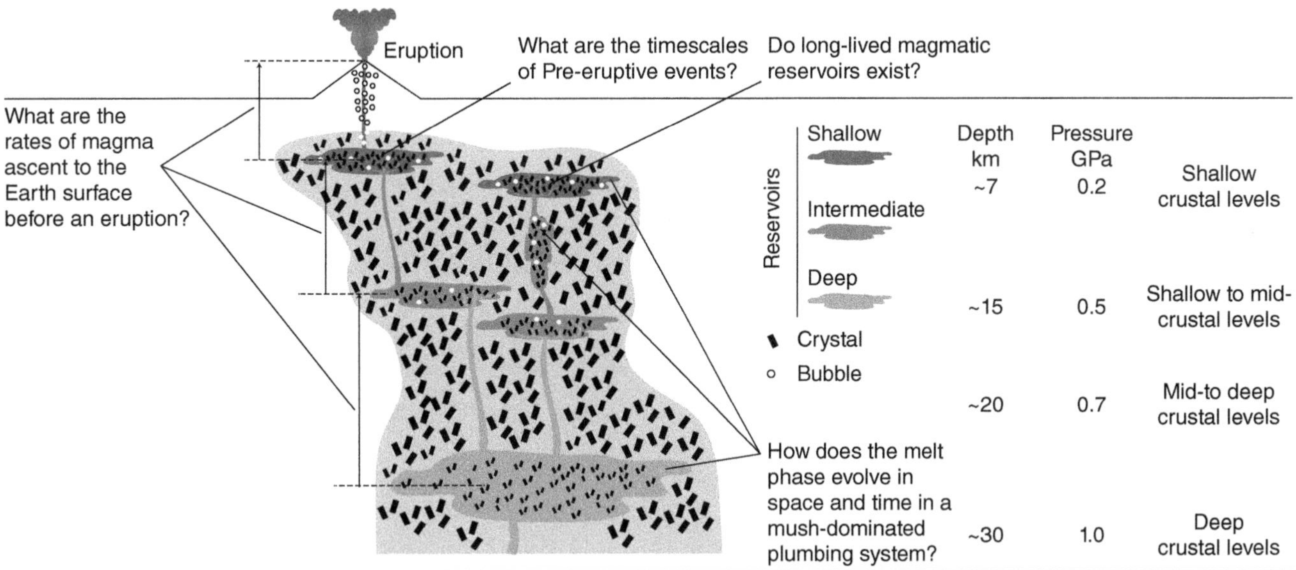

Figure 1.2 Simplified scheme of a mush-dominated volcanic plumbing system developing from deep crustal levels to shallow depthsand its relationship to the four main open questions addressed by this chapter.

volatile exsolution, and regional stresses (Dufek & Bachmann, 2010; Parmigiani et al., 2016; Petrelli et al., 2018; Ruprecht et al., 2008).

Many lines of evidence suggest, however, that the conceptual model of shallow magma chambers as melt-dominated domains is unrealistic (Cashman et al., 2017; Sparks & Cashman, 2017). Further, physical models based on thermal, mechanical, and dynamical principles suggest that long-lived and melt-dominated magma chambers are unlikely to develop or to be maintained, especially in the upper crust (Annen et al., 2015). As a consequence, magmatic reservoirs (i.e., the physical locus hosting magmas that may be melt-dominated or crystal-rich) have to be considered as complex systems formed mainly of crystal mushes, where liquid-dominated regions (i.e., volumes of eruptible magma) ephemerally develop from shallow to deep crustal levels (Bachmann & Huber, 2016; Cashman et al., 2017; K.M. Cooper & Kent, 2014; Edmonds et al., 2019; Sparks & Cashman, 2017; Zellmer & Annen, 2008), and where volatiles play a significant role (e.g., Bachmann & Bergantz, 2008; Edmonds & Wallace, 2017; Edmonds & Woods, 2018).

In this chapter, we present the methods and techniques currently utilized to unravel rates and timescales of magmatic processes. We then review the evidence that provides insights into magma ascent rates through the crust, timescales of magma storage, and the dynamic processes occurring in volcanic plumbing systems before eruptions. Our main goal is to link all the different clues that are currently utilized in the estimation of time-dependent magmatic processes, and to provide an inclusive vision of preeruptive dynamics with their associated rates and timescales. The questions that we will address are (Figure 1.2): How does the melt phase evolve in space and time in a mush-dominated magmatic plumbing system? Do long-lived magmatic reservoirs exist? What are the timescales of preeruptive events? What are the rates of magma ascent to the Earth surface before an eruption?

1.2. METHODS AND TECHNIQUES FOR ESTIMATION OF RATES AND TIMESCALES

In this section we present the methods and techniques currently utilized to unravel rates and timescales in magmatic systems, including: isotopic-decay dating, chemical-diffusion-based geospeedometers, crystal-size-distribution analysis, melt-inclusion timekeepers, thermal modeling of magmatic systems, mechanical modeling of magma ascent, and the modeling of the volatile phase. The fundamental principles of each technique are reported in Table 1.1, together with the main benefits, limitations, and uncertainties.

1.2.1. Isotopic-Decay Dating Techniques

Isotopic-decay dating techniques can be divided into those that allow the determination of eruption ages, and those that provide insight into preeruptive processes. The difference is principally due to the closure temperature of the elements and isotopes used. For the former, closure temperatures are significantly lower than magmatic temperatures, and the clock therefore starts ticking only at the point of rapid cooling due to eruption. For the latter, closure temperatures are at or close to magmatic temperatures, and preeruptive processes can therefore be timed if eruption ages are known or can be determined.

Table 1.1 Principles, Benefits, Limits, and Uncertainties of Methods and Techniques for Rates and Timescales Estimations. For the References, Please Refer to the Main Text

Method/ technique	Principles	Benefits	Limits	Uncertainties
K–Ar and Ar–Ar dating	Decay of ^{40}K to ^{40}Ar by electron capture; step heating allows ages to be obtained from partially altered rocks	Absolute dating of eruption age of K-bearing rocks, even if some parts have lost Ar due to alteration	Young rocks require high K contents to be datable; excess Ar problem, i.e. initial Ar content may be elevated	Relative uncertainties highly dependent on the age of the sample
Rb–Sr dating	Decay of ^{87}Rb to ^{87}Sr by electron emission	Absolute dating of eruption age of old samples	Rb and Sr are both fluid mobile large ion lithophile elements, and many systems do not remain closed for the long time periods measured	Variable, but typically millions of years for samples billions of years old
(U–Th)/He dating	Alpha decay from natural decay chains and ^{147}Sm producing ^{4}He	Determining rates and timescales of rock exhumation, mountain building and landscape evolution, with access to very young eruption ages in combination with U-series dating	When used to date volcanic eruptions, post-eruptive reheating events may significantly alter the results; also, radiation damage may influence the sensitivity of the (U-Th)/He system	Uncertainty of (U–Th)/He ages is typically in the range of 2–5% , expressed as 1σ
U-series dating	Radioactive decay from ^{238}U, ^{235}U, and ^{232}Th through a number of short-lived daughters to stable ^{206}Pb, ^{207}Pb and ^{208}Pb, respectively	Secular disequilibria introduced by geological processes can be dated; a range of intermediate isotopes with a range of half-lives provides access to a range of processes and ages; internal checks for consistency may be available through the use of different isotope pairs; inconsistencies may point to complexities	Typically, five times the half-life of the shortest-lived isotope involved; system needs to remain closed for this time; the geological processes introducing the disequilibria need to be short relative to the half-life of the isotope-pair investigated to provide age information without additional modeling	Dependent on degree of disequilibria displayed (high if close to the limit of five half-lives), and the half-life of the isotope-pair investigated
Geospeedometry	Diffusion of elements within or between crystals at magmatic temperatures	Dating high-T preeruptive processes; a range of elements with different diffusivities allows access to a range of timescales; provides constraints on magmatic temperatures when combined with radiometric determination of preeruptive ages	An initial concentration profile needs to be assumed; for dating purposes, temperatures need to be independently estimated	Absolute uncertainties are order of magnitudes, as diffusivities are typically not well enough constrained; relative uncertainties, i.e. difference between ages obtained, depend on age
Crystal-melt reactions	Reaction of crystals with surrounding melt due to crystal-melt disequilibrium; width of reaction rim provides yield time	Access to magma mixing processes that result in mineral–melt disequilibrium; access to very short timescales	Reaction rates need to be experimentally constrained, and are highly dependent on composition, temperature, and pressure of the system	Age-dependent, and unless experimental work done on system investigated, order-of-magnitude estimates
CSD analysis	Distribution of crystal sizes carries information on the timescales of crystallization	Simple systems provide access to time if growth rates can be constrained; further, CSDs can help reveal magmatic processes such as crystal fractionation, crystal accumulation, and magma mixing	To yield time information, growth rates need to be independently constrained, which is rarely the case	Order-of-magnitude or worse uncertainties, unless crystal growth rates are independently constrained

Concentration variance decay	Based on the concept of the time decay of the variance of chemical composition during magma mixing	Provides information about mixing-to-eruption timescales; statistically robust, since it considers multiple chemical elements, more than the few typically utilized by classical geospeedometers	Ability to track only the last mixing event before the eruption; requires "ad hoc" experiments	Constrained by experiments developed for each single case
Melt inclusion textural analysis	Based on quantitative three-dimensional textural analyses to calculate melt-inclusion faceting times	Determination of melt-inclusion faceting times provides information about the longevity of the magmatic system	Many different parameters must be independently constrained to provide significant results	On the order of 100%
Thermal modeling	Provides a wide range of analytical solutions and numerical models dealing with heat transfer during geological processes	Powerful tool based on a well known physical process; new numerical schemes and solutions are continuously developed	To provide reliable solutions, a careful selection of the adopted analytical solution and/or numerical scheme is required; initial and boundary conditions must be carefully constrained	Depending on the adopted analytical solution or numerical scheme
Mechanical modeling of volatile phase	Includes the numerical modeling of magma ascent and decompression, experimental constraints at high pressure and temperature, analogue modeling, and the measurement of bubble number density	Provides constraints on topical processes occurring during magma ascent and eruption	Behavior of volatile phases is still poorly understood, and requires further investigations	Depending on the approach adopted
Mechanical modeling of magma ascent	Models melt migration and magma ascent velocities	Provides insights about magma ascent patways, velocities and timescales; useful for inferring preeruptive dynamics	Requires knowledge of rheological properties of magma and surrounding hosting rocks; to achieve reliable results, knowledge of the provailing stress field is also required	Physical formulation and associated uncertainities are mainly related to the length-scale of the processes addressed
MND water exsolution rates	Based on the decompression-induced crystallization of microlites due to phase equilibria changes following water exsolution	Examining the link between the decompression process and the formation of microlites provides information on magma ascent and the resulting water exsolution processes	Main limitations are due to the disequilibrium vesiculation; kinetic constant must be determined accurately; not applicable to microlite-poor samples	Reproduces experimental MND by first order approximation
Seismic constraints	Spatial propagation of seismicity over time associated with magma movement	Direct access to rock fracturing in response to magma movement, and therefore direct access to ascent rates if earthquake locations can be constrained accurately	Constraining hypocentre locations accurately is dependent on good knowledge of subsurface seismic properties and geological structures; seismic signals of magma movement often too small to be detected	Dependent on the limits listed

1.2.1.1. Constraining Eruption Ages

1.2.1.1.1. K–Ar and Ar–Ar Dating Both K–Ar and Ar–Ar dating techniques are based on the decay of a naturally occurring isotope of potassium, ^{40}K, to ^{40}Ar (Kelley, 2002). The essential difference between K–Ar and Ar–Ar dating techniques lies in the measurement of potassium. In the K–Ar method, potassium is measured directly, whereas in the Ar–Ar technique, K is measured after the transformation of ^{39}K to ^{39}Ar by neutron bombardment (Kelley, 2002). The main advantage of the Ar–Ar technique over the K–Ar dating is that potassium and argon are effectively measured simultaneously (i.e., on the same aliquot of sample), providing a greater internal precision. Many efforts have been made to increase the precision of K–Ar dating (e.g., Gillot et al. 2006). Ar–Ar and K–Ar dating methods have been widely applied to estimate the ages of eruptive events (Bachmann, Oberli, et al., 2007; Bacon & Lanphere, 2006; Chenet et al., 2007; Deino et al., 2004; Iverson et al., 2014; Jicha et al, 2016; Lanphere, 2000; Muir et al., 2015; Petrosino et al., 2016; Reichow et al., 2002). As such, these methods are a prerequisite for determining preeruptive ages of crystallization and other magmatic processes through the use of radiometric clocks. However, dating young (< 20 ka) eruptive deposits remains a challenge, unless they are highly potassic, because of the long half-life of ^{40}K. As in all isotopic-decay dating techniques, the system needs to be closed to obtain reliable ages. Argon loss through alteration or high temperature events that can damage the mineral lattices will result in age estimates that are too young, while excess argon, which may be derived from degassing of deeper magmas or from the assimilation of old crustal xenoliths, will result in age estimates that are too old.

1.2.1.1.2. Rb–Sr Dating The Rb–Sr method is based on the decay of ^{87}Rb to ^{87}Sr (half-life of 4.88×10^{10} years). It is one of the oldest (Hahn & Walling, 1938) and widely utilized bulk-dating techniques in geology (Rink et al., 2015), finding extensive applications in igneous petrology (Glodny et al., 2002, 2003; Halama et al., 2018). Due to the low closure temperature of the Rb–Sr isotope system, this also is technique for determining eruption age—preeruption ages cannot be determined using this technique. Further, the long half-life of ^{87}Rb precludes high-resolution dating, but it is applicable for constraining the eruption ages of ancient volcanic rocks. In recent times, extensive application of the Rb–Sr dating method has been limited in situ techniques such as ion microprobe and laser ablation inductively coupled plasma mass spectrometry (LA-ICP-MS) are not possible with this method. Developments in new analytical techniques, however, have allowed the Rb–Sr method using laser ablation ICP-MS (Zack & Hogmalm, 2016), to be applied to the study of timescales of ancient subsolidus plutonic processes, such as pluton emplacement and hydrothermal alteration of plutonic rocks (Zellmer, Kimura, et al., 2018).

1.2.1.1.3. (U–Th)/He dating (U–Th)/He chronology relies on the radioactive decay of ^{238}U, ^{235}U, ^{232}Th, and ^{147}Sm to ^{4}He (Farley, 2002; Farley & Stockli, 2019). It is commonly applied to U- and Th-bearing minerals (e.g., apatite, titanite, and zircon) to unravel rates and timescales of rock exhumation, mountain building, and landscape evolution (Ehlers & Farley, 2003; House et al., 1998; Shuster et al., 2005). As an example, closure temperatures for apatite and zircon range from ~50 to ~200°C (Reiners, 2005). As a consequence, this technique is also suitable for dating strong cooling events in magmatic systems (i.e., eruptions; Farley et al., 2002; Schmitt et al. 2006, 2010). Although not yet widely utilized, the (U–Th)/He technique can successfully date the eruption age of young magmatic systems (i.e., < 1.5 Ma). For example, Danišík et al. (2017) recently showed how the combination of U–Th disequilibrium/U–Pb and (U–Th)/He methods can be applied successfully in dating zircons up to 2.5 ka, opening new perspectives for dating Quaternary magmatic systems. Also, the method has been applied successfully to refine the eruption chronology of the Ciomadul volcanic complex (e.g., Harangi, Lukács, et al., 2015; Molnár et al., 2018, 2019).

1.2.1.2. Constraining Preeruptive Crystallization Ages

Uranium-series dating methods are based on the decay of ^{232}Th, ^{238}U, and ^{235}U to ^{208}Pb, ^{206}Pb, and ^{207}Pb, respectively (Bourdon et al., 2003; K.M Cooper, 2015, 2019; K.M. Cooper & Reid, 2008; Ivanovich & Harmon, 1992; Schmitt, 2011; Schoene, 2013). These decay chains are widely utilized to provide crystallization ages of accessory minerals like zircon (Charlier & Zellmer, 2000; S.E. Jackson et al., 2004; Reid et al., 1997, 2011; Reid & Coath, 2000; Wilson & Charlier, 2016), monazite (Parrish, 1990), and titanite (Frost et al., 2001). Also, intermediate daughter nuclides are utilized in U-series disequilibrium dating. Closure temperatures are typically well above the solidus (e.g., above 900°C for Zircons; Lee et al., 1997), so that preeruptive crystallization ages can be determined through these dating methods. In detail, ^{238}U, ^{235}U, and ^{232}Th decay through a series of intermediate daughter nuclides (e.g., ^{234}U, ^{230}Th, ^{226}Ra, ^{222}Rn, and ^{210}Pb for ^{238}U) that are both radiogenic and radioactive, with half-lives ranging from seconds to 245 kyr. U-series disequilibrium dating is widely utilized for the estimation of the rates and timescales of magmatic systems (Black et al., 1998; Blundy et al., 2008; Bourdon et al., 2000; Cooper & Reid, 2003; Cunningham, Turner, Patia, et al., 2009; Landwehr et al., 2001; Pyle et al., 1988; Sims et al., 2013; M.B. Turner et al., 2013; Zellmer et al., 2008), providing a powerful tool in petrochronology. Example applications of uranium-series analyses to the study of eruptible magmas are: (a) the tracing of preeruptive magma degassing (Berlo et al., 2006; Kayzar et al., 2009; M.B. Turner et al., 2013); (b) timescale estimations of mixing events associated with eruptive events (Ruprecht & Cooper, 2012); (c) estimating the rates of magmatic processes, including cooling rates and residence times in magma reservoirs before eruptions (Bourdon et al., 2000; Charlier & Wilson, 2010; K.M. Cooper & Kent, 2014; Cunningham. Turner, Dosseto, et al., 2009; Harangi, Lukács, et al., 2015; Kuritani et al., 2007; Reid & Vazquez, 2017; Rogers et al., 2004; Schmitt et al., 2010, 2011; Wotzlaw et al., 2013); (d) constraining the genesis of volcanic rocks (Avanzinelli et al., 2007); (e) unraveling magma evolution and ascent at volcanic arcs (Zellmer et al., 2005); (f) determining the chronology of petrogenetic processes at mid-ocean ridges (Goldstein et al., 1993; K.H. Rubin et al., 2005).

As with other radiogenic dating methods, one of the key prerequisites of U-series dating is that the system remains closed to loss or gain of the specific isotopes used for age estimation of the event that established disequilibrium. In magmatic studies, the ($^{234}U/^{238}U$) activity ratio is frequently monitored for deviation from equilibrium, which would indicate interaction with fluids carrying fluid mobile elements (such as U and Ra). Recent work has shown that in some volcanic rocks, groundmasses in ^{234}U–^{238}U equilibrium can carry crystal cargo in ^{234}U–^{238}U disequilibrium (Zellmer, Freymuth, et al., 2014; Zellmer, Rubin, et al., 2015), precluding the determination of meaningful mineral isochrons. This demonstrates that care must be taken to obtain reliable mineral formation ages for magmatic systems on the basis of U-series dating.

1.2.2. Chemical-Diffusion-Based Geospeedometers

Chemical diffusion is a time-dependent process, widely used for the estimation of rates and timescales of magmatic processes (Costa et al., 2008; Costa & Morgan, 2010). It consists of the analyses of chemical heterogeneities in crystals and glasses, providing a record of magmatic processes. These heterogeneities progressively move to equilibrium profiles at rates that are mainly controlled by the diffusion coefficients. As a consequence, knowing suitable values for the diffusion coefficients, and measuring the chemical heterogeneities, makes it possible to obtain the timescales of the system using Fick's second law of diffusion.

Many studies focus on Fick's second-law-based geospeedometers to obtain timescales of open-system processes in magmatic reservoirs before an eruption, by investigating chemical heterogeneities in crystals (Costa et al., 2003, 2008; Costa & Dungan, 2005; Dohmen et al., 2018; Druitt et al., 2012; Flaherty et al., 2018; Garzanti et al., 2011; Petrone et al., 2016, 2018; Turner & Costa, 2007; Zellmer & Turner, 2007; Zellmer et al., 1999, 2011, 2012; Zellmer, Sakamoto, et al., 2016). Also, such geospeedometers have been applied to chemical heterogeneities in crystals for the study of magma cooling rates (Dohmen et al., 2018; Newcombe et al., 2014) and ascent rates (Demouchy et al., 2006; Ferriss et al., 2018; Lloyd et al., 2016).

It is noteworthy that Fick's second-law-based geospeedometers have not been used exclusively to study chemical heterogeneities in crystals. For example, they have been applied successfully to constrain ascent timescales using melt embayments (i.e., melt inclusions that have failed to become fully enclosed; Ferguson et al., 2016; Humphreys et al., 2008; Liu et al., 2007; Lloyd et al., 2014; Myers et al., 2018). During ascent and degassing, melt embayments will experience diffusive volatile loss due to direct connection to the host melt. This process has been modeled successfully by Fick's second law, providing constraints on the timescales of diffusive reequilibration, which can be related to the decompression rate of the magma (Humphreys et al., 2008; Lloyd et al., 2014).

1.2.3. Rates from Crystal-Melt Reactions During Magma Ascent

The rates of phenocryst-melt reactions produced by decompression can be utilized successfully to estimate the rates of magma ascent (Browne & Gardner, 2006). For example, the loss of volatiles from the melt during its ascent toward the surface produces the breakdown of volatile-bearing phenocrysts, such as amphibole and biotite. The kinetics of these reactions are complex functions of many variables, including mineral and glass compositions, melt viscosity, and temperature. As a consequence, they are not easily constrained theoretically, but they can be successfully studied experimentally (Rutherford & Devine, 2003). For example, the rates of reaction between amphibole phenocrysts and a coexisting rhyolitic melt have been studied (Rutherford & Hill, 1993), and these calibrations can be used to estimate ascent rates for dacitic and andesitic composition magmas. De Angelis et al. (2015) examined the process of amphibole decomposition through isobaric heating of magnesio-hornblende phenocrysts within a natural high-silica andesite glass, highlighting that the injection of new magma (i.e., heating) in a shallow magmatic reservoir produces amphibole reaction rims that have thicknesses, textures, and mineralogies consistent with many of the natural reaction rims seen at island-arc andesite volcanoes. De Angelis et al. (2015) also showed that heating-induced reaction rims are texturally consistent with experimental decompression reaction rims. As a consequence, it may be challenging to discern between decompression and heating mechanisms in nature (De Angelis et al., 2015). Phenocryst–melt reactions have been also applied to estimate magma ascent rates in alkali basalt and kimberlite magmas through studying the reactions occurring between the melt phase and garnets in garnet-bearing xenoliths (Canil & Fedortchouk, 1999). Finally, crystal–melt reaction rates may sometimes be determined from natural samples through geospeedometric techniques (section 1.2.2). The peritectic reaction of olivine and melt forming orthopyroxene is an example, which has recently been timed (Zellmer, Sakamoto, et al., 2016, 2018). This work also determined the rate of orthopyroxene crystal growth. Knowing crystal growth rates, in turn, is critical for the derivation of crystal residence times in magmatic systems through crystal-size-distribution (CSD) analysis (section 1.2.4).

1.2.4. Crystal-Size-Distribution Analysis

The analysis of crystal size distributions (CSD) is a valuable tool to estimate crystal growth rates and timescales of igneous systems (Cashman & Ferry, 1988; Castro et al., 2003; Higgins, 2000; Marsh, 1998, 1988; Mock & Jerram, 2005; Morgan et al., 2007). In the standard formulation, crystal-size-distribution analysis is performed through the "CSD diagram," in which the natural logarithm of the crystal population density is plotted against crystal size (Marsh, 1988, 1998). In the CSD diagram, the zero-size intercept is a measure of the nucleation density. In the same diagram, the slope obtained by data interpolation is a measure of the product between the mean linear growth rate and the average crystal residence time. Thus, to obtain crystal residence times from CSD analysis, crystal growth rates must be well known. Lack of reliable information on crystal growth rates, which is itself a function of the degree of undercooling, typically introduces large uncertainties in crystal residence time deductions from CSDs. Nevertheless, the analysis of CSDs is still widely used to unravel the fate and the timescales of magmas prior to eruption (Cashman & Ferry, 1988; Cashman & Marsh, 1988; Castro et al., 2003; Higgins, 2000, 2002; Lejeune & Richet, 1995; Mock & Jerram, 2005; Morgan & Jerram, 2006). As examples, CSD analysis has been used to constrain crystal growth mechanisms (Kile & Eberl, 2003), the cooling rates of magmatic systems (Cashman, 1993), the timescales of open-system events (Higgins, 1996a), magma ascent rates and crystal residence time in a magmatic mush (Brugger & Hammer, 2010; Higgins, 1996b; Mangan, 1990; Noguchi et al., 2006; Piochi et al., 2005; Salisbury et al., 2008). Semi-automatic measurement techniques now also allow the rapid determination of CSDs of thousands of microlites from glassy tephra samples (Lormand et al., 2018), allowing microlite CSDs to be investigated to gain insights on the timescales of magma ascent at the onset of eruption. Given the uncertainties, however, the application of CSD analysis is perhaps most powerful when undertaken in conjunction with other methods (Jerram & Martin, 2008), or when growth rates can be constrained independently (Zellmer, Sakamoto, et al., 2016, 2018).

1.2.5. Timescales of Magma Mixing Through Concentration Variance Decay and Process Modeling

Concentration variance decay is a geospeedometer based on the process of magma mixing (Perugini et al., 2013; Perugini et al., 2015). It derives from evidence that the mixing of magmas is a commonly observed phenomenon in explosive eruptions. In detail, it uses the decay with time of the concentration variance during the process of magma mixing to provide an estimate of the time lapse between an intrusion of new magma in a volcanic plumbing system and the eruption (Laeger et al., 2019; Perugini et al., 2013; Perugini et al. 2015; Rossi et al., 2017). Timescales of magma mixing have also been investigated by the use of numerical (Bergantz et al., 2015; Oldenburg et al., 1989; Perugini et al., 2004, 2006; Perugini, De Campos, Petrelli, Morgavi, et al., 2015; Petrelli et al., 2005, 2011; Ruprecht et al., 2008) and analogue (Jellinek et al., 1999; Jellinek & Kerr, 1999) modeling.

1.2.6. Timescale Estimations Through Melt-Inclusion Textural Analyses

Melt-inclusion timekeepers use textures of quartz-hosted melt inclusions to estimate quartz crystallization times and growth rates. They are based on quantitative three-dimensional textural analyses (e.g., propagation phase-contrast X-ray tomography) to calculate the melt-inclusion faceting time (i.e., the time required to develop the negative crystal shape of the host mineral). Melt inclusions are blebs of melt trapped in crystals during its growth. At time zero (i.e., just after trapping), the shape of a melt inclusion is typically circular. However, round melt inclusions are disequilibrium features, and equilibrium is approached as diffusion acts to fully facet them (Beddoe-Stephens et al., 1983; Chaigneau et al., 1980; Gualda, Pamukcu, et al., 2012; Manley, 1996). For example, Pamukcu et al. (2015) investigated large-volume, high-silica rhyolite eruptions obtaining decadal to centennial timescales, revealing that giant magma bodies can develop over notably short historical timescales.

1.2.7. Rates and Timescales from Thermal Modeling of Magmatic Systems

Timescale estimations by thermal numerical modeling consider heat exchange between magmatic bodies and their host rocks. The simplest models utilize a static formulation, where it is assumed that no motions occur within the system or, if motions occur, they can be neglected. These models can be totally static or based on incremental volume changes of the magmatic system. In contrast, dynamic models are typically governed by conservation equations (mass, momentum, and energy), may in addition involve the chemical evolution of the system (Oldenburg et al., 1989; Petrelli et al., 2016), and track the fate of crystals and volatiles (see also section 1.2.8; e.g., Petrelli et al., 2018) on the system. Thermal numerical modeling has been applied successfully to unravel the fate, timescales, and dynamics of volcanic plumbing systems (Blundy & Annen, 2016; Caricchi et al., 2012, 2014; Degruyter et al., 2016, 2017; Huber et al., 2009, 2011; Karakas et al., 2017; Leeman et al., 2008; Menand et al., 2015; Petrelli et al., 2018).

1.2.8. Rates and Timescales from Mechanical Modeling of the Volatile Phase in Magmas

Modeling of volatile phases (i.e., H_2O, CO_2, S, Cl, and F) provides useful information about the dynamics and timescales of magmatic systems (Cashman, 2004; Edmonds & Wallace, 2017; Wallace et al., 2015; Zellmer, Edmonds, et al., 2015). In particular, studies concerning the mechanical behavior of bubbles (i.e., exsolved volatiles) have been widely developed to infer magma ascent rates, and their implication on eruptive styles (Cashman, 2004; Degruyter et al., 2012; Edmonds & Wallace, 2017; Wallace et al., 2015). These studies include numerical modeling of magma ascent and decompression (Forestier-Coste et al., 2012; La Spina et al., 2017), experimental constraints at high pressure and temperature (Burgisser et al., 2017; Cluzel et al., 2008; Gardner et al., 2016; Mourtada-Bonnefoi & Laporte, 2004), analogue modeling (Spina et al., 2016), and the measurement of bubble number density (Toramaru, 2006).

1.2.9. Rates from the Mechanical Modeling of Magma Ascent

The mechanical modeling of magma ascent can be utilized proficiently for the estimation of melt-migration and magma-ascent velocities (Browne & Szramek, 2015; Rutherford, 2008). Physical formulation of the model is mainly related to the length scale at which the process occurs, ranging from grain-scale interactions up to crustal-length processes (Browne & Szramek, 2015; O'Neill & Spiegelman, 2010; Rutherford, 2008). For example, porous flow formulation has been utilized successfully for the study of magmatic solitary waves (Wiggins & Spiegelman, 1995), melt rising at mid-ocean ridges (Spiegelman, 1996), development of shear bands (Gebhardt & Butler, 2016), and reactive flow (Lissenberg & MacLeod, 2016; Parmigiani et al., 2011). Also, mechanical modeling has been utilized on mantle xenoliths to estimate the rates of magma ascent (O'Reilly & Griffin, 2010; Rutherford, 2008). Additional formulations are based on dike propagation (Dahm, 2000; Maimon et al., 2012; Rivalta et al., 2015; A.M. Rubin, 1995; Spence & Turcotte, 1990), vein network analysis (O'Neill & Spiegelman, 2010), and diapirism (Burov et al., 2003; Weinberg & Podladchikov, 1994) at the largest scale.

1.2.10. Microlite Number Density Water-Exsolution-Rate Meter

The basic assumption behind the microlite number density (MND) water-exsolution-rate meter (Toramaru et al., 2008) is an equivalence between the decompression-induced and the cooling-induced crystallization. As described by Toramaru et al. (2008), the method is based on the decompression-induced crystallization of microlites due to phase-equilibria changes following water exsolution. In practice, microlites in effusive or pyroclastic rocks are utilized to constrain water exsolution rates (Toramaru et al., 2008).

1.2.11. Ascent Rates from Seismic Data

The tracking of earthquake depths below a volcanic system over time may be used, in theory, to infer the rates of magma ascent (Rutherford, 2008). Early attempts of this have included linking volcanism in the New Hebrides to earthquakes at mantle depths beneath the volcanoes (Blot, 1972), and the tracking of earthquakes from the lower crust to shallower depth immediately prior to the Great Tolbachik Fissure eruption in Kamchatka (Fedotov et al., 1983). However, the direct association of earthquakes with magma migration is often ambiguous, providing misleading information (Druitt et al., 2012). This is because earthquake generation requires stressed and brittle rock, so aseismic subvolcanic regions may be unstressed and unfractured or hot and ductile. On the other hand, magma under pressure may generate earthquakes in brittle rock adjacent to magma-bearing conduits. Hence, magmatic zones may be either seismic or aseismic (Druitt et al., 2012; Rutherford, 2008). Despite these drawbacks, the analysis of seismic data frequently provides significant information and constraints on magma ascent below active volcanoes (De Siena et al., 2014; Rutherford, 2008)

1.2.12. Reliability of Rate and Timescale Estimations and Improved Strategies

Reliability of estimated rates and timescales mainly relies on the robustness of the utilized analytical techniques, experimental constraints, and theoretical models. Regarding this aspect, we can divide the reported methods into two main groups.

The first group contains rates and estimates directly obtained by an analytical technique. This group includes the K–Ar, Ar–Ar, Rb–Sr, (U–Th)/He, U–Th–Pb and U-series dating methods. Here, the uncertainties on estimated rates and timescales are related to the analytical technique, in terms of precision and accuracy (Schoene et al., 2013). For example, U–Pb dating on zircons can be performed by whole-grain bulk methods such as isotope-dilution thermal ionization mass spectrometry (ID-TIMS), or by in situ techniques like secondary ion mass spectrometry (SIMS) and laser ablation inductively coupled plasma mass spectrometry (LA-ICP-MS; Schaltegger et al., 2015). The achieved precision and accuracy, expressed as 2σ, are of the order of 0.1 %, 1–2 %, and 2–3 %, for the ID-TIMS, SIMS, and LA-ICP-MS, respectively (Schaltegger et al., 2015). The reported uncertainties highlight that ID-TIMS estimations provide about an order of magnitude more precise results than in situ techniques. However, for the SIMS and LA-ICP-MS techniques, the loss in precision and accuracy is balanced by the capability of obtaining age distributions and age heterogeneities within a single crystal with lateral resolutions approaching, in some cases, a few microns (Chew et al., 2017; Sas et al., 2019).

For the second group, containing methods where rate and timescale estimates involve multistep modeling of experimental, analytical, or geophysical data, the reliability of the results strongly depends on: (a) the uncertainties linked to the parameters utilized in the calculations; (b) the uncertainties related to the assumed initial and boundary condition, and the propagation of these uncertainties through each step of the model. For example, in Fick's second-law-based techniques, estimated timescales strongly rely on the parametrization utilized to constrain the diffusion coefficients, the selected initial and equilibrium chemical profiles, and on the propagation of the uncertainties through the utilized minimization technique or the implemented numerical scheme (Costa et al., 2008). Recently, many studies focused on error propagation in Fick's second-law-based geospeedometers (Bradshaw & Kent, 2017; Engi et al., 2017; Shea et al., 2015) demonstrate increasing attention to this critical topic. Rossi et al. (2017) and Gualda, Pamukcu, et al. (2012) have investigated the robustness of the concentration variance decay and of the textural analysis of melt inclusions, respectively. The robustness of CSD analysis has been widely discussed, and the method has been continuously improved over the time (Calzolaio et al., 2010; Higgins, 2002, 2006; Kile & Eberl, 2003; Marsh, 1998). Finally, the uncertainties related to thermal and fluid dynamics modeling have been investigated extensively (Blankenbach et al., 1989; King et al., 2010; Tosi et al., 2015; Van Keken et al., 1997). For these methods, uncertainties mainly derive from the assumptions utilized for the initial and boundary conditions and on the adopted numerical scheme (Blankenbach et al., 1989; King et al., 2010; Tosi et al., 2015; Van Keken et al., 1997). In all of these cases, a useful tool for the reduction of final uncertainties is analysis of the dominant errors (Hughes & Hase, 2010). Consider an estimate depending on the product of three parameters: A, B, C. Now, assume that these parameters are characterized by known uncertainties of 10%, 2.5%, and 2.5%, respectively. Performing a simple propagation of the uncertainties, we will discover that the resulting uncertainty is about 11%, dominated by the errors in A. As a consequence, spending efforts in reducing the uncertainties in B and C does not result in a significant improvement of our estimate. In contrast, focusing on the reduction of the uncertainties in A will improve the reliability of the results significantly. In the context of the subjects discussed here, to improve the reliability of our timescale estimates, focus should be on dominant errors like those associated with the parametrization of diffusion coefficients in Fick's second-law-based geospeedometers (Costa et al., 2008), the characterization of temperature gradients within the crust in thermal modeling (Annen et al., 2015), and the scaling laws in mechanical modeling (Rivalta et al., 2015), to provide a few examples.

The ability of unraveling different timescales associated with contiguous and, sometimes, overlapping processes (e.g., crystallization time in a magma chamber and the subsequent ascent time) associated with an eruptive event is also a challenge. To achieve this goal, the combination of multiple techniques is mandatory. For example, Arzilli et al. (2016) utilized: (a) textural data and mineral assemblages of synthetic samples; (b) experimental constraints of feldspar nucleation kinetics and growth rate; (c) textural data, mineral abundances, and CSD measurements on natural samples to unravel preeruptive magma conditions and unrest timescales during the Monte Nuovo eruption (CE 1538; Campi Flegrei, southern Italy). The combination of these different techniques highlights a shallow magmatic reactivation lasting a few months, and conduit dynamics timescales (i.e., ascent timescales) of the order of hours to days. Another example of the mutual application of multiple methods is provided by Ruth et al. (2018), who merged olivine diffusion chronometry and data on melt-inclusion volatiles to track the progressive ascent, mixing, and overall shallow accumulation of magma prior to the 2008 eruption of Llaima volcano, Chile. Ruth et al. (2018) elucidated prolonged magma mixing and ascent events followed by an eruption that tapped an extensive part of the vertically arranged, crystal-rich magma plumbing system.

As a further example, Petrelli et al. (2018) combined the application of thermal numerical investigations, volatile thermodynamic modeling, and mechanical formalisms to constrain the temporal evolution of a hydrous mafic magma stored at mid- to deep crustal levels and the ascent rates of the resulting residual melts. Results highlighted prolonged storages times at deep crustal levels, of the order of thousands of years, potentially followed by rapid ascents to the surface, of the order of days to months (Petrelli et al., 2018).

1.2.13. Summary of Timescales Accessible Through Different Dating Methods

Figure 1.3 shows an overview of the timescales, from hours to the time of Earth formation, that are accessible by using the different dating methods described above. Two of the methods have the potential to access the entire range of timescales of interest for geological applications: Fick's second-law-based geospeedometers and thermal modeling. Some techniques, mainly based on the decay of radioactive isotopes, are typically suitable for eruption age determinations and timescale estimations larger than 10^3 years, although some of the shorter lived isotopes may access more rapid processes (e.g., Sims et al., 2013). In contrast, other methods mainly focus on shorter timescales: the rates from crystal–melt reactions during magma ascent, the concentration variance decay method, the microlite number density water-exsolution-rate method, and the ascent rates from seismic data.

1.3. MECHANISMS, RATES, AND TIMESCALES OF MAGMA EMPLACEMENT, STORAGE, AND MIGRATION

In this section we discuss the mechanisms, rates, and timescales of emplacement, storage and migration of magmas in Earth's crust. We begin with the mechanisms of magma emplacement and storage at deep and mid-crustal levels (i.e., at depths of 15–20 km and below; section 1.3.1). Then, we investigate the rates of magma ascent from the deep and mid-crust to shallower levels (section 1.3.2), and the modes of magma storage at mid- to shallow crustal levels (section 1.3.3). Finally, we discuss the pre- to syneruptive dynamics and related timescales (section 1.3.4), including the influence of volatiles on the processes operating in this regime.

1.3.1. Mechanisms of Magma Emplacement and Storage at Deep and Mid-Crustal Levels

Many insights on the timescales of cooling, crystallization, chemical evolution, and volatile accumulation at mid- to deep crustal levels (i.e., at depths of 15–20 km and below) arise from the combination of thermal modeling (Annen, 2011; Annen et al., 2015; Blundy & Annen, 2016), fluid dynamics investigations (Petrelli et al., 2018), and geophysical data (Laumonier et al., 2017), taking into account boundary conditions provided by petrological studies (Melekhova et al., 2013; Nandedkar et al., 2014; Sisson & Grove, 1993).

Among these, thermal- and fluid-dynamics models provide significant insights (Annen, 2009, 2011; Petrelli et al., 2016, 2018), currently investigating two main scenarios: instantaneously emplaced systems, where the magmatic body remains isolated from external magmatic inputs (i.e., closed systems), and magmatic bodies that grow incrementally by periodic refilling of new parental magma (Annen, 2009, 2011). The latter hypothesis is currently favored (Annen, 2011). These models are often constrained by petrological experiments (Melekhova et al., 2013; Nandedkar et al., 2014; Sisson & Grove, 1993) and thermodynamic constraints (e.g., MELTS simulations; Gualda, Ghiorso, et al., 2012), providing fundamental information on the crystallization processes and the related release of latent heat. From the thermal point of view, a volume of magma contains sensible heat, defined as the heat capacity multiplied by the temperature difference between the magma and its surroundings, and latent heat, which is that released upon crystallization (Blundy & Annen, 2016). Modeling results highlight that mantle-derived magmas emplaced at mid- to deep crustal levels progressively build up a deep hot zone, where intermediate to felsic melts are generated by fractionation of the mafic magma and concomitant partial melting of previous intrusions and sometimes preexisting crust (Annen et al., 2006, 2015; Solano et al., 2012). For a quasi-instantaneously emplaced magmatic system, magma residence, cooling, and differentiation timescales are similar, mainly depending on the size of the magmatic body (Annen, 2011). In contrast, for magmatic systems that grow incrementally (i.e., where mafic magma is repeatedly injected), magma residence times depend on magma emplacement duration after the system was mature enough to accumulate melt (10^6 years), whereas differentiation times scale with the time each magma pulse takes to equilibrate with its environment (10^3 years; Annen et al., 2006, 2011; Blundy & Annen, 2016).

An additional parameter of paramount importance for constraining storage timescales is the magma supply rate. It is one of the main parameters governing the modulation of the geothermal gradient. At deep crustal levels,

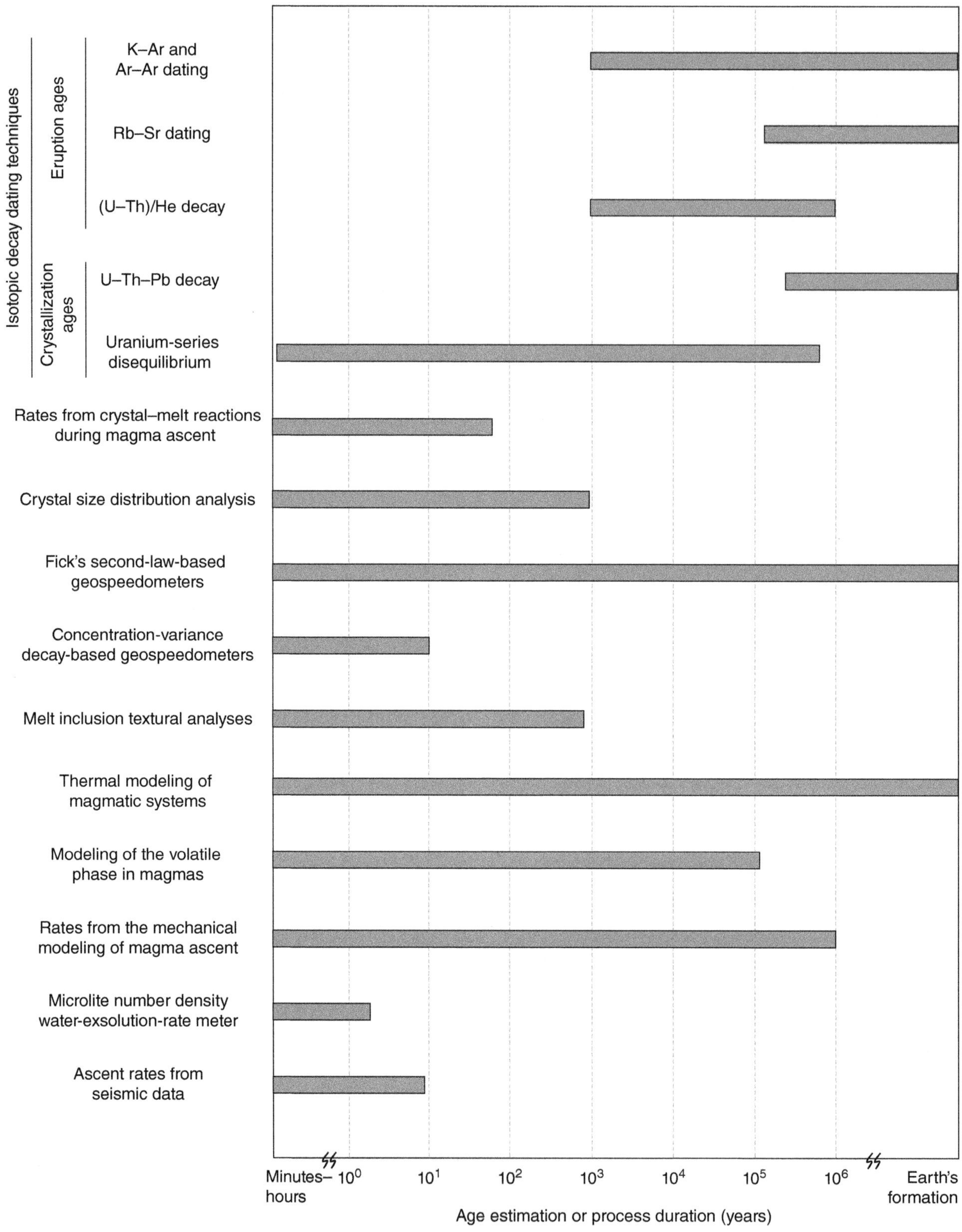

Figure 1.3 Range of timescales that can be determined using the reported analytical methods and modeling techniques. [Modified from Turner & Costa (2007).]

the supply rate is governed mainly by mantle magmas, with a possible additional contribution by crustal melting. Estimates for intrusion rates derived from thermal modeling and field observations range from 5×10^{-3} km^3 year^{-1} (Gelman et al., 2013) to ~10^{-2} km^3 year^{-1} (Annen, 2009), and from 2×10^{-4} to 2×10^{-1} km^3 year^{-1} (Dimalanta et al., 2002; Lipman & Bachmann, 2015), respectively. Any further constraints must take these estimates into account. Regarding this point, Karakas et al. (2017) investigated the thermal influence of lower crustal reservoirs on the upper continental crust. To achieve this goal, they used a model of stochastic and incremental injections of magma (as dikes and sills) over long timescales (≥ 10^5 years). They highlighted that continuous magma intrusion for extended timescales (i.e., 10^5 to 10^6 years) results in the development of large lower crustal mush zones, able to modify the thermal budget for upper crustal levels (Karakas et al., 2017). As a consequence, the total duration of magmatism should be combined with magma intrusion rate estimates to assess the capability of volcanic systems to form the largest explosive eruptions, in agreement with Karakas et al. (2017).

1.3.2. Rates of Magma Ascent from Deep and Mid-Crust to Shallower Depths

The unraveling of magma ascent rates from deep crustal levels or the lithospheric mantle is of paramount importance to constrain many processes, such as the incremental growth of shallow magmatic systems, degassing rates, eruptive styles, and, in general, the overall evolution of a volcanic feeding system.

The rate of magma ascent depends on several parameters like the physical properties of the magma (i.e., density, viscosity, and crystallinity), the pressure and overpressure conditions in the magma storage region, the rheological properties of the host rocks, and, in cases culminating in eruptive events, the resistance to flow in the conduit during the final part of the voyage to the Earth surface (Mastin & Ghiorso, 2001; Papale & Dobran, 1994; Pinkerton et al., 2002; Rutherford, 2008; Sparks et al., 2006). All of these variables are changing with time in response to: (a) petrological processes occurring within the ascending magma (e.g., magma degassing, groundmass crystallization, and phenocryst-melt reactions); (b) overpressure variations in the source region (i.e., drop in overpressure due to magma discharge); (c) rheological variation of the host rocks; (d) in the case of conduits, changes in the resistance to flow in the conduit. Also, these processes are nonlinearly related, making the exact unraveling of the time-dependent evolution of magma ascent a difficult task (Browne & Szramek, 2015; Gonnermann & Manga, 2013; O'Neill & Spiegelman, 2010; Rivalta et al., 2015; Rutherford, 2008). However, simplified mechanical and thermal models, average estimates, and the study of petrological features in the erupted rocks have been successfully utilized to unravel magma ascent rates (Jankovics et al., 2013; Rutherford, 2008).

For example, average magma-ascent velocities can be tentatively estimated by buoyancy-driven (Dahm, 2000) and magma-driven (A.M. Rubin, 1993, 1995) ascending dike theories (Rivalta et al., 2015). In the former approach, the dike is assumed to be isolated and its propagation is due to buoyancy forces only (Dahm, 2000). In the latter, propagation is driven not only by buoyancy forces, but also by the excess pressure of a reservoir connected to the dike (A.M. Rubin, 1995). Therefore, due to the force applied by the excess pressure, even dikes smaller than the critical length can propagate toward the surface. To provide some constraints, Figure 1.4 reports the ascent velocities (m s^{-1}) predicted by the buoyancy-driven (Figure 1.4a) and magma-driven (Figure 1.4b) ascending dike formulations, for a magma rising from a depth of 25 km. Typical dynamic viscosities for various types of melt and magma are also reported in Figure 1.4. The boundary conditions utilized in the calculations are in agreement with Petrelli et al. (2018). In detail, Figure 1.4a and Figure 1.4b report the dike ascent velocities as a function of three parameters:

1. the density contrast (i.e., the density differences between the ascending magma and the host rock) in the case of the buoyancy-driven formulation;
2. the pressure excess of the reservoir connected to the dike in the case of the magma-driven approach,
3. the dynamic viscosity (η) of the ascending magma.

As seen in Figure 1.4, ascent velocities strongly depend on magma viscosity, because they scale as η^{-1} in both investigated formulations. The strong dependence on viscosity becomes weaker when flow approaches the condition of turbulence, which typically occurs, in agreement with Sparks et al. (2006), for η values lower than about 10 Pa·s. Figure 1.4a indicates that the buoyancy-driven formulation predicts ascent velocities ranging between 10^{-2} and 6 m s^{-1} for a basaltic melt characterized by $10^1 < \eta < 10^2$ Pa·s and density contrasts with the host rock densities ranging from 50 to 500 kg m^{-3}. Higher density contrasts may be achieved by water-rich (e.g., H_2O ~8–10 wt%) magmas. The magma-driven formulation predicts, for the same basaltic melt, ascent velocities ranging from 0.02 to more than 10^2 m s^{-1}, for excess pressures ranging from 1 to 10 MPa (Figure 1.4b). The ascent velocities predicted for silicic melts (e.g., characterized by $10^7 < \eta < 10^8$ Pa·s) are significantly lower, ranging from 10^{-8} to 2×10^{-4} m s^{-1}. Volatile exsolution at shallow crustal levels can strongly affect the predicted ascent velocities. For example, Petrelli et al. (2018) calculated an increase of the velocities by a factor of ~2 and 10 at depths corresponding to ~4 and 1 km, respectively. All of the ascent velocities reported

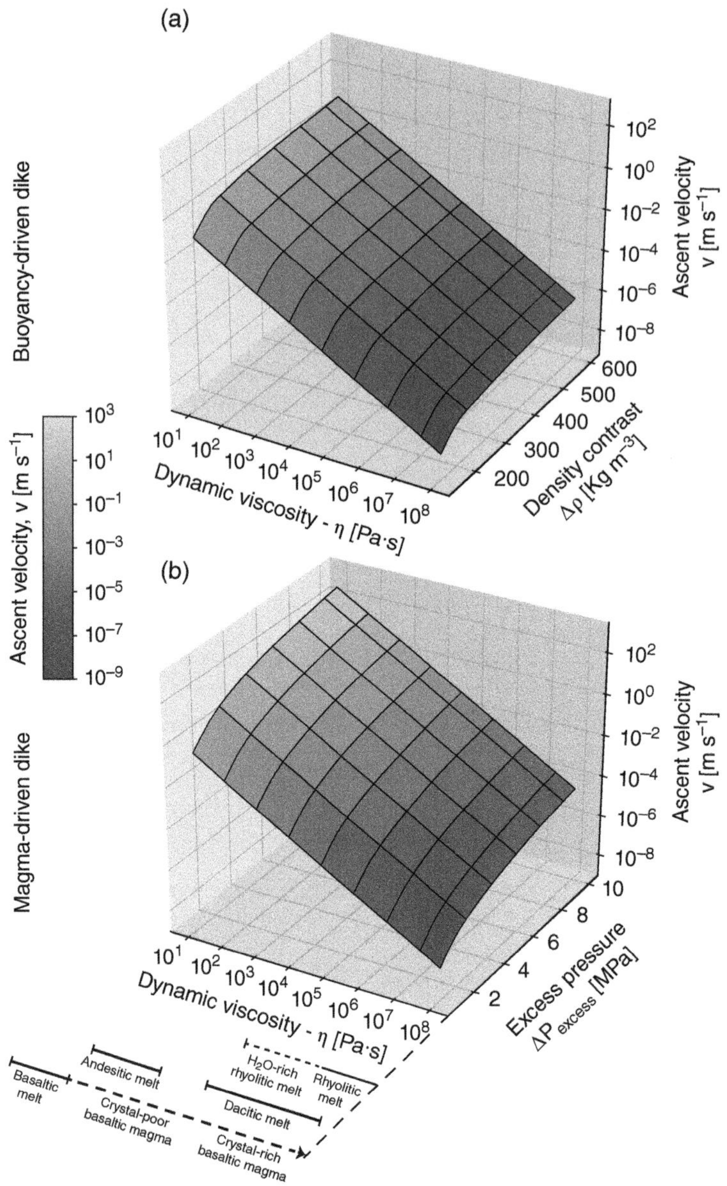

Figure 1.4 Dike ascent velocities (m s^{-1}) obtained by the (a) buoyancy-driven and (b) magma-driven ascending dike formulations. (a) Ascent velocities reported as a function of the density contrast ($\Delta\rho$) and the dynamic viscosity (η) according to Dahm (2000). (b) Ascent velocities reported as function of the excess of pressure (ΔP) and the dynamic viscosity (η) according to Rubin (1993).

above are based on robust theoretical and experimental formulations (Dahm, 2000; Rivalta et al., 2015; A.M. Rubin, 1993, 1995). However, they can be considered only a rough approximation of real ascent rates, because of the assumptions at the foundation of their formulations (e.g., constant rheological behavior of the ascending magma and of the host rocks) and the limited knowledge of the boundary conditions of the investigated physical problem (e.g., the values utilized for the fracture toughness and the shear modulus). Nevertheless, magma ascent rates obtained by using the buoyancy-driven (Dahm, 2000), and magma-driven (A.M. Rubin, 1995) ascending dikes formulations are in good agreement with ascent estimates for mantle-derived magmas from deep-crustal levels provided by petrological modeling, thermal constraints, and studies on mantle xenoliths reported in Table 1.2. For example, Ruprecht and Plank (2013), utilizing Ni diffusion profiles in olivine, reported ascent velocities from about 30 km depth ranging between 6×10^{-4} and 10^{-3} m s^{-1} for the andesitic magma feeding Irazù volcano. Also, they provide an independent ascent estimate larger than 10^{-3} m s^{-1} based on thermal considerations for dike propagation

Table 1.2 Review of the Ascent Velocities from Mid- to Deep-Crustal Levels as Function Magma Compositions

Reference	Magma	Observation	Minimum ascent rate (m s^{-1})	Average ascent rate (m s^{-1})	Maximum ascent rate (m s^{-1})
Rutherford (2008)	Alkali basalt	Hydrogen zoning in xenolith hosted olivines	0.2		0.5
Spera (1984)	Mantle derived melt	Kinetic and fluid dynamic constraints	0.01		10
O'Neill & Spiegelman (2010)	Mantle derived basaltic melt	Vein network analysis		0.24	
O'Neill & Spiegelman (2010); A.M. Rubin (1995)	Mantle derived magma	Dike ascent theoretical modeling	0.01		10
O'Reilly & Griffin (2010)	Mantle derived magma	Geochemical and microstructural constraints on xenoliths	0.2		2
Petrelli et al. (2018)	Water rich mafic magmas	Theoretical modeling uncompressible magma dike ascent	0.15		20
Rutherford (2008)	Kimberlitic magma	Garnet dissolution	1		30
Sparks et al. (2006)	Kimberlitic magma	Theoretical modeling	4		20
Spera (1984)	Kimberlitic magma	Kinetic and fluid dynamic constraints	10		30
Peslier et al. (2015)	Nephelinite magma	Water disequilibrium in xenolith hosted olivines	0.2		25
Brenna et al. (2018)	Alkali basalts	Mg/Fe, Ca, Mn, Ni, Li and H_2O diffusion in olivines	0.01		10
Sen (1983); Peslier et al. (2015)	Nephelinite magma	Density–viscosity calculation for a 20-cm diameter peridotite sphere		1	
Shaw (2004); Peslier et al. (2015)	Nephelinite/basanite magma	Fe–Mg diffusion in olivine from xenoliths	0.1		4
Denis et al. (2013); Peslier et al. (2015)	Nephelinite/basanite magma	Dehydration for the spinel-bearing xenoliths	3		12
Peslier & Luhr (2006)	Alkali basalts	Hydrogen loss from olivines in mantle xenoliths	0.2		0.5
Demouchy et al. (2006)	Alkali basalts	Dehydration profiles in olivine crystals	3		9
Ray et al. (2016)	Alkali basalts	Mantle xenolith–xenocryst combination of different approaches	0.1		8.4
Dégi et al. (2009)	Alkali basalts	Chemical profile of Fe-Ti oxides in granulite xenoliths	0.35		0.77
Szabó & Bodnar (1996)	Alkali basalts	Thickness of chemical zoning bands in spinel		0.93	
Harangi et al. (2013)	Alkali basalts	Ca profile of olivine xenocryst	0.14		0.19
Jankovics et al. (2013)	Alkali basalts	Dike propagation velocity and mantle xenoliths studies	0.1		3.9
Ruprecht & Plank (2013)	Andesitic magma	Ni diffusion profiles	6.0E-04		9.0E-04
Ruprecht & Plank (2013)	Andesitic magma	Thermal considerations on dike propagation		0.0012	
Annen et al. (2006)	Hydrous andesitic magma	Dike ascent theoretical modeling	1.4E-05		2.0E-01
O'Neill & Spiegelman (2010)	Diapiric ascent ductile crust	Theoretical modeling	1.0E-09		1.0E-07
O'Neill & Spiegelman (2010)	Diapiric ascent brittle crust	Theoretical modeling	1.0E-10		1.0E-09

(Ruprecht & Plank, 2013). Ascent rate estimates for mantle-derived basaltic melts range between 0.01 and 25 m s^{-1} on the basis of theoretical modeling (O'Neill & Spiegelman, 2010; Rutherford, 2008; Spera, 1984), geochemical and microstructural constraints on mantle xenoliths (O'Reilly & Griffin, 2010), and water disequilibrium in olivine (Peslier et al., 2015; Rutherford, 2008). The ascent rates of kimberlitic magmas, being characterized by very low viscosities (i.e., $10^0 < \eta < 10^1$ Pa·s), record the highest values and range from 1 to 30 m s^{-1} (Rutherford, 2008; Sparks et al., 2006; Spera, 1984). Finally, it is interesting to note that the ascent velocities predicted for deep-seated silicic magmas by dike formulations and diapiric ascent are partially overlapping, since they range from 10^{-8} to 10^{-4} and from 10^{-10} to 10^{-7} m s^{-1}, respectively (O'Neill & Spiegelman, 2010).

1.3.3. The Nature of Magma Storage at Mid- and Shallow Crustal Levels

The study of mid- to shallow magmatic systems (i.e., shallower than ~15 km) can be successfully performed utilizing multifaceted modeling approaches that combine petrological, geophysical, and observations in the field (e.g., Bachmann, Miller, et al., 2007; Glazner et al., 2004; Lipman, 1988, 2007; Lipman & Bachmann, 2015). As an example, Lipman (1988) provided a detailed study of the emplacement and petrological evolution of the mid-Tertiary Latir volcanic field and its cogenetic granitic batholith (northern New Mexico, United States), elucidating the evolution of a silicic magmatic system in the upper crust. In detail, Lipman (1988) highlighted that multiple differentiation processes were active during the evolution of silicic magmatic systems in the upper crust. They are crystal fractionation, replenishment by mantle and lower crustal melts of varying chemical and isotopic character, mixing of evolved with more primitive magmas, upper crustal assimilation, and, perhaps, volatile-transfer processes (Lipman, 1988).

The combination of absolute crystallization ages (e.g., U–Th–Pb geochronology on zircon; e.g., Bourdon et al., 2000; Kuritani et al., 2007; Rogers et al., 2004; Schmitt et al., 2011) and chemical diffusion data (i.e., Fick's second-law-based geospeedometers; e.g., Costa et al., 2008; Costa & Morgan, 2010) provides significant constraints on the thermal evolution of magma storage episodes (e.g., K.M. Cooper & Kent, 2014; A.E. Rubin et al., 2017). An apparent dichotomy of absolute ages based on radiometric dating and geospeedometric ages based on diffusion data was previously recognized two decades ago at Santorini, Greece (Zellmer et al., 1999). Over the next several years, this evidence was corroborated by studies of many other arc volcanoes (e.g., Zellmer et al., 2005), indicating that magma storage is typically characterized by temperatures low enough for chemical diffusion to essentially cease, so that geospeedometry records only intermittent high-temperature episodes of crustal magma residence (Zellmer, 2009). More recently, the thermal histories of magma bodies were more formally quantified (K.M. Cooper, 2019; K.M. Cooper & Kent, 2014; Kent & Cooper, 2018; Ruprecht & Cooper, 2012;), initially using a compilation of published U-series data and Zellmer et al.'s (2003) geospeedometric modeling approach, and later through a combination of U-series dating and Li diffusion in single zircon crystals (A.E. Rubin et al., 2017).

Additional constraints come from crystallization temperature estimation through petrologic thermometers (e.g., Ti in zircons; Fu et al., 2008), allowing the tracking of preeruptive magmatic timescales, thermal states, and magma eruptibility (Barboni et al., 2016; K.M. Cooper, 2019; Kaiser et al., 2017; Kent & Cooper, 2018; Klemetti & Clynne, 2014; Tierney et al., 2016). However, interpretation of the results obtained is inconclusive, and still widely debated (Barboni & Schoene, 2014; K.M. Cooper & Kent, 2014; Kent & Cooper, 2018; A.E. Rubin et al., 2017). For example, natural zircon crystallization ages often show distributions that can be successfully modeled assuming extended magma storage (> 100 kyr; "long-lived magma reservoirs") at relatively elevated temperatures (supersolidus $T > 750°C$; "warm storage"; Barboni et al., 2016). However, Kent and Cooper (2018) showed that the same distributions also could be produced by subsolidus ($T < 700°C$; "cold storage") magma storage periods interrupted by a limited number of short thermal rejuvenation events. As an overall survey, zircon ages and chemical signatures suggest that many silicic magmatic bodies aggregate and compositionally evolve over timescales of 10^4 to 10^6 years (Bachmann, Miller, et al., 2007; Bacon & Lowenstern, 2005; Charlier & Wilson, 2010; Claiborne et al., 2010; Coleman et al., 2004; Eddy et al., 2016; Lipman & Bachmann, 2015; J.S. Miller et al., 2007; Reid & Vazquez, 2017; Spera & Bohrson, 2018; Vazquez & Reid, 2004; Wotzlaw et al., 2013). However, shorter timescales are also reported (Bragagni et al., 2014; Pamukcu et al., 2015; Schmitt, 2006).

Since the available data are not fully conclusive, the current debate focuses on two main interpretations: the first supports the idea of a magma storage in a "warm" eruptible state (i.e., where eruptible lenses of melt survive for extended periods; Barboni et al., 2016); the second preferring a "cold," dominantly immobile storage, where the eruptible state can be achieved only for brief periods (K.M. Cooper & Kent, 2014), but relatively rapidly (Michaut & Jaupart, 2006). In general, studies arguing for long-lived magma reservoirs often support the idea of large, highly crystalline regions in the upper crust where melt-rich, eruptible pockets, i.e., warm mush zones, survive for extended time periods, and from which melt can be extracted (Barboni et al., 2016; Hildreth, 2004).

In contrast, other studies suggest cold storage (K.M. Cooper & Kent, 2014; A.E. Rubin et al., 2017), where mobile magma bodies are ephemeral features that grow incrementally, fed by deep sources and erupting on short timescales (Menand et al., 2015; Wotzlaw et al., 2015).

Further constraints come from Szymanowski et al. (2017), who combined U–Pb geochronological investigations on zircon and titanite with mineral thermometry (Ferry & Watson, 2007; Hayden et al., 2008) on the dated crystals to constrain both the time and temperature of magma storage within large-volume magma reservoirs. Results highlight that zircon and titanite crystals record an extended period of magma assembly (i.e., more than 600 kyr) under dominant cold storage conditions (i.e., 680–730°C), suggesting that protracted low-temperature storage culminating in late-stage reheating is a widespread feature of large crystal-rich eruptions (Szymanowski et al., 2017). A plausible synthesis of the two reported end-members (i.e., warm and cold storage of a magmatic body) is the development of a thermally heterogeneous reservoir at near-solidus conditions where high temperature eruptible magmas are ephemeral, dispersed, and rapidly mobilized prior to eruptions (Cooper, 2019). In particular, M.D. Jackson et al. (2018) indicated reactive flow, rather than fractional crystallization, as a critical process in crustal mush reservoirs, able to produce differentiated, low-crystallinity, eruptible magmas. These magmas typically host older crystals, stored at low temperatures (i.e., cold conditions; M.D. Jackson et al., 2018).

Thermal modeling provides significant constraints on the thermophysical conditions and timescales of magma storage at mid- to shallow crustal levels (Annen & Sparks, 2002; Blundy & Annen, 2016; Karakas et al., 2017). One of the most debated issues regards the minimal magma intrusion rate required to maintain one or more melt-rich lenses of magma (i.e., in warm conditions) in the crust (Bachmann et al., 2002; Molloy et al., 2008; Spera & Bohrson, 2018). Further insights come from mechanical modeling through analogue experiments (Jellinek & DePaolo, 2003), highlighting that the large intrusion rates implied by thermal models ($\geq 10^{-2}$ km^3 year^{-1}) are unlikely to sustain large magma bodies in the upper crust. However, as suggested by Karakas et al. (2017), modified thermal budgets at mid- to shallow crustal levels, produced by prolonged magma emplacements even at low intrusion rates (i.e., 10^{-3} to 10^{-2} km^3 year^{-1}), could support the development of significant volumes of eruptible magmas, potentially leading to large eruptions. As a consequence, young magmatic systems may not be able to support a so-called "super eruption." In contrast, long-lived systems (i.e., $>10^6$ years) may accumulate an amount of eruptible magma potentially able to feed a super eruption (Karakas et al., 2017).

Geophysical techniques (e.g., seismic tomography, magnetotelluric investigations, and gravimetric surveys) have been utilized in the attempt to provide a real-time image of where magma is, or may be, stored within the crust (Pritchard & Gregg, 2016; Stelten et al., 2014). As reported by C.F. Miller (2016), results are intriguing but frustrating (Stelten et al., 2014). Geophysical techniques point to the presence of molten material beneath many volcanoes, for example, Uturuncu (Bolivia), Ciomadul (SE Carpathians), Yellowstone (Wyoming, United States), St. Helens, Rainier, and Adams (Washington, United States) volcanoes (Bedrosian et al., 2018; Comeau et al., 2015, 2016; Harangi, Novák, et al., 2015; Hill et al., 2009; Huang et al., 2015; Laumonier et al., 2017; C.F. Miller, 2016), and in NE Japan (K.X. Chen et al., 2018; Zellmer et al., 2019). However, melt domains appear to be only a very small portion of the magmatic plumbing systems, largely or entirely uneruptible (C.F. Miller, 2016). In addition, no large volumes of eruptible magma have yet been identified anywhere (C.F. Miller, 2016; Pritchard & Gregg, 2016). The evidence from geophysical observations corroborates thermophysical considerations, highlighting that the development of large liquid-rich magma domains (i.e., crystal-poor magma chambers) for extended time windows is not likely in the upper crust (Blundy & Annen, 2016). On the contrary, there is an increasing number of studies suggesting a rapid assembly of melt-dominated eruptible magma domains just prior to eruption (e.g., Burgisser & Bergantz, 2011; Gualda & Sutton, 2016; Till et al., 2015).

Regarding the mechanical behavior of melt lenses in magmatic mushes, Seropian et al. (2018) investigated the timescales required for the development of Rayleigh–Taylor instabilities (i.e., the instability of the interface between two magmas characterized by different densities), and the subsequent rising of melt towards the Earth surface. They found that the growth rate of an instability depends mainly on the difference in density between the two magmas, their viscosity, and the dimension of the system (Seropian et al., 2018). Results highlight a wide range of calculated timescales for the development of Rayleigh–Taylor instabilities, ranging between 10^1 and 10^4 years (Seropian et al., 2018).

Additional constraints on the behavior of crystal-rich magmatic systems during storage at upper crustal levels come from the development of numerical fluid-dynamics models (Bergantz et al., 2017; Dufek & Bachmann, 2010; Parmigiani et al., 2016). Among these, the definition of the mechanical "lock up" of crystal mushes, the definition of crystallinity conditions at which crystal–liquid separation takes place most efficiently, and the investigation of the role of volatiles are of paramount importance. Regarding mechanical "lock up" of crystal mushes, Bergantz et al. (2017) showed that it is not uniquely associated with a particular crystal-volume fraction, and that distinct mechanical behaviors can emerge simultaneously within a crystal-rich system.

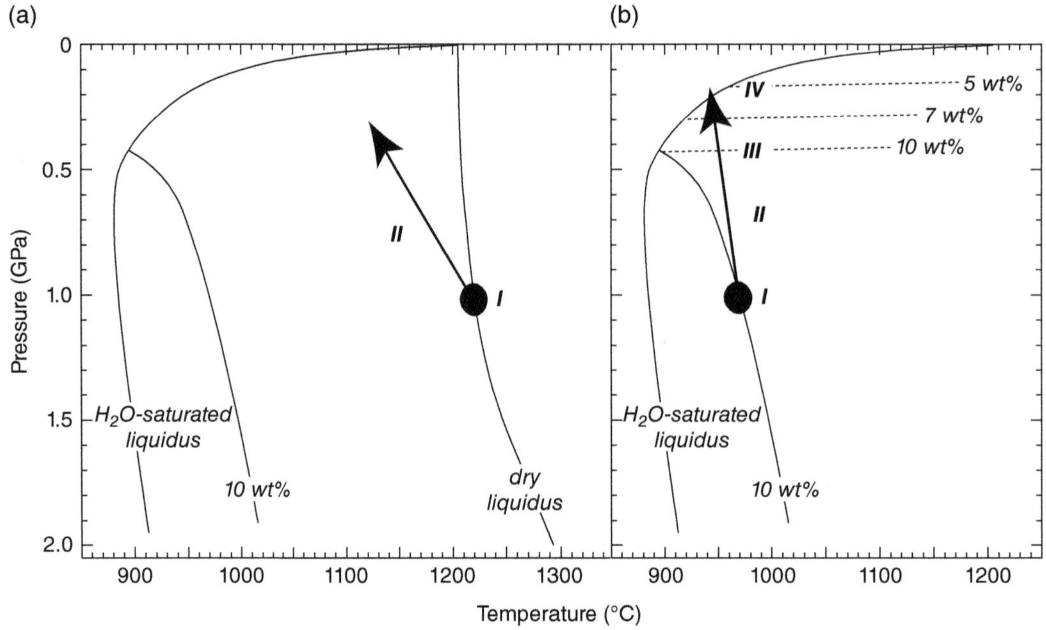

Figure 1.5 Pressure–temperature diagram showing liquidi of andesitic melts that are water-saturated, or water-undersaturated (10 wt% H_2O), or dry. The ascent paths for (a) dry and (b) hydrous but H_2O-understurated andesitic magmas from a pressure of 1 GPa are provided (modified from Annen et al., 2005). In the diagram, the processes occurring at different stages of magma ascent are also reported: (a) segregation (I) and progressive crystallization (II) for a dry melt; (b) segregation (I), crystal resorption (II), water saturation (III), and crystallization (IV) for a water-rich melt.

In particular, the "unlocking" of a crystal mush can be understood as a process that produces an increase in the viscous number (i.e., a measure of the connectedness of the particle network with higher values indicating a diminished contiguity) and a decrease of the coordination number (i.e., the average number of particle–particle contacts per particle throughout the granular system), whether initiated by a thermal event (Burgisser & Bergantz, 2011; Huber et al., 2011) or overpressure from a new injection (Bergantz et al., 2015). Significant findings are also reported by Dufek & Bachmann (2010), who showed that crystal–liquid separation takes place most efficiently within a crystallinity window of ~50–70 vol% crystals. Finally, Parmigiani et al. (2016) investigated, numerically, how the magmatic volatile phase behaves in crystal-rich and crystal-poor magmas, showing that volatiles tend to migrate efficiently in crystal-rich parts of a magma reservoir and accumulate in crystal-poor regions. All these results are of paramount importance to define preeruptive dynamics and related timescales of these complex crystal mushes.

1.3.4. Eruption Run-up and Associated Timescales

The approach to the investigation of pre- to synneruptive timescales is largely controlled by the architecture of the magmatic system under investigation: mafic to intermediate-composition melts may rapidly traverse the differentiated crust, while intermediate-composition to felsic melts are generally thought to originate from magmatic mushes that require unlocking before melt can be extracted.

1.3.4.1. Mafic to Intermediate Systems

For mafic to intermediate-composition melts, two competing endmember scenarios may be envisaged. One, in which magmas ascent relatively slowly and crystallize mostly in response to concomitant cooling (Figure 1.5a). Here, melt segregation (I) is immediately followed by crystallization (II). This scenario may most closely resemble melts that carry low amounts of volatiles and that ascend through a dense, mafic crust, e.g., in rift settings. In this case, most of the crystal cargo may have formed from the magma that carries the crystals (Field et al., 2013), i.e., constituent of phenocrysts (*sensu strictu*), and dating the crystal cargo will thus provide direct insights into the timescales of magma evolution and ascent leading up to eruption. For example, in mid-ocean ridge lavas, residence times of crystals grown during melt ascent in dikes range from weeks to months (Zellmer et al., 2011, 2012). In contrast, the other scenario reflects rapid magma ascent and crystallization in response to decompression-induced degassing (Figure 1.5b). Here, melt segregation (I) leads to the melt entering superliquidus space, and any crystals within the melt or encountered by the melt will be resorbed (II). Water saturation and degassing (III) will in this case commence at above 0.4 GPa. Crystallization (IV), however, will only occur at shallower depths (~0.2 GPa). This scenario may most closely resemble melts that carry large amounts of volatiles and thus ascend rapidly due to their low density, as is typically the case in subduction-zone environments. Here, magma ascent from the deep crust all the way up to the surface is infrequent, as most magmas will stall due to viscosity increase by degassing-induced crystallization during ascent (Zellmer, Pistone, et al., 2016). However, eruptions of highly undercooled melts may occur if magma ascent is extremely fast; in this case, there will not be enough time for the kinetic processes of nucleation and crystallization to operate. Eruptive products will then be very sparsely phyric, and much of the crystal cargo may be xenocrysts (unrelated to the magmatic system) or antecrysts (related to a consanguineous, but earlier magmatic episode) (Jerram & Martin, 2008), i.e., taken up from one of the many previous small intrusive bodies formed from magmas that stalled during ascent (Zellmer, Freymuth, et al., 2014; Zellmer, Pistone, et al., 2016; Zellmer, Sakamoto, et al., 2014). Here, it is the absence of the phenocrysts (sensu strictu) that may allow us to constrain extremely rapid magma ascent rates of up to tens of meters per second, which will be challenging in terms of hazard mitigation (Petrelli et al., 2018). It is critical to recognize that, in this case, the study of the xenocrystic and antecrystic crystal cargo, including estimations of its formation depth, temperature, and age, may not inform the magmatic processes that resulted in the actual eruption. Instead, these parameters may provide information about the protracted magma assembly before eruptions of basaltic magmas (e.g., Jankovics et al., 2015, 2019), and they may deliver insights into present-day plutonic processes (Zellmer, Pistone, et al., 2016). Quantification of melt-ascent timescales may be accomplished, however, by studying diffusion of volatile elements from melt inclusions and melt embayments in the crystal cargo (Y. Chen et al., 2013; Lloyd et al., 2014), revealing very rapid magma transfer from depth to the surface (Table 1.3).

1.3.4.2. Intermediate to Felsic Systems

The estimation of unlocking timescales of magmatic mushes stored at shallow crustal levels is one of the main challenges of modern petrological and volcanological investigations in intermediate to felsic systems (Bachmann et al., 2005; Bachmann & Bergantz, 2003; G.F. Cooper, Wilson, et al., 2016; Sliwinski et al., 2017; Spera & Bohrson, 2018; Till et al., 2015). Again, thermal, fluid-dynamic, and thermodynamic modeling provide significant constraints on unlocking timescales. For example, Spera & Bohrson (2018) performed thermodynamic and heat transfer calculations for the unlocking of near-solidus rhyolitic mush, heated by the recharge of basaltic magma. Results highlight a broad range of unlocking timescales, ranging from ~10^{-2} to ~10^{6} years, calculated for magma volumes from between 0.1 and 5000 km^3, mainly scaling with the dimension of the system and its dynamics (e.g., stretching and folding). Similar timescales also have been reported by other studies (Bachmann & Huber, 2016; Huber et al., 2010) utilizing fluid-dynamics investigations.

Further constraints on the timescales of preeruptive dynamics mainly come from the study of crystal residence times by diffusion modeling (Costa et al., 2003, 2008; Costa & Dungan, 2005; Dohmen et al., 2018; Druitt et al., 2012; Flaherty et al., 2018; Garzanti et al., 2011; Petrone et al., 2016, 2018), melt-inclusion timekeepers (Pamukcu et al., 2015), concentration variance decay (Perugini et al., 2013; Perugini, De Campos, et al., 2015; Rossi et al., 2017), and CSD analysis (Cashman & Ferry, 1988; Cashman & Marsh, 1988; Castro et al., 2003; Higgins, 2000, 2002; Lejeune & Richet, 1995; Mock & Jerram, 2005; Morgan & Jerram, 2006).

The modeling of chemical diffusion (i.e., Fick's second-law-based geospeedometry) provides significant constraints on the timescales of open-system processes, such as magma assimilation, magma mixing, and the unlocking of magmatic mushes (Costa & Morgan, 2010; Turner & Costa, 2007). For example, the timescales of magma mixing have been widely investigated by chemical diffusion modeling (Costa et al., 2008; Costa & Chakraborty, 2004; Petrone et al., 2016). Table 1.4 (updated from Costa & Chakraborty, 2004) reports some estimates of timescales of mixing-triggered eruptions. Table 1.4 highlights a broad range of timescales, varying across 10 orders of magnitude from minutes (i.e., ~10^{-6} years) to 10^{3} years. These estimates must be interpreted with care, keeping in mind that different techniques often provide timescale estimates of different, often nonlinear, processes operating in the volcanic plumbing system before an eruption. For example, diffusion estimates based on the mixing-induced extreme outward rims of crystals or on concentration variance decay provide insights about processes and timescales associated with the final intrusion event before an eruption. As a consequence, these methods usually yield the shortest timescales. Diffusion estimates based on whole crystals provide information about their residence times within the magmatic system, but only at magmatic temperatures, i.e., not including periods of cool storage close to or below the solidus of the system. In contrast, crystallization ages typically provide the longest timescales, including periods of cool storage, although some of these estimates may be biased by incorporation of antecrystic materials into the sample, which may not be easily distinguishable from true phenocrysts.

Table 1.3 Review of Ascent Velocities from Shallow Reservoirs for Different Volcanic Systems

Reference	Volcanic system	Observation	Minimum ascent rate (m s^{-1})	Average ascent rate (m s^{-1})	Maximum ascent rate (m s^{-1})
Rutherford (2008)	Mount St. Helens, USA	Groundmass crystallization	0.01		>0.2
Rutherford (2008); Rutherford & Hill (1993)	Mount St. Helens, USA	Hornblende rims	0.04		>0.18
Geschwind & Rutherford (1995); Scandone & Malone (1985); Rutherford (2008)	Mount St. Helens, USA	Calculation from mass eruption rate		3.00	
Rutherford (2008)	Mount St. Helens, USA	Seismicity	0.07		0.6
Humphreys et al. (2008)	Mount St. Helens, USA	H_2O diffusion	37		64
Papale et al. (1998); Carey & Sigurdsson (1985); Rutherford (2008)	Mount St. Helens, USA	Numerical modeling	4		15
Rutherford & Devine (2003); Rutherford (2008)	Soufriere Hills, Montserrat	Amphibole rims	0.001		>0.2
Rutherford (2008)	Soufriere Hills, Montserrat	Magnetite	0.003		>0.2
Nakamura (1995); Rutherford (2008); Venezky & Rutherford (1999)	Mount Unzen, Japan	Magnetite zonation	0.03		0.007
Kagiyama et al. (1999); Rutherford (2008)	Mount Unzen, Japan	Seismicity		0.002	
McCanta et al. (2007); Rutherford (2008)	Mount Shasta, USA	Amphibole rims and plagioclase growth	0.004		0.006
Liu et al. (2007); Rutherford (2008)	Taupo Volcanic Zone, New Zealand	CO_2 and H_2O diffusion	0.05		0.35
Liu et al. (2007); Rutherford (2008)	Taupo Volcanic Zone, New Zealand	Theoretical modeling (Papael et al., 2008)	5		8
Myers et al. (2018)	Taupo Volcanic Zone, New Zealand	CO_2 and H_2O	0.06		2.6
Anderson (1991); Rutherford (2008)	Yellowstone Caldera, USA	Diffusion of volatiles		40	
Myers et al. (2018)	Yellowstone Caldera, USA	CO_2 and H_2O gradients	0.3		4
Myers et al. (2016)	Yellowstone Caldera, USA	CO_2 and H_2O gradients	0.3		1.5
Myers et al. (2018)	Long Valley Caldera, USA	CO_2 and H_2O gradients	0.6		13
Scott et al. (2012)	Santiaguito, Guatemala	Amphibole rims	0.007		0.023
Miwa et al. (2009)	Sakurajima, Japan	MND water exsolution rate meter	0.11		0.35
Cassidy et al. (2015)	Colima, Mexico	MND water exsolution rate meter	0.0001		0.068
Lloyd et al. (2016)	Volcán de Fuego, Guatemala	Nanosims results from olivine-hosted melt embayments	11		17

Table 1.4 Summary of Timescales of Open Magmatic System Experiencing Magma Refilling from Deeper Crustal Levels. Timescales for Closed System Differentiation are also Reported for Reference

References	Time (prior to eruption)	Magmatic system	Method	Process
Mashima (2004)	Hours to days	Saga–Futagoyama, Japan	Reaction coronas, plagioclase dissolution	Basalt dacite mixing in conduit
White (1996)	Days to weeks	Pinatubo, Philippines	Seismological data	Basaltic intrusion in silicic reservoir
Nakamura (1995); Venezky & Rutherford (1999)	Weeks to months	Mount Unzen, Japan	Fe–Ti oxide zoning	Dacite–dacite mixing
Chertkoff & Gardner (2004)	Days to 6 months	Ceboruco, Mexico	Fe–Ti oxide zoning, plagioclase overgrowth	Rhyodacite–dacite–mafic mixing
Coombs et al. (2000)	Days to 2 years	Trident, Alaska	Fe–Ti oxide and Fe–Mg olivine zoning	Dacite–andesite mixing
Zellmer et al. (2003)	10–1200 years	Soufriere Hills, Montserrat	Sr and Ba plagioclase zoning	Andesite remobilization by mafic intrusions
Devine et al. (2003)	Days to weeks	Soufriere Hills, Montserrat	Fe–Ti oxide zoning	Andesite remobilization by mafic intrusions
Costa & Chakraborty (2004)	Year to decades	San Pedro, Chile	Fe–Mg olivine zoning	Basaltic andesite–dacite mixing at depth
Tomiya et al. (2013)	Hours to days	Shinmoedake volcano, Japan	Mg, Al, Ti magnetite zoning	Basalt–dacite or basaltic andesite–dacite mixing
Petrone et al. (2016, 2018)	10–50 years	Stromboli volcano, Italy	Non-isothermal BSE grayscale zoning in clinopyroxenes	Shoshonitic basalts–shoshonites mixing
Matthews et al. (2012)	10–60 years	Taupo Volcanic Zone, New Zealand	Ti in quartz	Mafic magma intrusion in silicic reservoir
Perugini, De Campos, Petrelli, & Dingwell (2015)	Minutes to hours	Campi Flegrei caldera, Italy	Concentration variance decay	Shoshonite injection in phonolitic reservoir
Montagna et al. (2015)	Hours	Campi Flegrei caldera, Italy	Numerical modeling	Shoshonite injection in phonolitic reservoir
Astbury et al. (2018)	Hours to days	Campi Flegrei caldera	Thickness of final recharge zones and outermost rims	Shoshonite injection in phonolitic reservoir
Arzilli er al. (2016)	Hours to 2 days	Campi Flegrei caldera, Italy	Feldspar nucleation kinetics and growth rate estimates	Shoshonite injection in phonolitic reservoir
Iovine et al. (2017)	2–60 years	Campi Flegrei caldera, Italy	Diffusion chronometry performed on sanidine crystals	Shoshonite injection in phonolitic reservoir
Chamberlain et al. (2014)	150–500 years	Long Valley caldera, USA	Diffusion of Ti in quartz, Ba and Sr in sanadine, and Fe–Mg in orthopyroxene	Bishop Tuff rhyolite rejuvination by mafic injection
Wark et al. (2007)	< 100 years	Long Valley caldera, USA	Ti in quartz	Bishop Tuff rhyolite rejuvination by mafic injection
Till et al. (2015)	~10 months	Yellowstone caldera, USA	Nanometer-scale trace element diffusion in sanidine crystals	Ryolitic magma rejuvination by silicic intrusion
Druitt et al. (2012)	Less than 100 years	Santorini volcano, Greece	Mg, Sr and Ti diffusion in plagioclase	Recharge of the magma reservoir by silicic magma

(Continued)

Table 1.4 (Continued)

References	Time (prior to eruption)	Magmatic system	Method	Process
Flaherty et al. (2018)	Less than 2 years	Santorini volcano, Greece	Mg–Fe diffusion profiles in ortho- and clinopyroxenes	Recharge of the magma reservoir by silicic magma
G.F. Cooper et al. (2017)	Decades to centuries	Taupo Volcanic Zone, New Zealand	Fe–Mg diffusion in orthopyroxenes and Ti diffusion in quartz	Silicic mush rejuvination by a less evolved magma
K.M. Cooper, Sims et al. (2016)	Decades to a few centuries	Krafla volcano, Iceland	U-series and diffusion data	Complex interactions between melts and crystallized material
Hartley et al. (2016)	6–10 days	Laki, Iceland	Mg diffusion in Olivine	Open system basaltic feeding system
Kilgour et al. (2014)	3–5 months	Taupo Volcanic Zone, New Zealand	BSE zoning in pyroxenes as Fe–Mg proxy	Andesitic magma-mush and magma–magma interaction
Rae et al. (2016)	Months to years	Kīlauea Volcano, Hawaii	Fe–Mg olivine zoning	Mafic reservoir refilled by primitive magmas from depth
Ubide & Kamber (2018)	2–12 days	Mount Etna, Italy	Clinopyroxene growth rates	Mafic recharge into mafic resident mushes
Singer et al. (2016)	Weeks	Novarupta vent, Alaska	Fe–Mg diffusion in orthopyroxene and Mg in plagioclase rims	Silicic eruption primed by mafic recharge
Giuffrida & Viccaro (2017)	< 5 months	Mount Etna, Italy	Fe-Mg diffusion modeling on olivine	Mafic recharge into mafic resident mushes
Snyder (2000); Sparks & Walker (1977); Folch & Martí (1998)	Days to weeks	Various	Numerical modeling	Basaltic intrusion in silicic reservoir
Hawkesworth et al. (2000); Turner et al. (2003); Condomines et al. (2003); Reagan et al. (2003); Hawkesworth et al. (2004); George et al. (2004); Matthews et al. (2012); Reid (2003)	10^3 to 10^5 years	Various	Th–U, Ra–Th disequilibria, Ti diffusion in quartz	Mafic to silicic closed-system differentiation

Note. Modified from Costa & Chakraborty (2004). BSE, back-scattered electrons

Crystal-size-distribution analyses provide estimations about crystal growth processes and crystal residence times, but the latter critically depend on independent constraints on crystal growth rates. In practice, most studies estimate growth rates without independent constraints, which are typically challenging to obtain (cf. Zellmer, Sakamoto, et al., 2016, 2018). Thus, residence-time estimates based on CSD studies likely have order-of-magnitude uncertainties, although this is rarely communicated.

Table 1.4 also reports estimated timescales for mafic to silicic closed-system differentiation using U–Th and Ra–Th disequilibria, of the order 10^3 to 10^5 years. The latter are strikingly similar to the timescales of melt differentiation in the deep crust on the basis of thermal modeling (cf. section 1.4), and support the notion of Annen et al. (2005) that magmatic differentiation from mafic to intermediate compositions may dominantly occur in the deep crust, prior to magma ascent. In contrast, differentiation from intermediate to felsic compositions of erupted material is more likely occurring at mid- to shallow crustal levels (Adam et al., 2016; Dosseto et al., 2008), consistent with the higher viscosities of silica-rich magmas, which impede their ascent from great depths, and with their typically lower temperatures compared to mafic eruptives, although there are notable exceptions in some subduction-zone environments (Straub et al., 2014).

Table 1.3 reports ascent rates of intermediate to felsic melts from their storage region at the onset of eruption. Many of these estimates employ studies of volatile diffusion and of mineral reaction-rim dimensions, although numerical modeling of groundmass crystallization, and seismic constraints, have also been employed (Anderson, 1991; Carey & Sigurdsson, 1985; Cassidy et al., 2015; Cox et al., 2012; Geschwind & Rutherford, 1995; Humphreys et al., 2008; Kagiyama et al., 1999; Liu et al., 2007; McCanta et al., 2007; Miwa et al., 2009; Myers et al., 2016, 2018; Nakamura, 1995; Papale et al., 1998; Rutherford, 2008; Rutherford & Devine, 2003; Rutherford & Hill, 1993; Scandone & Malone, 1985; Stein & Spera, 1996; Venezky & Rutherford, 1999). With the exception of magmas feeding large, Plinian eruption columns, ascent rates are slower than those of hydrous mafic magmas. This is consistent with the higher viscosity of intermediate to felsic melts, which impedes their flow through the volcanic conduit.

1.3.4.3. Interplay Between Ascent Rates and Evolution of Volatiles in Modulating Styles and Timescales of Eruptions

The role of volatiles in modulating styles and timescales of eruption has been widely discussed in the literature (e.g., Cassidy et al., 2016; Edmonds & Woods, 2018; Gonnermann & Manga, 2013). Here, we summarize the main achievements and recent advances (i.e., Edmonds & Woods, 2018). As widely known, volatiles modulate many magma reservoir processes (e.g., Cassidy et al., 2016, 2018; Edmonds & Woods, 2018; Gonnermann & Manga, 2013). As an example, exsolved volatile species increase the buoyancy of magmas (e.g., Edmonds & Woods, 2018; Petrelli et al., 2018), affect phase equilibria (e.g., Ghiorso & Gualda, 2015), and may promote convective dynamics (e.g., Cardoso & Woods, 1999; Huppert et al., 1982; Ruprecht et al., 2008). Also, the exsolved volatile phase may accumulate in the roof zones of crystal-rich magma reservoirs (e.g., Parmigiani et al., 2016), potentially activating hydrothermal systems (Chiodini et al., 2012), and triggering explosive eruptive dynamics (e.g., Bachmann & Bergantz, 2008; Edmonds & Wallace, 2017; Edmonds & Woods, 2018).

However, the role of volatiles always must be considered in the light of many other physical parameters affecting the evolution of a magma reservoir (e.g., Cassidy et al., 2018). In terms of magma storage, Petrelli et al. (2018) investigated the timescales of volatiles accumulation, measured as the times needed to achieve twice the initial water content in the melt phase due to the crystallization of mainly anhydrous phases. Results show that timescales of water accumulation are a function of the dynamics occurring within the magmatic system, with convection-dominated scenarios highlighting accumulation times that are faster than static cases by a factor of ~2. For example, a magmatic system characterized by a width equal to ~5 km and 3 km height, in static conditions (i.e., heat transfer dominated by conduction), will accumulate within ~100 kyr. In contrast, magmatic bodies in which convective motions dominate the transfer of heat within the system will double the water concentration in the melt phase in less than ~50 kyr (Petrelli et al., 2018).

In terms of magma ascent, volatiles are highly effective in reducing melt viscosity, and this will impact ascent rates. Dry melts will rise less rapidly and crystallize due to concomitant cooling (Figure 1.5a), while hydrous melts will rise rapidly and reach superliquidus conditions during ascent (Figure 1.5b). If pathways for rapid magma ascent are available, e.g., promoted by tectonic fracturing of the crust (Zellmer et al., 2019), direct melt ascent from deep crustal levels to the surface is feasible in both cases (Figure 1.6a). If not, more slowly ascending hydrous melts will crystallize in response to decompression-induced degassing, and the concomitant increase in their viscosity may result in preeruptive stalling (Zellmer, Edmonds, et al., 2015) and cooling, leading to the formation of magma-mush reservoirs (Figure 1.6b). The initial volatile content of the melt will impact the depth of degassing-induced crystallization, concomitant melt-viscosity increases, and therefore stalling depths. For example, Zellmer, Sakamoto, et al. (2016) have shown that magma reservoir depths increase from west to east along the Trans-Mexican Volcanic Belt (TMVB), likely due to a concomitant eastwards increase in volatiles of primary magmas released from the mantle wedge. Shallower ponding levels of magmas in the western TMVB are supported by an increase in surface heat flux westwards along the arc.

Summarizing, despite the range of different processes governing eruption styles, Cassidy et al. (2018) proposed a simplified model accounting for ascent rates from storage to surface and for outgassing efficiency. The ascent rate depends mainly on the crustal stress field, overpressure, conduit radius, viscosity, and magma buoyancy (Cassidy et al., 2018). In contrast, outgassing efficiency is mainly governed by the viscosity, pressure or solubility, time (linked inherently to ascent rate), bubble nucleation/coalescence, permeability of magma and conduit wall rock, and fragmentation processes (Cassidy et al., 2018). Magma buoyancy and ascent speeds are increased by the occurrence of exsolved volatiles, potentially leading to explosive eruptions (e.g., Petrelli et al., 2018). However, Degruyter et al. (2017) suggested that, during recharge events, exsolved volatiles may lower the explosivity due to an increase in magma reservoir compressibility, dampening pressurization (Cassidy et al., 2018). It is straightforward that the interplay between exsolved volatiles and the many other parameters governing ascent rates and eruptive behavior requires further studies by direct measurements (e.g., Edmonds et al., 2014), petrological experiments (e.g., Pistone et al., 2015, 2017), and studies using analogue (e.g., Spina et al., 2016) and numerical modeling (Degruyter et al., 2017).

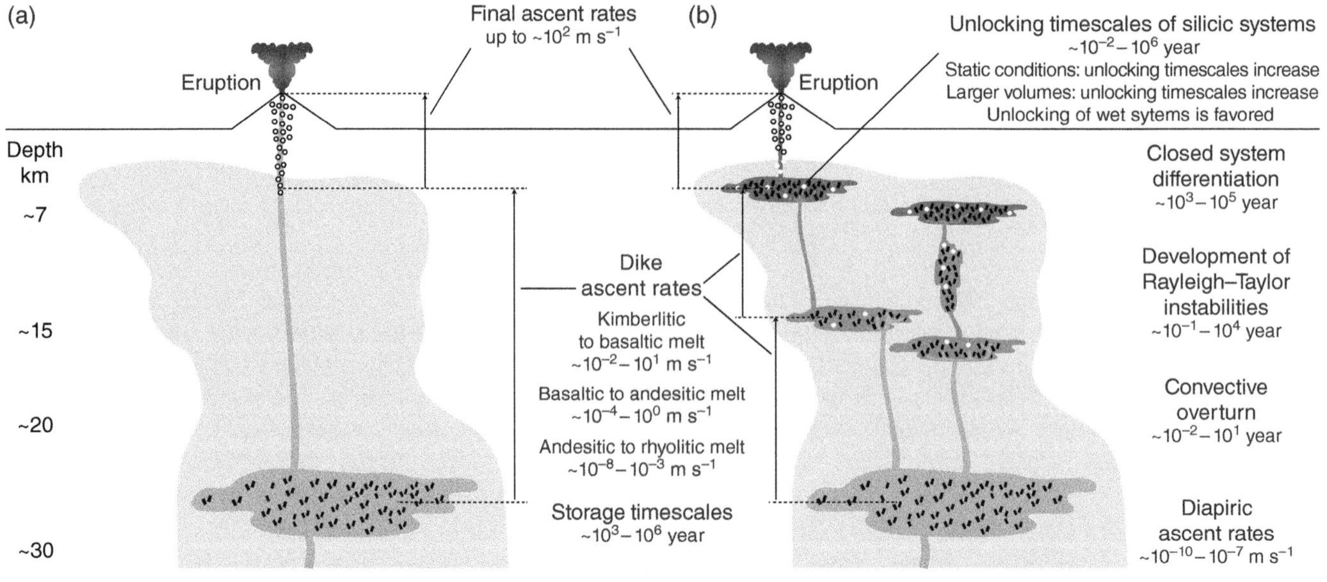

Figure 1.6 Summary of rates and timescales of magma transfer, storage, emplacement, and eruption. (a) Direct magma ascent from deep crustal levels, facilitated by the availability of ascent pathways. (b) Where direct ascent pathways to the surface do not exist, intermittent magma and mush reservoirs will develop at mid- to shallow crustal levels. Magmatic volatile contents likely have a strong influence on ascent rates in (a), as well as the depth of storage in (b). See text for discussion.

1.4. CONCLUSIONS AND OUTLOOK

In this chapter we have posed the following questions (Figure 1.2):

How does the melt phase evolve in space and time in a mush-dominated magmatic plumbing system?

Do long-lived magmatic reservoirs exist?

What are the timescales of preeruptive events?

What are the rates of magma ascent to the Earth surface before an eruption?

We have reviewed the large number of diverse geochemical, petrological, and geophysical techniques that are at our disposal to address these issues, and have presented a summary of our current understanding of the rates and timescales in Figure 1.6.

We distinguish two endmember scenarios, namely direct magma ascent from deep crustal levels to the surface (Figure 1.6a), and intermittent storage and magmatic evolution in mid- to shallow crustal magma and mush reservoirs (Figure 1.6b). Besides the control of magma composition, critical factors in determining the ascent of magmas from the mantle, their storage in the crust, and the eventual propagation to the surface, is the role of volatiles, as well as tectonic controls that may enhance or hinder magma ascent to the surface. A better understanding of these processes may ultimately provide significant insights into potential links between subvolcanic processes and the style of volcanic eruptions.

Clearly, future work on this subject will significantly benefit from determination of rates and timescales of magmatism in the context of magmatic volatiles and their cycles, and in relation to crustal stress fields, to achieve a more systemic understanding of the parameters that control magmatic processes and their timescales on Earth.

ACKNOWLEDGMENTS

We would like to acknowledge all colleagues who have contributed to the advances in our understanding of the rates and timescales of magma transfer, storage, emplacement and eruption over the last few decades. GFZ acknowledges support through the Ministry of Business, Innovation, and Employment (MBIE) grant MAUX1507 and a Massey University Research Fund. MP thanks Agata, Anna, and Caterina for being a precious source of motivation. The editorial handling and the reviewers are gratefully acknowledged.

REFERENCES

Adam, J., Turner, S., & Rushmer, T. (2016). The genesis of silicic arc magmas in shallow crustal cold zones. *Lithos*, *264*, 472–494. https://doi:10.1016/j.lithos.2016.07.036

Anderson, A. (1991). The chronology of colonization in New Zealand. *Antiquity*, *65*(249), 767–795. https://doi:10.1017/S0003598X00080510

Annen, C. (2009). From plutons to magma chambers: Thermal constraints on the accumulation of eruptible silicic magma in the upper crust. *Earth and Planetary Science Letters*, *284*, 409–416. https://doi:10.1016/j.epsl.2009.05.006

Annen, C. (2011). Implications of incremental emplacement of magma bodies for magma differentiation, thermal aureole dimensions and plutonism–volcanism relationships. *Tectonophysics*, *500*, 3–10. https://doi:10.1016/j.tecto.2009.04.010

Annen, C., Blundy, J. D., Leuthold, J., & Sparks, R. S. J. (2015). Construction and evolution of igneous bodies: Towards an integrated perspective of crustal magmatism. *Lithos*, *230*, 206–221. https://doi:10.1016/j.lithos.2015.05.008

Annen, C., Blundy, J. D., & Sparks, R. S. J. (2006). The sources of granitic melt in deep hot zones. *Transactions of the Royal Society of Edinburgh, Earth Sciences*, *97*. https://doi:10.1017/S0263593300001462

Annen, C., Blundy, J. D., & Sparks, R. S. J. (2005). The genesis of intermediate and silicic magmas in deep crustal hot zones. *Journal of Petrolology*, *47*, 505–539. https://doi:10.1093/petrology/egi084

Annen, C., & Sparks, R. S. J. (2002). Effects of repetitive emplacement of basaltic intrusions on thermal evolution and melt generation in the crust. *Earth and Planetary Science Letters*, *203*, 937–955. https://doi:10.1016/S0012-821X(02)00929-9

Arzilli, F., Morgavi, D., Petrelli, M., Polacci, M., Burton, M., Di Genova, D., et al.. (2019). The unexpected explosive sub-Plinian eruption of Calbuco volcano (22–23 April 2015; southern Chile): Triggering mechanism implications. *Journal of Volcanology and Geothermal Research*, *378*, 35–50. https://doi: 10.1016/j.jvolgeores.2019.04.006

Arzilli, F., Piochi, M., Mormone, A., Agostini, C., & Carroll, M. R. (2016). Constraining pre-eruptive magma conditions and unrest timescales during the Monte Nuovo eruption (1538 AD; Campi Flegrei, Southern Italy): Integrating textural and CSD results from experimental and natural trachyphonolites. *Bulletin of Volcanology*, *78*, 72. https://doi:10.1007/s00445-016-1062-z

Astbury, R. L., Petrelli, M., Ubide, T., Stock, M. J., Arienzo, I., D'Antonio, M., & Perugini, D. (2018). Tracking plumbing system dynamics at the Campi Flegrei caldera, Italy: High-resolution trace element mapping of the Astroni crystal cargo. *Lithos*, *318–319*, 464–477. https://doi:10.1016/j.lithos.2018.08.033

Avanzinelli, R., Elliott, T., Tommasini, S., & Conticelli, S. (2007). Constraints on the genesis of potassium-rich Italian volcanic rocks from U/Th disequilibrium. *Journal of Petrolology*, *49*, 195–223. https://doi:10.1093/petrology/egm076

Bachmann, O. (2010). Timescales associated with large silicic magma bodies. In A. Dosseto, S. P. Turner, J. A. Van Orman (Eds.), *Timescales of magmatic processes: From core to atmosphere* (pp. 212–230). Chichester, UK: J. Wiley & Sons. https://doi:10.1002/9781444328509.ch10

Bachmann, O., & Bergantz, G. W. (2003). Rejuvenation of the Fish Canyon magma body: A window into the evolution of large-volume silicic magma systems. *Geology*, *31*, 789. https://doi:10.1130/G19764.1

Bachmann, O., & Bergantz, G. W. (2008). The magma reservoirs that feed supereruptions. *Elements*, *4*, 17–21. https://doi: 10.2113/GSELEMENTS.4.1.17

Bachmann, O., Dungan, M. A., & Bussy, F. (2005). Insights into shallow magmatic processes in large silicic magma bodies: the trace element record in the Fish Canyon magma body, Colorado. *Contributions to Mineralogy and Petrology*, *149*, 338–349. https://doi:10.1007/s00410-005-0653-z

Bachmann, O., Dungan, M. A., & Lipman, P. W. (2002). The Fish Canyon magma body, San Juan volcanic field, Colorado: Rejuvenation and eruption of an upper-crustal batholith. *Journal of Petrolology*, *43*, 1469–1503.

Bachmann, O., & Huber, C. (2016). Silicic magma reservoirs in the Earth's crust. *American Mineralogist*, *101*, 2377–2404. https://doi:10.2138/am-2016-5675

Bachmann, O., Oberli, F., Dungan, M. A., Meier, M., Mundil, R., & Fischer, H. (2007). $^{40}Ar/^{39}Ar$ and U–Pb dating of the Fish Canyon magmatic system, San Juan Volcanic field, Colorado: Evidence for an extended crystallization history. *Chemical Geology*, *236*, 1–2, 34–166.

Bachmann, O., Miller, C. F., & de Silva, S. L. (2007). The volcanic–plutonic connection as a stage for understanding crustal magmatism. *Journal of Volcanology and Geothermal Research*, *167*, 1–4, 1–23.

Bacon, C. R., & Lanphere, M. A. (2006). Eruptive history and geochronology of Mount Mazama and the Crater Lake region, Oregon. *Bulletin of the Geological Society of America*, *118*, 1331–1359.

Bacon, C. R., & Lowenstern, J. B. (2005). Late Pleistocene granodiorite source for recycled zircon and phenocrysts in rhyodacite lava at Crater Lake, Oregon. *Earth and Planetary Science Letters*, *233*, 277–293. https://doi:10.1016/j.epsl.2005.02.012

Barboni, M., Boehnke, P., Schmitt, A. K., Harrison, T. M., Shane, P., Bouvier, A.-S., & Baumgartner, L. (2016). Warm storage for arc magmas. *Proceedings of the National Academy of Sciences*, *113*, 13959–13964. https://doi:10.1073/pnas.1616129113

Barboni, M., & Schoene, B. (2014). Short eruption window revealed by absolute crystal growth rates in a granitic magma. *Nature Geoscience*, *7*, 524–528. https://doi:10.1038/ngeo2185

Bedrosian, P. A., Peacock, J. R., Bowles-Martinez, E., Schultz, A., & Hill, G. J. (2018). Crustal inheritance and a top-down control on arc magmatism at Mount St Helens. *Nature Geoscience*, *1*. https://doi:10.1038/s41561-018-0217-2

Beddoe-Stephens, B., Aspden, J. A., & Shepherd, T. J. (1983). Glass inclusions and melt compositions of the Toba Tuffs, Northern Sumatra. *Contributions to Mineralogy and Petrology*, *83*, 278–287.

Bergantz, G. W., Schleicher, J. M., & Burgisser, A. (2015). Open-system dynamics and mixing in magma mushes. *Nature Geoscience*, *8*, 793–796. https://doi:10.1038/ngeo2534

Bergantz, G. W., Schleicher, J. M., & Burgisser, A. (2017). On the kinematics and dynamics of crystal-rich systems. *Journal of Geophysical Research-Solid Earth*, *122*, 6131–6159. https://doi:10.1002/2017JB014218

Berlo, K., Turner, S., Blundy, J., Black, S., & Hawkesworth, C. (2006). Tracing pre-eruptive magma degassing using ($^{210}Pb/^{226}Ra$) disequilibria in the volcanic deposits of the 1980–1986 eruption of Mount St. Helens. *Earth and Planetary Science Letters*, *249*, 337–349. https://doi:10.1016/j.epsl.2006.07.018

Black, S., Macdonald, R., DeVivo, B., Kilburn, C. R. J., & Rolandi, G. (1998). U-series disequilibria in young (AD 1944) Vesuvius rocks: preliminary implications for magma residence times and volatile addition. *Journal of Volcanology and Geothermal Research, 82*, 97–111. https://doi:10.1016/S0377-0273(97)00059-0

Blankenbach, B., Busse, F., Christensen, U., Cserepes, L., Gunkel, D., Hansen, U., et al. (1989). A benchmark comparison for mantle convection codes. *Geophysical Journal International,98*,23–38. https://doi:10.1111/j.1365-246X.1989.tb05511.x

Blot, C. (1972). Volcanisme et séismes du manteau supérieur dans l'Archipel des Nouvelles-Hébrides. *Bulletin of Volcanology, 36*, 446–461. https://doi:10.1007/BF02597120

Blundy, J., Cashman, K. V., & Berlo, K. (2008). Evolving magma storage conditions beneath Mount St. Helens inferred from chemical variations in melt inclusions from the 1980–1986 and current (2004–2006) eruptions. *US Geological Survey Professional Paper, 1750-33*, 755–790.

Blundy, J. D., & Annen, C. (2016). Crustal magmatic systems from the perspective of heat transfer. *Elements, 12*. https://doi:10.2113/gselements.12.2.115

Bourdon, B., Turner, S., Henderson, G. M., & Lundstrom, C. C. (2003). Introduction to U-series geochemistry. *Reviews in Mineralogy and Geochemistry, 52*(1), 1–21.

Bourdon, B., Worner, G., & Zindler, A. (2000). U-series evidence for crustal involvement and magma residence times in the petrogenesis of Parinacota volcano, Chile. *Contributions to Mineralogy and Petrology, 139*, 458–469. https://doi:10.1007/s004100000150

Bradshaw, R. W., & Kent, A. J. R. (2017). The analytical limits of modeling short diffusion timescales. *Chemical Geology, 466*, 667–677. https://doi:10.1016/j.chemgeo.2017.07.018

Bragagni, A., Avanzinelli, R., Freymuth, H., & Francalanci, L. (2014). Recycling of crystal mush-derived melts and short magma residence times revealed by U-series disequilibria at Stromboli volcano. *Earth and Planetary Science Letters, 404*, 206–219. https://doi:10.1016/j.epsl.2014.07.028

Brenna, M., Cronin, S. J., Smith, I. E. M., Tollan, P. M. E., Scott, J. M., Prior, D. J., Bambery, K., & Ukstins, I. A. (2018). Olivine xenocryst diffusion reveals rapid monogenetic basaltic magma ascent following complex storage at Pupuke Maar, Auckland Volcanic Field, New Zealand. *Earth and Planetary Science Letters, 499*, 13–22. https://doi:10.1016/j.epsl.2018.07.015

Browne, B. L., & Gardner, J. E. (2006). The influence of magma ascent path on the texture, mineralogy, and formation of hornblende reaction rims. *Earth and Planetary Science Letters, 246*, 3–4, 161–176.

Browne, B., & Szramek, L. (2015). Rates of magma ascent and storage. In H. Sigurdsson (Ed.), *The encyclopedia of volcanoes (pp.203–214)*. Elsevier. https://doi:10.1016/B978-0-12-385938-9.00009-2

Burchardt, S. (Ed.) (2018). *Volcanic and igneous plumbing systems: Understanding magma transport, storage, and evolution in the Earth's crust*. Elsevier. https://doi: 10.1016/C2015-0-06837-X

Brugger, C. R., & Hammer, J. E. (2010). Crystal size distribution analysis of plagioclase in experimentally decompressed hydrous rhyodacite magma. *Earth and Planetary Science Letters, 300*, 246–254. https://doi:10.1016/j.epsl.2010.09.046

Burgisser, A., & Bergantz, G. W. (2011). A rapid mechanism to remobilize and homogenize highly crystalline magma bodies. *Nature, 471*, 212–215. https://doi:10.1038/nature09799

Burgisser, A., Chevalier, L., Gardner, J. E., & Castro, J. M. (2017). The percolation threshold and permeability evolution of ascending magmas. *Earth and Planetary Science Letters, 470*, 37–47. https://doi:10.1016/J.EPSL.2017.04.023

Burgisser, A., & Degruyter, W. (2015). Magma ascent and degassing at shallow levels. In H. Sigurdsson (Ed.), *The encyclopedia of volcanoes* (pp. 225–236). Elsevier. https://doi:10.1016/B978-0-12-385938-9.00011-0

Burov, E., Jaupart, C., & Guillou-Frottier, L. (2003). Ascent and emplacement of buoyant magma bodies in brittle-ductile upper crust. *Journal of Geophysical Research, 108*(B4). https://doi.org/10.1029/2002JB001904

Calzolaio, M., Arzilli, F., & Carroll, M. R. (2010). Growth rate of alkali feldspars in decompression-induced crystallization experiments in a trachytic melt of the Phlegraean Fields (Napoli, Italy). *European Journal of Mineralogy, 22*, 485–493. https://doi:10.1127/0935-1221/2010/0022-2012

Canil, D., Fedortchouk, Y. (1999). Garnet dissolution and the emplacement of kimberlites. *Earth and Planetary Science Letters, 167*, 227–237. https://doi:10.1016/S0012-821X(99)00019-9

Carey, S., & Sigurdsson, H. (1985). The May 18, 1980 eruption of Mount St. Helens: 2. Modeling of dynamics of the Plinian phase. *Journal of Geophysical Research, 90*(B4), 294–2958. https://doi.org/10.1029/JB090iB04p02948

Cardoso, S., & Woods, A. W. (1996). Interfacial turbulent mixing in stratified magma reservoirs. *Journal of Volcanology and Geothermal Research, 73*, 157–175.

Caricchi, L., Annen, C., Blundy, J., Simpson, G., & Pinel, V. (2014). Frequency and magnitude of volcanic eruptions controlled by magma injection and buoyancy. *Nature Geoscience, 7*, 126–130. https://doi:10.1038/ngeo2041

Caricchi, L., Annen, C., Rust, A., & Blundy, J. (2012). Insights into the mechanisms and timescales of pluton assembly from deformation patterns of mafic enclaves. *Journal of Geophysical Research, 117*, B11206. https://doi:10.1029/2012JB009325

Cashman, K. V. (1993). Relationship between plagioclase crystallization and cooling rate in basaltic melts. *Contributions to Mineralogy and Petrology, 113*, 126–142. https://doi:10.1007/BF00320836

Cashman, K. V. (2004). Volatile controls on magma ascent and eruption. In R. S. J. Sparks, C. J. Hawkesworth (Eds.), *The State of the Planet: Frontiers and Challenges in Geophysics, Geophysical Monograph Series* (Vol. *150*). Washington, DC: American Geophysical Union. https://doi:10.1029/150GM10

Cashman, K. V., & Ferry, J. M. (1988). Crystal size distribution (CSD) in rocks and the kinetics and dynamics of crystallization. *Contributions to Mineralogy and Petrology, 99*, 401–415. https://doi:10.1007/BF00371933

Cashman, K. V., & Marsh, B. D. (1988). Crystal size distribution (CSD) in rocks and the kinetics and dynamics of crystallization II: Makaopuhi lava lake. *Contributions to Mineralogy and Petrology, 99*, 292–305. https://doi:10.1007/BF00375363

Cashman, K. V., Sparks, R. S. J., & Blundy, J. D. (2017). Vertically extensive and unstable magmatic systems: A unified view of igneous processes. *Science*, *80*(355), eaag3055, https://doi:10.1126/science.aag3055

Cassidy, M., Watt, S. F. L., Talling, P. J., Palmer, M. R., Edmonds, M., Jutzeler, M., et al. (2015). Rapid onset of mafic magmatism facilitated by volcanic edifice collapse. *Geophysical Research Letters*, *42*, 4778–4785. https://doi:10.1002/2015GL064519

Cassidy, M., Castro, J. M., Helo, C., Troll, V. R., Deegan, F. M., Muir, D., Neave, D. A., & Mueller, S. P. (2016). Volatile dilution during magma injections and implications for volcano explosivity. *Geology*, *44*(12), 1027–1030.

Cassidy, M., Manga, M., Cashman, K., & Bachmann, O. (2018). Controls on explosive–effusive volcanic eruption styles. *Nature Communications*, *9*(1), 2839.

Castro, J. M., Cashman, K. V., & Manga, M. (2003). A technique for measuring 3D crystal-size distributions of prismatic microlites in obsidian. *American Mineralogist*, *88*, 1230–1240.

Castro, J. M., & Dingwell, D. B. (2009). Rapid ascent of rhyolitic magma at Chaitén volcano, Chile. *Nature*, *461*, 780–783. https://doi:10.1038/nature08458

Chaigneau, M., Massare, D., & Clocchiatti, R. (1980). Contribution à l'étude des inclusions vitreuses et des éléments volatils contenus dans les phénocristaux de quartz de roches volcaniques acides. *Bulletin of Volcanology*, *43*, 233–240.

Chamberlain, K. J., Morgan, D. J., & Wilson, C. J. N. (2014). Timescales of mixing and mobilisation in the Bishop Tuff magma body: perspectives from diffusion chronometry. *Contributions to Mineralogy and Petrology*, *168*, 1034. https://doi:10.1007/s00410-014-1034-2

Charlier, B., & Zellmer, G. (2000). Some remarks on U–Th mineral ages from igneous rocks with prolonged cystallisation histories. *Earth and Planetary Science Letters*, *183*, 457–469. https://doi:10.1016/S0012-821X(00)00298-3

Charlier, B. L. A., & Wilson, C. J. N. (2010). Chronology and evolution of caldera-forming and post-caldera magma systems at Okataina volcano, New Zealand from zircon U–Th model-age spectra. *Journal of Petrology*, *51*, 1121–1141.

Chen, K. X., Gung, Y., Kuo, B.-Y., & Huang, T. Y. (2018). Crustal magmatism and deformation fabrics in northeast Japan revealed by ambient noise tomography. *Journal of Geophysical Research-Solid Earth*. https://doi:10.1029/2017JB015209

Chen, Y., Provost, A., Schiano, P., & Cluzel, N. (2013). Magma ascent rate and initial water concentration inferred from diffusive water loss from olivine-hosted melt inclusions. *Contributions to Mineralogy and Petrology*, *165*, 525–541. https://doi:10.1007/s00410-012-0821-x

Chenet, A. L., Quidelleur, X., Fluteau, F., Courtillot, V., & Bajpai, S. (2007). ^{40}K–^{40}Ar dating of the main Deccan large igneous province: Further evidence of KTB age and short duration. *Earth and Planetary Science Letters*, *263*, 1–15. https://doi:10.1016/j.epsl.2007.07.011

Chew, D. M., Petrus, J. A., Kenny, G., & McEvoy, N. (2017). Rapid high-resolution U–Pb LA-Q-ICPMS age mapping of zircon. *Journal of Analytical Atomic Spectrometry*, *32*, 262–276. https://doi:10.1039/c6ja00404k

Chertkoff, D. G., & Gardner, J. E. (2004). Nature and timing of magma interactions before, during, and after the caldera-forming eruption of Volcán Ceboruco, Mexico. *Contributions to Mineralogy and Petrology*, *146*, 715–735. doi:10.1007/s00410-003-0530-6

Chiodini, G., Caliro, S., De Martino, P., Avino, R., & Gherardi, F. (2012). Early signals of new volcanic unrest at Campi Flegrei caldera? Insights from geochemical data and physical simulations. *Geology*, *40*(10), 943–946.

Claiborne, L. L., Miller, C. F., Flanagan, D. M., Clynne, M. A., & Wooden, J. L. (2010). Zircon reveals protracted magma storage and recycling beneath Mount St. Helens. *Geology*, *38*, 1011–1014. https://doi:10.1130/G31285.1

Cluzel, N., Laporte, D., Provost, A., & Kannewischer, I. (2008). Kinetics of heterogeneous bubble nucleation in rhyolitic melts: implications for the number density of bubbles in volcanic conduits and for pumice textures. *Contributions to Mineralogy and Petrology*, *156*, 745–763. https://doi:10.1007/s00410-008-0313-1

Coleman, D. S., Gray, W., & Glazner, A. F. (2004). Rethinking the emplacement and evolution of zoned plutons: Geochronologic evidence for incremental assembly of the Tuolumne Intrusive Suite, California. *Geology*, *32*, 433–436. https://doi:10.1130/G20220.1

Comeau, M. J., Unsworth, M. J., Ticona, F., & Sunagua, M. (2015). Magnetotelluric images of magma distribution beneath Volcán Uturuncu, Bolivia: Implications for magma dynamics. *Geology*, *43*(3), 243–246.

Comeau, M. J., Unsworth, M. J., & Cordell D. (2016). New constraints on the magma distribution and composition beneath Volcán Uturuncu and the southern Bolivian Altiplano from magnetotelluric data. *Geosphere*, *12*(5), 1391–1421.

Condomines, M., Gauthier, P. J., & Sigmarsson, O. (2003). Timescales of magma chamber processes and dating of young volcanic rocks, *Reviews in Mineralogy and Geochemistry*, *52*(1), 125–174. doi:10.2113/0520125

Coombs, M. L., Eichelberger, J. C., & Rutherford, M. J. (2000). Magma storage and mixing conditions for the 1953–1974 eruptions of Southwest Trident volcano, Katmai National Park, Alaska. *Contributions to Mineralogy and Petrology*, *140*, 99–118. doi:10.1007/s004100000166

Cooper, G. F., Morgan, D. J., & Wilson, C. J. N. (2017). Rapid assembly and rejuvenation of a large silicic magmatic system: Insights from mineral diffusive profiles in the Kidnappers and Rocky Hill deposits, New Zealand. *Earth and Planetary Science Letters*, *473*, 1–13. https://doi:10.1016/j.epsl.2017.05.036

Cooper, G.F., Wilson, C.J.N., Millet, M.A., Baker, J.A. (2016). Generation and rejuvenation of a supervolcanic magmatic system: A case study from Mangakino volcanic centre, New Zealand. *Journal of Petrolology*, *57*, 1135–1170. https://doi:10.1093/petrology/egw035

Cooper, K. M. (2015). Timescales of crustal magma reservoir processes: Insights from U-series crystal ages. In L. Caricchi, J. D. Blundy (Eds.), *Chemical, physical and temporal evolution of magmatic systems* (Special Publication 422, pp. 141–174). London: Geological Society.

Cooper, K. M. (2019). Time scales and temperatures of crystal storage in magma reservoirs: Implications for magma reservoir dynamics. *Philosophical Transactions of the Royal*

Society A Mathematical, Physical and Engineering Sciences, *377*(2139). https://doi.org/10.1098/rsta.2018.0009

Cooper, K. M., & Kent, A. J. R. (2014). Rapid remobilization of magmatic crystals kept in cold storage. *Nature*, *506*, 480–3. https://doi:10.1038/nature12991

Cooper, K.M., & Reid, M. R. (2003). Re-examination of crystal ages in recent Mount St. Helens lavas: Implications for magma reservoir processes. *Earth and Planetary Science Letters*, *213*, 149–167. https://doi:10.1016/S0012-821X(03)00262-0

Cooper, K. M., & Reid, M. R. (2008). Uranium-series crystal ages. *Reviews in Mineralogy and Geochemistry*, *69*, 479–544. https://doi:10.2138/rmg.2008.69.13

Cooper, K. M., Sims, K. W. W., Eiler, J. M., & Banerjee, N. (2016). Timescales of storage and recycling of crystal mush at Krafla Volcano, Iceland. *Contributions to Mineralogy and Petrology*, *171*. https://doi:10.1007/s00410-016-1267-3

Costa, F., & Chakraborty, S. (2004). Decadal time gaps between mafic intrusion and silicic eruption obtained from chemical zoning patterns in olivine. *Earth and Planetary Science Letters*, *227*, 517–530 https://doi:10.1016/j.epsl.2004.08.011

Costa, F., Chakraborty, S., & Dohmen, R. (2003). Diffusion coupling between trace and major elements and a model for calculation of magma residence times using plagioclase. *Geochimica et Cosmochimica Acta*, *67*, 2189–2200. https://doi:10.1016/S0016-7037(02)01345-5

Costa, F., Dohmen, R., & Chakraborty, S. (2008). Time scales of magmatic processes from modelling the zoning patterns of crystals, *Reviews in Mineralogy and Geochemistry*, *69*(1), 545–594 . https://doi:10.2138/rmg.2008.69.14

Costa, F., & Dungan, M. (2005). Short time scales of magmatic assimilation from diffusion modeling of multiple elements in olivine. *Geology*, *33*, 837–840. https://doi:10.1130/G21675.1

Costa, F., & Morgan, D. (2010). Time Constraints from Chemical Equilibration in Magmatic Crystals ,In A. Dosseto, S. P. Turner, J. A. Van Orman (Eds.), *Timescales of magmatic processes: From core to atmosphere* (pp. 125–159). Chichester, UK: J. Wiley & Sons. https://doi:10.1002/9781444328509

Cox, S., Rutter, H., Sims, A., Manga, M., Weir, J., Ezzy, T., et al. (2012). Hydrological effects of the M_W 7.1 Darfield (Canterbury) earthquake, 4 September 2010, New Zealand. *New Zealand Journal of Geology and Geophysics*, *55*, 231–247. https://doi:10.1080/00288306.2012.680474

Crisp, J. A. (1984). Rates of magma emplacement and volcanic output. *Journal of Volcanology and Geothermal Research*, *20*, 177–211. https://doi:10.1016/0377-0273(84)90039-8

Cunningham, H. S., Turner, S. P., Dosseto, A., Patia, H., Eggins, S. M., & Arculus, R. J. (2009). Temporal variations in U-series disequilibria in an active caldera, Rabaul, Papua New Guinea. *Journal of Petrolology*, *50*, 507–529. https://doi:10.1093/petrology/egp009

Cunningham, H. S., Turner, S. P., Patia, H., Wysoczanski, R., Nichols, A. R. L., Eggins, S., & Dosseto, A. (2009). ($^{210}Pb/^{226}Ra$) variations during the 1994–2001 intracaldera volcanism at Rabaul Caldera. *Journal of Volcanology and Geothermal Research*, *184*, 416–426. https://doi:10.1016/j.jvolgeores.2009.04.018

Dahm, T. (2000). On the shape and velocity of fluid-filled fractures in the Earth. *Geophysical Journal International*, *142*, 181–192. https://doi:10.1046/j.1365-246x.2000.00148.x

Daines, M. J., & Pec, M. (2015). Migration of melt. In H. Sigurdsson (Ed.), *The encyclopedia of volcanoes* (pp. 49–64). https://doi:10.1016/B978-0-12-385938-9.00002-X

Danišík, M., Schmitt, A. K., Stockli, D. F., Lovera, O. M., Dunkl, I., & Evans, N. J. (2017). Application of combined U–Th-disequilibrium/U–Pb and (U–Th)/He zircon dating to tephrochronology. *Quaternary Geochronology*, *40*, 23–32. https://doi:10.1016/J.QUAGEO.2016.07.005

De Angelis, S. H., Larsen, J., Coombs, M., Dunn, A., & Hayden, L. (2015). Amphibole reaction rims as a record of pre-eruptive magmatic heating: An experimental approach. *Earth and Planetary Science Letters*, *426*, 235–245.

De Siena, L., Thomas, C., Waite, G.P., Moran, S.C., & Klemme, S. (2014). Attenuation and scattering tomography of the deep plumbing system of Mount St. Helens. *Journal of Geophysical Research-Solid Earth, 119*, 8223–8238. https://doi:10.1002/2014JB011372

Dégi, J., Abart, R., Török, K., Rhede, D., & Petrishcheva, E. (2009). Evidence for xenolith-host basalt interaction from chemical patterns in Fe–Ti-oxides from mafic granulite xenoliths of the Bakony-Balaton Volcanic field (W-Hungary). *Mineralogy and Petrology*, *95*, 219–234. https://doi:10.1007/s00710-008-0035-0

Degruyter, W., Bachmann, O., Burgisser, A., & Manga, M. (2012). The effects of outgassing on the transition between effusive and explosive silicic eruptions. *Earth and Planetary Science Letters*, *349–350*, 161–170. https://doi:10.1016/j.epsl.2012.06.056

Degruyter, W., Huber, C., Bachmann, O., Cooper, K. M., & Kent, A. J. R. (2016). Magma reservoir response to transient recharge events: The case of Santorini volcano (Greece). *Geology*, *44*(1), 23–26.

Degruyter, W., Huber, C., Bachmann, O., Cooper, K. M., & Kent, A. J. R. (2017). Influence of exsolved volatiles on reheating silicic magmas by recharge and consequences for eruptive style at Volcán Quizapu (Chile). *Geochemistry, Geophysics, Geosystems*, *18*, 11, 4123–4135.

Deino, A. L., Orsi, G., de Vita, S., & Piochi, M. (2004). The age of the Neapolitan Yellow Tuff caldera-forming eruption (Campi Flegrei caldera – Italy) assessed by $^{40}Ar/^{39}Ar$ dating method. *Journal of Volcanology and Geothermal Research*, *133*, 157–170. https://doi:10.1016/S0377-0273(03)00396-2

Demouchy, S., Jacobsen, S. D., Gaillard, F., & Stem, C. R. (2006). Rapid magma ascent recorded by water diffusion profiles in mantle olivine. *Geology*, *34*, 429–432. https://doi:10.1130/G22386.1

Denis, C. M., Demouchy, S., & Shaw, C. S. J. (2013). Evidence of dehydration in peridotites from Eifel Volcanic Field and estimates of the rate of magma ascent. *Journal of Volcanology and Geothermal Research*, *258*, 85–99. https://doi:10.1016/j.jvolgeores.2013.04.010

Devine, J. D., Rutherford, M. J., Norton, G. E., & Young, S. R. (2003). Magma storage region processes inferred from geochemistry of Fe–Ti oxides in andesitic magma, Soufrière Hills Volcano, Montserrat, W.I. *Journal of Petrolology*, *44*, 1375–1400.

Dimalanta, C., Taira, A., Yumul, G., Tokuyama, H., & Mochizuki, K. (2002). New rates of western Pacific island arc magmatism from seismic and gravity data. *Earth and*

Planetary Science Letters, *202*, 105–115. https://doi:10.1016/S0012-821X(02)00761-6

Dohmen, R., Faak, K., & Blundy, J. D. (2018). Chronometry and speedometry of magmatic processes using chemical diffusion in olivine, plagioclase and pyroxenes. In M. J. Kohn, M. Engi, P. Lanari (Eds.), *Petrochronology: Methods and applications, Reviews in mineralogy and geochemistry* (Vol. *83*, pp. 535–575). Chantilly, VA: Mineralogical Society of America. https://doi:10.2138/rmg.2017.83.16

Dosseto, A., Turner, S. P., Sandiford, M., & Davidson, J. (2008). Uranium-series isotope and thermal constraints on the rate and depth of silicic magma genesis. In C. Annen, G. F. Zellmer (Eds.), *Dynamics of crustal magma transfer, storage and differentiation* (Special Publication 304, pp. 169–181). London: Geological Society.. https://doi:10.1144/SP304.9

Dosseto, A., Turner, S. P., & Van Orman, J. A. (Eds.) (2010). *Timescales of magmatic processes: from core to atmosphere.* Chichester, UK: John Wiley & Sons. https://doi:10.1002/9781444328509

Druitt, T. H., Costa, F., Deloule, E., Dungan, M., & Scaillet, B. (2012). Decadal to monthly timescales of magma transfer and reservoir growth at a caldera volcano. *Nature*, *482*, 77–80. https://doi:10.1038/nature10706

Dufek, J., & Bachmann, O. (2010). Quantum magmatism: Magmatic compositional gaps generated by melt-crystal dynamics. *Geology*, *38*, 687–690. https://doi:10.1130/G30831.1

Eddy, M. P., Bowring, S. A., Miller, R. B., & Tepper, J. H. (2016). Rapid assembly and crystallization of a fossil large-volume silicic magma chamber. *Geology*, *44*, 331–334. https://doi:10.1130/G37631.1

Edmonds, M., Cashman, K. V., Holness, M., & Jackson, M. (2019). Architecture and dynamics of magma reservoirs. *Philosophical Transactions of the Royal Society A: Mathematical, Physical and Engineering Sciences*, *377*(2139). https://doi.org/10.1098/rsta.2018.0298

Edmonds, M., Humphreys, M. C. S., Hauri, E. H., Herd, R. A., Wadge, G., Rawson, H., et al. (2014). Pre-eruptive vapour and its role in controlling eruption style and longevity at Soufrière Hills Volcano. *Geological Society, London, Memoirs*, *39*, 291–315. https://doi.org/10.1144/M39.16

Edmonds, M., & Wallace, P. J. (2017). Volatiles and exsolved vapor in volcanic systems. *Elements*, *13*(1), 29–34. https://doi.org/10.2113/gselements.13.1.29

Edmonds, M., & Woods, A. W. (2018). Exsolved volatiles in magma reservoirs. *Journal of Volcanology and Geothermal Research*, *368*, 13–30.

Ehlers, T. A., & Farley, K. A. (2003). Apatite (U–Th)/He thermochronometry: methods and applications to problems in tectonic and surface processes. *Earth and Planetary Science Letters*, *206*, 1–14. https://doi:10.1016/S0012-821X(02)01069-5

Engi, M., Lanari, P., & Kohn, M. J. (2017). Significant ages—an introduction to petrochronology. *Reviews in mineralogy and geochemistry* (Vol. 83, pp. 1–12). Chantilly, VA: Mineralogical Society of America. https://doi:10.2138/rmg.2017.83.1

Farley, K.A. (2002). (U-Th)/He Dating: techniques, calibrations, and applications. *Reviews in Mineralogy and Geochemistry*, *47*, 819–844. https://doi:10.2138/rmg.2002.47.18

Farley, K. A., Kohn, B. P., & Pillans, B. (2002). The effects of secular disequilibrium on (U–Th)/He systematics and dating of Quaternary volcanic zircon and apatite. *Earth and Planetary Science Letters*, *201*(1), 117–125.

Farley, K. A., & Stockli, D. F. (2019). (U–Th)/He dating of phosphates: Apatite, monazite, and xenotime. *Reviews in Mineralogy and Geochemistry*, *48*(1), 559–577. https://doi.org/10.2138/rmg.2002.48.15

Fedotov, S. A., Gorel'chik, V. I., Stepanov, V. V., & Garbuzova, V. T. (1983). The development of the great Tolbachik fissure eruption in 1975 from seismological data. In S. A. Fedotov, Y. K. Markhinin (Eds.), *The Great Tolbachik Fissure Eruption* (pp. 189–203). Cambridge University Press.

Ferguson, D. J., Gonnermann, H. M., Ruprecht, P., Plank, T., Hauri, E. H., Houghton, B. F., & Swanson, D. A. (2016). Magma decompression rates during explosive eruptions of Kīlauea volcano, Hawaii, recorded by melt embayments. *Bulletin of Volcanology*, *78*(10). 10.1007/s00445-016-1064-x

Ferriss, E., Plank, T., Newcombe, M., Walker, D., & Hauri, E. (2018). Rates of dehydration of olivines from San Carlos and Kilauea Iki. *Geochimica et Cosmochimica Acta*, *242*, 165–190.

Ferry, J. M., & Watson, E. B. (2007). New thermodynamic models and revised calibrations for the Ti-in-zircon and Zr-in-rutile thermometers. *Contributions to Mineralogy and Petrology*, *154*, 429–437. https://doi:10.1007/s00410-007-0201-0

Field, L., Blundy, J., Calvert, A., & Yirgu, G. (2013). Magmatic history of Dabbahu, a composite volcano in the Afar Rift, Ethiopia. *Geological Society of America Bulletin*, *125*, 128–147. https://doi:10.1130/B30560.1

Flaherty, T., Druitt, T. H., Tuffen, H., Higgins, M. D., Costa, F., & Cadoux, A. (2018). Multiple timescale constraints for high-flux magma chamber assembly prior to the Late Bronze Age eruption of Santorini (Greece). *Contributions to Mineralogy and Petrology*, *173*. https://doi:10.1007/s00410-018-1490-1

Folch, A., Martí, J. (1998). The generation of overpressure in felsic magma chambers by replenishment. *Earth and Planetary Science Letters*, *163*, 301–314. https://doi:10.1016/S0012-821X(98)00196-4

Forestier-Coste, L., Mancini, S., Burgisser, A., & James, F. (2012). Numerical resolution of a mono-disperse model of bubble growth in magmas. *Applied Mathematical Modelling*, *36*, 5936–5951. https://doi:10.1016/j.apm.2012.01.031

Frost, B. R., Chamberlain, K. R., & Schumacher, J. C. (2001). Sphene (titanite): Phase relations and role as a geochronometer. *Chemical Geology*, *172*, 131–148. https://doi:10.1016/S0009-2541(00)00240-0

Fu, B., Page, F. Z., Cavosie, A. J., Fournelle, J., Kita, N. T., Lackey, J. S., Wilde, S. A., & Valley, J. W. (2008). Ti-in-zircon thermometry: applications and limitations. *Contributions to Mineralogy and Petrology*, *156*, 197–215. https://doi:10.1007/s00410-008-0281-5

Gardner, J. E., Jackson, B. A., Gonnermann, H., & Soule, S. A. (2016). Rapid ascent and emplacement of basaltic lava during the 2005–06 eruption of the East Pacific Rise at ca. 9°51′N as inferred from CO_2. *Earth and Planetary Science Letters*, *453*, 152–160. https://doi:10.1016/j.epsl.2016.08.007

Garzanti, E., Andó, S., France-Lanord, C., Censi, P., Vignola, P., Galy, V., & Lupker, M. (2011). Mineralogical and chemical variability of fluvial sediments 2. Suspended-load silt (Ganga–Brahmaputra, Bangladesh). *Earth and Planetary Science Letters*, *302*, 107–120. https://doi:10.1016/j.epsl.2010.11.043

Gebhardt, D. J., & Butler, S. L. (2016). Linear analysis of melt band formation in a mid-ocean ridge corner flow. *Geophysical Research Letters*, *43*, 3700–3707. https://doi:10.1002/2016GL068688

Gelman, S. E., Deering, C. D., Gutierrez, F. J., & Bachmann, O. (2013). Evolution of the Taupo Volcanic Center, New Zealand: petrological and thermal constraints from the Omega dacite. *Contributions to Mineralogy and Petrology*, *166*, 1355–1374. https://doi:10.1007/s00410-013-0932-z

George, R., Turner, S., Hawkesworth, C., Bacon, C. R., Nye, C., Stelling, P., & Dreher, S. (2004). Chemical versus temporal controls on the evolution of tholeiitic and calc-alkaline magmas at two volcanoes in the Alaska–Aleutian arc. *Journal of Petrology*, *45*, 203–219.

Geschwind, C. H., & Rutherford, M. J. (1995). Crystallization of microlites during magma ascent: the fluid mechanics of 1980–1986 eruptions at Mount St Helens. *Bulletin of Volcanology*, *57*, 356–370. https://doi:10.1007/BF00301293

Ghiorso, M. S., & Gualda, G. A. R. (2015). An H_2O–CO_2 mixed fluid saturation model compatible with rhyolite-MELTS. *Contributions to Mineralogy and Petrology*, *169*(6), 53.

Gillot, P.Y., Hildebrand, A., Lefèvre, J.,C., Albore-Livadie, C. (2006). The K/Ar Dating Method : Principle, Analytical Techniques, and Application to Holocene Volcanic Eruptions in Southern Italy. *Acta Vulcanologica*, *18*, 1–2, https://doi:10.1400/93820

Giuffrida, M., & Viccaro, M. (2017). Three years (2011–2013) of eruptive activity at Mt. Etna: Working modes and timescales of the modern volcano plumbing system from micro-analytical studies of crystals. *Earth-Science Reviews*, *171*, 289–322. doi:10.1016/j.earscirev.2017.06.003

Glazner, A. F., Bartley, J. M., Coleman, D. S., Gray, W., & Taylor, R. Z. (2004). Are plutons assembled over millions of years by amalgamation from small magma chambers? *GSA Today*, *14*(4–5), 4–11.

Glodny, J., Austrheim, H., Molina, J. F., Rusin, A. I., & Seward, D. (2003). Rb/Sr record of fluid–rock interaction in eclogites: The Marun-Keu complex, Polar Urals, Russia. *Geochimica et Cosmochimica Acta*, *67*, 4353–4371. https://doi:10.1016/S0016-7037(03)00370-3

Glodny, J., Bingen, B., Austrheim, H., Molina, J. F., & Rusin, A. (2002). Precise eclogitization ages deduced from Rb/Sr mineral systematics: The Maksyutov complex, Southern Urals, Russia. *Geochimica et Cosmochimica Acta*, *66*, 1221–1235. https://doi:10.1016/S0016-7037(01)00842-0

Goldstein, S. J., Murrell, M. T., & Williams, R. W. (1993). ^{231}Pa and ^{230}Th chronology of mid-ocean ridge basalts. *Earth and Planetary Science Letters*, *115*, 151–159. https://doi:10.1016/0012-821X(93)90219-Y

Gonnermann, H. M., & Manga, M. (2013). Dynamics of magma ascent in the volcanic conduit. In S. A. Fagents, T. K. P. Gregg, R. M. C. Lopes (Eds.), *Modeling volcanic processes: The physics and mathematics of volcanism* (pp. 55–84). Cambridge University Press. https://doi:10.1017/CBO9781139021562.004

Gonnermann, H. M., & Taisne, B. (2015). Magma transport in dikes. In H. Sigurdsson (Ed.), *The encyclopedia of volcanoes* (pp.215–224). Elsevier. https://doi:10.1016/B978-0-12-385938-9.00010-9

Gualda, G. A. R., Ghiorso, M. S., Lemons, R. V., & Carley, T. L. (2012). Rhyolite-MELTS: a modified calibration of MELTS optimized for silica-rich, fluid-bearing magmatic systems. *Journal of Petrolology*, *53*, 875–890. https://doi:10.1093/petrology/egr080

Gualda, G. A. R., Pamukcu, A. S., Ghiorso, M.S., Anderson, A. T., Sutton, S. R., & Rivers, M. L. (2012). Timescales of quartz crystallization and the longevity of the Bishop giant magma body. *PLoS One*, *7*, e37492. https://doi:10.1371/journal.pone.0037492

Gualda, G. A.R., & Sutton, S. R. (2016). The year leading to a supereruption. *PLoS One*, *11*(7), e0159200.

Hahn, O., & Walling, E. (1938). Über die Möglichkeit geologischer Altersbestimmungen rubidiumhaltiger Mineralien und Gesteine. *Zeitschrift für anorganische und allgemeine Chemie*, *236*, 78–82. https://doi:10.1002/zaac.19382360109

Halama, R., Glodny, J., Konrad-Schmolke, M., & Sudo, M. (2018). Rb–Sr and in situ ^{40}Ar/^{39}Ar dating of exhumation-related shearing and fluid-induced recrystallization in the Sesia zone (Western Alps, Italy). *Geosphere*, *14*, 1425–1450. https://doi:10.1130/GES01521.1

Harangi, S., Sági, T., Seghedi, I., Ntaflos, T. (2013). Origin of basaltic magmas of Perşani volcanic field, Romania: A combined whole rock and mineral scale investigation. *Lithos*, *180–181*, 43–57. https://doi:10.1016/j.lithos.2013.08.025

Harangi, S., Lukács, R., Schmitt, A. K., Dunkl, I., Molnár, K., Kiss, B., et al. (2015). Constraints on the timing of Quaternary volcanism and duration of magma residence at Ciomadul volcano, east-central Europe, from combined U–Th/He and U–Th zircon geochronology. *Journal of Volcanology and Geothermal Research*, *301*, 66–80.

Harangi, S., Novák, A., Kiss, B., Seghedi, I., Lukács, R., Szarka, L., et al. (2015). Combined magnetotelluric and petrologic constrains for the nature of the magma storage system beneath the Late Pleistocene Ciomadul volcano (SE Carpathians). *Journal of Volcanology and Geothermal Research*, *290*, 82–96.

Hartley, M. E., Morgan, D. J., Maclennan, J., Edmonds, M., & Thordarson, T. (2016). Tracking timescales of short-term precursors to large basaltic fissure eruptions through Fe–Mg diffusion in olivine. *Earth and Planetary Sciences Letters*, *439*, 58–70. doi:10.1016/J.EPSL.2016.01.018

Hawkesworth, C. J., Blake, S., Evans, P., Hughes, R., Macdonald, R., Thomas, L. E., Turner, S. P., & Zellmer, G. (2000). Time scales of crystal fractionation in magma chambers—integrating physical, isotopic and geochemical perspectives. *Journal of Petrology*, *41*, 991–1006. doi:10.1093/petrology/41.7.991

Hawkesworth, C., George, R., Turner, S., Zellmer, G., Rhiannon, G., Turner, S., & Zellmer, G. (2004). Time scales of magmatic processes. *Earth and Planetary Science Letters*, *218*, 1–16. https://doi:10.1016/S0012-821X(03)00634-4

Hayden, L. A., Watson, E. B., & Wark, D. A. (2008). A thermobarometer for sphene (titanite). *Contributions to Mineralogy and Petrology*, *155*, 529–540. https://doi:10.1007/s00410-007-0256-y

Higgins, M. D. (1996a). Magma dynamics beneath Kameni volcano, Thera, Greece, as revealed by crystal size and shape measurements. *Journal of Volcanology and Geothermal Research, 70*, 37–48. https://doi:10.1016/0377-0273(95)00045-3

Higgins, M. D. (1996b). Crystal size distributions and other quantitative textural measurements in lavas and tuff from Egmont volcano (Mt. Taranaki), New Zealand. *Bulletin of Volcanology, 58*, 194–204. https://doi:10.1007/s004450050135

Higgins, M. D. (2000). Measurement of crystal size distributions. *American Mineralogist, 85*, 1105–1116. https://doi:10.2138/am-2000-8-901

Higgins, M. D. (2002). Closure in crystal size distributions (CSD), verification of CSD calculations, and the significance of CSD fans. *American Mineralogist, 87*, 171–175. https://doi:10.2138/am-2002-0118

Higgins, M. D. (2006). Verification of ideal semi-logarithmic, lognormal or fractal crystal size distributions from 2D datasets. *Journal of Volcanology and Geothermal Research, 154*, 8–16. https://doi:10.1016/j.jvolgeores.2005.09.015

Hildreth, W. (2004). Volcanological perspectives on Long Valley, Mammoth Mountain, and Mono Craters: several contiguous but discrete systems. *Journal of Volcanology and Geothermal Research, 136*, 169–198. https://doi:10.1016/J.JVOLGEORES.2004.05.019

Hill, G. J., Caldwell, T. G., Heise, W., Chertkoff, D. G., Bibby, H. M., Burgess, M. K., Cull, J. P., & Cas, R. A. F. (2009). Distribution of melt beneath Mount St Helens and Mount Adams inferred from magnetotelluric data. *Nature Geoscience, 2*, 785–789. https://doi:10.1038/ngeo661

House, M. A., Wernicke, B. P., & Farley, K. A. (1998). Dating topography of the Sierra Nevada, California, using apatite (U–Th)/He ages. *Nature, 396*, 66–69. https://doi:10.1038/23926

Huang, H. H., Lin, F. C., Schmandt, B., Farrell, J., Smith, R. B., & Tsai, V. C. (2015). Volcanology. The Yellowstone magmatic system from the mantle plume to the upper crust. *Science, 348*, 773–6. https://doi:10.1126/science.aaa5648

Huber, C., Bachmann, O., & Dufek, J. (2010). The limitations of melting on the reactivation of silicic mushes. *Journal of Volcanology and Geothermal Research, 195*, 97–105. https://doi:10.1016/j.jvolgeores.2010.06.006

Huber, C., Bachmann, O., & Dufek, J. (2011). Thermo-mechanical reactivation of locked crystal mushes: Melting-induced internal fracturing and assimilation processes in magmas. *Earth and Planetary Science Letters, 304*, 443–454. https://doi:10.1016/j.epsl.2011.02.022

Huber, C., Bachmann, O., & Manga, M. (2009). Homogenization processes in silicic magma chambers by stirring and mushification (latent heat buffering). *Earth and Planetary Science Letters, 283*, 38–47. https://doi:10.1016/j.epsl.2009.03.029

Hughes, I., & Hase, T. (2010). *Measurements and their uncertainties: A practical guide to modern error analysis*. Oxford University Press.

Humphreys, M. C. S., Menand, T., Blundy, J. D., & Klimm, K. (2008). Magma ascent rates in explosive eruptions: Constraints from H_2O diffusion in melt inclusions. *Earth and Planetary Science Letters, 270*, 25–40. https://doi:10.1016/j.epsl.2008.02.041

Huppert, H. E., Sparks, R. S. J., & Turner, J. S. (1982). Effects of volatiles on mixing in calc-alkaline magma systems. *Nature, 297*, 554–557. https://doi:10.1038/297554a0

Iovine, R.S., Fedele, L., Mazzeo, F.C., Arienzo, I., Cavallo, A., Wörner, G., et al. (2017). Timescales of magmatic processes prior to the ~4.7 ka Agnano-Monte Spina eruption (Campi Flegrei caldera, Southern Italy) based on diffusion chronometry from sanidine phenocrysts. *Bulletin of Volcanology, 79*. https://doi:10.1007/s00445-017-1101-4

Ivanovich, M., & Harmon, R. S. (1992). *Uranium-series disequilibrium: applications to earth, marine, and environmental sciences*, 2nd edn. Oxford, UK: Clarendon Press.

Iverson, N. A., Kyle, P. R., Dunbar, N. W., McIntosh, W. C., & Pearce, N. J. G. (2014). Eruptive history and magmatic stability of Erebus volcano, Antarctica: Insights from englacial tephra. *Geochemistry, Geophysics, Geosystems, 15*, 4180–4202. https://doi:10.1002/2014GC005435

Jackson, M. D., Blundy, J., & Sparks, R. S. J. (2018). Chemical differentiation, cold storage and remobilization of magma in the Earth's crust. *Nature, 564*, 405–409.

Jackson, S. E., Pearson, N. J., Griffin, W. L., & Belousova, E. A. (2004). The application of laser ablation-inductively coupled plasma-mass spectrometry to in situ U–Pb zircon geochronology. *Chemical Geology, 211*, 47–69. https://doi:10.1016/j.chemgeo.2004.06.017

Jankovics, M. É., Dobosi, G., Embey-Isztin, A., Kiss, B., Sági, T., Harangi, S., & Ntaflos, T. (2013). Origin and ascent history of unusually crystal-rich alkaline basaltic magmas from the western Pannonian Basin. *Bulletin of Volcanology, 75*, 1–23. https://doi:10.1007/s00445-013-0749-7

Jankovics, M. É., Harangi, S., Németh, K., Kiss, B., & Ntaflos, T. (2015). A complex magmatic system beneath the Kissomlyó monogenetic volcano (western Pannonian Basin): Evidence from mineral textures, zoning and chemistry. *Journal of Volcanology and Geothermal Research, 301*, 38–55.

Jankovics, M. É., Sági, T., Astbury, R. L., Petrelli, M., Kiss, B., Ubide, T., et al. (2019). Olivine major and trace element compositions coupled with spinel chemistry to unravel the magmatic systems feeding monogenetic basaltic volcanoes. *Journal of Volcanology and Geothermal Research, 369*, 203–223.

Jellinek, A. M., & DePaolo, D. J. (2003). A model for the origin of large silicic magma chambers: precursors of caldera-forming eruptions. *Bulletin of Volcanology, 65*, 363–381. https://doi:10.1007/s00445-003-0277-y

Jellinek, A. M., & Kerr, R. C. (1999). Mixing and compositional stratification produced by natural convection 2. Applications to the differentiation of basaltic and silicic magma chambers and komatiite lava flows. *Journal of Geophysical Research-Solid Earth, 104*, 7203–7218.

Jellinek, A. M., Kerr, R. C., & Griffiths, R. W. (1999). Mixing and compositional stratification produced by natural convection 1. Experiments and their application to Earth's core and mantle. *Journal of Geophysical Research-Solid Earth, 104*, 7183–7201

Jerram, D. A., & Martin, V. M. (2008). Understanding crystal populations and their significance through the magma plumbing system. In C. Annen, G. F. Zellmer (Eds.), *Dynamics of crustal magma transfer, storage and differentiation* (Special Publication 304, pp. 133–148). London: Geological Society. https://doi:10.1144/SP304.7

Jicha, B. R., Singer, B. S., & Sobol, P. (2016). Re-evaluation of the ages of $^{40}Ar/^{39}Ar$ sanidine standards and supereruptions in the western U.S. using a Noblesse multi-collector mass spectrometer. *Chemical Geology, 431*, 54–66.

Kagiyama, T., Utada, H., & Yamamoto, T. (1999). Magma ascent beneath Unzen Volcano, SW Japan, deduced from the electrical resistivity structure. *Journal of Volcanology and Geothermal Research, 89*, 35–42. https://doi:10.1016/S0377-0273(98)00120-6

Kaiser, J. F., de Silva, S., Schmitt, A. K., Economos, R., & Sunagua, M. (2017). Million-year melt–presence in monotonous intermediate magma for a volcanic–plutonic assemblage in the Central Andes: Contrasting histories of crystal-rich and crystal-poor super-sized silicic magmas. *Earth and Planetary Science Letters, 457*, 73–86. https://doi:10.1016/j.epsl.2016.09.048

Karakas, O., Degruyter, W., Bachmann, O., & Dufek, J. (2017). Lifetime and size of shallow magma bodies controlled by crustal-scale magmatism. *Nature Geoscience, 10*, 446–450. https://doi:10.1038/ngeo2959

Kayzar, T. M., Cooper, K. M., Reagan, M. K., & Kent, A. J. R. (2009). Gas transport model for the magmatic system at Mount Pinatubo, Philippines: Insights from $(^{210}Pb)/(^{226}Ra)$. *Journal of Volcanology and Geothermal Research, 181*, 124–140.

Kelley, S. (2002). K–Ar and Ar–Ar Dating. *Reviews in Mineralogy and Geochemistry, 47*, 785–818. https://doi:10.2138/rmg.2002.47.17

Kent, A. J. R., Cooper, K. M. (2018). How well do zircons record the thermal evolution of magmatic systems? *Geology, 46*, 111–114. https://doi:10.1130/G39690.1

Kile, D. E., & Eberl, D. D. (2003). On the origin of size-dependent and size-independent crystal growth: Influence of advection and diffusion. *American Mineralogist, 88*, 1514–1521. https://doi:10.2138/am-2003-1014

Kilgour, G. N., Saunders, K. E., Blundy, J. D., Cashman, K. V., Scott, B. J., & Miller, C. A. (2014). Timescales of magmatic processes at Ruapehu volcano from diffusion chronometry and their comparison to monitoring data. *Journal of Volcanology and Geothermal Research, 288*, 62–75. https://doi:10.1016/j.jvolgeores.2014.09.010

King, S. D., Lee, C., van Keken, P. E., Leng, W., Zhong, S., Tan, E., Tosi, N., & Kameyama, M. C. (2010). A community benchmark for 2-D Cartesian compressible convection in the Earth's mantle. *Geophysical Journal International, 180*, 73–87. https://doi:10.1111/j.1365-246X.2009.04413.x

Klemetti, E. W., & Clynne, M. A. (2014). Localized rejuvenation of a crystal mush recorded in zircon temporal and compositional variation at the Lassen volcanic center, Northern California. *PLoS One, 9*. https://doi:10.1371/journal.pone.0113157

Kohn, M. J., Engi, M., & Lanari, P. (2018). *Petrochronology: Methods and applications. Reviews in mineralogy and geochemistry* (Vol. 83). Chantilly, VA: Mineralogical Society of America. https://doi:10.2138/rmg.2017.83

Kuritani, T., Yokoyama, T., & Nakamura, E. (2007). Rates of thermal and chemical evolution of magmas in a cooling magma chamber: A chronological and theoretical study on basaltic and andesitic lavas from Rishiri Volcano, Japan. *Journal of Petrolology, 48*, 1295–1319.

Laeger, K., Petrelli, M., Morgavi, D., Lustrino, M., Pimentel, A., Paredes-Mariño, J., et al. (2019). Pre-eruptive conditions and triggering mechanism of the ~ 16 ka Santa Bárbara explosive eruption of Sete Cidades Volcano (São Miguel, Azores). *Contributions to Mineralogy and Petrology, 174*(2), 11, https://doi: 10.1007/s00410-019-1545-y

La Spina, G., de' Michieli Vitturi, M., & Clarke, A. B. (2017). Transient numerical model of magma ascent dynamics: application to the explosive eruptions at the Soufrière Hills Volcano. *Journal of Volcanology and Geothermal Research, 336*, 118–139. https://doi:10.1016/j.jvolgeores.2017.02.013

Landwehr, D., Blundy, J., Chamorro-Perez, E. M., Hill, E., & Wood, B. (2001). U-series disequilibria generated by partial melting of spinel lherzolite. *Earth and Planetary Science Letters, 188*. https://doi:10.1016/S0012-821X(01)00328-4

Lanphere, M. A. (2000). Duration of sedimentation of the Creede Formation from $^{40}Ar/^{39}Ar$ ages. *Geological Society of America Special Paper, 346*, 71–76.

Laumonier, M., Gaillard, F., Muir, D., Blundy, J., & Unsworth, M. (2017). Giant magmatic water reservoirs at mid-crustal depth inferred from electrical conductivity and the growth of the continental crust. *Earth and Planetary Science Letters, 457*, 173–180. https://doi:10.1016/j.epsl.2016.10.023

Lee, J. K. W., Williams, I. S., & Ellis, D. J. (1997). Pb, U and Th diffusion in natural zircon. *Nature, 390*(6656), 159–162, https://doi: 10.1038/36554

Leeman, W. P., Annen, C., & Dufek, J. (2008). Snake River Plain – Yellowstone silicic volcanism: implications for magma genesis and magma fluxes. In C. Annen, G. F. Zellmer (Eds.), *Dynamics of crustal magma transfer, storage and differentiation* (Special Publication 304, pp. 235–1259). London: Geological Society. https://doi:10.1144/SP304.12

Lejeune, A. M., & Richet, P. (1995). Rheology of crystal-bearing silicate melts: an experimental study at high viscosities. *Journal of Geophysical Research, 100*, B3, 4215–4229, https://doi: 10.1029/94JB02985

Lipman, P. W. (1988). Evolution of silicic magma in the upper crust: The mid-Tertiary Latir volcanic field and its cogenetic granitic batholith, northern New Mexico, U.S.A. *Transactions of the Royal Society of Edinburgh: Earth Sciences, 79*(2–3), 265–288.

Lipman, P. W. (2007). Incremental assembly and prolonged consolidation of Cordilleran magma chambers: Evidence from the Southern Rocky Mountain volcanic field. *Geosphere, 3*(1), 42–70.

Lipman, P. W., & Bachmann, O. (2015). Ignimbrites to batholiths: Integrating perspectives from geological, geophysical, and geochronological data. *Geosphere, 11*, 705–743. https://doi:10.1130/GES01091.1

Lissenberg, C. J., & MacLeod, C. J. (2016). A Reactive Porous Flow Control on Mid-ocean Ridge Magmatic Evolution. *Journal of Petrolology, 57*, 2195–2220. https://doi:10.1093/petrology/egw074

Liu, Y., Anderson, A. T., & Wilson, C. J. N. (2007). Melt pockets in phenocrysts and decompression rates of silicic magmas before fragmentation. *Journal of Geophysical Research, 112*, B06204. https://doi:10.1029/2006JB004500

Lloyd, A. S., Ruprecht, P., Hauri, E. H., Rose, W., Gonnermann, H. M., & Plank, T. (2014). NanoSIMS results from olivine-hosted melt embayments: Magma ascent rate during explosive basaltic eruptions. *Journal of Volcanology and Geothermal Research*, *283*, 1–18. https://doi:10.1016/j.jvolgeores.2014.06.002

Lloyd, A. S., Ferriss, E., Ruprecht, P., Hauri, E. H., Jicha, B. R., & Plank, T. (2016). An assessment of clinopyroxene as a recorder of magmatic water and magma ascent rate. *Journal of Petrology*, *57*(10), 1865–1886.

Lormand, C., Zellmer, G. F., Németh, K., Kilgour, G., Mead, S., Palmer, A.S., et al. (2018). Weka Trainable Segmentation plugin in ImageJ: a semi-automatic tool applied to Crystal Size Distributions of microlites in volcanic rocks. *Microscopy and Microanalysis*, *24*, 667–675. https://doi.org/10.1017/S1431927618015428

Maimon, O., Lyakhovsky, V., Melnik, O., & Navon, O. (2012). The propagation of a dyke driven by gas-saturated magma. *Geophysical Journal International*, *189*, 956–966. https://doi:10.1111/j.1365-246X.2012.05342.x

Mangan, M. T. (1990). Crystal size distribution systematics and the determination of magma storage times: The 1959 eruption of Kilauea volcano, Hawaii. *Journal of Volcanology and Geothermal Research*, *44*, 295–302. https://doi:10.1016/0377-0273(90)90023-9

Manley, C. R. (1996). Morphology and maturation of melt inclusions in quartz phenocrysts from the Badlands rhyolite lava flow, southwestern Idaho. *American Mineralogist*, *81*, 158–168.

Marsh, B. D. (1988). Crystal size distribution (CSD) in rocks and the kinetics and dynamics of crystallization – I. Theory. *Contributions to Mineralogy and Petrology*, *99*, 277–291. https://doi:10.1007/BF00375362

Marsh, B. D. (1998). On the interpretation of crystal size distributions in magmatic systems. *Journal of Petrolology*, *39*, 553–599. https://doi:10.1093/petroj/39.4.553

Mashima, H. (2004). Time scale of magma mixing between basalt and dacite estimated for the Saga-Futagoyama volcanic rocks in northwest Kyushu, southwest Japan. *Journal of Volcanology and Geothermal Research*, *131*, 333–349. doi:10.1016/S0377-0273(03)00412-8

Mastin, L. G.,& Ghiorso, M. S. (2001). Adiabatic temperature changes of magma–gas mixtures during ascent and eruption. *Contributions to Mineralogy and Petrology*, *141*, 307–321.

Matthews, N. E., Pyle, D. M., Smith, V. C., Wilson, C. J. N., Huber, C., & van Hinsberg, V. (2012). Quartz zoning and the pre-eruptive evolution of the ~340-ka Whakamaru magma systems, New Zealand. *Contributions to Mineralogy and Petrology*, *163*, 87–107. https://doi:10.1007/s00410-011-0660-1

McCanta, M. C., Rutherford, M. J., & Hammer, J. E. (2007). Pre-eruptive and syn-eruptive conditions in the Black Butte, California dacite: Insight into crystallization kinetics in a silicic magma system. *Journal of Volcanology and Geothermal Research*, *160*, 263–284. https://doi:10.1016/j.jvolgeores.2006.10.004

Melekhova, E., Annen, C., & Blundy, J. D. (2013). Compositional gaps in igneous rock suites controlled by magma system heat and water content. *Nature Geoscience*, *6*, 385–390. https://doi:10.1038/ngeo1781

Menand, T., Annen, C., & de Saint Blanquat, M. (2015). Rates of magma transfer in the crust: Insights into magma reservoir recharge and pluton growth. *Geology*, *43*, 199–202. https://doi:10.1130/G36224.1

Michaut, C., & Jaupart, C. (2006). Ultra-rapid formation of large volumes of evolved magma. *Earth and Planetary Science Letters*, *250*, 38–52.

Miller, C. F. (2016). Eruptible magma. *Proceedings of the National Academy of Sciences*, *113*, 13941–13943. https://doi:10.1073/pnas.1617105113

Miller, J. S., Matzel, J. E. P., Miller, C. F., Burgess, S. D., & Miller, R. B. (2007). Zircon growth and recycling during the assembly of large, composite arc plutons. *Journal of Volcanology and Geothermal Research*, *167*, 282–299. https://doi:10.1016/j.jvolgeores.2007.04.019

Miwa, T., Toramaru, A., & Iguchi, M. (2009). Correlations of volcanic ash texture with explosion earthquakes from vulcanian eruptions at Sakurajima volcano, Japan. *Journal of Volcanology and Geothermal Research*, *184*, 473–486. https://doi:10.1016/j.jvolgeores.2009.05.012

Mock, A., & Jerram, D. A. (2005). Crystal size distributions (CSD) in three dimensions: Insights from the 3D reconstruction of a highly porphyritic rhyolite. *Journal of Petrolology*, *46*, 1525–1541. https://doi:10.1093/petrology/egi024

Molloy, C., Shane, P., & Nairn, I. (2008). Pre-eruption thermal rejuvenation and stirring of a partly crystalline rhyolite pluton revealed by the Earthquake Flat Pyroclastics deposits, New Zealand. *Journal of the Geological Society, London*, *165*, 435–447. https://doi:10.1144/0016-76492007-071

Molnár, K., Harangi, S., Lukács, R., Dunkl, I., Schmitt, A. K., Kiss, B., Garamhegyi, T., & Seghedi, I. (2018). The onset of the volcanism in the Ciomadul Volcanic Dome Complex (Eastern Carpathians): Eruption chronology and magma type variation. *Journal of Volcanology and Geothermal Research*, *354*, 39–56. https://doi:10.1016/J.JVOLGEORES.2018.01.025

Molnár, K., Lukács, R., Dunkl, I., Schmitt, A.K., Kiss, B., Seghedi, I., Szepesi, J., & Harangi, S. (2019). Episodes of dormancy and eruption of the Late Pleistocene Ciomadul volcanic complex (Eastern Carpathians, Romania) constrained by zircon geochronology. *Journal of Volcanology and Geothermal Research*, *373*, 133–147.

Montagna, C. P., Papale, P., & Longo, A. (2015). Timescales of mingling in shallow magmatic reservoirs. In L. Caricchi, J. D. Blundy (Eds.), *Chemical, physical and temporal evolution of magmatic systems* (Special Publication 422, pp. 131–140). London: Geological Society. https://doi:10.1144/SP422.6

Morgan, D. J., & Jerram, D. A. (2006). On estimating crystal shape for crystal size distribution analysis. *Journal of Volcanology and Geothermal Research*, *154*, 1–7. https://doi:10.1016/j.jvolgeores.2005.09.016

Morgan, D. J., Jerram, D. A., Chertkoff, D. G., Davidson, J. P., Pearson, D. G., Kronz, A., & Nowell, G. M. (2007). Combining CSD and isotopic microanalysis: Magma supply and mixing processes at Stromboli Volcano, Aeolian Islands, Italy. *Earth and Planetary Science Letters*, *260*, 419–431. https://doi:10.1016/j.epsl.2007.05.037

Mourtada-Bonnefoi, C. C., & Laporte, D. (2004). Kinetics of bubble nucleation in a rhyolitic melt: an experimental study of the effect of ascent rate. *Earth and Planetary Science Letters*, *218*, 521–537. https://doi:10.1016/S0012-821X(03)00684-8

Muir, D. D., Barfod, D. N., Blundy, J. D., Rust, A. C., Sparks, R. S. J., & Clarke, K. M. (2015). The temporal record of magmatism at Cerro Uturuncu, Bolivian Altiplano. In L. Caricchi,

J. D. Blundy (Eds.), *Chemical, physical and temporal evolution of magmatic systems* (Special Publication 422, pp. 57–83). London: Geological Society. https://doi:10.1144/SP422.1.

Myers, M. L., Wallace, P. J., Wilson, C. J. N., Morter, B. K., & Swallow, E. J. (2016). Prolonged ascent and episodic venting of discrete magma batches at the onset of the Huckleberry Ridge supereruption, Yellowstone. *Earth and Planetary Science Letters*, *451*, 285–297. https://doi:10.1016/j.epsl.2016.07.023

Myers, M. L., Wallace, P. J., Wilson, C. J. N., Watkins, J. M., & Liu, Y. (2018). Ascent rates of rhyolitic magma at the onset of three caldera-forming eruptions. *American Mineralogist*, *103*, 952–965. https://doi:10.2138/am-2018-6225

Nakamura, M. (1995). Continuous mixing of crystal mush and replenished magma in the ongoing Unzen eruption. *Geology*, *23*, 807–810. https://doi:10.1130/0091-7613(1995)023<0807:CMOCMA>2.3.CO;2

Nandedkar, R. H., Ulmer, P., & Müntener, O. (2014). Fractional crystallization of primitive, hydrous arc magmas: An experimental study at 0.7 GPa. *Contributions to Mineralogy and Petrology*, *167*, 1–27. https://doi:10.1007/s00410-014-1015-5

Newhall, C., Self, S., & Robock, A. (2018). Anticipating future Volcanic Explosivity Index (VEI) 7 eruptions and their chilling impacts. *Geosphere*, *14*(2), 1–32, https://doi:10.1130/GES01513.1

Newcombe, M.E., Fabbrizio, A., Zhang, Y., Ma, C., Le Voyer, M., Guan, Y., et al. (2014). Chemical zonation in olivine-hosted melt inclusions. *Contributions to Mineralogy and Petrology*, *168*(1), 1–26.

Noguchi, S., Toramaru, A., & Shimano, T. (2006). Crystallization of microlites and degassing during magma ascent: Constraints on the fluid mechanical behavior of magma during the Tenjo Eruption on Kozu Island, Japan. *Bulletin of Volcanology*, *68*, 432–449. https://doi:10.1007/s00445-005-0019-4

O'Neill, C., & Spiegelman, M. (2010). Formulations for simulating the multiscale physics of magma ascent. In A. Dosseto, S. P. Turner, J. A. Van Orman (Eds.), *Timescales of magmatic processes: From core to atmosphere* (pp. 87–101). Chichester, UK: J. Wiley & Sons.. https://doi:10.1002/9781444328509.ch4

O'Reilly, S. Y., & Griffin, W. L. (2010). Rates of magma ascent: constraints from mantle-derived xenoliths. In A. Dosseto, S. P. Turner, J. A. Van Orman (Eds.), *Timescales of magmatic processes: From core to atmosphere* (pp. 116–124). Chichester, UK: J. Wiley & Sons.ttps://doi:10.1002/9781444328509.ch6

Oldenburg, C. M., Spera, F. J., Yuen, D. A., & Sewell, G. (1989). Dynamic mixing in magma bodies: theory, simulations, and implications. *Journal of Geophysical Research*, *94*, 9215–9236. https://doi:10.1029/JB094iB07p09215

Pamukcu, A. S., Gualda, G. A. R., Bégué, F., Gravley, & D. M. (2015). Melt inclusion shapes: Timekeepers of short-lived giant magma bodies. *Geology*, *43*, 947–950. https://doi:10.1130/G37021.1

Papale, P., & Dobran, F. (1994). Magma flow along the volcanic conduit during the plinian and pyroclastic flow phases of the May 18, 1980, Mount St. Helens eruption. *Journal of Geophysical Research*, *99*, 4355–4373.

Papale, P., Neri, A., & Macedonio, G. (1998). The role of magma composition and water content in explosive eruptions: 1. Conduit ascent dynamics. *Journal of Volcanology and Geothermal Research*, *87*, 75–93. https://doi:10.1016/S0377-0273(98)00101-2.

Parmigiani, A., Faroughi, S., Huber, C., Bachmann, O., & Su, Y. (2016). Bubble accumulation and its role in the evolution of magma reservoirs in the upper crust. *Nature*, *532*, 492–495. https://doi:10.1038/nature17401

Parmigiani, A., Huber, C., Bachmann, O., & Chopard, B. (2011). Pore-scale mass and reactant transport in multiphase porous media flows. *Journal of Fluid Mechanics*, *686*, 40–76. https://doi:10.1017/jfm.2011.268.

Parrish, R.R. (1990). U–Pb dating of monazite and its application to geological problems. *Canadian Journal of Earth Sciences*, *27*, 1431–1450. https://doi:10.1139/e90-152

Perugini, D., De Campos, C. P., Dingwell, D. B., & Dorfman, A. (2013). Relaxation of concentration variance: A new tool to measure chemical element mobility during mixing of magmas. *Chemical Geology*, *335*, 8–23. https://doi:10.1016/j.chemgeo.2012.10.050

Perugini, D., De Campos, C. P., Petrelli, M., & Dingwell, D. B. (2015). Concentration variance decay during magma mixing: a volcanic chronometer. *Nature Scientific Reports*, *5*, 14225. https://doi:10.1038/srep14225

Perugini, D., De Campos, C. P., Petrelli, M., Morgavi, D., Vetere, F.P., & Dingwell, D. B. (2015). Quantifying magma mixing with the Shannon entropy: Application to simulations and experiments. *Lithos*, *236–237*, 299–310. https://doi:10.1016/j.lithos.2015.09.008

Perugini, D., Petrelli, M., & Poli, G. (2006). Analysis of concentration patterns in volcanic rocks: Insights into dynamics of highly explosive volcanic eruptions. *Physica A: Statistical Mechanics and its Applications*, *370*, 741–746. https://doi:10.1016/j.physa.2006.02.023

Perugini, D., Ventura, G., Petrelli, M., & Poli, G. (2004). Kinematic significance of morphological structures generated by mixing of magmas: a case study from Salina Island (southern Italy). *Earth and Planetary Science Letters*, *222*, 1051–1066. https://doi:10.1016/j.epsl.2004.03.038

Peslier, A. H., Bizimis, M., & Matney, M. (2015). Water disequilibrium in olivines from Hawaiian peridotites: Recent metasomatism, H diffusion and magma ascent rates. *Geochimica et Cosmochimica Acta*, *154*, 98–117. https://doi:10.1016/j.gca.2015.01.030

Peslier, A. H., & Luhr, J. F. (2006). Hydrogen loss from olivines in mantle xenoliths from Simcoe (USA) and Mexico: Mafic alkalic magma ascent rates and water budget of the sub-continental lithosphere. *Earth and Planetary Science Letters*, *242*, 302–319. https://doi:10.1016/j.epsl.2005.12.019

Petrelli, M., El Omari, K., Le Guer, Y., & Perugini, D. (2016). Effects of chaotic advection on the timescales of cooling and crystallization of magma bodies at mid crustal levels. *Geochemistry, Geophysics, Geosystems*, *17*, 425–441. https://doi:10.1002/2015GC006109

Petrelli, M., El Omari, K., Spina, L., Le Guer, Y., La Spina, G., & Perugini, D. (2018). Timescales of water accumulation in magmas and implications for short warning times of explosive eruptions. *Nature Communications*, *9*, 770. https://doi:10.1038/s41467-018-02987-6

Petrelli, M., Perugini, D., & Poli, G. (2005). Time-scales of hybridisation of magmatic enclaves in regular and chaotic flow fields: Petrologic and volcanologic implications.

Bulletin of Volcanology, *68*, 285–293. https://doi:10.1007/s00445-005-0007-8

Petrelli, M., Perugini, D., & Poli, G. (2011). Transition to chaos and implications for time-scales of magma hybridization during mixing processes in magma chambers. *Lithos*, *125*, 211–220. https://doi:10.1016/j.lithos.2011.02.007

Petrone, C. M., Braschi, E., Francalanci, L., Casalini, M., & Tommasini, S. (2018). Rapid mixing and short storage timescale in the magma dynamics of a steady-state volcano. *Earth and Planetary Science Letters*, *492*, 206–221. https://doi:10.1016/j.epsl.2018.03.055

Petrone, C. M., Bugatti, G., Braschi, E., & Tommasini, S. (2016). Pre-eruptive magmatic processes re-timed using a non-isothermal approach to magma chamber dynamics. *Nature Communications*, *7*. https://doi:10.1038/ncomms12946

Petrosino, P., Morabito, S., Jicha, B. R., Milia, A., Sprovieri, M., & Tamburrino, S. (2016). Multidisciplinary tephrochronological correlation of marker events in the eastern Tyrrhenian Sea between 48 and 105 ka. *Journal of Volcanology and Geothermal Research*, https://doi:10.1016/j.jvolgeores.2016.02.001

Pinkerton, H., Wilson, L., & Macdonald, R. (2002). The transport and eruption of magma from volcanoes: a review. *Contemporary Physics*, *43*,197–210.

Piochi, M., Mastrolorenzo, G., & Pappalardo, L. (2005). Magma ascent and eruptive processes from textural and compositional features of Monte Nuovo pyroclastic products, Campi Flegrei, Italy. *Bulletin of Volcanology*, *67*, 663–678. https://doi:10.1007/s00445-005-0410-1

Pistone, M., Caricchi, L., Fife, J. L., Mader, K., & Ulmer, P. (2015). In situ X-ray tomographic microscopy observations of vesiculation of bubble-free and bubble-bearing magmas. *Bulletin of Volcanology*, *77*, 1–15.

Pistone, M., Whittington, A. G., Andrews, B. J., & Cottrell, E. (2017). Crystal-rich lava dome extrusion during vesiculation: an experimental study. *Journal of Volcanology and Geothermal Research*, *347*, 1–14., https://doi: 10.1016/j.jvolgeores.2017.06.018

Pritchard, M. E., & Gregg, P. M. (2016). Geophysical evidence for silicic crustal melt in the continents: Where, what kind, and how much? *Elements*, *12*, 121–127. https://doi:10.2113/gselements.12.2.121

Pyle, D. M., Ivanovich, M., & Sparks, R. S. J. (1988). Magma-cumulate mixing identified by U-Th disequilibrium dating. *Nature*, *331*, 157–159.

Rae, A. S. P., Edmonds, M., Maclennan, J., Morgan, D., Houghton, B., Hartley, M. E., & Sides, I. (2016). Time scales of magma transport and mixing at Kīlauea Volcano, Hawai'i. *Geology*, *44*, 463–466. https://doi:10.1130/G37800.1

Ray, A., Hatui, K., Paul, D. K., Sen, G., Biswas, S. K., & Das, B. (2016). Mantle xenolith-xenocryst-bearing monogenetic alkali basaltic lava field from Kutch Basin, Gujarat, Western India: Estimation of magma ascent rate. *Journal of Volcanology and Geothermal Research*, *312*, 40–52. https://doi:10.1016/J.JVOLGEORES.2016.01.015

Reagan, M. K., Sims, K. W. W., Erich, J., Thomas, R. B., Cheng, H., Edwards, R. L., Layne, G., & Ball, L., 2003. Time-scales of differentiation from mafic parents to rhyolite in North American continental arcs. *Journal of Petrology*, *44*, 1703–1726.

Reichow, M. K., Saunders, A. D., White, R. V., Pringle, M. S., Al'Mukhamedov, A. I., Medvedev, A. I., & Kirda, N. P. (2002). $^{40}Ar/^{39}Ar$ dates from the West Siberian Basin: Siberian flood basalt province doubled. *Science*, *80*(296), 1846–1849. https://doi:10.1126/science.1071671

Reid, M. R. (2003). Timescales of Magma Transfer and Storage in the Crust, Treatise on Geochemistry. In *The crust, Treatise on geochemistry* (Vol. *3*–9, pp. 1–27). Elsevier Inc.. https://doi.org/10.1016/B0-08-043751-6/03022-X

Reid, M. R., & Coath, C. D. (2000). In situ U–Pb ages of zircons from the Bishop Tuff: No evidence for long crystal residence times. *Geology*, *28*(5), 443–446.

Reid, M. R., & Vazquez, J. A. (2017). Fitful and protracted magma assembly leading to a giant eruption, Youngest Toba Tuff, Indonesia. *Geochemistry, Geophysics, Geosystems*, *18*(1), 156–177.

Reid, M. R., Coath, C. D., Harrison, T. M., & McKeegan, K. D. (1997). Prolonged residence times for the youngest rhyolites associated with Long Valley Caldera: $^{230}Th–^{238}U$ ion microprobe dating of young zircons. *Earth and Planetary Science Letters*, *150*(1–2), 27–39.

Reid, M. R., Vazquez, J. A., & Schmitt, A. K. (2011). Zircon-scale insights into the history of a Supervolcano, Bishop Tuff, Long Valley, California, with implications for the Ti-in-zircon geothermometer. *Contributions to Mineralogy and Petrology*, *161*(2), 293–311.

Reiners, P. W. (2005). Zircon (U–Th)/He thermochronometry. *Reviews in Mineralogy and Geochemistry*, *58*, 151–179.

Rink, W. J., Thompson, J. W., Heaman, L. M., Jull, A. J. T., & Paces, J. B. (2015). *Encyclopedia of scientific dating methods*. Springer.

Rivalta, E., Taisne, B., Bunger, A. P., & Katz, R. F. (2015). A review of mechanical models of dike propagation: Schools of thought, results and future directions. *Tectonophysics*, *638*, 1–42. https://doi:10.1016/j.tecto.2014.10.003

Rogers, N. W., Evans, P. J., Blake, S., Scott, S. C., & Hawkesworth, C. J. (2004). Rates and timescales of fractional crystallization from$^{238}U–^{230}Th–^{226}Ra$ disequilibria in trachyte lavas from Longonot volcano, Kenya. *Journal of Petrolology*, *45*, 1747–1776. https://doi:10.1093/petrology/egh032

Romero, J. E., Morgavi, D., Arzilli, F., Daga, R., Caselli, A., Reckziegel, F., et al. (2016). Eruption dynamics of the 22–23 April 2015 Calbuco Volcano (Southern Chile): Analyses of tephra fall deposits. *Journal of Volcanology and Geothermal Research*, *317*, 15–29. https://doi:10.1016/J.JVOLGEORES.2016.02.027

Rossi, S., Petrelli, M., Morgavi, D., González-García, D., Fischer, L. A., Vetere, F., & Perugini, D. (2017). Exponential decay of concentration variance during magma mixing: Robustness of a volcanic chronometer and implications for the homogenization of chemical heterogeneities in magmatic systems. *Lithos*, *286–287*, 396–407. https://doi:10.1016/j.lithos.2017.06.022

Rubin, A. E., Cooper, K. M., Till, C. B., Kent, A. J. R., Costa, F., Bose, M., et al. (2017). Rapid cooling and cold storage in a silicic magma reservoir recorded in individual crystals. *Science*, *80*(356), 1154–1157. https://doi:10.1126/science.aam8720

Rubin, A. M. (1993). On the thermal viability of dikes leaving magma chambers. *Geophysical Research Letters*, *20*, 257–260. https://doi:10.1029/92GL02783

Rubin, A. M. (1995). Propagation of magma-filled cracks. *Annual Reviews in Earth Planetary Sciences*, *23*, 287–336. https://doi:10.1146/annurev.ea.23.050195.001443

Rubin, K. H., Van Der Zander, I., Smith, M. C., & Bergmanis, E. C. (2005). Minimum speed limit for ocean ridge magmatism from ^{210}Pb–^{226}Ra–^{230}Th disequilibria. *Nature*, *437*, 534–538. https://doi:10.1038/nature03993

Ruprecht, P., Bergantz, G. W., & Dufek, J. (2008). Modeling of gas-driven magmatic overturn: Tracking of phenocryst dispersal and gathering during magma mixing. *Geochemistry, Geophysics, Geosystems*, *9*(7). https://doi:10.1029/2008GC002022

Ruprecht, P., & Cooper, K. M. (2012). Integrating the uranium-series and elemental diffusion geochronometers in mixed magmas from Volcan Quizapu, Central Chile. *Journal of Petrolology*, *53*, 841–871. https://doi:10.1093/petrology/egs001

Ruprecht, P., & Plank, T. (2013). Feeding andesitic eruptions with a high-speed connection from the mantle. *Nature*, *500*, 68–72. https://doi:10.1038/nature12342.

Ruth, D. C. S., Costa, F., Bouvet de Maisonneuve, C., Franco, L., Cortés, J. A., & Calder, E. S. (2018). Crystal and melt inclusion timescales reveal the evolution of magma migration before eruption. *Nature Communications*, *9*, 2657. https://doi:10.1038/s41467-018-05086-8

Rutherford, M. J. (2008). Magma ascent rates. *Reviews in Mineralogy and Geochemistry*, *69*(1), 241–271. https://doi:10.2138/rmg.2008.69.7

Rutherford, M. J., & Devine, J. D. (2003). Magmatic conditions and magma ascent as indicated by hornblende phase equilibria and reactions in the 1995–2002 Soufrière Hills magma. *Journal of Petrolology*, *44*, 1433–1453. https://doi:10.1093/petrology/44.8.1433

Rutherford, M. J., & Hill, P. M. (1993). Magma ascent rates from amphibole breakdown: an experimental study applied to the 1980–1986 Mount St. Helens eruptions. *Journal of Geophysical Research*, *98*(B11), 19667–19685.

Salisbury, M. J., Bohrson, W. A., Clynne, M. A., Ramos, F. C., & Hoskin, P. (2008). Multiple plagioclase crystal populations identified by crystal size distribution and in situ chemical data: Implications for timescales of magma chamber processes associated with the 1915 eruption of Lassen Peak, CA. *Journal of Petrolology*, *49*, 1755–1780. https://doi:10.1093/petrology/egn045

Sas, M., Kawasaki, N., Sakamoto, N., Shane, P., Zellmer, G. F., Kent, A. J. R., & Yurimoto, H. (2019). The ion microprobe as a tool for obtaining strontium isotopes in magmatic plagioclase: A case study at Okataina Volcanic Centre, New Zealand. *Chemical Geology, 513*, 153–166.

Scandone, R., Cashman, K. V., & Malone, S. D. (2007). Magma supply, magma ascent and the style of volcanic eruptions. *Earth and Planetary Science Letters*, *253*, 513–529. https://doi:10.1016/j.epsl.2006.11.016

Scandone, R., & Malone, S. D. (1985). Magma supply, magma discharge and readjustment of the feeding system of mount St. Helens during 1980. *Journal of Volcanology and Geothermal Research*, *23*, 239–262. https://doi:10.1016/0377-0273(85)90036-8

Schaltegger, U., Schmitt, A. K., & Horstwood, M. S. A. (2015). U–Th–Pb zircon geochronology by ID-TIMS, SIMS, and laser ablation ICP-MS: Recipes, interpretations, and opportunities. *Chemical Geology*, *402*, 89–110. https://doi:10.1016/J.CHEMGEO.2015.02.028

Schmitt, A. K. (2011). Uranium series accessory crystal dating of magmatic processes. *Annual Reviews in Earth and Planetary Sciences*, *39*, 321–49.

Schmitt, A. K. (2006). Laacher See revisited: High-spatial-resolution zircon dating indicates rapid formation of a zoned magma chamber. *Geology*, *34*, 597–600. https://doi:10.1130/G22533.1

Schmitt, A. K., Perfit, M. R., Rubin, K. H., Stockli, D. F., Smith, M. C., Cotsonika, L. A., et al. (2011). Rapid cooling rates at an active mid-ocean ridge from zircon thermochronology. *Earth and Planetary Science Letters*, *302*, 349–358. https://doi:10.1016/j.epsl.2010.12.022

Schmitt, A. K., Stockli, D. F., & Hausback, B. P. (2006). Eruption and magma crystallization ages of Las Tres Vírgenes (Baja California) constrained by combined $^{230}Th/^{238}U$ and (U–Th)/He dating of zircon. *Journal of Volcanology and Geothermal Research*, *158*, 3–4, 281–295.

Schmitt, A. K., Stockli, D. F., Niedermann, S., Lovera, O. M., & Hausback, B. P. (2010). Eruption ages of Las Tres Vírgenes volcano (Baja California): A tale of two helium isotopes. *Quaternary Geochronology*, *5*(5), 503–511.

Schoene, B. (2013). U–Th–Pb geochronology. In K. Turekian, H. Holland (Eds.), *Treatise on geochemistry* (2nd edn, pp. 341–378). Elsevier. https://doi:10.1016/B978-0-08-095975-7.00310-7

Schoene, B., Condon, D. J., Morgan, L., & McLean, N. (2013). Precision and accuracy in geochronology. *Elements*, *9*, 19–24. https://doi:10.2113/gselements.9.1.19

Scott, J. A. J., Mather, T. A., Pyle, D. M., Rose, W. I., & Chigna, G. (2012). The magmatic plumbing system beneath Santiaguito Volcano, Guatemala. *Journal of Volcanology and Geothermal Research*, *237–238*, 54–68.

Sen, G. (1983). A petrologic model for the constitution of the upper mantle and crust of the Koolau shield, Oahu, Hawaii, and Hawaiian magmatism. *Earth and Planetary Science Letters*, *62*, 215–228. https://doi:10.1016/0012-821X(83)90085-7

Seropian, G., Rust, A. C., & Sparks, R. S. J. (2018). The gravitational stability of lenses in magma mushes: Confined Rayleigh–Taylor instabilities. *Journal of Geophysical Research-Solid Earth, 123*, 3593–3607. https://doi:10.1029/2018JB015523

Shaw, C. S.J . (2004). The temporal evolution of three magmatic systems in the West Eifel volcanic field, Germany. *Journal of Volcanology and Geothermal Research*, *131*, 213–240. https://doi:10.1016/S0377-0273(03)00363-9

Shea, T., Costa, F., Krimer, D., & Hammer, J. E. (2015). Accuracy of timescales retrieved from diffusion modeling in olivine: A 3D perspective. *American Mineralogist*, *100*, 2026–2042. https://doi:10.2138/am-2015-5163

Shuster, D. L., Vasconcelos, P. M., Heim, J. A., & Farley, K. A. (2005). Weathering geochronology by (U–Th)/He dating of goethite. *Geochimica et Cosmochimica Acta*, *69*, 659–673. https://doi:10.1016/J.GCA.2004.07.028

Siebert, L., Cottrell, E., Venzke, E., & Andrews, B. (2015). Earth's volcanoes and their eruptions: An overview. In H. Sigurdsson (Ed.), *The encyclopedia of volcanoes* (pp. 239–255). Elsevier. https://doi:10.1016/B978-0-12-385938-9.00012-2

Siebert, L., Simkin, T., & Kimberly, P. (2011). *Volcanoes of the world*. University of California Press.
Sigmundsson, F., Hreinsdóttir, S., Hooper, A., Árnadóttir, T., Pedersen, R., Roberts, M.J., et al. (2010). Intrusion triggering of the 2010 Eyjafjallajökull explosive eruption. *Nature*, *468*, 426–430. https://doi:10.1038/nature09558
Sigurdsson, H. (Ed.) (2015). *Encyclopedia of volcanoes*, 2nd edn. Elsevier. https://doi:10.1016/B978-0-12-385938-9.00007-9
Sims, K. W. W., Pichat, S., Reagan, M. K., Kyle, P. R., Dulaiova, H., Dunbar, N. W., et al. (2013). On the time scales of magma genesis, melt evolution, crystal growth rates and magma degassing in the Erebus volcano magmatic system using the ^{238}U, ^{235}U and ^{232}Th decay series. *Journal of Petrolology*, *54*, 235–271. https://doi:10.1093/petrology/egs068
Singer, B. S., Costa, F., Herrin, J. S., Hildreth, W., & Fierstein, J. (2016). The timing of compositionally-zoned magma reservoirs and mafic 'priming' weeks before the 1912 Novarupta-Katmai rhyolite eruption. *Earth and Planetary Science Letters*, *451*, 125–137. doi:10.1016/j.epsl.2016.07.015
Sisson, T. W., & Grove, T. L. (1993). Experimental investigations of the role of H_2O in calc-alkaline differentiation and subduction zone magmatism. *Contributions to Mineralogy and Petrology*, *113*, 143–166. https://doi:10.1007/BF00283225
Sliwinski, J. T., Bachmann, O., Dungan, M. A., Huber, C., Deering, C. D., Lipman, P. W., Martin, L. H. J., & Liebske, C. (2017). Rapid pre-eruptive thermal rejuvenation in a large silicic magma body: the case of the Masonic Park Tuff, Southern Rocky Mountain volcanic field, CO, USA. *Contributions to Mineralogy and Petrology*, *172*. https://doi:10.1007/s00410-017-1351-3
Snyder, D. (2000). Thermal effects of the intrusion of basaltic magma into a more silicic magma chamber and implications for eruption triggering. *Earth and Planetary Science Letters*, *175*, 257–273. https://doi:10.1016/S0012-821X(99)00301-5
Solano, J. M. S., Jackson, M. D., Sparks, R. S. J., Blundy, J. D., & Annen, C. (2012). Melt segregation in deep crustal hot zones: A mechanism for chemical differentiation, crustal assimilation and the formation of evolved magmas. *Journal of Petrolology*, *53*, 1999–2026. https://doi:10.1093/petrology/egs041
Sparks, R. S. J., Baker, L., Brown, R. J., Field, M., Schumacher, J., Stripp, G., & Walters, A. (2006). Dynamical constraints on kimberlite volcanism. *Journal of Volcanology and Geothermal Research*, *155*, 18–48. https://doi:10.1016/j.jvolgeores.2006.02.010
Sparks, R. S. J., & Cashman, K. (2017). Dynamic magma systems: Implications for forecasting volcanic activity. *Elements*, *13*(1), 35–40.
Sparks, R. S. J., & Walker, G. P. L. (1977). The significance of vitric-enriched air-fall ashes associated with crystal-enriched ignimbrites. *Journal of Volcanology and Geothermal Research*, *2*, 329–341. https://doi:10.1016/0377-0273(77)90019-1
Spence, D. A., & Turcotte, D. L. (1990). Buoyancy-driven magma fracture: A mechanism for ascent through the lithosphere and the emplacement of diamonds. *Journal of Geophysical Research*, *95*, 5133. https://doi:10.1029/JB095iB04p05133
Spera, F. J. (1984). Carbon dioxide in petrogenesis III: role of volatiles in the ascent of alkaline magma with special reference to xenolith-bearing mafic lavas. *Contributions to Mineralogy and Petrology*, *88*, 217–232. https://doi:10.1007/BF00380167
Spera, F. J., & Bohrson, W. A. (2018). Rejuvenation of crustal magma mush: A tale of multiply nested processes and timescales. *American Journal of Science*, *318*, 90–140. https://doi:10.2475/01.2018.05
Spiegelman, M. (1996). Geochemical consequences of melt transport in 2-D: The sensitivity of trace elements to mantle dynamics. *Earth and Planetary Science Letters*, *139*, 115–132.
Spina, L., Scheu, B., Cimarelli, C., Arciniega-Ceballos, A., & Dingwell, D. B. (2016). Time scales of foam stability in shallow conduits: Insights from analogue experiments. *Geochemistry, Geophysics, Geosystems*, *17*, 4179–4194. https://doi:10.1002/2016GC006455
Stein, D. J., & Spera, F. J. (1996). Molecular dynamics simulations of liquids and glasses in the system $NaAlSiO_4$–SiO_2: Physical properties and transport mechanisms. *American Mineralogist*, *81*, 284–302. https://doi:10.2138/am-1996-3-403
Stelten, M. E., Cooper, K. M., Vazquez, J. A., Calvert, A. T., & Glessner, J. J. G. (2014). Mechanisms and timescales of generating eruptible rhyolitic magmas at Yellowstone Caldera from zircon and sanidine geochronology and geochemistry. *Journal of Petrolology*, *56*, 1607–1642. https://doi:10.1093/petrology/egv047
Stothers, R. B. (1984). The great tambora eruption in 1815 and its aftermath. *Science*, *224*, 1191–1198. https://doi:10.1126/science.224.4654.1191
Straub, S. M., Zellmer, G. F., Gómez-Tuena, A., Espinasa-Pereña., R., Martin-Del Pozzo, A. L., Stuart, F. M., & Langmuir, C. H. (2014). A genetic link between silicic slab components and calc-alkaline arc volcanism in central Mexico. In A. Gómez-Tuena, S. M. Straub, G. F. Zellmer, *Orogenic andesites and crustal growth* (Special Publication 385, pp. 31–64). London: Geological Society. https://doi:10.1144/SP385.14
Szabó, C., & Bodnar, R. J. (1996). Changing magma ascent rates in the Nógrád-Gömör Volcanic Field northern Hungary/southern Slovakia: Evidence from CO_2-rich fluid inclusions in metasomatized upper mantle xenoliths. *Petrologiya*, *4*, 221–230.
Szymanowski, D., Wotzlaw, J. F, Ellis, B. S., Bachmann, O., Guillong, M., & von Quadt, A. (2017). Protracted near-solidus storage and pre-eruptive rejuvenation of large magma reservoirs. *Nature Geoscience*, *10*, 777–782. https://doi:10.1038/ngeo3020
Tierney, C. R., Schmitt, A .K., Lovera, O. M., & de Silva, S.L. (2016). Voluminous plutonism during volcanic quiescence revealed by thermochemical modeling of zircon. *Geology*, *44*, 683–686. https://doi:10.1130/G37968.1
Till, C. B., Vazquez, J. A., & Boyce, J. W. (2015). Months between rejuvenation and volcanic eruption at Yellowstone caldera, Wyoming. *Geology*, *43*, 695–698. https://doi:10.1130/G36862.1
Tomiya, A., Miyagi, I., Saito, G., & Geshi, N. (2013). Short time scales of magma-mixing processes prior to the 2011 eruption of Shinmoedake volcano, Kirishima volcanic group, Japan. *Bulletin of Volcanology*, *75*, 750. https://doi:10.1007/s00445-013-0750-1
Toramaru, A. (2006). BND (bubble number density) decompression rate meter for explosive volcanic eruptions. *Journal of Volcanology and Geothermal Research*, *154*, 303–316. https://doi:10.1016/j.jvolgeores.2006.03.027

Toramaru, A., Noguchi, S., Oyoshihara, S., & Tsune, A. (2008). MND (microlite number density) water exsolution rate meter. *Journal of Volcanology and Geothermal Research*, *175*, 156–167. https://doi:10.1016/j.jvolgeores.2008.03.035

Tosi, N., Stein, C., Noack, L., Hüttig, C., Maierová, P., Samuel, H., et al.. (2015). A community benchmark for viscoplastic thermal convection in a 2-D square box. *Geochemistry, Geophysics, Geosystems*, *16*, 2175–2196. https://doi:10.1002/2015GC005807

Turekian, K., & Holland, H. (Eds.) (2013). *Treatise on Geochemistry*, 2nd edn. Elsevier.

Turner, M. B., Reagan, M. K., Turner, S. P., Sparks, R. S. J., Handley, H. K., Girard, G., & Suh, C. E. (2013). Timescales of magma degassing —insights from U-series disequilibria, Mount Cameroon, West Africa. *Journal of Volcanology and Geothermal Research*, *262*, 38–46. https://doi:10.1016/j.jvolgeores.2013.06.003

Turner, S., & Costa, F. (2007). Measuring timescales of magmatic evolution. *Elements*, *3*, 267–272. https://doi:10.2113/gselements.3.4.267

Turner, S., George, R., Jerram, D. A., Carpenter, N., & Hawkesworth, C. (2003). Case studies of plagioclase growth and residence times in island arc lavas from Tonga and the Lesser Antilles, and a model to reconcile discordant age information. *Earth and Planetary Science Letters*, *214*, 279–294. https://doi:10.1016/S0012-821X(03)00376-5

Ubide, T., & Kamber, B. S. (2018). Volcanic crystals as time capsules of eruption history. *Nature Communications*, *9*, 326. https://doi:10.1038/s41467-017-02274-w

Van Keken, P. E., King, S. D., Schmeling, H., Christensen, U. R., Neumeister, D., & Doin, M. P. (1997). A comparison of methods for the modeling of thermochemical convection. *Journal of Geophysical Research, B Solid Earth*, *102*, 22477–22495.

Vazquez, J. A., & Reid, M. R. (2004). Probing the accumulation history of the voluminous Toba magma. *Science*, *80*(305), 991–994. https://doi:10.1126/science.1096994

Venezky, D. Y., & Rutherford, M. J. (1999). Petrology and Fe–Ti oxide reequilibration of the 1991 Mount Unzen mixed magma. *Journal of Volcanology and Geothermal Research*, *89*, 213–230. https://doi:10.1016/S0377-0273(98)00133-4

Wallace, P. J., Plank, T., Edmonds, M., & Hauri, E. H. (2015). Volatiles in magmas. In H. Sigurdsson (Ed.), *The encyclopedia of volcanoes* (pp. 163–183). Elsevier. https://doi:10.1016/B978-0-12-385938-9.00007-9

Wark, D. A., Hildreth, W., Spear, F. S., Cherniak, D. J., & Watson, E. B. (2007). Pre-eruption recharge of the Bishop magma system. *Geology*, *35*, 235–238. https://doi:10.1130/G23316A.1

Weinberg, R. F., & Podladchikov, Y. (1994). Diapiric ascent of magmas through power law crust and mantle. *Journal of Geophysical Research*, *99*, 9543–9559. https://doi:10.1029/93JB03461

White, R.A, (1996). Precursory deep long-period earthquakes at Mount Pinatubo spatio-temporal link to a basalt trigger. In C. G. Newhall, R. S. Punonngbayan (Eds.), *Fire and mud: Eruptions and lahars of Mount Pinatubo, Philippines* (pp. 233-246. Seattle: University of Washington Press.

Wicks, C., de la Llera, J. C., Lara, L.E., & Lowenstern, J. (2011). The role of dyking and fault control in the rapid onset of eruption at Chaitén volcano, Chile. *Nature*, *478*, 374–377. https://doi:10.1038/nature10541

Wiggins, C., & Spiegelman, M. (1995). Magma migration and magmatic solitary waves in 3-D. *Geophysical Research Letters*, *22*, 1289–1292. https://doi:10.1029/95GL00269

Wilson, C. J. N., & Charlier, B. L. A. (2016). The life and times of silicic volcanic systems. *Elements*, *12*, 103–108. https://doi:10.2113/gselements.12.2.103

Wotzlaw, J. F., Schaltegger, U., Frick, D. A., Dungan, M. A., Gerdes, A. G., & Günther, D. (2013). Tracking the evolution of large-volume silicic magma reservoirs from assembly to supereruption. *Geology*, *41*, 8, 867–870.

Wotzlaw, J. F., Bindeman, I. N., Stern, R. A., D'Abzac, F.-X., & Schaltegger, U. (2015). Rapid heterogeneous assembly of multiple magma reservoirs prior to Yellowstone supereruptions. *Nature Scientific Reports*, *5*, 14026. https://doi:10.1038/srep14026

Zack, T., & Hogmalm, K. J. (2016). Laser ablation Rb/Sr dating by online chemical separation of Rb and Sr in an oxygen-filled reaction cell. *Chemical Geology*, *437*, 120–133. https://doi:10.1016/J.CHEMGEO.2016.05.027

Zellmer, G. F. (2009). Petrogenesis of Sr-rich adakitic rocks at volcanic arcs: Insights from global variations of eruptive style with plate convergence rates and surface heat flux. *Journal of the Geological Society, London*, *166*, 725–734. https://doi:10.1144/0016-76492008-0721

Zellmer, G.F., & Annen, C. (2008). An introduction to magma dynamics. In C. Annen, G. F. Zellmer (Eds.), *Dynamics of crustal magma transfer, storage and differentiation* (Special Publication 304, pp. 1–13). London: Geological Society. https://doi:10.1144/SP304.1

Zellmer, G.F., Annen, C., Charlier, B. L. A., George, R. M. M., Turner, S. P., & Hawkesworth, C. J. (2005). Magma evolution and ascent at volcanic arcs: constraining petrogenetic processes through rates and chronologies. *Journal of Volcanology and Geothermal Research*, *140*, 171–191. https://doi:10.1016/j.jvolgeores.2004.07.020

Zellmer, G. F., Blake, S., Vance, D., Hawkesworth, C., & Turner, S. (1999). Plagioclase residence times at two island arc volcanoes (Kameni Islands, Santorini, and Soufriere, St. Vincent) determined by Sr diffusion systematics. *Contributions to Mineralogy and Petrology*, *136*, 345–357. https://doi:10.1007/s004100050543

Zellmer, G. F., Chen, K.-X., Gung, Y., Kuo, B.-Y., & Yoshida, T. (2019). Magma transfer processes in the NE Japan arc: Insights from crustal ambient noise tomography combined with volcanic eruption records. *Frontiers in Earth Science*, *7*. https://doi.org/10.3389/feart.2019.00040

Zellmer, G. F., Dulski, P., Iizuka, Y., & Perfit, M. R. (2012). Rates and processes of crystallization in on-axis and off-axis MOR basaltic melts. *Lithos*, *154*, 1–15. https://doi:10.1016/j.lithos.2012.07.019

Zellmer, G. F., Edmonds, M., & Straub, S. M. (2015). Volatiles in subduction zone magmatism. In G. F. Zellmer, M. Edmonds, S. M. Straub (Eds.), *The role of volatiles in the genesis, evolution, and eruption of arc magmas* (Special Publication 410, pp. 1–17). London: Geological Society.

Zellmer, G. F., Freymuth, H., Cembrano, J. M., Clavero, J. E., Veloso, E. A. E., & Sielfeld, G. G. (2014). Altered mineral uptake into fresh arc magmas: Insights from U-Th isotopes of samples from Andean volcanoes under differential crustal stress regimes. In A. Gomez-Tuena, S. M. Straub, G. F. Zellmer (Eds.), *Orogenic andesites and crustal growth* (Special Publication 385, pp. 185–208). London: Geological Society. https://doi:10.1144/SP385.9

Zellmer, G. F., Kimura, J.-I., Chang, Q., Shellnutt, J. G., Sas, M., & Shane, P. (2018). Rapid determination of initial $^{87}Sr/^{86}Sr$ and estimation of the Rb–Sr age of plutonic rocks by LA-ICPMS of variably altered feldspars: An example from the 1.14 Ga Great Abitibi Dyke, Ontario, Canada. *Lithos, 314–315*, 52–58. https://doi:10.1016/j.lithos.2018.05.024

Zellmer, G. F., Pistone, M., Iizuka, Y., Andrews, B. J., Gómez-Tuena, A., Straub, S. M., & Cottrell, E. (2016). Petrogenesis of antecryst-bearing arc basalts from the Trans-Mexican Volcanic Belt: Insights into along-arc variations in magma-mush ponding depths, H_2O contents, and surface heat flux. *American Mineralogist*, *101*, 2405–2422. https://doi:10.2138/am-2016-5701

Zellmer, G. F., Rubin, K. H., Dulski, P., Iizuka, Y., Goldstein, S. L., & Perfit, M. R. (2011). Crystal growth during dike injection of MOR basaltic melts: Evidence from preservation of local Sr disequilibria in plagioclase. *Contributions to Mineralogy and Petrology*, *161*, 153–173. https://doi:10.1007/s00410-010-0518-y

Zellmer, G. F., Rubin, K. H., Grönvold, K., & Jurado-Chichay, Z. (2008). On the recent bimodal magmatic processes and their rates in the Torfajökull-Veidivötn area, Iceland. *Earth and Planetary Science Letters*, *269*, 387–397. https://doi:10.1016/j.epsl.2008.02.026

Zellmer, G. F., Rubin, K. H., Miller, C. A., Shellnutt, J. G., Belousov, A., & Belousova, M. (2015). Resolving discordant U–Th–Ra ages: Constraints on petrogenetic processes of recent effusive eruptions at Tatun Volcano Group, northern Taiwan. In L. Caricchi, J. D. Blundy (Eds.), *Chemical, physical and temporal evolution of magmatic systems* (Special Publication 422, pp. 175–188). London: Geological Society. https://doi:10.1144/SP422.3

Zellmer, G. F., Sakamoto, N., Iizuka, Y., Miyoshi, M., Tamura, Y., Hsieh, H.-H., & Yurimoto, H. (2014). Crystal uptake into aphyric arc melts: Insights from two-pyroxene pseudo-decompression paths, plagioclase hygrometry, and measurement of hydrogen in olivines from mafic volcanics of SW Japan. In A. Gomez-Tuena, S. M. Straub, G. F. Zellmer (Eds.), *Orogenic andesites and crustal growth* (Special Publication 385, pp. 161–184). London: Geological Society. https://doi:10.1144/SP385.3

Zellmer, G. F., Sakamoto, N., Matsuda, N., Iizuka, Y., Moebis, A., & Yurimoto, H. (2016). On progress and rate of the peritectic reaction Fo + SiO_2 → En in natural andesitic arc magmas. *Geochimica et Cosmochimica Acta*, *185*, 383–393. https://doi:10.1016/j.gca.2016.01.005.

Zellmer, G. F., Sakamoto, N., Matsuda, N., Iizuka, Y., Moebis, A., & Yurimoto, H. (2018). Corrigendum to "On progress and rate of the peritectic reaction Fo + SiO_2 → En in natural andesitic arc magmas" [*Geochimica et Cosmochimica Acta, 185* (2016) 383–393]. *Geochimica et Cosmochimica Acta*, *220*, 552–553. https://doi:10.1016/j.gca.2017.09.043.

Zellmer, G. F., Sparks, R. S. J., Hawkesworth, C. J., & Wiedenbeck, M. (2003). Magma emplacement and remobilization timescales beneath Montserrat: Insights from Sr and Ba zonation in plagioclase phenocrysts. *Journal of Petrolology*, *44*, 1413–1431.

Zellmer, G. F., & Turner, S. P. (2007). Arc dacite genesis pathways: evidence from mafic enclaves and their hosts in Aegean lavas. *Lithos*, *95*, 346–362.

2

Boundary-Layer Melts Entrapped as Melt Inclusions? The Case of Phosphorus- and CO_2-Rich Spinel-Hosted Melt Inclusions from El Hierro, Canary Islands

Marc-Antoine Longpré[1], John Stix[2], and Nobumichi Shimizu[3]

ABSTRACT

Melt inclusions provide unique information on the volatile budgets and compositional diversity of magmatic systems, but are susceptible to synentrapment and postentrapment compositional modifications. The possible formation of chemically anomalous boundary layers around growing crystals and their subsequent entrapment as melt inclusions have been demonstrated experimentally and theoretically, but are thought to be of negligible importance in natural melt-inclusion suites. Here we report on the major, trace, and volatile element compositions of spinel-hosted melt inclusions from El Hierro, Canary Islands, that show anomalous departures in FeO_t–SiO_2 space and high concentrations of P_2O_5, a slow diffuser in silicate melts, with respect to olivine-hosted melt inclusions, the matrix glass, and bulk rock from the same samples. These inclusions also display extremely high CO_2 concentrations and high S/H_2O and Cl/F ratios. With the rapid growth textures of spinels, these observations suggest that spinel-hosted melt inclusions represent boundary-layer melts, the compositions of which were controlled by synentrapment crystal growth and incomplete diffusive relaxation. Our results document a rare case of entrapment of boundary-layer melt in natural magmas, indicate that fast-grown spinel may be a poor melt-inclusion host target, and highlight a means to flag potential boundary-layer melts in melt-inclusion data sets.

2.1. INTRODUCTION

Melt inclusions are minute droplets of silicate melt trapped in crystals that can provide important information on the magmatic systems within which they formed (see reviews by Danyushevsky et al., 2002; Frezzotti, 2001; Kent, 2008; Lowenstern, 1995, 2003; Métrich & Wallace, 2008; Roedder, 1984; Schiano, 2003; Sobolev, 1996). Particularly, melt inclusions provide unique insights into (a) the volatile element (CO_2, H_2O, S, F, Cl) concentrations in magmas prior to syneruptive degassing at low pressure (e.g., Métrich & Wallace, 2008) and (b) the diversity of silicate liquid composition in magmatic systems that may be partially obscured by mixing during mobilization of magma batches (e.g., Kent, 2008).

It is well established, however, that melt inclusions are susceptible to processes that can modify their original composition at the time of entrapment, so care must be taken when interpreting melt-inclusion data sets. Upon cooling, crystallization of the host mineral will occur on the walls of the inclusion, a process referred to as postentrapment crystallization (PEC), which most dramatically affects compatible elements in the host mineral (Danyushevsky et al., 2000; Kress & Ghiorso, 2004). Postentrapment crystallization may also be accompanied by diffusive reequilibration of the inclusion with its host; for example, there is strong evidence that the combination

[1] *School of Earth and Environmental Sciences, Queens College, City University of New York, NY, USA*

[2] *Earth and Planetary Sciences, McGill University, Montreal, Quebec, Canada*

[3] *Woods Hole Oceanographic Institution, Woods Hole, MA, USA*

Dynamic Magma Evolution, Geophysical Monograph 254, First Edition. Edited by Francesco Vetere.

of these processes commonly leads to "iron loss" in olivine-hosted melt inclusions (Danyushevsky et al., 2000). Relatively simple correction schemes have been developed for olivine-hosted melt inclusions, assuming that the extent of iron loss can be estimated (Danyushevsky et al., 2000; Danyushevsky & Plechov, 2011). Equivalent models do not yet exist for melt inclusions hosted in other mineral phases, although Kress and Ghiorso (2004) proposed a thermodynamic approach for PEC correction for olivine-, orthopyroxene-, and feldspar-hosted melt inclusions.

Postentrapment processes may also affect volatile elements in melt inclusions. Notably, it has been demonstrated that H^+ can rapidly diffuse in or out of melt inclusions through the host mineral to equilibrate with the external melt (Bucholz et al., 2013; Gaetani et al., 2012; Hartley et al., 2015; Lloyd et al., 2013; Portnyagin et al., 2008; Qin et al., 1992). Similar open-system behavior may also characterize fluorine (Koleszar et al., 2009; Le Voyer et al., 2014). In addition, CO_2 initially dissolved in the melt inclusion may be subsequently preferentially sequestrated into "shrinkage" bubbles (Hartley et al., 2014; Moore et al., 2015; Steele-Macinnis et al., 2011; Wallace et al., 2015). Finally, sulfur may be affected by postentrapment formation of a sulfide bleb, an expected side effect of iron loss in olivine-hosted melt inclusions (Danyushevsky et al., 2000; Longpré et al., 2017).

Aside from postentrapment modification, the possibility that melt-inclusion compositions might reflect disequilibrium processes occurring at the crystal–melt interface prior to and during entrapment (and therefore be unrepresentative of the bulk magmatic liquid) has long been recognized (Baker, 2008; Faure & Schiano, 2005; Lu et al., 1995). Indeed, compositional gradients may arise in a boundary layer surrounding a growing crystal due to preferential incorporation or exclusion of certain elements in the mineral. Enrichment of incompatible elements in such boundary layers may lead to local saturation of accessory minerals, such as apatite and pyrrhotite, that can then be entrapped as inclusions in the growing phenocrysts (Bacon, 1989; Harrison & Watson, 1984). Baker (2008) performed melt-inclusion-formation experiments and calculations simulating the generation and relaxation of chemical gradients at the crystal–melt interface, and he concluded that at crystal growth rates high enough ($\sim 10^{-8}$–10^{-9} m s^{-1}) to form melt inclusions then slow diffusers, particularly phosphorus, might be enriched in melt inclusions with respect to the bulk melt. However, natural melt-inclusion suites generally bear little, if any, evidence for entrapment of boundary-layer melt and diffusive fractionation (Lowenstern, 1995; Lu et al., 1995; Zajacz & Halter, 2007).

This chapter focuses on naturally quenched lava balloons of basanite bulk composition (Martí et al., 2013; Meletlidis et al., 2015) produced during the submarine eruption that took place in 2011–2012 just south of El Hierro, in the Canary Islands. Olivine- and spinel-hosted melt inclusions from these lava balloons were previously shown to differ in their major element compositions, which together with the occurrence of normal and reverse zoning patterns in olivine and clinopyroxene crystals, was interpreted to reflect mixing of a dominant basanite magma with a more evolved magma at depth (Longpré et al., 2014). Longpré et al. (2017) discussed the volatile-rich and oxidized nature of the dominant basanite magma, represented by olivine-hosted melt inclusions, with CO_2, H_2O, and S concentrations reaching 3420 ppm, 3.0 wt% and 5080 ppm, respectively. Beaudry et al. (2018) subsequently presented S isotope data on both melt and sulfide inclusions from the same samples that reveal the recycled nature of the sulfur signal once degassing is accounted for. Here we present new volatile-element and trace-element data on the spinel-hosted melt inclusions and reevaluate their significance in the context of previously published work. We also provide supporting data on spinel compositions. We find that, compared to olivine-hosted melt inclusions, the matrix glass and bulk rock from the same samples, spinel-hosted inclusions show drastic differences in major, trace, and volatile elements that are best explained by synentrapment modification of melt composition via host mineral crystallization and diffusive fractionation, causing elevated concentrations of slow diffusing elements such as phosphorus and carbon. Both regional and general implications of these results are then discussed.

2.2. SAMPLES AND METHODS

We extracted spinel grains from lava balloon fragments erupted in seawater on five different dates (15 October, 31 October, 27 November, and 6 December 2011, and 28 January 2012) spanning much of the 2011–2012 eruptive sequence at El Hierro; Longpré et al. (2014, 2017) have described these samples. Because spinel is opaque, melt inclusions were exposed by progressive grinding of the mineral grains and frequent petrographic observation under reflected light. We commonly, though not always, intersected vapor bubbles during grinding and occasionally observed clinopyroxene or sulfide microlites within the spinel-hosted melt inclusions. All but one (# 4.67) of the selected inclusions lacked visible microlites and we generally avoided inclusions that appeared leaked or connected to the bulk melt via capillaries. Prior to analysis, melt inclusions and host grains were polished and transferred to an indium mount following established sample preparation protocols (for details see, e.g., Helo et al., 2011).

The concentrations of dissolved volatiles (CO_2, H_2O, F, S, Cl) in 20 spinel-hosted melt inclusions were measured by secondary ion mass spectrometry (SIMS) on a CAMECA IMS 1280 instrument at Woods Hole Oceanographic

Institution (Table SM4; see link in Supplementary Material (SM) section). Calibration curves for $^{12}C/^{30}Si$, $^{16}O^{1}H/^{30}Si$, $^{19}F/^{30}Si$, $^{32}S/^{30}Si$, and $^{35}Cl/^{30}Si$ versus the respective volatile component were constructed using 10 basaltic to basaltic andesite standard glasses. The calibration curves and their slopes' small standard errors (<3.7%), the low analytical backgrounds, the good reproducibility of our in-run secondary standard glass (P1326-2), and the error propagation method are fully described by Longpré et al. (2017; see also Table SM1). Figure SM1 shows the good reproducibility of replicate analyses on three high-CO_2 spinel-hosted melt inclusions, further confirming the robustness of the volatile analyses.

Following SIMS work, we analyzed the major element compositions of the melt inclusions and host minerals by electron probe microanalysis (EPMA) on a JEOL 8900 instrument at McGill University (Tables SM3 and SM4). Longpré et al. (2014) have outlined details of the analytical procedures. Line scans across melt inclusions were performed utilizing an acceleration voltage of 15 kV, a current of 50 nA, a focused beam, a dwell time of 1000 ms and a step size of 0.5 μm (Table SM5). Finally, trace-element concentrations in a subset of 11 melt inclusions were measured by laser ablation inductively coupled plasma mass spectrometry (LA-ICP-MS) at Lamont-Doherty Earth Observatory, using a New Wave UP193-FX laser coupled to a Thermo VG PQ ExCell mass spectrometer (see Longpré et al., 2017 for details; Tables SM2 and SM4).

2.3. RESULTS AND DISCUSSION

2.3.1. Spinel Morphology and Composition

Spinel macrocrysts (≥ 500 μm across) are present in the samples from the 31 October, 27 November, and 6 December 2011 lava balloons, but absent in the 15 October 2011 and 28 January 2012 samples. However, spinel also occurs as inclusions in olivine and clinopyroxene macrocrysts (in all but the 15 October samples) and microlites in the glassy groundmass (in all samples). Both macrocrysts and microlites typically display skeletal to dendritic morphologies, suggesting late-stage rapid growth from the carrier melt, whereas spinel inclusions show euhedral to rounded morphologies (Figure 2.1a–d;Haggerty, 1991).

On average, the 2011–2012 spinels have a composition approaching $Fe^{2+}_{1.0}Mg^{2+}_{0.4}Fe^{3+}_{0.9}Al^{3+}_{0.3}Ti^{4+}_{0.4}O^{2-}_{4}$ (Longpré et al., 2014), although some—particularly inclusions—also contain significant Cr (up to 0.67 cation units, corresponding to 26 wt% Cr_2O_3). Figure 2.2 summarizes the observed compositional variations in terms of sample eruption date and spinel type (whether macrocryst, inclusion, or microlite). Generally, the spinels follow the Fe–Ti trend described by Barnes and Roeder (2001), which is expressed by initially increasing then decreasing TiO_2 with increasing $Fe^{3+}/(Fe^{3+} + Al + Cr)$ (Figure 2.2b). This trend is common in alkali basalt suites and has been interpreted to reflect spinel crystallization concomitant with olivine and clinopyroxene fractionation (Barnes & Roeder, 2001).

In detail, there appears to be systematic spinel compositional variations according to eruption date. This is primarily reflected as higher Ti content at a given Cr content or a $Fe^{3+}/(Fe^{3+} + Al + Cr)$ ratio for later samples, and it is particularly evident when focusing on the microlites only (Figure 2.2b inset). Increasing Ti content in spinel may reflect lower fO_2 conditions (Frost & Lindsley, 1991), suggesting that the El Hierro magma became progressively more reduced as the eruption proceeded.

2.3.2. Uncorrected Composition of Spinel-Hosted Melt Inclusions

2.3.2.1. Major and Trace Elements

Figure 2.3 compares the compositions of uncorrected spinel- and olivine-hosted melt inclusions, the average matrix glass, and the average bulk rock from the 2011–2012 eruption with the El Hierro liquid line of descent defined by published bulk rock and glass compositions (Carracedo et al., 2001; Hausen, 1972; Pellicer, 1979; Stroncik et al., 2009). Corrected compositions of spinel- and olivine-hosted melt inclusions are also shown but will be discussed in section 2.3.3. Uncorrected spinel-hosted melt inclusions show a wide range of compositions, spanning 42.8–53.6 wt% SiO_2, 1.2–4.6 wt% TiO_2, 9.4–15.9 wt% FeO_t, 1.7–6.7 wt% MgO, 5.2–12.0 wt% CaO, 4.4–7.1 Na_2O, 1.5–2.6 K_2O and 0.8–2.0 P_2O_5. These compositions are significantly more evolved than the 2011–2012 bulk rock and matrix glass and fall close to or on the high-SiO_2 side of the El Hierro liquid line of descent, with two striking exceptions: (a) they plot distinctly above the well-defined bulk-rock–glass FeO_t–SiO_2 array ($R^2 = 0.95$); and (b) they typically have higher P_2O_5 than any known local magma composition (Figure 2.3a and h). In comparison, uncorrected olivine-hosted melt inclusions display a more restricted compositional range, with 41.7–45.1 wt% SiO_2, 4.3–5.5 wt% TiO_2, 8.7–13.7 wt% FeO_t, 2.7–9.6 wt% MgO, 10.3–14.5 wt% CaO, 4.2–5.4 Na_2O, 1.5–1.9 K_2O and 0.6–1.2 P_2O_5, and plot near the 2011–2012 bulk rock and matrix glass. Notably, in contrast to spinel-hosted melt inclusions, several uncorrected olivine-hosted melt inclusions fall below the FeO_t–SiO_2 array (Figure 2.3a).

In a primitive-mantle-normalized multielement diagram (Figure 2.4), spinel-hosted melt inclusions show the most enriched composition within the 2011–2012 El Hierro samples, with, for example, 110 ±17 (1σ) ppm Nb (167 × primitive mantle). Overall, however, the general shape of their trace-element profile is very similar to that of olivine-hosted melt inclusions, the matrix glass, and bulk rock.

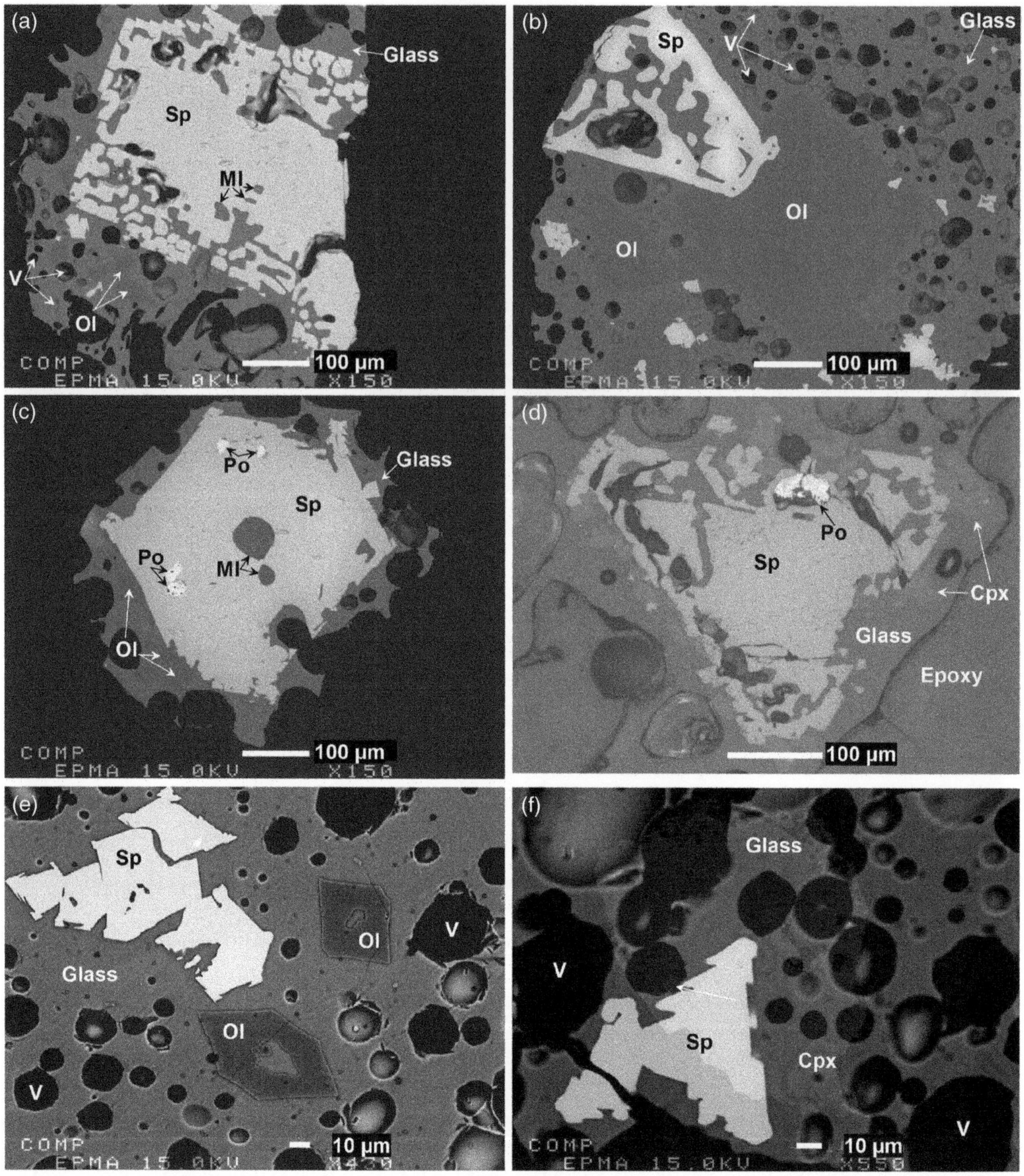

Figure 2.1 Spinel textures in the El Hierro basanite. All panels but (d) (reflected light) show backscatter electron images. Both macrocrysts (a–d) and microlites (e, f) show skeletal to dendritic habits indicating rapid growth. Panels (c) and (d) show evidence for the association of sulfide saturation with spinel growth. Cpx, clinopyroxene; MI, melt inclusion; Sp, spinel; Ol, olivine; Po, pyrrhotite; V, vesicle.

2.3.2.2. Volatile Elements

Figure 2.5 summarizes the volatile element systematics of spinel-hosted melt inclusions, highlighting significant differences with olivine-hosted melt inclusions. Most strikingly, several spinel-hosted inclusions show much higher CO_2 concentrations, with 10 values above 4000 ppm and a maximum of 12,000 ppm! In addition, H_2O contents in spinel-hosted melt inclusions are comparatively low, with most values around 0.5–1.0 wt% and all below 2.0 wt%, whereas olivine-hosted melt inclusions show higher H_2O concentrations up to 3.0 wt%. Sulfur concentrations reach similarly high values in spinel-hosted and olivine-hosted melt inclusions (4870 and 5080 ppm, respectively), but are higher in spinel-hosted inclusions at a given H_2O

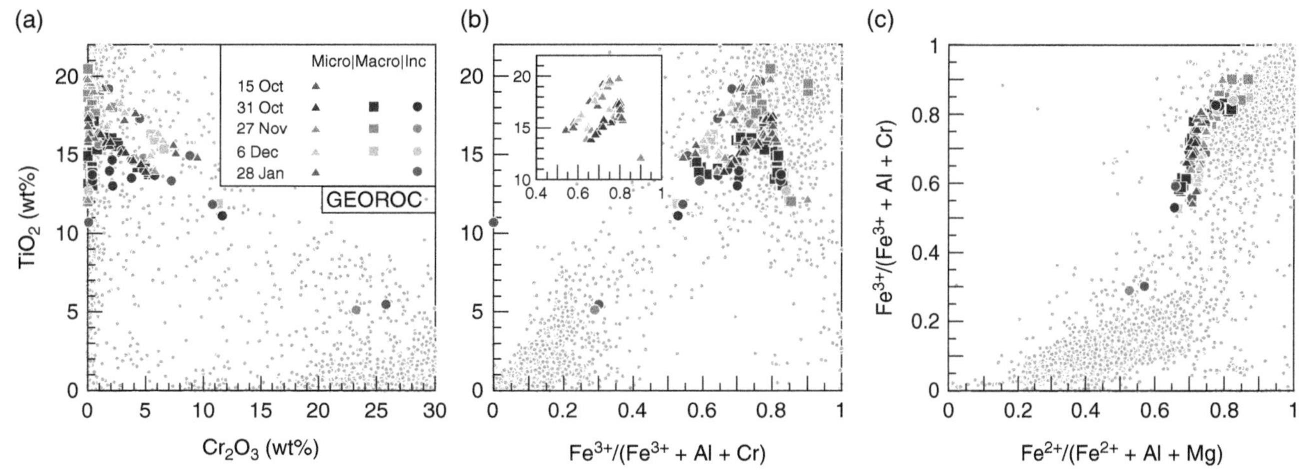

Figure 2.2 Composition of spinel macrocrysts (Macro; squares), microlites (Micro; triangles), and macrocryst-hosted inclusions (Inc; circles) in the 2011–2012 El Hierro lava balloons color-coded by eruption date. Plots show TiO_2 content as a function of (a) Cr_2O_3 content and (b) $Fe^{3+}/(Fe^{3+} + Al + Cr)$. Panel (c) shows $Fe^{3+}/(Fe^{3+} + Al + Cr)$ versus $Fe^{2+}/(Fe^{2+} + Mg)$. Note the generally increasing TiO_2 with time during the eruption, which may reflect a decrease in fO_2 with time (Frost & Lindsley, 1991), as best shown by the microlites (inset in (b)). The composition of spinels from a global compilation of basaltic and related rocks extracted from the GEOROC database (http://georoc.mpch-mainz.gwdg.de/georoc/Start.asp) is shown for comparison.

content. Moreover, unlike for olivine-hosted melt inclusions, fluorine and chlorine are not correlated in spinel-hosted melt inclusions, which show generally higher Cl at similar F concentrations and thus higher Cl/F.

A comparison of volatiles with lithophile elements showing similar geochemical behavior during partial melting and crystallization is insightful. Despite some extremely high CO_2 concentrations, the measured CO_2/Nb ratios of spinel-hosted melt inclusions only reach modest values, slightly higher than those of olivine-hosted inclusions but still dramatically short of all published estimates for undegassed melts derived from Earth's mantle (Figure 2.5d). In contrast, H_2O/Ce and S/Dy reach greater values in olivine-hosted melt inclusions relative to spinel-hosted melt inclusions, with the highest ratios meeting or exceeding mantle estimates (Figure 2.5e). Finally, whereas El Hierro samples have similar Cl/K ratios overlapping with primitive mantle estimates, olivine-hosted melt inclusions and the matrix glass show F/Nd ratios typically in excess of mantle estimates, whereas spinel-hosted melt inclusions extend to lower values (Figure 2.5f).

2.3.3. Synentrapment or Postentrapment Modification of Spinel-Hosted Melt Inclusions

The departure of melt-inclusion compositions from the FeO_t–SiO_2 array defined by El Hierro bulk rocks and matrix glasses is anomalous and strongly suggests melt-inclusion modification, either by synentrapment or postentrapment processes. In the case of uncorrected olivine-hosted melt inclusions, which have SiO_2 contents similar to 2011–2012 bulk rocks and matrix glasses, their relatively low FeO_t can be readily explained by iron loss during postentrapment crystallization and diffusive reequilibration with the host olivine (Danyushevsky et al., 2000) and can be corrected for using Petrolog3 (Danyushevsky & Plechov, 2011) assuming an initial FeO_t dictated by the bulk-rock–glass FeO_t–SiO_2 relationship (Longpré et al., 2014). Following this routine correction, olivine-hosted melt inclusions fall on the El Hierro liquid line of descent for all major elements (Figure 2.3). In contrast, potential synentrapment or postentrapment modifications of spinel-hosted melt inclusions are poorly understood. Kamenetsky (1996) argued postentrapment effects to be minimal in chromium spinel due to the low Cr content of the silicate melt, but this argument does not hold for Fe–Ti-rich spinels. Therefore, in order to constrain the cause of the offset of spinel-hosted melt inclusions in FeO_t–SiO_2 space and explain other characteristic features of spinel-hosted melt inclusions, particularly their high P_2O_5, incompatible trace-element and CO_2 concentrations, as well as their high S/H_2O and Cl/F ratios, we consider the following processes:

1. postentrapment crystallization of spinel on the melt-inclusion walls;
2. postentrapment dissolution of spinel into the melt inclusions;
3. postentrapment diffusion of iron into the melt inclusions;
4. formation and entrapment of a boundary-layer melt.

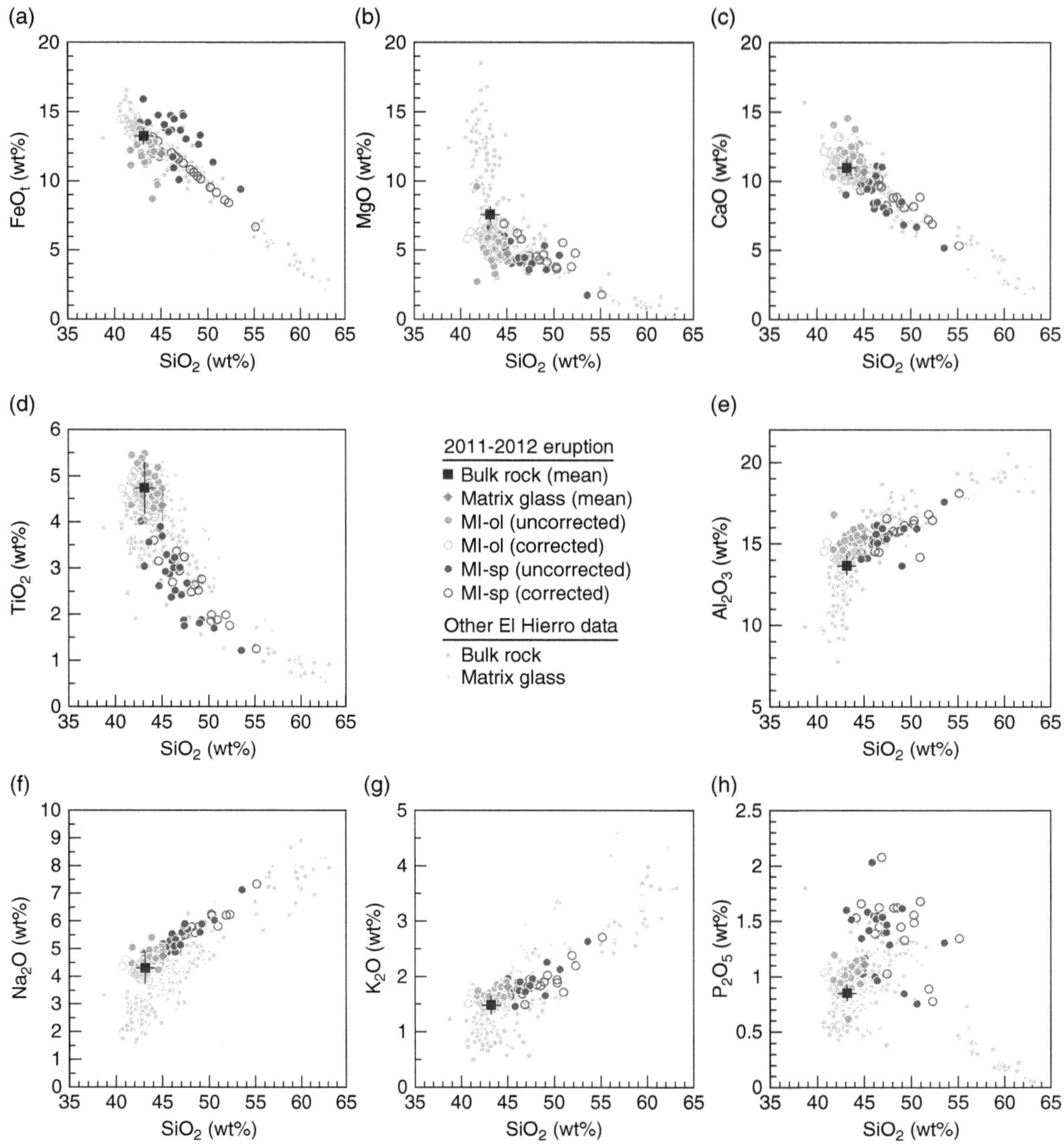

Figure 2.3 Major element composition of spinel-hosted melt inclusions (MI-sp, uncorrected and corrected) compared to that of olivine-hosted melt inclusions (MI-ol, uncorrected and corrected) and the average bulk rock and matrix glass of the 2011–2012 eruption products (data from Longpré et al. (2014), except for bulk rocks from Martí et al. (2013)). Also shown for comparison are published bulk-rock and glass data for prehistoric eruptions (Carracedo et al., 2001; Hausen, 1972; Pellicer, 1979; Stroncik et al., 2009), which define a liquid line of descent for El Hierro magmas.

Potential postentrapment volatile loss is discussed separately in sections 2.3.5 and 2.3.6.

2.3.3.1. Postentrapment Crystallization

Figure 2.6 shows backscatter electron (BSE) images and Si and Fe signal-intensity transects (acquired by EPMA) for two olivine-hosted and two spinel-hosted melt inclusions. Additional BSE images and X-ray elemental maps taken at higher magnification at the spinel–melt-inclusion interface are presented in Figures SM2 and SM3 for the inclusions shown in Figure 2.6c and d. Although not as obvious as for olivine-hosted melt inclusions, there is evidence for some postentrapment crystallization of spinel-hosted melt inclusions. For example, melt-inclusion sample MI-sp 4.66

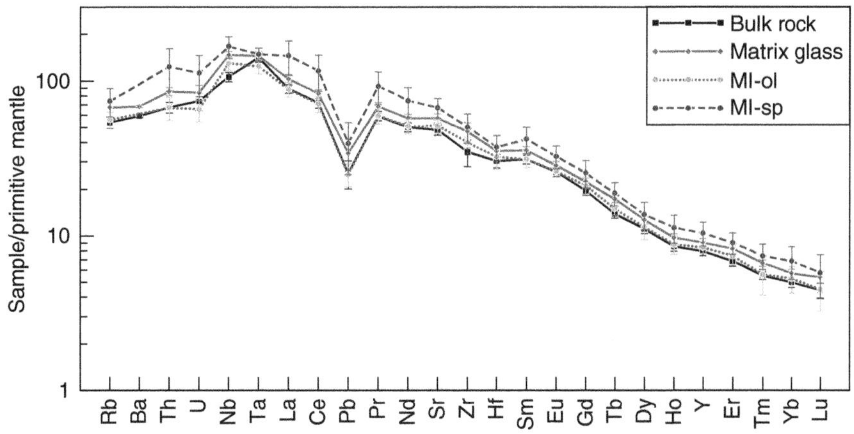

Figure 2.4 Primitive-mantle normalized (McDonough & Sun, 1995) multielement diagram showing the average composition of spinel-hosted melt inclusions (n = 11, this work), olivine-hosted melt inclusions (n = 15), the matrix glass (n = 18) (Longpré et al., 2017) and the bulk rock (n = 10) (Martí et al., 2013). Spinel-hosted melt inclusions display the most enriched concentrations in incompatible trace elements, but otherwise similar trace element profile patterns.

displays a band ~5 μm wide enriched in Si at the contact with spinel (Figure 2.6d), which in turn shows a halo ~1 μm wide chiefly characterized by higher Mg concentration (Figure SM2). If significant, we would expect postentrapment crystallization of spinel to lead to a strong depletion of elements compatible in spinel (e.g., Fe and Ti) and an enrichment of elements incompatible in spinel (e.g., Si and Ca) in the melt inclusions. In FeO_t–SiO_2 space, this should cause a sharp decrease in FeO_t with increasing SiO_2, with a slope steeper than that of the bulk-rock–glass trend (Figure 2.7). Assuming that El Hierro melt compositions must fall on the FeO_t–SiO_2 array, postentrapment spinel crystallization would tend to bring modified melt-inclusion compositions below the bulk-rock–glass trend and, therefore, it cannot explain data falling above it. In addition, while this process could be consistent with elevated incompatible trace element concentrations (e.g., P_2O_5, Nb, CO_2), it should not a priori fractionate S/H_2O and Cl/F ratios, as these elements are presumably all similarly incompatible in spinel. Therefore, although we cannot rule out its possible contribution to data variability in spinel-hosted melt inclusions, postentrapment crystallization alone seems incapable to reproduce some key features of our data set.

2.3.3.2. Postentrapment Dissolution

Dissolution of spinel into the melt inclusions should have effects opposite to that of postentrapment crystallization. In Figure 2.7, starting from the average 2011–2012 bulk rock composition, this should lead to an increase in FeO_t with a decrease in SiO_2. This process could thus apparently translate melt compositions falling on the bulk-rock–glass trend to the positions occupied by uncorrected spinel-hosted melt inclusions. However, spinel dissolution would be expected to deplete the melt in incompatible elements, such as P_2O_5, Nb, and CO_2, which is not observed. Furthermore, dissolution does not provide an obvious mechanism to affect S/H_2O and Cl/F ratios.

2.3.3.3. Postentrapment Iron Gain

Iron gain by postentrapment diffusion from the host spinels to the melt inclusions would yield a steep increase in FeO_t accompanied by decreasing SiO_2 in the modified melt (Figure 2.7). This could effectively displace melts from the bulk-rock–glass array to the high FeO_t side of it, and we previously hypothesized that the El Hierro spinel-hosted melt inclusions had been affected by this process (Longpré et al., 2014). Corrected spinel-hosted melt-inclusion compositions shown in Figure 2.3 were thus obtained by removing FeO_t until the FeO_t–SiO_2 relationship was satisfied (Figure 2.3a). This procedure involved an average subtraction of 2% FeO_t, with a maximum subtraction of 6% FeO_t, from the melt-inclusion composition. After this correction, the major element compositions of melt inclusions largely fall within the published data range for El Hierro lavas. However, elevated P_2O_5 contents remain a notable exception that remains unexplained in this scheme. An additional problem with iron gain is that, like spinel dissolution, it should deplete the affected melt inclusions in incompatible elements, contrary to observations.

2.3.3.4. Entrapment of Boundary-Layer Melts and Diffusive Fractionation

A boundary-layer melt forming next to a rapidly growing spinel crystal would be predicted to initially be depleted in spinel-compatible elements and enriched in incompatible elements with respect to the melt far away

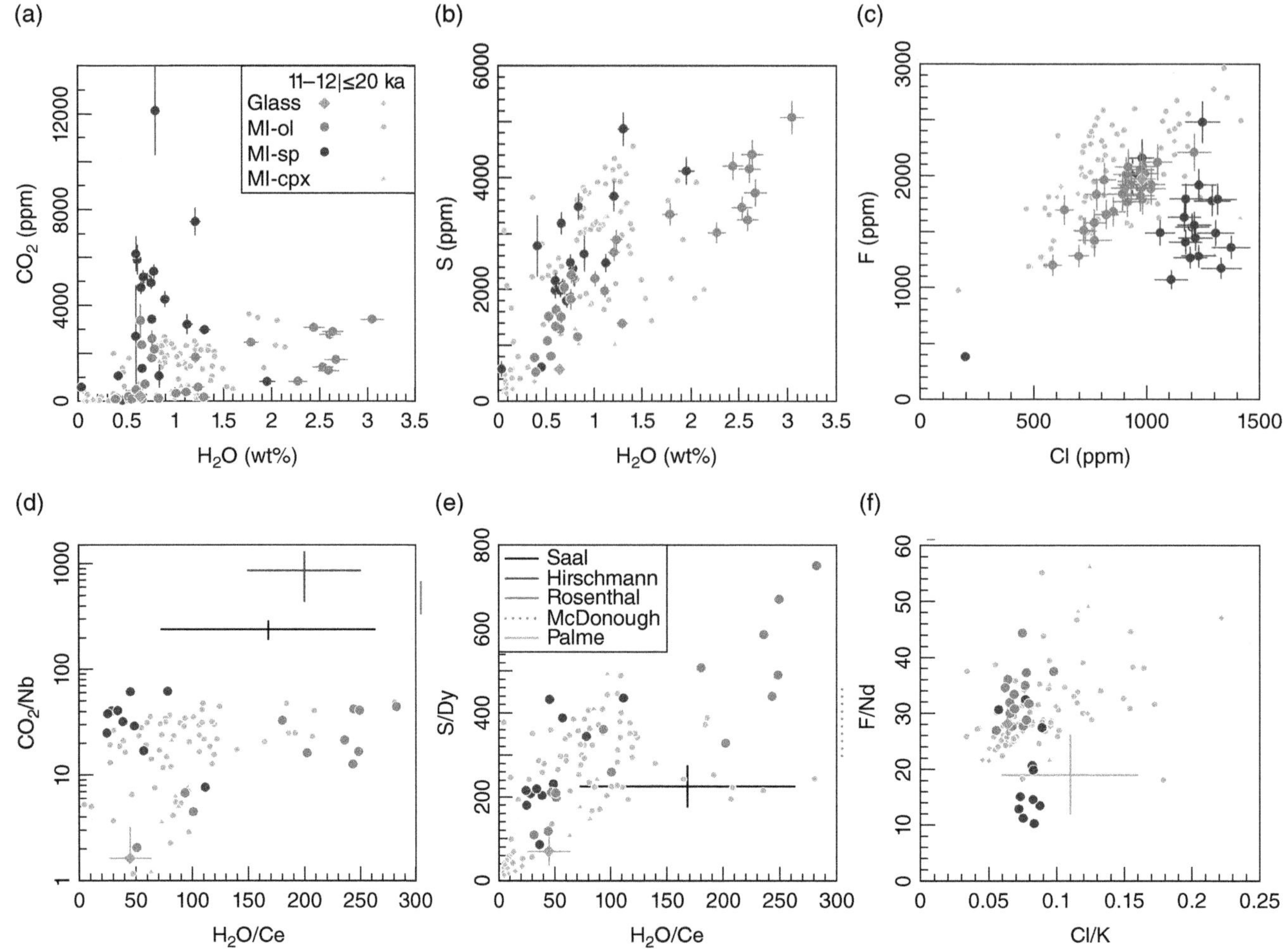

Figure 2.5 Volatile and volatile/lithophile element systematics of spinel-hosted melt inclusions (this work) compared to olivine-hosted melt inclusions and the matrix glass (Longpré et al., 2017) from the 2011–2012 samples. The data of Taracsák et al. (2019) for olivine- and clinopyroxene-hosted melt inclusions and matrix glasses for El Hierro tephras younger than ~20 ka are also shown (≤ 20 ka). (a) Several spinel-hosted inclusions show high CO_2 concentrations at moderate H_2O content. (b) At a given H_2O content, spinel-hosted melt inclusions typically display higher S concentrations than their olivine-hosted counterparts. (c) Unlike olivine-hosted melt inclusions, spinel-hosted melt inclusions do not show a F versus Cl correlation and are characterized by high Cl over a range of F concentrations. (d) Some CO_2/Nb ratios (note *y*-axis log scale) are slightly higher in spinel-hosted melt inclusions than in olivine-hosted melt inclusions but still much lower than expected for undegassed mantle melts. (e) High S/Dy ratios indicate that the sulfur content of the El Hierro mantle exceeds that of the depleted MORB mantle (see Taracsák et al., 2019). (f) Most El Hierro samples, except the 2011–2012 spinel-hosted melt inclusions, show F/Nd values in excess of the primitive mantle estimate. Plotted mantle volatile/lithophile elemental ratios are: for CO_2/Nb, 239 ± 46 (Saal et al., 2002), 505 ± 168 (Rosenthal et al., 2015), and 860 ± 420 (Hirschmann, 2018); for H_2O/Ce, 168 ± 95 (Saal et al., 2002) and 200 ± 50 (Hirschmann, 2018); for S/Dy, 225 ± 49 (Saal et al., 2002) and 371 ± 86 (McDonough & Sun, 1995); for F/Nd, 19 ± 7 (Palme & O'Neill, 2003); and for Cl/K, 0.0075 ± 0.0025 (Saal et al., 2002) and 0.11 ± 0.05 (Palme & O'Neill, 2003). Note that when a reference constrained only one of the two ratios plotted, the bar is placed just outside the plot area (to the right or the top).

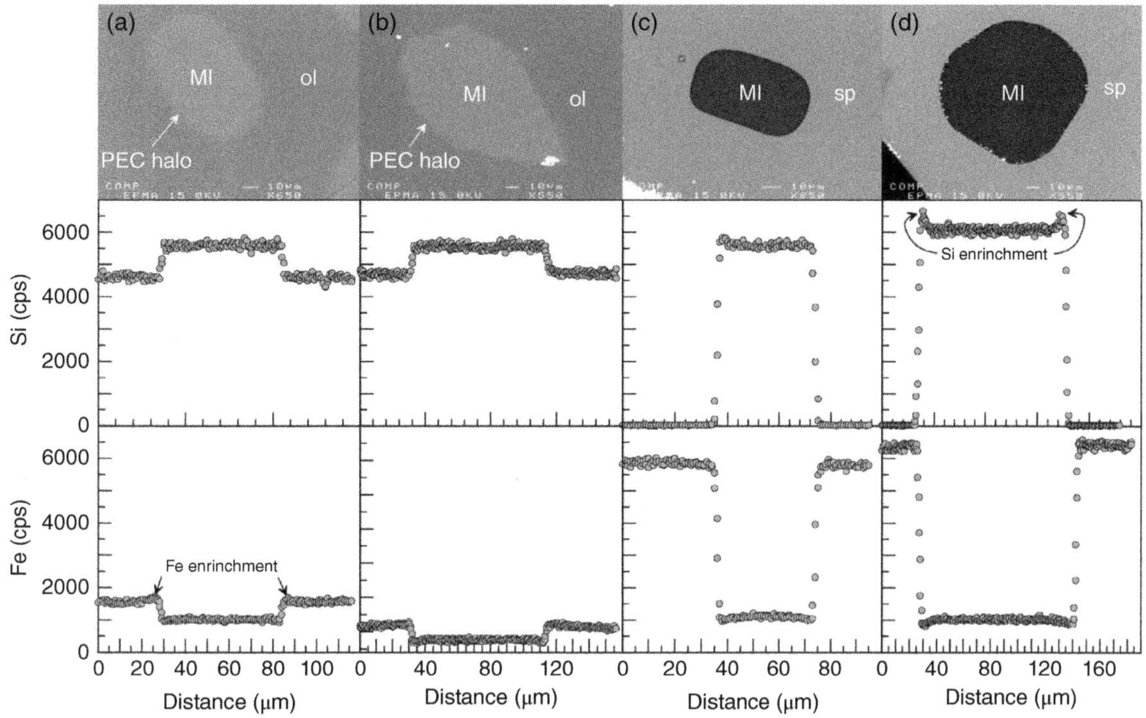

Figure 2.6 Backscattered electron images and Si and Fe intensity line scans (in counts per second, cps) of (a, b) olivine-hosted and (c, d) spinel-hosted melt inclusions from the 2011–2012 El Hierro samples. *Y*-axis scales are the same for all panels. A compositional halo due to postentrapment crystallization (PEC) of the mineral host is visible in olivine-hosted melt inclusions; such a feature appears only at higher magnification for spinel-hosted melt inclusions (Figures S2 and S3). Fe enrichment is commonly observed in olivine near the interface with a melt inclusion. Spinel composition does not seem significantly influenced by the presence of a melt inclusion (see minor effects observed in Figures S2 and S3), but Si enrichment in the melt inclusion shown in (d) is observed near the interface with its host spinel.

from crystals (hereafter referred to as the far-field melt). At first glance this would appear inconsistent with the high FeO_t of spinel-hosted melt inclusions at a given SiO_2. However, given time, the chemical gradient at the crystal–melt interface will be reduced by diffusion at magmatic temperature, with fast-diffusing elements reequilibrating faster with the far-field melt than slow-diffusing ones. If a boundary-layer melt is entrapped as an inclusion before full diffusive reequilibration, the concentration of a given element in the melt inclusion will be a complex function of its partition coefficient in the host and its diffusivity in the melt—the FeO_t–SiO_2 systematics in a hypothetical boundary layer at the spinel–melt interface may thus be difficult to predict due to the contrasting partitioning behavior of Fe and Si in spinel. On the other hand, because it is one of the slowest diffusers in silicate melts, phosphorus should be particularly sensitive to such synentrapment diffusive fractionation (Figure 2.8) (e.g., Baker, 2008). Coupled with the incompatibility of P_2O_5 in spinel and evidence for rapid spinel growth (Figure 2.1), this provides an attractive mechanism to explain the elevated P_2O_5 concentrations in spinel-hosted melt inclusions.

The boundary-layer-melt hypothesis predicts a correlation between melt-inclusion size and composition because, proportionally, smaller inclusions should sample melt closer to the crystal–melt interface and thus should deviate more strongly from the far-field melt than larger inclusions (Lu et al., 1995; Baker, 2008). Thus, if the 2011–2012 El Hierro spinel-hosted melt inclusions represent boundary-layer melts, we would expect to observe greater enrichment of slow diffusers with respect to fast diffusers for smaller inclusions. To test this, we compare P_2O_5 and K_2O, which we take as the most reliable tracers of potential boundary-layer effects as both should be incompatible in most host-mineral phases but have diffusivities orders of magnitude apart in the melt (Figure 2.8). Figure 2.9a shows that most spinel-hosted melt inclusions smaller than ~50 μm in their shortest dimension are characterized by elevated P_2O_5/K_2O ratios compared to larger spinel-hosted inclusions, olivine-hosted melt inclusions, the matrix glass, and bulk rock. This is strong evidence of boundary-layer effects in these spinel-hosted melt inclusions. In detail, surprisingly, we find a slight but noticeable

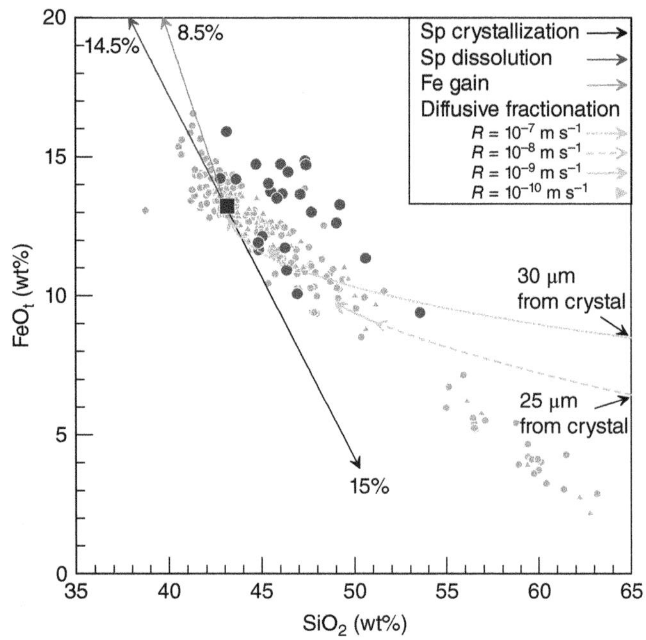

Figure 2.7 Synentrapment and postentrapment modification scenarios for spinel-hosted melt inclusions in FeO_t–SiO_2 space. The compositions of uncorrected spinel-hosted melt inclusions and the average bulk rock are shown for the 2011–2012 samples, together with bulk rock and glass data for prehistoric samples (symbols and data sources as in Figure 2.3). Blue arrows show the predicted effects on melt composition of postentrapment spinel crystallization, spinel dissolution, and iron gain, as well as boundary-layer-melt entrapment coupled to diffusive fractionation for different crystal growth rates (*R*) (see text for details). The starting melt composition is taken as the average bulk rock (navy square). The percentages of postentrapment spinel crystallization, spinel dissolution, and iron gain are indicated, represented by the length of their respective arrow. For boundary-layer-melt entrapment and diffusive fractionation scenarios, the melt composition shows a greater departure from the starting composition near the crystal–melt interface and tends to approach this composition farther from this interface (see also Figure 2.11). We modeled the composition of the melt from 0 μm (beginning of arrow) to 200 μm (at arrow-head tip) from the crystal–melt interface. For $R = 10^{-7}$ m s^{-1} and $R = 10^{-8}$ m s^{-1}, the composition of the melt at 0 μm plots off the graph to the right and thus we indicate the distance from the crystal–melt interface at which SiO_2 decreases to 65 wt%. For $R = 10^{-9}$ m s^{-1} and $R = 10^{-10}$ m s^{-1}, the melt composition from 0–200 μm from the crystal–melt interface is also significantly different from the starting composition, but falls within El Hierro's liquid line of descent (in the case of $R = 10^{-10}$ m s^{-1}, the melt composition gradient from 0–200 μm is smaller than the arrow head).

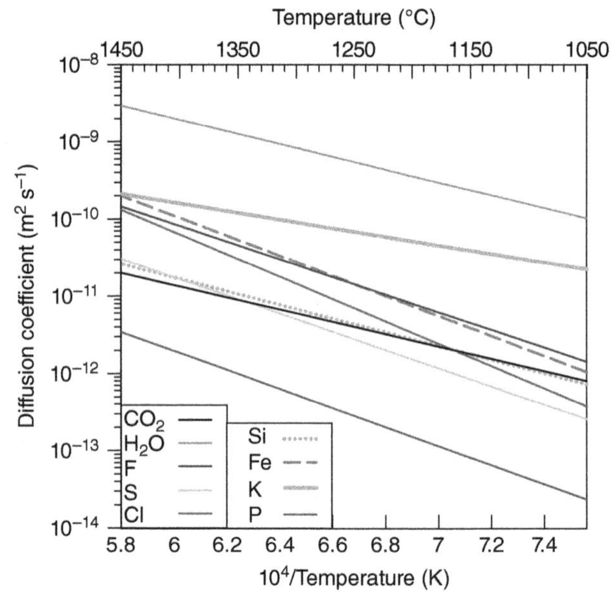

Figure 2.8 Comparison of diffusion coefficients for Si (Lesher et al., 1996), Fe (Zhang et al., 2010, equation 86), K (Zhang et al., 2010, equation 11), P (Baker, 2008), CO_2 (Zhang & Ni, 2010, equation 41), H_2O (Zhang & Ni, 2010, equation 22), S (Freda et al., 2005), F, and Cl (Alletti et al., 2007) in anhydrous (except for water diffusion, which is calculated here for 1 wt% H_2O) basaltic melts. Under hydrous conditions, diffusion coefficients are expected to increase in all cases. The diffusion expressions for P, S, Cl, and F are based on experiments using the same starting material—an alkaline basalt from Etna—and are thus directly comparable; this may not be the case for Si, Fe, K, and CO_2.

negative correlation between P_2O_5/K_2O and size for olivine-hosted melt inclusions also, translating into mean P_2O_5/K_2O ratios of 0.62 ± 0.06, which is a bit higher than for the mean matrix glass (0.59 ± 0.04) and bulk rock (0.59 ± 0.08) (Figure 2.9a). This indicates that even the olivine-hosted melt inclusions may record minor boundary-layer effects.

The rest of Figure 2.9 (b–e) shows other key slow diffuser/fast diffuser ratios as a function of size of the 2011–2012 melt inclusions. While no clear trends emerge for SiO_2/FeO_t (see section 2.3.4), most small spinel-hosted melt inclusions show markedly higher CO_2/K and S/H_2O ratios than larger spinel-hosted melt inclusions and olivine-hosted melt inclusions of any size, consistent with the boundary layer hypothesis. However, we note that, in contrast to P_2O_5/K_2O, both CO_2/K and S/H_2O are sensitive to degassing, as illustrated by the low matrix glass values. Despite this it appears that CO_2 is even more enriched with respect to potassium than P_2O_5 in spinel-hosted melt inclusions, which would not be expected based on their diffusivities alone (Figure 2.8). A possible explanation for this intriguing observation, potentially linked to melt-inclusion decrepitation, is discussed in section 2.3.5. Also puzzling is the much higher Cl/F ratio of spinel-hosted melt inclusions, regardless of size, with respect to olivine-hosted melt inclusions and the matrix glass. As for P_2O_5/K_2O, we note a hint of a negative trend between Cl/F and

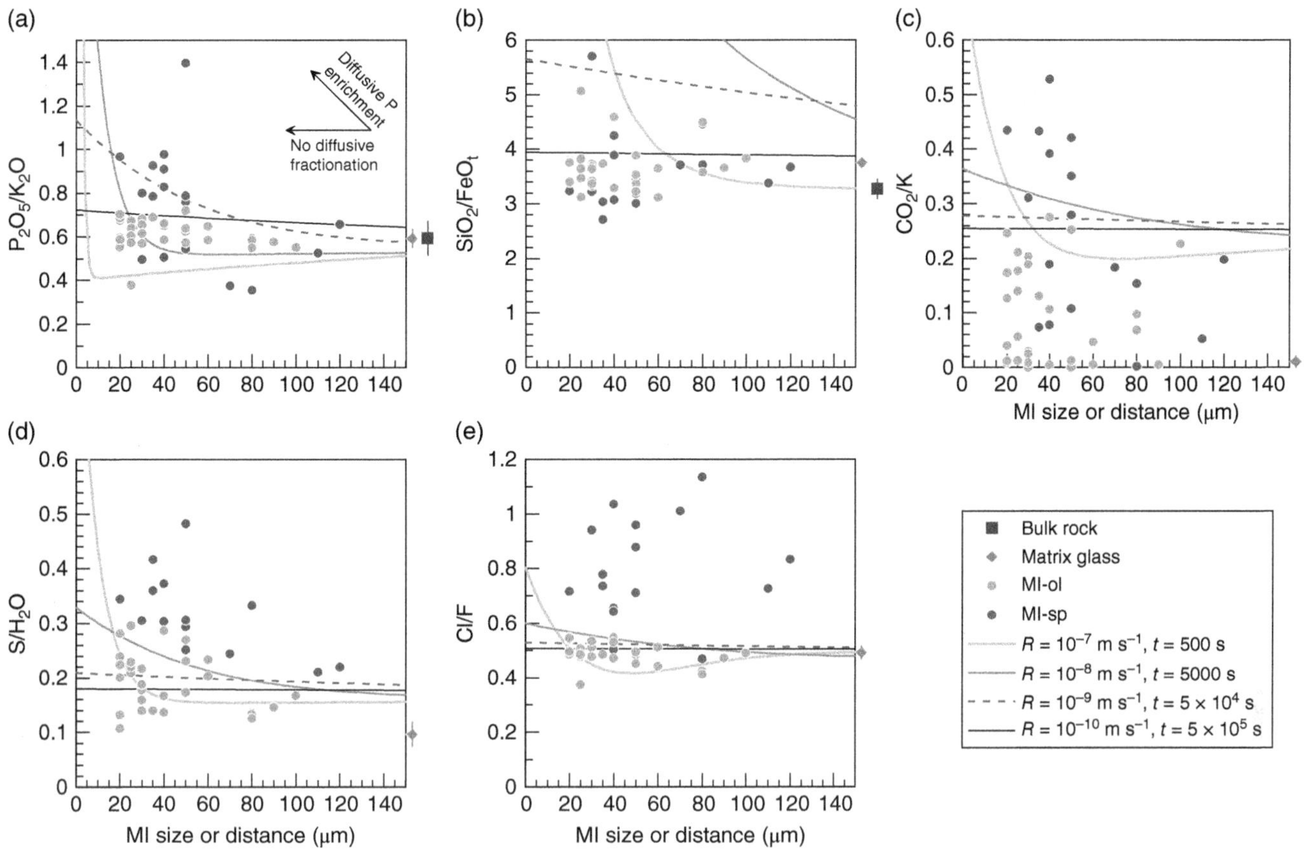

Figure 2.9 Slow-diffuser/fast-diffuser ratios as a function of melt-inclusion shortest dimension. (a) P_2O_5/K_2O, which we take as the most reliable measure of potential diffusive pileup at the crystal–melt interface, is elevated for small spinel-hosted melt inclusions relative to other 2011–2012 samples, which supports the hypothesis that these melt inclusions represent entrapped boundary-layer melts. A faint negative correlation between this ratio and size of olivine-hosted melt inclusions appears to be present. In all panels, the calculated composition of the boundary-layer melt as a function of distance from the crystal–melt interface is shown for four end-member crystal growth rate (R) and time step (t) scenarios as indicated in the legend and discussed in the text. The average bulk rock was used as far-field melt composition for major elements (Table 2.1), whereas far-field CO_2, H_2O, S, Cl and F concentrations were taken as 3000 ppm, 3 wt%, 5000 ppm, 1000 ppm and 2000 ppm, respectively, based on high-end (least degassed) values for olivine-hosted melt inclusions. Average bulk rock and matrix glass compositions, if applicable, are shown just outside of the plot area to the right.

size for olivine-hosted melt inclusions, which may indicate minor diffusive fractionation of halogens. The behavior of halogens is further discussed in section 2.3.6.

Diffusive fractionation of trace elements is also predicted to occur in crystal–melt interface boundary layers (Holycross & Watson, 2016). An attempt to test this for our samples is shown in Figure 2.10, where elements are arranged in order of increasing diffusion coefficient—i.e., the slowest diffusers on the left, the fastest on the right (table 3A of Holycross & Watson, 2016). Primitive-mantle normalized concentrations are divided by that of potassium in order to damp the effect of partitioning and isolate that of diffusivity. On such a diagram samples affected by boundary layer effects should be more enriched (with respect to K) in the slow diffusers than samples unaffected by diffusive fractionation. Observations are somewhat consistent with this prediction; although slow-diffusing Hf, Zr, and Ta do not show appreciable difference between samples, spinel-hosted melt inclusions are indeed significantly more enriched in U, Th, Nb, La, and Ce with respect to K than other samples.

In summary, several observations, particularly the rapid growth textures of spinel and negative correlations between slow diffuser/fast diffuser ratios (P_2O_5/K_2O, CO_2/K, and S/H_2O) and melt-inclusion size, are qualitatively consistent with boundary layer effects having played an important role in the composition of the 2011–2012 spinel-hosted melt inclusions. We thus quantitatively test the boundary-layer model against our data in the following section.

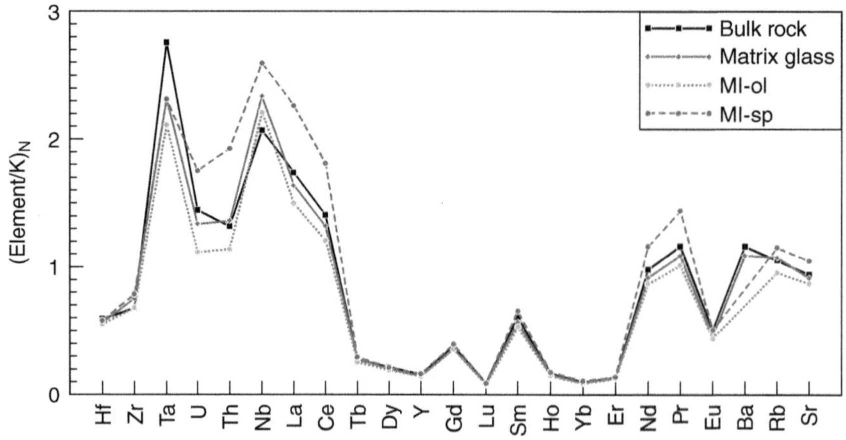

Figure 2.10 Primitive-mantle normalized (McDonough & Sun, 1995) trace element enrichment relative to fast-diffusing potassium for the El Hierro sample averages (as in Figure 2.4). The sequence of elements is in order of increasing diffusivity from left to right as per table 2.3A of Holycross and Watson (2016). Melt inclusions representing boundary-layer melts would be predicted to show greater enrichment of slow diffusers (left) with respect to samples unaffected by diffusive fractionation. This is consistent with observations for U, Th, Nb, La, and Ce, which show higher mean enrichment in spinel-hosted melt inclusions than in other samples, whereas slow diffusing Hf and Zr do not show appreciable difference between samples.

2.3.4. Modeling the Boundary-Layer-Melt Composition

Using the approach of Baker (2008; after Smith et al., 1955), we modeled the composition of the silicate liquid ($C_l(x,t)$) (for SiO_2, FeO_t, K_2O, P_2O_5, CO_2, H_2O, S, F, and Cl) at time t (s) and distance x (m) from the crystal–melt interface with equation (2.1) at fixed but various crystal growth rates R (m s^{-1}):

$$C_l(x,t) = C_0 \left\{ \begin{array}{l} 1 + \dfrac{1-K}{2K} e^{-\frac{R}{D}x} \operatorname{erfc}\left[\dfrac{1}{2}\sqrt{\dfrac{1}{Dt}}(x - Rt)\right] \\ -\dfrac{1}{2}\operatorname{erfc}\left[\dfrac{1}{2}\sqrt{\dfrac{1}{Dt}}(x + Rt)\right] \\ +\dfrac{(1-K)}{2}\left(\dfrac{1}{1-K} - \dfrac{1}{K}\right) e^{-(1-K)\left(\frac{R}{D}\right)(x+KRt)} \\ \operatorname{erfc}\left[\dfrac{1}{2}\sqrt{\dfrac{1}{Dt}}(x + (2K-1)Rt\right] \end{array} \right\} \tag{2.1}$$

where C_0 is the composition of the melt at infinite distance (far-field melt) from the crystal–melt interface, K is the mineral–melt partition coefficient (here, concentration of the element of interest in spinel divided by its concentration in the far-field melt), D (m^2 s^{-1}) is the diffusion coefficient of the element in the melt, and erfc is the error function complement. We used the average bulk-rock composition (Martí et al., 2013) and representative volatile element concentrations (Longpré et al., 2017) as far-field melt composition and the average spinel composition (Longpré et al., 2014; this work) to derive empirical spinel–melt partition coefficients for SiO_2 and FeO_t. The concentrations of K_2O, P_2O_5, CO_2, H_2O, S, F, and Cl were not measured in spinel and we used $K = 0.001$ for these incompatible elements (cf. Baker, 2008). The diffusivity of the elements of interest in basaltic melt is compared in Figure 2.8; for our calculations, we used diffusion coefficients at a temperature of 1150°C (Table 2.1). We considered crystal growth rates R of 10^{-7}, 10^{-8}, 10^{-9}, and 10^{-10} m s^{-1}, spanning a wide range of conditions relevant to natural magmas (see compilation by Baker, 2008). For each growth rate, the time step was chosen to allow the growth of 50 μm of crystal, based on the typical dimension of melt inclusions selected for analysis (e.g., Lowenstern, 1995). Table 2.1 lists input parameters for the calculations.

It is clear from Figure 2.11 that, at rapid growth rates (10^{-7}–10^{-9} m s^{-1}) thought to be required for melt-inclusion formation (Faure and Schiano, 2005; Baker, 2008), calculations predict strong enrichment of slow-diffusing incompatible elements at significant distances from the crystal–melt interface during the growth step. In contrast, water, with its significantly higher diffusivity, shows little to no enrichment. Iron, as a compatible element, shows strong depletions in the modeled boundary-layer melt. At the slower growth rate of 10^{-10} m s^{-1}, only P_2O_5 (elevated) and FeO_t (depleted) are significantly affected.

Because of the low partition coefficient of SiO_2 in spinel, the boundary-layer melt may become more enriched in SiO_2 than it is depleted in FeO_t, which may produce compositions falling on the high side of the bulk rock–glass FeO_t–SiO_2 array, similar to uncorrected spinel-hosted melt inclusions. To illustrate this possibility,

Table 2.1 Input Parameters in Diffusion Calculations of Boundary-Layer Melt

Element	$[C]_{melt}$	$[C]_{spinel}$	K	D (m^2 s^{-1})	Reference
SiO_2	43.17	0.04	0.001	2.1E-12	Lesher et al. (1996)
TiO_2	4.74	15.67	–	–	
Al_2O_3	13.65	6.17	–	–	
FeO	13.22	66.38	5.02	5.1E-12	Zhang et al. (2010, equation 86)
MnO	0.24	0.37	–	–	
MgO	7.58	6.92	–	–	
CaO	10.95	0.06	–	–	
Na_2O	4.31	0.00	–	–	
K_2O	1.48	0.00	0.001	4.4E-11	Zhang et al. (2010, equation 11)
P_2O_5	0.85	0.01	0.001	1.1E-13	Baker (2008)
Total	100.20	95.62			
CO_2	3000	–	0.001	2.1E-12	Zhang & Ni (2010, equation 41)
H_2O	1	–	0.001	8.5E-10	Zhang & Ni (2010, equation 22)
S	5000	–	0.001	1.1E-12	Freda et al. (2005)
Cl	1000	–	0.001	2.2E-12	Alletti et al. (2007)
F	2000	–	0.001	5.8E-12	Alletti et al. (2007)

Note. $[C]_{melt}$ is the composition of bulk melt (major and volatile elements given as wt% and ppm values, respectively); $[C]_{spinel}$ is the composition of spinel; K is the spinel/melt partition coefficients of the elements of interest; D is their diffusion coefficient at 1150°C, with associated source; –, not applicable.

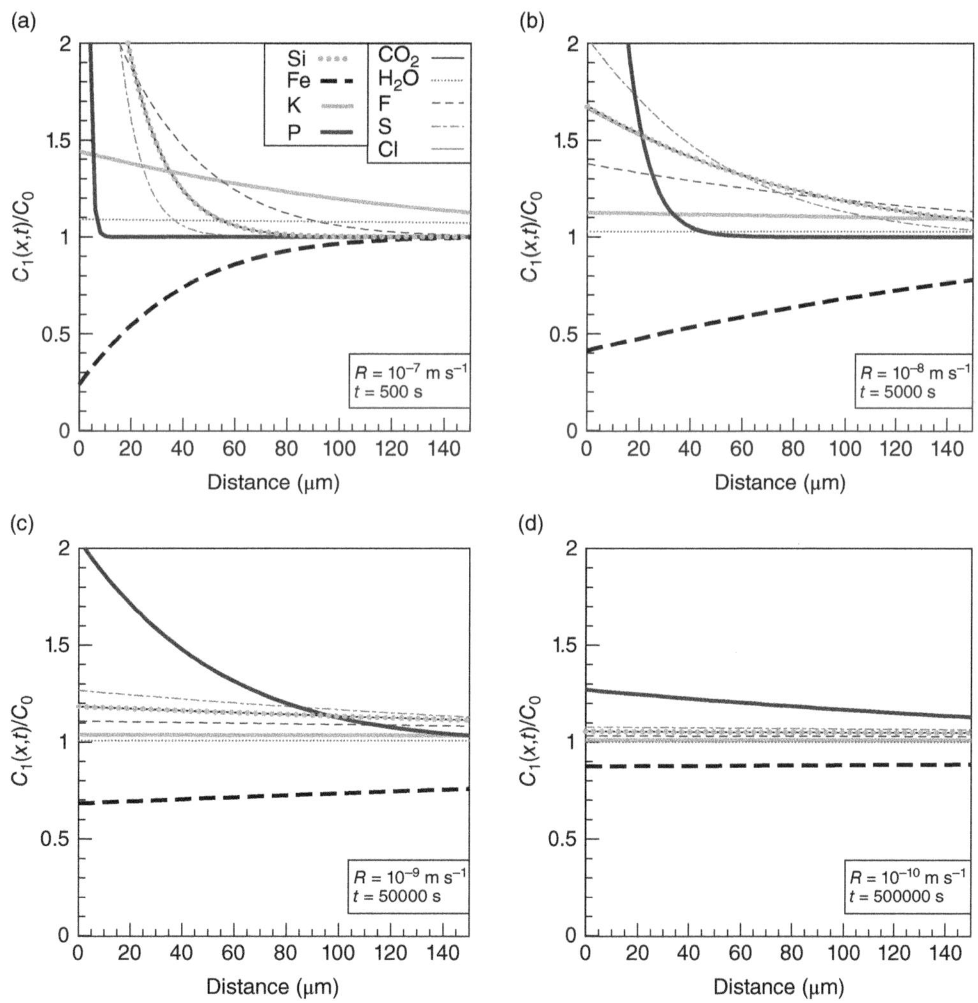

Figure 2.11 Modeling of boundary-layer-melt composition in front of a growing spinel crystal at varying crystal growth rate and a temperature of 1150°C. $C_1(x,t)/C_0$ is the ratio of the melt composition at distance x from the crystal–melt interface and time t to the far-field melt composition, which is here taken as the average bulk-rock composition (see also Table 2.1). For each crystal growth rate (R), the growth step modeled allows for the growth of 50 μm of crystal. Curves for Si, CO_2, and Cl essentially plot over each other in all panels due to identical measured/assumed spinel–melt partition coefficient and similar calculated diffusivity at 1150°C (Table 2.1).

we plotted FeO_t and SiO_2 contents and SiO_2/FeO_t ratios of modeled boundary-layer melts at varying distance from the crystal–melt interface against the El Hierro data in Figures 2.7 and 2.9b. We find that, indeed, the composition of the boundary-layer melt at distances of 0–200 μm from the crystal–melt interface for growth rates of 10^{-7} and 10^{-8} m s^{-1} is displaced to the right of the FeO_t–SiO_2 trend, although these compositions have lower FeO_t than most uncorrected spinel-hosted melt inclusions (Figure 2.7). Predicted boundary-layer-melt SiO_2/FeO_t ratios plotted in Figure 2.9b are generally moderately to strongly elevated with respect to that of the far-field melt, but spinel- and olivine-hosted melt inclusions do not show significantly different SiO_2/FeO_t as a function of size. Note that we did not attempt to model the diffusive relaxation step that is likely to occur prior to boundary-layer-melt entrapment (Baker, 2008). This step may or may not bring the boundary-layer-melts closer to uncorrected spinel-hosted melt inclusions in Figures 2.7 and 2.9b.

Aside from SiO_2/FeO_t, our calculations are generally consistent with the observations of higher slow diffuser/fast diffuser ratios, particularly P_2O_5/K_2O, CO_2/K and S/H_2O, in smaller spinel-hosted melt inclusions shown in Figure 2.9. We thus conclude that boundary-layer-melt entrapment and diffusive fractionation during rapid crystal growth best explains the anomalous composition of the El Hierro spinel-hosted inclusions. In detail, the match of our computations to the data is far from perfect but that is to be expected when comparing complex natural processes and simplified models. Limitations in our model may include the simplified single step approach, neglecting the aforementioned relaxation step during melt-inclusion sealing at low crystal growth rate (Baker, 2008), and uncertainties in the diffusion coefficients for Si, Fe, K and CO_2 in alkaline mafic melts (Zhang and Ni, 2010; Zhang et al., 2010). In addition, as discussed in section 2.3.3.1, postentrapment crystallization likely contributed to data variability in spinel-hosted melt inclusions, and we now discuss the possibility of postentrapment volatile loss further affecting CO_2 and halogen systematics.

2.3.5. Origin of High CO_2 Concentrations

Of all measured elements, CO_2 displays, on average, the greatest enrichment factor (2.8) in spinel-hosted melt inclusions (mean 3930 ± 2880 ppm) with respect to olivine-hosted melt inclusions (mean 1390 ± 1270 ppm). In comparison, P_2O_5 shows a mean enrichment factor of 1.3. The extremely high CO_2 concentrations in spinel-hosted melt inclusions thus cannot be explained by diffusive fractionation alone. A possible solution to this conundrum may lie in the preservation potential of original CO_2 concentrations as a function of melt-inclusion host mineral. Indeed, Tait (1992) proposed that the propensity of melt inclusions

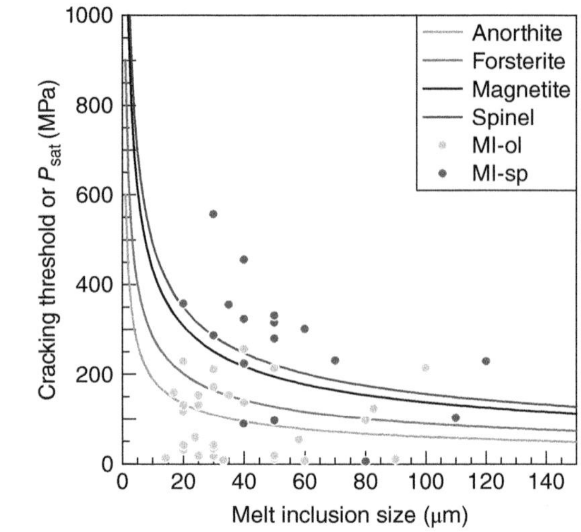

Figure 2.12 Cracking threshold estimates as a function of size and host mineral for melt inclusions. Cracking threshold is computed using equation (23) of Tait (1992), with tensile-elastic-modulus and fracture-surface-energy values for minerals taken from Tromans and Meech (2002; Alexander Lloyd, personal communication, 2019). In Tait's (1992) equation, melt-inclusion size is the radius of a spherical inclusion. Volatile saturation pressures (P_{sat}) for olivine-hosted and spinel-hosted melt inclusions calculated with the model of Iacono-Marziano et al. (2012) are shown for comparison. For our inclusions, size refers to their shortest measured dimension.

to decrepitate and thus leak volatiles is a function of host mineral strength and melt-inclusion size. In addition, Maclennan (2017) argued that decrepitation regulates global systematics (excluding arcs) of CO_2 contents in olivine-hosted melt inclusions. At El Hierro, we previously suggested that the relatively low volatile-saturation pressures (<260 MPa) recorded by olivine-hosted melt inclusions, which are at odds with other barometric methods and seismic evidence indicating deeper magma plumbing, are best explained by decrepitation (Longpré et al., 2017). What if spinel-hosted melt inclusions were less susceptible to bubble formation and decrepitation—and therefore postentrapment CO_2 loss—relative to olivine-hosted melt inclusions? In Figure 2.12, we used the model of Tait (1992) and the data of Tromans and Meech (2002) to derive coarse estimates of cracking threshold as a function of host mineral and melt-inclusion size. This calculation indicates that, indeed, magnetite- and spinel-hosted melt inclusions should be able to withstand greater overpressures than olivine-hosted inclusions, providing a plausible mechanism to preserve high CO_2 concentrations and volatile saturation pressures.

We note that, despite high CO_2 concentrations, spinel-hosted melt inclusions show modest CO_2/Nb ratios, with a maximum of 62 measured for a melt inclusion containing 7500 ppm CO_2. In comparison, the highest CO_2/Nb ratio we measured in olivine-hosted inclusions is 42 (Longpré

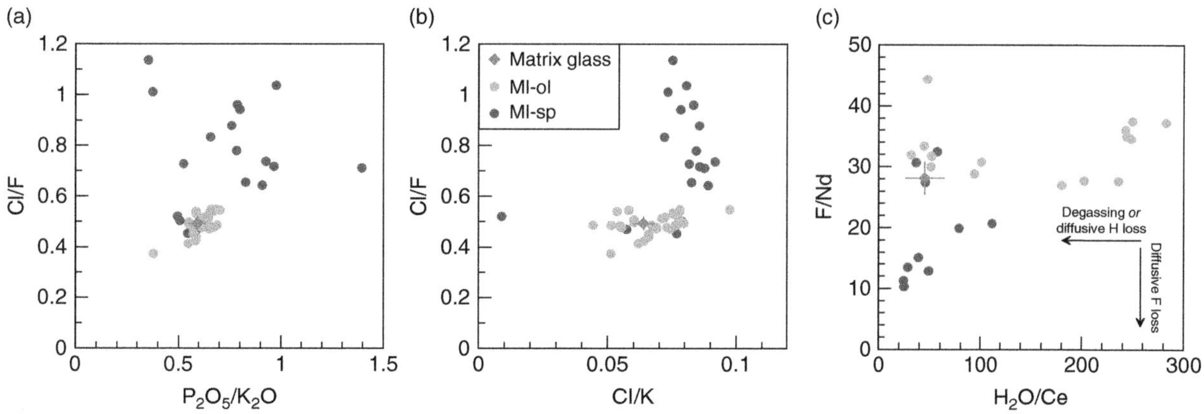

Figure 2.13 Behavior of halogens. (a) Most spinel-hosted melt inclusions have a much higher Cl/F ratio than olivine-hosted melt inclusions. A slight positive correlation ($R^2 = 0.51$) between Cl/F and P_2O_5/K_2O for olivine-hosted melt inclusions suggests diffusive fractionation of Cl/F. (b) While olivine-hosted melt inclusions have near constant Cl/F with varying Cl/K, most spinel-hosted melt inclusions have varying Cl/F at near constant Cl/K. (c) No correlation between F/Nd and H_2O/Ce emerges for olivine-hosted melt inclusions, but a subset of spinel-hosted melt inclusions with anomalously low F/Nd (for El Hierro) show a positive correlation of these two ratios, which could be explained by (postentrapment?) diffusive loss of both F and H_2O.

et al., 2017). These CO_2/Nb values are far below any reasonable estimate for undegassed mantle melts that should have CO_2/Nb greater than at least ~240 (Saal et al., 2002; Figure 2.5d). Even the melt inclusion containing 12,000 ppm CO_2, assuming it has a Nb concentration of ~120 ppm similar to our highest measured values (this inclusion is too small to place a 30 μm diameter laser beam in it and thus it was not analyzed for trace elements), would have a rather low CO_2/Nb ratio of ~100. Therefore, even the highest measured CO_2 concentrations most likely represent significantly degassed values—we emphasize that undegassed CO_2 concentrations in the 2011–2012 basanite, and other El Hierro magmas, likely reached weight percent levels prior to deep-seated degassing (Longpré et al., 2017; Taracsák et al., 2019).

2.3.6. Behavior of Halogens

The markedly different behavior of halogens in spinel- and olivine-hosted melt inclusions merits further discussion. We first highlight that, based on the high Cl and F concentrations of the matrix glass (Figure 2.5c and f), any significant role of degassing in producing the observed halogen systematics can be ruled out. While somewhat elevated Cl/F ratios would be expected in boundary-layer melts, most spinel-hosted melt inclusions show Cl/F values that appear too high to be explained by synentrapment diffusive fractionation alone—our end-member model calculations do not produce Cl/F ratios that approach observed values (Figure 2.9e). Yet, Cl/F is positively correlated, albeit weakly, with P_2O_5/K_2O in olivine-hosted melt inclusions (Figure 2.13a), hinting that the process producing high P_2O_5/K_2O ratios in spinel-hosted melt inclusions might also be responsible for their high Cl/F. Could our model dramatically underestimate Cl/F diffusive fractionation in the boundary-layer melt? An alternative solution is to call upon postentrapment diffusive F loss, which has been implicated in a few melt-inclusion data sets (Koleszar et al., 2009; Le Voyer et al., 2014). This would be consistent with the anomalously low (for El Hierro) F/Nd ratios that are positively correlated with H_2O/Ce for several spinel-hosted melt inclusions (Figure 2.13c), which could reflect concomitant postentrapment diffusive F and H^+ loss. However, in the absence of F degassing here, it is unclear what would drive F loss out of the melt inclusions. In brief, the data suggest diffusive fractionation of halogens either synentrapment or postentrapment, or both, but remaining discrepancies prevent a more definitive conclusion.

2.3.7. Implications for the 2011–2012 El Hierro Eruption

This chapter presents a reinterpretation of the origin of the spinel-hosted melt inclusions, and a discussion of associated implications for the 2011–2012 magmatic system is warranted. Indeed, Longpré et al. (2014) initially postulated that spinel-hosted melt inclusions recorded the presence of evolved melts (up to basaltic trachyandesite) in the 2011–2012 plumbing system. In this model, these melts hybridized with the dominant body of magma, represented by the olivine-hosted melt inclusions, to yield the intermediate composition matrix glass. This scenario assumed that spinel-hosted melt inclusions suffered postentrapment iron gain, which, as outlined in section 2.3.3.2, we now argue is unlikely based on their elevated P_2O_5 and incompatible trace-element concentrations. Instead, the reinterpretation of spinel-hosted melt inclusions as boundary-layer melts affected by synentrapment diffusive fractionation removes the need to invoke the presence of evolved melts in the 2011–2012 magmatic system (this is not to say that evolved melts may not be present under El Hierro). In this context, olivine-hosted melt inclusions and the matrix glass likely account for the full range of 2011–2012 basanite melt compositions (cf. Longpré et al., 2017; Meletlidis et al., 2015).

Despite this reevaluation, the main conclusions of Longpré et al. (2014) remain valid. In our revised model, mixing of closely related basanite melts, rather than mixing of significantly different mafic and evolved melts, occurred in mantle magma reservoirs to produce the bulk magma. This is recorded by variably enriched olivine-hosted melt inclusions and the presence of both normally and reversely zoned olivine and clinopyroxene crystals (Longpré et al., 2014, 2017). Here, the matrix glass composition was produced by late-stage crystallization of olivine, clinopyroxene, and spinel in the bulk magma, rather than by hybridization of mafic and evolved melts. The broad implications of the diffusion chronometry results, recording deep mixing and mobilization of the 2011–2012 magma shortly prior to and during eruption, are essentially unchanged (Longpré et al., 2014).

2.3.8. General Implications

To our knowledge, our results provide the first documented case of boundary-layer-melt entrapment and associated diffusive fractionation of volatiles in natural melt inclusions. This accidental finding suggests that spinel crystals, particularly if displaying rapid growth textures, are not good targets for melt-inclusion analysis. In line with Baker (2008), we propose that anomalous P_2O_5 concentrations—or, perhaps better, P_2O_5/K_2O ratios—may flag potentially spurious melt-inclusion compositions. A systematic comparison of P_2O_5/K_2O ratios in melt inclusions and their host matrix glasses and bulk rocks in other data sets may help further test the relevance of entrapment of boundary-layer melt, or lack thereof, in natural magmas.

2.4. CONCLUSIONS

The main results and conclusions of this study are as follows:

1. Spinel macrocrysts and microlites in the 2011–2012 El Hierro basanite show rapid growth textures.

2. Spinel-hosted melt inclusions in these samples are characterized by anomalous FeO_t–SiO_2 systematics and high P_2O_5 and CO_2 contents (as well as high S/H_2O and Cl/F ratios) compared to olivine-hosted-melt inclusions, the matrix glass, and bulk rock.

3. Slow diffuser/fast diffuser ratios (e.g., P_2O_5/K_2O) are elevated in the smallest spinel-hosted melt inclusions with respect to other samples.

4. Entrapment of boundary-layer melts affected by synentrapment diffusive enrichment of slow-diffusing elements best explains the characteristics of spinel-hosted melt inclusions. Postentrapment crystallization and a differential susceptibility of spinel- and olivine-hosted melt inclusions to postentrapment volatile loss via decrepitation (CO_2) and diffusive reequilibration (F? H_2O?) also may have contributed to data variability.

5. These results indicate that spinel-hosted melt inclusions do not represent evolved melts present in the 2011–2012 magmatic system as we previously proposed.

6. This work documents one of the first cases of entrapment of boundary-layer melt in natural samples, cast doubts on the reliability of fast-grown, Fe–Ti-rich spinel as melt-inclusion hosts, and provides a framework to test for potential boundary layer effects in other melt-inclusion data sets.

ACKNOWLEDGMENTS

Instituto Geográfico Nacional provided lava balloon samples and support during fieldwork. Sampling permits were delivered by the Cabildo de El Hierro—Medio Ambiente. We are grateful to Brian Monteleone, Lang Shi, and Louise Bolge, who provided help during SIMS, electron microprobe and LA-ICP-MS analysis, respectively. We thank Patrick Beaudry, Don Baker, Andreas Klügel, and Alexander Lloyd for fruitful discussions, Zoltán Zajacz and an anonymous referee for constructive reviews, and Francesco Vetere and Adrian Fiege for editorial handling. Alexander Lloyd kindly shared his ideas and calculations on the cracking threshold of melt inclusions. Financial support was provided by Queens College, partially through the Paula and Jeffrey Gural Endowed Professorship in Geology, to M-AL and by NSERC Discovery and Accelerator grants to JS.

SUPPLEMENTARY MATERIAL

Tables SM1–SM5 are available as supplementary material at http://sees.qc.cuny.edu/files/2020/03/Longpré_etal_Supplementary_Tables_1-5.xlsx and Figures SM1–SM3 are available as supplementary material at http://sees.qc.cuny.edu/files/2020/03/Longpré_etal_Supplementary_Figures_1-3.pdf

REFERENCES

Alletti, M., Baker, D. R., & Freda, C. (2007). Halogen diffusion in a basaltic melt. *Geochimica et Cosmochimica Acta*, *71*(14), 3570–3580.

Bacon, C. R. (1989). Crystallization of accessory phases in magmas by local saturation adjacent to phenocrysts. *Geochimica et Cosmochimica Acta*, *53*(5), 1055–1066.

Baker, D. (2008). The fidelity of melt inclusions as records of melt composition. *Contributions to Mineralogy and Petrology*, *156*(3), 377–395.

Barnes, S. J., & Roeder, P. L. (2001). The range of spinel compositions in terrestrial mafic and ultramafic rocks. *Journal of Petrology*, *42*(12), 2279–2302.

Beaudry, P., Longpré, M.-A., Economos, R., Wing, B. A., Bui, T. H., & Stix, J. (2018). Degassing-induced fractionation of multiple sulphur isotopes unveils post-Archaean recycled oceanic crust signal in hotspot lava. *Nature Communications*, *9*(1), 5093.

Bucholz, C. E., Gaetani, G. A., Behn, M. D., & Shimizu, N. (2013). Post-entrapment modification of volatiles and oxygen fugacity in olivine-hosted melt inclusions. *Earth and Planetary Science Letters*, *374*, 145–155.

Carracedo, J. C., Rodríguez Badiola, E., Guillou, H., de la Nuez, J., & Pérez Torrado, F. J. (2001). Geology and volcanology of La Palma and El Hierro, Western Canaries. *Estudios Geológicos*, *57*, 175–273.

Danyushevsky, L. V., Della-Pasqua, F. N., & Sokolov, S. (2000). Re-equilibration of melt inclusions trapped by magnesian olivine phenocrysts from subduction-related magmas: petrological implications. *Contributions to Mineralogy and Petrology*, *138*(1), 68–83.

Danyushevsky, L. V., McNeill, A. W., & Sobolev, A. V. (2002). Experimental and petrological studies of melt inclusions in phenocrysts from mantle-derived magmas: an overview of techniques, advantages and complications. *Chemical Geology*, *183*(1–4), 5–24.

Danyushevsky, L. V., & Plechov, P. (2011). Petrolog3: Integrated software for modeling crystallization processes. *Geochemistry, Geophysics, Geosystems*, *12*(7), Q07021.

Faure, F., & Schiano, P. (2005). Experimental investigation of equilibration conditions during forsterite growth and melt inclusion formation. *Earth and Planetary Science Letters*, *236*(3), 882–898.

Freda, C., Baker, D. R. & Scarlato, P. (2005). Sulfur diffusion in basaltic melts. *Geochimica et Cosmochimica Acta*, *69*(21), 5061–5069.

Frezzotti, M.-L. (2001). Silicate-melt inclusions in magmatic rocks: applications to petrology. *Lithos*, *55*(1–4), 273–299.

Frost, B. R., & Lindsley, D. H. (1991). Occurrence of iron-titanium oxides in igneous rocks. *Reviews in Mineralogy and Geochemistry*, *25*(1), 433–468.

Gaetani, G. A., O'Leary, J. A., Shimizu, N., Bucholz, C. E., & Newville, M. (2012). Rapid reequilibration of H_2O and oxygen fugacity in olivine-hosted melt inclusions. *Geology*, *40*(10), 915–918.

Haggerty, S. E. (1991). Oxide textures; a mini-atlas. *Reviews in Mineralogy and Geochemistry*, *25*(1), 129–219.

Harrison, T. M., & Watson, E. B. (1984). The behavior of apatite during crustal anatexis: Equilibrium and kinetic considerations. *Geochimica et Cosmochimica Acta*, *48*(7), 1467–1477.

Hartley, M. E., Maclennan, J., Edmonds, M., & Thordarson, T. (2014). Reconstructing the deep CO_2 degassing behaviour of large basaltic fissure eruptions. *Earth and Planetary Science Letters*, *393*, 120–131.

Hartley, M. E., Neave, D. A., Maclennan, J., Edmonds, M., & Thordarson, T. (2015). Diffusive over-hydration of olivine-hosted melt inclusions. *Earth and Planetary Science Letters*, *425*, 168–178.

Hausen, H. (1972). Outlines of the geology of El Hierro. *Commentationes Physico-Mathematicae*, *43*, 65–167.

Helo, C., Longpré, M.-A., Shimizu, N., Clague, D. A., & Stix, J. (2011). Explosive eruptions at mid-ocean ridges driven by CO_2-rich magmas, *Nature Geoscience*, *4*(4), 260–263.

Hirschmann, M. M. (2018). Comparative deep Earth volatile cycles: The case for C recycling from exosphere/mantle fractionation of major (H_2O, C, N) volatiles and from H_2O/Ce, CO_2/Ba, and CO_2/Nb exosphere ratios. *Earth and Planetary Science Letters*, *502*, 262–273.

Holycross, M. E., & Watson, E. B. (2016). Diffusive fractionation of trace elements in basaltic melt. *Contributions to Mineralogy and Petrology*, *171*(10), 80.

Iacono-Marziano, G., Morizet, Y., Le Trong, E., & Gaillard, F. (2012). New experimental data and semi-empirical parameterization of H_2O–CO_2 solubility in mafic melts. *Geochimica et Cosmochimica Acta*, *97*, 1–23.

Kamenetsky, V. (1996). Methodology for the study of melt inclusions in Cr-spinel, and implications for parental melts of MORB from FAMOUS area. *Earth and Planetary Science Letters*, *142*(3–4), 479–486.

Kent, A. J. R. (2008). Melt inclusions in basaltic and related volcanic rocks. *Reviews in Mineralogy and Geochemistry*, *69*(1), 273–331.

Koleszar, A. M., Saal, A. E., Hauri, E. H., Nagle, A. N., Liang, Y., & Kurz, M. D. (2009). The volatile contents of the Galapagos plume; evidence for H_2O and F open system behavior in melt inclusions. *Earth and Planetary Science Letters*, *287*(3–4), 442–452.

Kress, V. C., & Ghiorso, M. S. (2004). Thermodynamic modeling of post-entrapment crystallization in igneous phases. *Journal of Volcanology and Geothermal Research*, *137*(4), 247–260.

Le Voyer, M., Asimow, P. D., Mosenfelder, J. L., Guan, Y., Wallace, P. J., Schiano, P., Stolper, E. M., & Eiler, J. M. (2014). Zonation of H_2O and F concentrations around melt inclusions in olivines. *Journal of Petrology*, *55*(4), 685–707.

Lesher, C., Hervig, R., & Tinker, D. (1996). Self diffusion of network formers (silicon and oxygen) in naturally occurring basaltic liquid. *Geochimica et Cosmochimica Acta*, *60*(3), 405–413.

Lloyd, A. S., Plank, T., Ruprecht, P., Hauri, E. H., & Rose, W. (2013). Volatile loss from melt inclusions in pyroclasts of differing sizes. *Contributions to Mineralogy and Petrology*, *165*(1), 129–153.

Longpré, M.-A., Klügel, A., Diehl, A., & Stix, J. (2014). Mixing in mantle magma reservoirs prior to and during the 2011–2012 eruption at El Hierro, Canary Islands. *Geology*, *42*(4), 315–318.

Longpré, M.-A., Stix, J., Klügel, A., & Shimizu, N. (2017). Mantle to surface degassing of carbon- and sulphur-rich alkaline magma at El Hierro, Canary Islands. *Earth and Planetary Science Letters*, *460*, 268–280.

Lowenstern, J. B. (1995). Applications of silicate-melt inclusions to the study of magmatic volatiles. In J. F. H. Thompson, (Ed.), *Magmas, fluids and ore deposits* (pp. 71–99). Quebec City: Mineralogical Association of Canada.

Lowenstern, J. B. (2003). Melt inclusions come of age: volatiles, volcanoes, and Sorby's legacy. *Developments in Volcanology*, *5*, 1–21.

Lu, F., A. T. Anderson, & A. M. Davis (1995). Diffusional gradients at the crystal/melt interface and their effect on the compositions of melt Inclusions, *The Journal of Geology*, *103*(5), 591–597.

Maclennan, J. (2017). Bubble formation and decrepitation control the CO_2 content of olivine-hosted melt inclusions. *Geochemistry, Geophysics, Geosystems*, *18*(2), 597–616.

Martí, J., Castro, A., Rodríguez, C., Costa, F., Carrasquilla, S., Pedreira, R., & Bolos, X. (2013). Correlation of magma evolution and geophysical monitoring during the 2011–2012 El Hierro (Canary Islands) submarine eruption. *Journal of Petrology*, *54*(7), 1349–1373.

McDonough, W. F., & Sun, S.-S. (1995). The composition of the Earth. *Chemical Geology*, *120*(3–4), 223–253.

Meletlidis, S., Di Roberto, A., Cerdeña, I. D., Pompilio, M., Bertagnini, A., Benito-Saz, M. A., Del Carlo, P., & Aparicio, S. S.-M. (2015). New insight into the 2011–2012 unrest and eruption of El Hierro Island (Canary Islands) based on integrated geophysical, geodetical and petrological data. *Annals of Geophysics*, *58*(5), S0546.

Métrich, N., & Wallace, P. (2008). Volatile abundances in basaltic magmas and their degassing paths tracked by melt inclusions. *Reviews in Mineralogy and Geochemistry*, *69*(1), 363–402.

Moore, L., Gazel, E., Tuohy, R., Lloyd, A., Esposito, R., Steele-MacInnis, M., et al. (2015). Bubbles matter: An assessment of the contribution of vapor bubbles to melt inclusion volatile budgets. *American Mineralogist*, *100*(4), 806–823.

Palme, H., & O'Neill, H. S. C. (2003). Cosmochemical estimates of mantle composition. *Treatise on Geochemistry*, *2*, 1–38.

Pellicer, J. M. (1979). Estudio geoquímico del vulcanismo de la isla de Hierro Archipiélago Canario. *Estudios Geológicos*, *35*, 15–29.

Portnyagin, M., Almeev, R., Matveev, S., & Holtz, F. (2008). Experimental evidence for rapid water exchange between melt inclusions in olivine and host magma, *Earth and Planetary Science Letters*, *272*(3–4), 541–552.

Qin, Z., Lu, F., & Anderson, A. T. (1992). Diffusive reequilibration of melt and fluid inclusions. *American Mineralogist*, *77*, 565–576.

Roedder, E. (1984). *Fluid inclusions*. Washington, DC: Mineralogical Society of America.

Rosenthal, A., Hauri, E. H., & Hirschmann, M. M. (2015). Experimental determination of C, F, and H partitioning between mantle minerals and carbonated basalt, CO_2/Ba and CO_2/Nb systematics of partial melting, and the CO_2 contents of basaltic source regions. *Earth and Planetary Science Letters*, *412*(0), 77–87.

Saal, A. E., Hauri, E. H., Langmuir, C. H., & Perfit, M. R. (2002). Vapour undersaturation in primitive mid-ocean-ridge basalt and the volatile content of Earth's upper mantle. *Nature*, *419*(6906), 451–455.

Schiano, P. (2003). Primitive mantle magmas recorded as silicate melt inclusions in igneous minerals. *Earth-Science Reviews*, *63*(1–2), 121–144.

Smith, V. G., Tiller, W. A., & Rutter, J. (1955). A mathematical analysis of solute redistribution during solidification. *Canadian Journal of Physics*, *33*(12), 723–745.

Sobolev, A. V. (1996). Melt inclusions in minerals as a source of principle petrological information. *Petrology*, *4*(3), 209–220.

Steele-Macinnis, M., Esposito, R., & Bodnar, R. J. (2011). Thermodynamic model for the effect of post-entrapment crystallization on the H_2O–CO_2 systematics of vapor-saturated, silicate melt inclusions. *Journal of Petrology*, *52*(12), 2461–2482.

Stroncik, N. A., Klügel, A., & Hansteen, T. H. (2009). The magmatic plumbing system beneath El Hierro (Canary Islands): constraints from phenocrysts and naturally quenched basaltic glasses in submarine rocks. *Contributions to Mineralogy and Petrology*, *157*(5), 593–607.

Tait, S. (1992). Selective preservation of melt inclusions in igneous phenocrysts. *American Mineralogist*, *77*(1–2), 146–155.

Taracsák, Z., Hartley, M. E., Burgess, R., Edmonds, M., Iddon, F., & Longpré, M.-A. (2019). High fluxes of deep volatiles from ocean island volcanoes: Insights from El Hierro, Canary Islands. *Geochimica et Cosmochimica Acta*, *258*, 19–36.

Tromans, D., & Meech, J. A. (2002). Fracture toughness and surface energies of minerals: theoretical estimates for oxides, sulphides, silicates and halides. *Minerals Engineering*, *15*(12), 1027–1041.

Wallace, P. J., Kamenetsky, V. S., & Cervantes, P. (2015). Melt inclusion CO_2 contents, pressures of olivine crystallization, and the problem of shrinkage bubbles. *American Mineralogist*, *100*(4), 787–794.

Zajacz, Z., & Halter, W. (2007). LA-ICPMS analyses of silicate melt inclusions in co-precipitated minerals: Quantification, data analysis and mineral/melt partitioning. *Geochimica et Cosmochimica Acta*, *71*(4), 1021–1040.

Zhang, Y., & Ni, H. (2010). Diffusion of H, C, and O components in silicate melts. *Reviews in Mineralogy and Geochemistry*, *72*(1), 171–225.

Zhang, Y., Ni, H., & Chen, Y. (2010). Diffusion data in silicate melts. *Reviews in Mineralogy and Geochemistry*, *72*(1), 311–408.

3

Apatite as a Monitor of Dynamic Magmatic Evolution at Torfajökull Volcanic Center, Iceland

Lissie Connors[1], Tamara L. Carley[1], and Adrian Fiege[2]

ABSTRACT

Apatite, a powerful geochemical tool for understanding the evolution of volatile and trace elements in magmatic systems, is a common accessory phase in Icelandic rocks. Torfajökull, a historically active Icelandic volcano erupting apatite-bearing magmas, exhibits a unique compositional history; from more-evolved peralkaline rhyolites (Pleistocene) to less-evolved metaluminous rhyolites (Holocene). Geochemical analyses of apatites and volcanic glasses from five eruptions spanning Torfajökull's rhyolitic history (384 ka to 1477 CE) contribute new insights into the volatile (H-F-Cl-S) and rare earth element (REE) evolution of the system. Some of the apatites contain exceptionally high REE compositions (≤20 wt% REE_2O_3), with concentrations decreasing over time. This temporal trend corresponds to changing magma compositions and saturation of different REE-bearing phases, as mafic–felsic magma mingling becomes more prominent. Volcanic glass and apatite compositions indicate increasing volatile abundances in magmas through time. Sulfur concentrations are low in apatite (20–100 μg g^{-1}), as is expected in reduced systems, with the exception of elevated sulfur in apatite from the 67 ka eruption (20–570 μg g^{-1}). Magma oxidation induced by volcanic degassing (further exemplified by Cl–OH contents in apatite), perhaps explains this signature. The success of this apatite-focused investigation of Torfajökull reveals the potential of apatite to elucidate magmatic and volcanic processes in Iceland.

3.1. INTRODUCTION

Apatite—commonly ($Ca_5(PO_4)_3(F,Cl,OH)$)—is distinctive in its ability to incorporate all major magmatic volatiles in its crystal structure as major (F, Cl, OH) or trace elements (S, C, Br) (Hughes & Rankovan, 2002; Marks et al., 2012; Piccoli & Candela, 2002). Here, the structural incorporation of sulfur (S) is influenced by the oxygen fugacity of the system (e.g., S^{2-} or S^{6+}; Konecke et al, 2017b). Experimental studies of apatite have constrained volatile partitioning in apatite–melt and apatite–melt–fluid systems for a continually expanding series of magmatic conditions and compositions (e.g. Doherty et al., 2014; Konecke et al, 2017b; Mathez & Webster, 2005; Webster et al., 2009; Webster et al., 2017). Apatite can also incorporate significant amounts of rare earth elements (REE), reaching total contents of > 1 wt% oxides (Hughes & Rakovan, 2002) and rarely even > 10 wt% oxides (Rønsbo, 1989). These features make apatite a powerful geochemical tool for investigating the volatile, redox, and REE evolution of magmas (e.g. Belousova et al., 2002; Fleet & Pan, 2002; Konecke et al, 2017b; Piccoli & Candela, 2002; Rønsbo, 1989; Webster & Piccoli, 2015).

The geochemistry of apatite has been used to study a range of processes in magmatic systems from a variety of tectonic regions, and from terrestrial to extraterrestrial environments (e.g. Boyce et al., 2010; McCubbin et al., 2011; Parat et al., 2002; Piccoli & Candela, 2002; Streck & Dilles, 1998). In conjunction with matrix glass compositions, apatite geochemistry can be used to quantify pre-eruptive volatile contents, and estimate gas emissions of volcanic systems,

[1] *Department of Geology and Environmental Geosciences, Lafayette College, Easton, PA, USA*

[2] *Department of Earth and Planetary Sciences, American Museum of Natural History, New York, NY, USA*

Dynamic Magma Evolution, Geophysical Monograph 254, First Edition. Edited by Francesco Vetere.

providing new perspectives on volcanic degassing (e.g., Boyce & Hervig, 2008a; Scott et al., 2015; Stock et al., 2016).

Despite the valuable contributions that apatite studies have made globally, and despite apatite's ubiquity as an accessory phase in Icelandic magmas (e.g., Carmichael, 1964; Moorbath et al., 1968; Wood, 1978), Icelandic apatite has rarely been investigated (c.f. Bergþórsdóttir, 2018). This study focuses on Torfajökull, an active central volcano in southern Iceland. Torfajökull has ~400 kyrkyr eruptive history (the longest documented activity of any historically active Icelandic volcano; McGarvie et al., 2006), and prior studies indicated pronounced changes in geochemistry over time (from peralkaline to metaluminous; Gunnarsson et al., 1998; Jónasson, 2007; McGarvie, 1984; McGarvie et al., 2006; Zellmer et al., 2008). Apatites from five eruptions spanning Torfajökull's history, from 384 ka to its most recent eruption in 1477 CE, are used to examine the volatile and trace elemental evolution of this rhyolitic system and to explore the possibility of redox effects as a function of degassing.

3.2. GEOLOGICAL BACKGROUND

3.2.1. Silicic Volcanism in Iceland

Iceland has been constructed by productive magmatism and volcanism at the intersection of the Mid-Atlantic Ridge and the Icelandic hotspot (Pálmason & Sæmundsson, 1974; Schilling, 1973; Vink, 1984; Vogt, 1971). Icelandic crust is significantly thicker than typical oceanic crust (25–40 km vs. 3–7 km; Bjarnason, 2008), and contains an unusual abundance of silicic material for an oceanic island (≥ 10% of exposed crust contain >6 wt% SiO_2; Jonasson, 2007; Walker, 1963). The processes resulting in the formation of this overly thickened silicic crust has long been studied and debated (Bindeman et al., 2012; Carley et al., 2014; Carmichael, 1964; Furman et al., 1992; Jónasson, 1994; Macdonald et al., 1987, 1990; Martin & Sigmarsson, 2007, 2010; Nicholson et al., 1991; Prestvik et al., 2001; Sigmarsson et al., 1991; Sigurdsson & Sparks, 1981; Wood, 1978).

Icelandic rhyolites are concentrated at central volcanoes, i.e., localized, polygenetic features that erupt magmas of diverse compositions (e.g., Thordarson & Larsen, 2007). Fissures often accompany central volcanoes, i.e., laterally extensive mafic eruption features following the strike of the rift. Central volcanoes typically consist of ~20–30% rhyolite, but there are rare examples of central volcanoes with substantially greater abundances of rhyolite (e.g., >80% at Torfajökull; Gunnarsson et al., 1998; Walker, 1963).

The local tectonic setting of a central volcano (e.g., on-rift, off-rift) may have significant influence on its particular rhyolite generation process. Martin and Sigmarsson (2007) hypothesized that the anatexis or extensive assimilation dominates silicic petrogenesis at hot, on-rift volcanoes, while fractional crystallization from a basaltic parent magma dominates at cool, off-rift volcanoes. This hypothesis has been supported and refined in subsequent, isotope-based studies (e.g., Carley et al., 2020; Martin & Sigmarsson, 2010; Schattel et al., 2014). The concept of localized tectonics is complicated, including an ongoing history of rift relocations and rift jumps, through Iceland's thickened crust. For example, there is evidence of extinct abandoned rifts dating to 26–15 Ma in the Westfjords (Hardarson & Fitton, 1997; Martin & Sigmarsson, 2010; Sæmundsson, 1979), and the propagating Eastern Volcanic Zone (EVZ) is thought to be establishing the future location of the Mid-Atlantic Ridge in south-central Iceland. This suggests that throughout its life cycle, changing tectonic conditions of a central volcano may impact silicic petrogenesis and other magmatic processes.

3.2.2. Torfajökull Volcanic Complex

Torfajökull, a historically active volcano in south-central Iceland (Figure 3.1), is unique among Icelandic volcanoes. Its tectonic setting is transitional and dynamic, located on the margin of the nonrifting Southern Flank Zone (SFZ)—at the convergence of the South Iceland Seismic Zone (SISZ) and the southward-propagating EVZ. During its >400 kyr eruptive history (McGarvie et al., 2006), Torfajökull (Figure 3.2) has erupted significant volumes of rhyolitic material (~250 km^3; Gunnarsson et al., 1998), making it the greatest concentration of exposed silicic rock in Iceland (Jónasson, 2007) and thus in the modern ocean (Gunnarsson et al., 1998).

Individual rhyolitic eruptions contributing to this large silicic center span across the Pleistocene–Holocene boundary. Older (≥ 67 ka), larger volume (> 1 km^3), rhyolites erupted subglacially to form ridges and tuyas, while the younger (< 67 ka), smaller volume (< 1 km^3), rhyolites erupted subaerially as lava flows (McGarvie et al., 2006). Torfajökull rhyolites exhibit dramatic, well-documented, compositional changes with time (Figure 3.2). The oldest, Pleistocene rhyolites have the highest SiO_2 and alkalinity. The Pleistocene rhyolites have undergone more extensive post-emplacement alteration compared to the younger units. The younger, Holocene rhyolites are progressively less evolved, with lower SiO_2 and alkalinity as well as increasing evidence of interaction with mafic magmas (e.g., Blake, 1984; Gunnarsson et al., 1998; Larsen, 1984; McGarvie, 1984; McGarvie et al., 1990; McGarvie et al., 2006; Mørk, 1984; Zellmer et al., 2008). This compositional "devolution" observed at Torfajökull is thought to be a function of Torfajökull's transitional tectonic setting, with increasing influence of the southward-propagating EVZ at ~70 ka. Specifically, the younger Torfajökull rhyolites have experienced magma mixing with tholeiitic basalt

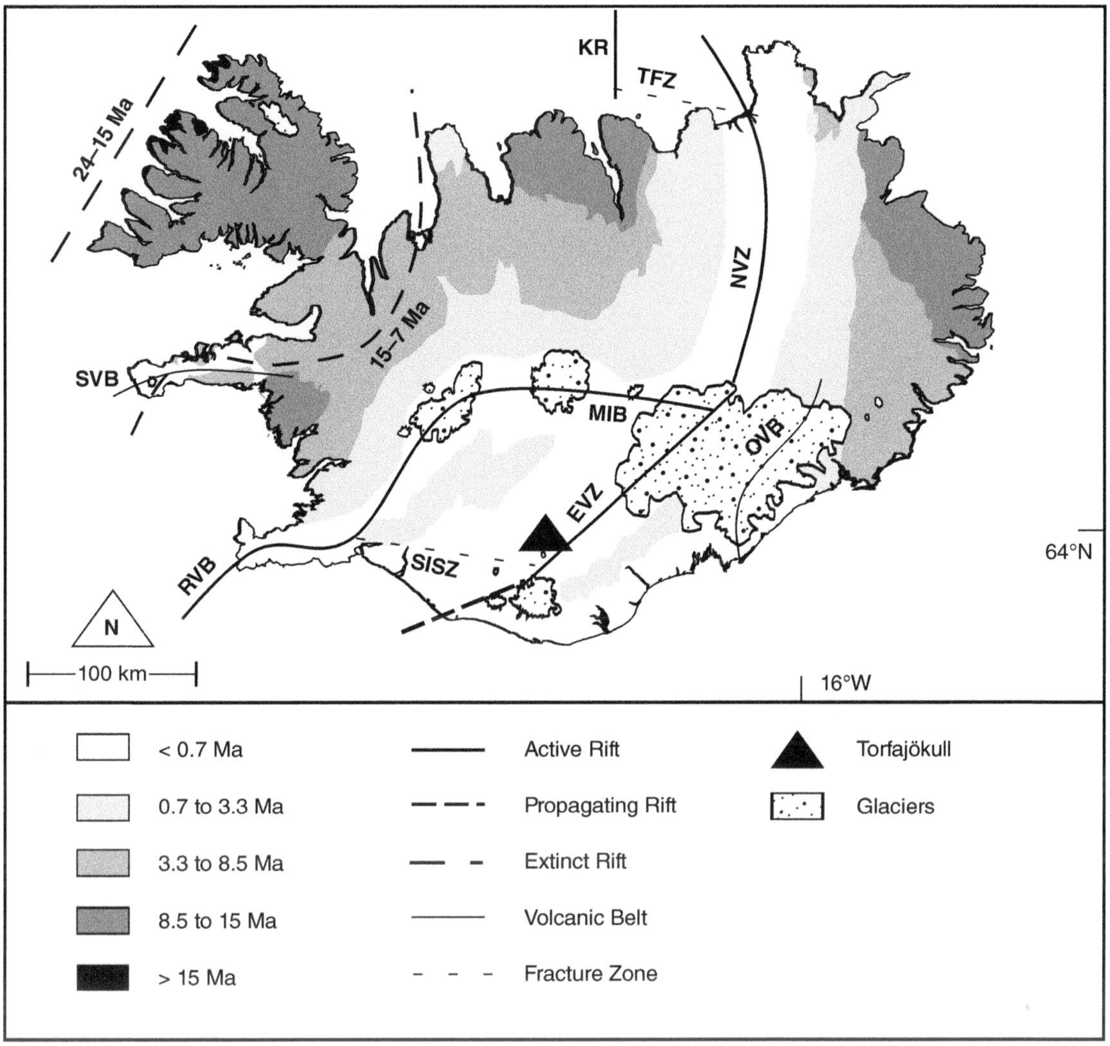

Figure 3.1 Geologic map of Iceland. Torfajökull is located in southern Iceland at the junction of the Eastern Volcanic Zone (EVZ) and the South Iceland Seismic Zone (SISZ). Ages of volcanics are indicated by shades of gray (i.e., white youngest and black oldest). Major tectonic features: WVZ, Western Volcanic Zone; NVZ, Northern Volcanic Zone; RVB, Reykjanes Volcanic Belt; KR, Kolbeinsey Ridge; SVB, Snæfellsness Volcanic Belt; OVB, Öræfi Volcanic Belt; MIB, Mid-Iceland Belt; TFZ, Tjörnes Fracture Zone. [Modified from Thordarson & Hoskuldsson (2002).]

from the rift-related Veiðivötn-Bárdarbunga fissure swarm (e.g., Blake, 1984; Larsen, 1984; McGarvie, 1984; Zellmer et al., 2008).

3.2.3. The Role of Volatiles at Torfajökull

Volatiles within magmatic systems can influence magma evolution at depth, dictate eruption styles, and control the environmental impacts of volcanic eruptions (e.g., De Vivo et al., 2005; Oppenheimer et al., 2003; Wallace et al., 2015). Understanding the abundance and composition of these volatile constituents within a magma body is important, as they are the drivers of the ascent, effervescence, expansion, and fragmentation of magma in explosive eruptions (e.g., Lowenstern, 2003)—all of which are still poorly constrained in subglacial rhyolitic eruptions (e.g., Owen et al., 2012, 2013).

Icelandic rhyolites have traditionally been characterized as H_2O-poor (e.g., Jónasson, 2007; Macdonald et al., 1990; Sigurdsson, 1977), but feldspar-hosted melt inclusions (MIs) from the 67 ka subglacial rhyolites from Torfajökull contain up to 4.8 wt% H_2O (Owen et al., 2013). Chlorine concentrations measured from the same volcanic edifice indicate the degassing of a relatively Cl-rich magma (Owen et al., 2013). These findings suggest that Icelandic rhyolites (at least at Torfajökull) are not as volatile-depleted as previously thought. However, it remains unclear how these volatiles have evolved, degassed, or been replenished throughout Torfajökull's ~400 kyr rhyolitic history. There is even greater uncertainty

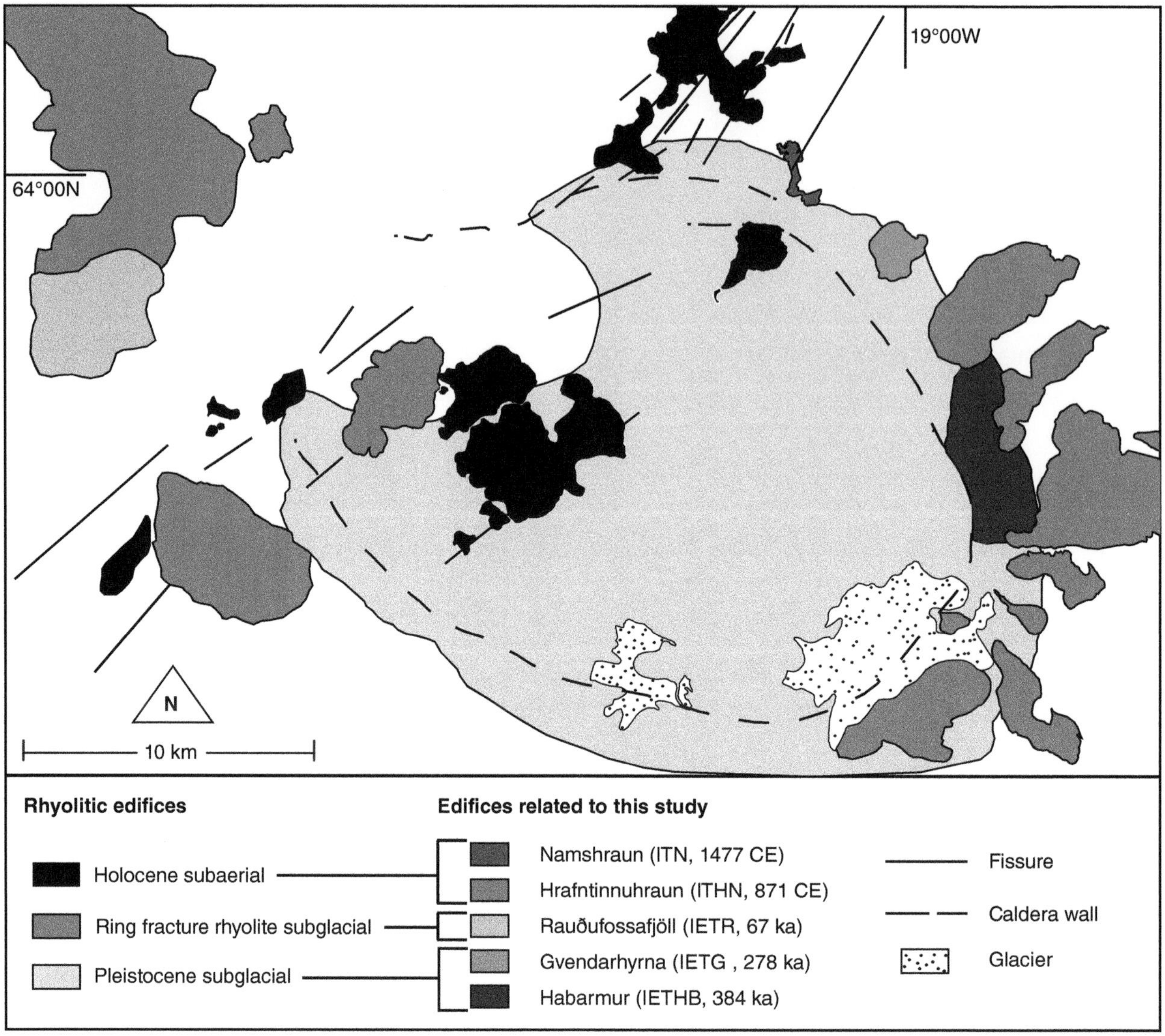

Figure 3.2 Geologic map of Torfajökull volcano, showing the distribution of rhyolites: light gray, older rhyolitic subglacial eruptives; gray, ring fracture rhyolites; black, recent postglacial lava flows. Fissures indicated on the western region of the map illustrate the intersection of Torfajökull with the Veidivötn Fissure Swarm. [Modified from McGarvie et al. (2006), Sæmundsson & Friðleifsson (2001), and Owen et al. (2013).]

regarding the S budget and evolution of the system, as the S contents of these MIs were all below the analytical detection limit (Owen et al., 2013).

3.2.4. Apatite as a Tracer of Volatile Evolution

Apatite is a common accessory phosphate mineral that has been observed in a variety of igneous and metamorphic rocks globally (Piccoli & Candela, 2002; Webster & Piccoli, 2015; and references therein). Minerals within the apatite group have a general formula of $A_{10}(XO_4)_6Z_2$ where the *A*-site is primarily occupied by Ca^{2+}, but can also accommodate Sr^{2+}, Pb^{2+}, Ba^{2+}, Mg^{2+}, Mn^{2+}, Re^{2+}, REE^{3+}, Eu^{2+}, Cd^{2+}, and Na^{2+} (among others). The *X*-site is typically occupied by P^{5+} and exhibits a four-fold coordination, but the *X*-site can also accommodate cations such as Si^{4+}, C^{4+}, S^{6+}, As^{5+}, and V^{5+} (Hughes & Rankovan, 2002; Riker et al., 2018). The *Z*-site, or anion-column site, is occupied by the halogens F^-, Cl^- and OH^-, but can also contain reduced S species (e.g., HS^-, S^{2-}; Kim et al., 2017; Konecke et al, 2017b). The wide variety of cations able to substitute within the apatite structure yields diverse apatite compositions in natural systems (e.g., Hughes & Rakovan, 2002). Apatite is a developing geochemical tool in tracing the evolution of F, OH, and Cl in magmatic and volcanic systems (eg., Boyce & Hervig, 2008b; Scott et al., 2015). Studies of S speciation in the apatite structure (e.g., S^{2-} or S^{6+}) enable the use of

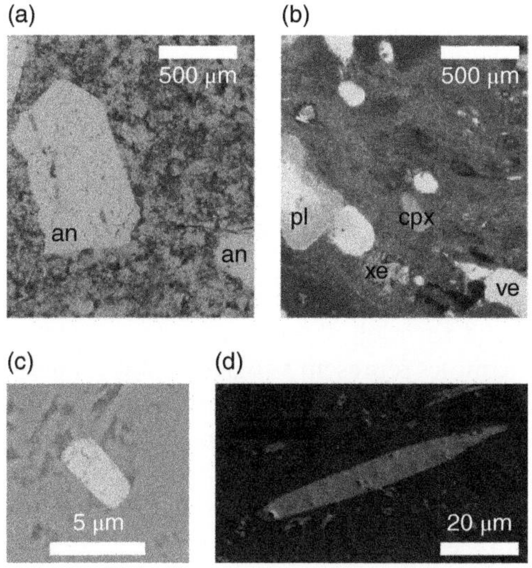

Figure 3.3. Example thin-section photographs of eruptive units IETHB and ITHN and backscatter electron (BSE) images of apatites. Thin-section photographs in plane polarized light of (a) peralkaline rhyolitic units older than 67 ka (IETHB, IETG) with euhedral anorthoclase (an) phenocrysts and homogeneous matrix and (b) metalluminous rhyolitic units 67 ka and younger (IETR, ITHN, ITN) with a diverse mineral assemblage and evidence of mingling between rhyolitic and basaltic magmas (pl, plagioclase; ol, olivine; cpx, clinopyroxene; xe, xenolith; vs, vesicle). (c) BSE image of a small anorthoclase-hosted inclusion. (d) BSE image of a large matrix-hosted apatite.

apatite as an oxybarometer in magmatic and hydrothermal systems (Kim et al., 2017; Konecke et al, 2017a; Konecke et al, 2017b; Sadove et al., 2019). Future advances in measuring and understanding C in apatite may make this mineral an even more powerful tracer for volatiles in magmas (Riker et al., 2018).

Apatite is a particularly useful geochemical tracer in felsic systems, where it saturates early and crystallizes over a wide temperature range (often >200°C; Harrison & Watson, 1984; Piccoli & Candela, 2002; Pichavant et al., 1992; Webster & Piccoli, 2015).

Most volcanic apatites are compositionally unzoned with respect to halogens (e.g., Piccoli & Candela, 2002; Webster & Piccoli, 2015). However, intracrystalline variability with respect to trace elements and halogens is sometimes identified in volcanic apatite (e.g., Boyce & Hervig, 2008a, 2008b; Parat et al., 2002). Experiments performed by Brenan (1993) show relatively fast timescales of halogen diffusion in the apatite structure. Halogen zoning in apatite is therefore interpreted to be evidence of short residence times of apatite in preeruptive magmas with heterogeneous halogen compositions (Boyce & Hervig, 2008). For other volatiles, such as S, processes of diffusion in apatite are comparably slower (Peng et al., 1997). Complex zonation of S is thus relatively common in volcanic apatites (Peng et al., 1997; Streck & Dilles, 1998; Scott et al., 2015). This suggests that the S signature of a magma is better retained in the apatite structure relative to the halogens, and zonation of S may correspond to a record of changing T, P, and fO_2 within the system (see also Konecke et al, 2017a; Konecke et al, 2017b).

Previous studies of volatiles in magmatic systems (including Torfajökull) have been largely based on the compositions of silicate MIs (e.g., Head et al., 2011; Johnson et al., 2011; Owen et al., 2013; Portnyagin et al., 2007). Melt inclusions are vulnerable to post-entrapment modifications such as alteration, leakage, devitrification, and shrinkage bubble formation (Baker et al., 2005; Bucholz et al., 2013; Esposito et al., 2014; Lloyd et al., 2013; Lowenstern, 1995; Moore et al., 2015; Wallace et al., 2015), which may compromise their primary compositional signatures. The formation of vapor bubbles within slower cooling MIs, as well as diffusion processes causing the migration of H_2O towards the melt–crystal interface, can lead to misrepresentation of primary volatile contents (Audétat & Lowenstern, 2014; Baker et al., 2005).

Apatite is relatively more resistant to weathering and alteration compared to volcanic glasses (Piccoli & Candela, 2002). Therefore, it may preserve a more representative record of volatile and trace elemental concentrations within a magma. Considering the significant influence of temperature on apatite saturation (Harrison & Watson, 1984; Piccoli & Candela, 2002; Pichavant et al., 1992; Webster & Piccoli, 2015), the timing of apatite crystallization can differ from the MI entrapment, recording potentially an alternative range of post-entrapment and syneruptive volatile contents. Furthermore, chemical zonation is often preserved in apatite phenocrysts and may record temporal volatile changes not preserved in MIs. Hence, even for magmatic systems with well-characterized MI compositions, apatite can contribute a valuable new temporal perspective.

3.2.4.1. Equilibrium Dissociation Constant versus Nernst Partitioning Coefficient

The compositional relationship between a mineral phase and a coexisting melt under equilibrium conditions can be described by either equilibrium dissociation constants (K_d) or by Nernst-type partitioning coefficients ($\boldsymbol{D}$). Nernst partitioning coefficients describe a ratio between an elemental constituent in a melt versus a mineral phase, while equilibrium dissociation constants (K_d) describe the exchange relationship between multiple elements found at the same structural site. As a first-order approximation, K_d values are applied and required to describe apatite–melt chemical relationships for major structural components that compete with each other.

This includes F, OH, and Cl on the anion column site (e.g., $K_{dOH\text{–}Cl} = Cl_{melt} \times OH_{ap}/OH_{melt} \times Cl_{ap}$; for details see Webster et al., 2017). Thus, the relative concentrations of F, OH, and Cl in the melt dictate the F–OH–Cl composition of a crystallizing apatite. For example, the proportion of F in apatite greatly exceeds F in the melt, because F is compatible in the apatite structure (Webster & Piccoli, 2015). The relative concentrations of F, OH, and Cl cannot serve as a direct proxy for volatiles in the melt because the relative compatibilities of F, OH, and Cl vary as a function of melt compositions and conditions (Boyce et al. 2014; Webster et al., 2017).

In contrast, as a first-order approximation, Nernst partitioning coefficients only apply for trace, and sometimes minor, elements. Their incorporation should not affect other (major) elements, or this effect should at least be negligible (i.e., $\boldsymbol{D}(X) = C(X)_{ap}/C(X)_{melt}$), where $C(X)$ is the concentration the element of interest in the apatite or melt). The $\boldsymbol{D}$ values are more appropriate for trace elements in apatite, such as S, whose incorporation (on the order of 100–1000s of $\mu g\ g^{-1}$; e.g., Konecke et al, 2017b; Parat et al., 2011) has minimal effects on the melt in comparison to major elements, such as F and Cl (Boyce et al., 2014).

Experimentally derived equilibrium dissociation constants and partition coefficients for rhyolitic apatites (i.e. Doherty et al., 2014; Konecke et al, 2017a; Webster et al., 2009, 2017) can be used to approximate the conditions of a magma body in a setting such as Torfajökull. By comparing apatite estimates of volatiles in the melt with measured glass compositions, the applicability of apatite partition coefficients can be assessed for Icelandic systems. Differences in volatile components in measured glass compositions versus modeled melt compositions can then be used to investigate degassing processes during an eruptive event (i.e., Scott et al., 2015).

3.3. METHODS

3.3.1. Sample Selection and Preparation

Five samples are used to represent the ~400 kyr silicic history of the Torfajökull volcanic system (384 ka to 1477 CE). The samples represent subglacial and subaerial eruptions, as well as peralkaline and metaluminous magma compositions. Unit locations are shown in Figure 3.2, and exact coordinates of sample sites, and a brief overview of sample characteristics, are provided in Table 3.1.

Rock chips with dimensions of approximately 2 × 2 cm were mounted in epoxy and polished for subsequent analyses. For scanning electron microscopy (SEM) characterization, the polished samples were coated in gold using a Denton Desk V sputter coater. For electron probe microanalysis (EPMA), the gold coating was removed and replaced with carbon using a Denton Vacuum Benchtop Turbo Carbon Coater.

3.3.2. Bulk Rock Geochemistry

Samples of silicic lava from the five units of interest were sent to Activation Laboratories (ActLabs, Ancaster, Ontario, Canada) for commercial lithogeochemical analysis. Upon arrival at ActLabs, samples were pulverized in a steel mill. Major and trace element concentrations were then measured by multiple techniques (analytical package code: 4E-Research+ICP/MS): inductively coupled plasma

Table 3.1 Sample Names, Locations, and Ages

Sample	Unit	Epoch	Age[a]	Location	Composition	ASI[b]
ITN	Namshraun	Holocene	1477 CE	27W 0595465 710064 Hjorsey 1955	Metaluminous rhyolite	0.87
ITHN	Hrafntinnuhraun	Holocene	870 CE	27W 0583860 7094213 Hjorsey 1955	Metaluminous rhyolite	0.92
IETR	Rauðufossafjöll	Pleistocene	67 ka	27W 0576365 7095290 WGS 1984	Peralkaline rhyolite	0.88
IETG	Gvendarhyrna	Pleistocene	278 ka	27W 0597859 7097451 WGS 1984	Peralkaline rhyolite	0.87
IETHB	Harbarmur	Pleistocene	384 ka	27W 0600092 7094232 WGS 1984	Peralkaline rhyolite	0.87

Note. [a]Ages from McGarvie et al. (2006). [b]Aluminum Saturation Index, calculated from Frost and Frost (2008).

optical emission spectrometry (ICP-OES), instrumental neutron activation analysis (INAA), inductively coupled plasma mass spectrometry (ICP-MS) and X-ray fluorescence spectrometry (XRF). Results, which were verified by ActLabs using a suite of internationally recognized standards, are summarized in Table SM1 (see link in Supplementary Material (SM) section). The results for two units (ITN, ITHn) have been published previously (Carley et al., 2011).

3.3.3. Scanning Electron Microscopy

Samples were characterized by using a Phenom ProX Desktop SEM at the Acopian Engineering Center at Lafayette College. Apatite crystals suitable for EPMA were located and identified through a combination of backscattered electron (BSE) imaging and energy dispersive spectroscopy (EDS; used for elemental mapping and spot analyses). Phosphorus element maps were created at an acceleration voltage of 10 kV using a focused beam. Phases rich in P were further investigated by using EDS spot analyses (15 kV, focused beam) to accurately identify apatite crystals.

3.3.4. Electron Probe Microanalysis

Samples were analyzed using a Cameca SX100 electron microprobe at the American Museum of Natural History in New York City. We followed the procedure applied by Konecke et al, 2017a for the apatite analyses. The electron microprobe was calibrated to measure 17 elements in apatites, in major and trace quantities (P, Ca, F, Cl, S, Ba, Ce, La, Nd, Fe, Mg, Mn, Al, K, Na, Si, Ti, and Sr). The reference materials used for calibration were Durango apatite (P, Ca), MgF_2 (F), scapolite (Cl), barite (Ba, S), $CePO_4$ (Ce), $LaPO_4$ (La), $NdPO_4$ (Nd), fayalite (Fe), rhodonite (Mn), orthoclase (Al, K), jadeite (Na), diopside (Si, Mg), rutile (Ti), and strontianite (Sr). Elements in apatite were analyzed by using an acceleration voltage of 15 kV, a beam current of 10 nA and a beam size of 2 µm for all elements besides S. For S, a beam current of 40 nA and a beam size of 1 µm was used on two spectrometers. The peak count times were 5 s for F, 10 s for P and Ca, 20 s for Ce, La, Nd, Fe, Mn, K, Si, Ti, Ba, and Sr, 60 s for Cl, as well as 300 s for S. Fluorine, along with major elements Ca and P, were analysed first. This procedure, in combination with the low beam current and low count time minimized the effect of F migration during analyses (e.g., Goldoff et al., 2012; Konecke et al, 2017b; Stock et al., 2015; Stormer et al., 1993). The reliability and precision of the apatite analyses was confirmed by frequent monitoring of Durango apatite; e.g., the F content was typically reproduced within 1σ uncertainty of ±0.3 wt%. However, a linear, time-interpolated drift correction had to be applied for one session, where F in Durango apatite relatively increased by 2.6% during the correction. This correction affected 36 out of 109 analyses, clearly indicated in Table SM2. The settings for S result in a detection limit of ~25 µg g^{-1} when applying the Ancey et al. (1979) method. Line transects of point analyses were programmed to measure chemical variation across some larger (e.g., > 20 µm diameter) apatite grains, using on average a 2 µm step size. Multiple point analyses were also collected at the rims and cores of larger crystals. In smaller crystals (e.g., < 10 µm) only core measurements were possible.

For the matrix glass analyses, the electron microprobe was calibrated to measure Na, K, Fe, Mg, Al, Si, Ca, Ti, Mn, P, F, Cl, and S. The standards used for the calibration were MgF_2 (F), jadeite (Na), orthoclase (Al, K), scapolite (Cl), rutile (Ti), diopside (Si, Ca, Mg), rhodonite (Mn), fayalite (Fe), FeS (S), and Durango apatite (P). The glasses were analyzed with an acceleration voltage of 15 kV and a beam size of 2 µm. Fluorine, Na, K, and Fe were analyzed using a beam current of 2 nA. The beam current was increased to 10 nA for the analyses of Mg, Al, Si, Ca, Ti, Mn, and P, and to 40 nA for the analyses of S and Cl. Peak count times were 5 s for volatiles and trace elements, and 300 s for S. The reliability and precision of the glass analyses was verified by measuring reference glasses (e.g., VG-2; Jarosewich et al., 1980). Sodium migration was negligible; i.e., the Na contents in the reference glasses were reproduced within 1σ uncertainty of ±0.15 wt%. Representative EPMA data are listed in Table 3.2 (apatites) and Table 3.3 (glasses). All EPMA data are available in Table SM2 (apatites) and Table SM3 (glasses).

Table 3.2 Average Major and Volatile Element Compositions of Torfajökull Glasses

Elements (wt%)	Sample ITN	ITHN	IETR	IETG	IETHB
SiO_2	64.83	69.33	73.95	78.93	73.48
Al_2O_3	14.85	14.08	12.14	10.26	12.80
K_2O	3.54	4.75	4.25	4.32	5.22
Na_2O	5.72	5.16	5.15	4.32	5.53
CaO	2.23	1.06	0.42	0.03	0.05
FeO	4.10	2.09	1.69	1.78	2.57
MgO	0.42	0.21	0.07	0.21	0.01
MnO	0.09	0.04	0.07	0.02	0.04
TiO_2	0.55	0.47	0.41	0.10	0.09
P_2O_5	0.12	0.09	0.04	0.02	0.02
F	0.11	0.25	0.23	0.04	0.05
Cl	0.14	0.20	0.10	0.01	0.02
SO_2	0.11	0.10	0.08	0.02	0.02
Total	96.80	97.82	98.62	99.86	99.88
AST (°C)[a]	845	890	865	830	779

Note. [a] Apatite saturation temperature calculated from average glass compositions (Piccoli & Candela, 1994).

Table 3.3 Representative Analyses of Torfajökull Apatites (25th, 50th, and 75th Percentiles Based on Fluorine Compositions) for Each Eruptive Unit (ITN, ITHN, IETR, IETG, IETHB)

Elements	ITN			ITHN			IETR			IETG			IETHB		
(wt %)	25th	50th	75th	25th	50th	75th	25th	50th	75th	25th	50th	75th	25th	50th	75th
P_2O_5	41.11	39.82	41.82	41.78	39.80	40.28	37.07	40.80	38.30	32.01	31.57	30.27	38.28	32.12	33.72
CaO	50.60	51.17	50.19	52.10	52.14	50.76	54.44	49.94	51.71	43.87	41.52	41.15	51.76	42.88	42.95
SiO_2	1.64	1.45	1.40	1.17	1.47	1.46	0.97	1.56	1.14	6.35	5.81	7.26	2.32	7.24	5.00
Na_2O	0.99	0.03	0.30	0.02	0.04	0.07	0.05	0.07	0.21	0.44	0.42	0.63	0.01	0.46	0.36
K_2O	0.14	0.14	0.08	0.12	0.13	0.10	0.21	0.38	0.20	0.20	0.23	0.20	0.28	0.27	0.19
Al_2O_3	0.10	0.07	0.11	0.09	0.14	0.15	0.03	0.10	0.04	0.60	0.30	0.62	0.05	0.33	0.33
FeO	0.56	0.49	0.57	0.43	0.61	0.59	0.43	0.41	0.36	BDL	BDL	BDL	0.19	BDL	BDL
MgO	0.12	0.10	0.14	0.14	0.12	0.19	0.15	0.06	0.14	0.04	0.28	0.16	0.02	0.00	0.00
MnO	BDL	BDL	0.02	BDL	BDL	BDL	BDL	BDL	BDL	BDL	BDL	BDL	BDL	BDL	BDL
TiO_2	0.06	0.03	0.11	0.05	0.04	0.05	0.17	0.08	0.21	BDL	BDL	BDL	BDL	BDL	BDL
SrO	0.03	BDL	0.05	0.00	BDL	BDL	0.06	BDL	BDL	0.07	0.05	0.10	BDL	0.02	0.03
BaO	BDL	BDL	BDL	0.03	BDL	0.02	0.06	BDL	BDL	0.05	0.07	0.04	0.03	0.04	0.02
La_2O_3	0.45	0.49	0.05	0.37	0.57	0.67	0.50	0.52	0.43	3.10	2.84	2.57	1.07	2.38	2.50
Ce_2O_3	1.23	1.18	0.35	1.05	1.20	1.36	1.23	1.61	1.33	7.81	6.88	7.25	2.66	5.93	6.52
Nd_2O_3	0.49	0.51	0.05	0.45	0.53	0.52	0.53	0.73	0.61	3.92	3.45	3.63	1.26	3.15	3.51
Total REE*	2.16	2.18	4.35	1.87	2.29	2.55	2.27	2.87	2.37	14.83	13.17	13.45	4.98	11.46	12.53
SO_2	BDL	BDL	0.06	0.01	0.01	0.01	0.07	0.04	BDL	BDL	0.02	BDL	BDL	0.00	0.00
F	2.40	2.84	3.35	3.31	3.46	3.63	2.72	3.30	3.63*	2.43	3.33	3.67	1.91	3.15	3.64
Cl	0.34	0.24	0.28	0.21	0.27	0.27	0.12	0.19	0.10	0.02	0.03	0.02	0.01	0.01	0.01
OH*	0.34	0.24	0.28	0.58	0.30	0.10	1.17	0.52	0.00	1.36	0.30	0.00	2.17	0.88	0.06
Total	100.33	98.58	98.93	101.31	100.52	100.12	98.82	99.89	98.78	100.92	96.82	97.57	99.89	98.65	98.85
O = F,Cl	1.12	1.25	1.48	1.44	1.52	1.59	1.17	1.47	1.71	1.03	1.41	1.55	0.83	1.33	1.33
Total	100.67	98.29	97.90	100.44	99.30	98.63	98.82	98.94	97.06	101.26	95.71	96.02	101.23	98.20	97.38

Note. BDL, below detection limit.
*Total REE, La_2O_3+Ce_2O_3+NdO_3
*OH calculated stoichiometrically using method from Ketcham (2015)

3.3.5. Data Analysis

The EPMA data were filtered to ensure only "clean" (i.e., single phase) measurements were included in the final data assessment. Mixed analyses (i.e., measurements of apatite grains contaminated with signals from adjacent phases) were identified by elevated Si, Na, and K contents. These elements are rarely present in high abundance in apatite (i.e., above detection limit), and thus these analyses were removed from the data set. To further evaluate the validity of measurements, the Ca, P, and anion (F–OH–Cl) site occupation was verified. Here, atoms per formula unit (apfu) were calculated on the basis of 26 anion sites in the apatite structure ($Ca_{10}(PO_4)_6(F,Cl,OH)_2$; see Ketcham, 2015) and are listed in Table SM1. The total apfu values for the P-site were calculated using the sum of P, Si, and Ti values. Data points with P-site totals lower than 5 or higher than 7 apfu were excluded from the data set. The total apfu values for the Ca-site were calculated using the sum of the Ca, La, Ce, Nd, Fe, Mn, Na, Ba, Sr, and Al values (for common substitution mechanisms, see Pan & Fleet, 2002). Contents of Mn (detection limit: ~900 μg g^{-1}), Ba (~2000 μg g^{-1}), Al (~250 μg g^{-1}), and Sr (~1400 μg g^{-1}) in apatite were typically below their measured detection limit, having negligible contributions to the Ca-site totals. All Ca-site totals lower than 9 and higher than 11 were excluded from the data set. We highlight that the remaining analyses show the P/Ca apfu ratio typical for apatite (Figure 3.4). Thus, possible contributions by the host phase (glass or feldspar) to the analyses are negligible as their P/Ca apfu ratios are vastly different. Similarly, accessory REE phases such as britholite, chevkinite, and deloneite are characterized by significantly different P/Ca apfu ratios (0.18, 0, and 1.5, respectively). Analyses with anomalously high apfu F values (>2.45 apfu) were removed, as these values greatly exceed what is stoichiometrically possible for an apatite crystal. The filtered data set contains 138 apatite analyses. Besides signal contamination from nearby phases, we suggest that uneven surfaces and cracks (both common in the samples from Torfajökull and relating to the fragile nature of the samples) are the main reason for nonstoichiometric nature of some discarded analyses.

High F measurements with values between 2.15 and 2.45 apfu were corrected, calculating a maximum F value assuming an OH concentration of zero. This was done by subtracting the Cl values from 2 apfu, the maximum occupancy of the halogen sites within apatite. Data points with corrected F values (25 out of 129 analyses) are indicated by an asterisk in Table SM2. The high F contents measured in several samples indicate some beam damage (Goldoff et al., 2012), resulting in an overestimation of F and an underestimation of OH, where the latter was calculated based on stoichiometry. However, while the absolute values may be affected, the F–Cl–H trends recorded by the apatites should remain intact.

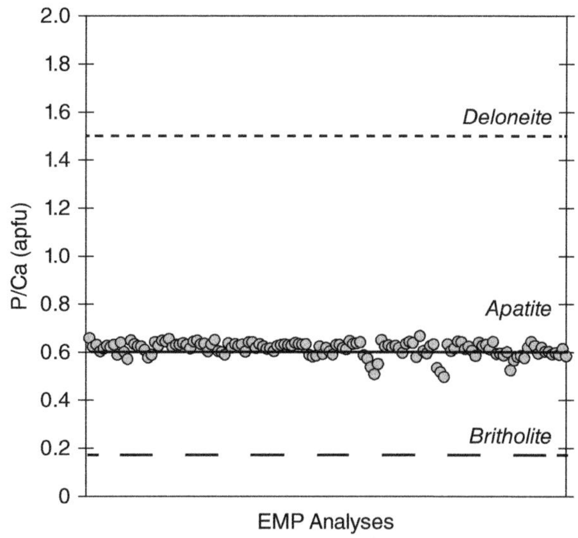

Figure 3.4 Evaluation of the EPMA apatite data. The P/Ca apfu ratios of the filtered data set are shown in comparison to P/Ca apfu ratios of end-member apatite ($Ca_5(PO_4)_3(F,Cl,OH)$), britholite ($(Ce,Ca,Th,La,Nd)_5(SiO_4,PO_4)_3(OH,F)$), and deloneite ($NaCa_2SrCe(PO_4)_3F$).

The OH contents within the apatites were estimated based on stoichiometry, due to the inability of the electron microprobe to measure H concentrations. This method is applied frequently in the literature (e.g., Doherty et al., 2014; Goldoff et al., 2012; Piccoli & Candela, 2002; Stoppa & Liu, 1995) and was recently refined by Ketcham et al. (2015). The calculations assume that all halogen sites within the crystal structure are occupied, allowing use of the measured F and Cl contents for the estimation of OH.

Glass analyses with unreasonably low totals for (mostly) degassed matrix glass (e.g., <96%) were excluded. Glass data points with apparent contamination from other phases, as determined by characteristics such as high Fe (>6%), Ca (>2%), or Al (>15%) were removed from the final data set, which contains 48 analyses (Table SM1).

3.4. RESULTS

3.4.1. Petrography

The oldest samples (IETHB, 384 ka; IETG, 278 ka) are highly altered porphyritic rhyolites with devitrified matrices. Samples from the subglacial Pleistocene rhyolites are similar to one another in texture and mineralogy, although IETHB is more visibly altered than IETG. Matrix-hosted phenocryst phases include large anorthoclase crystals (≤5 mm in diameter) and augite. In both of these units, apatites exist in low abundances and are almost exclusively hosted in large anorthoclase phenocrysts (Figure 3.3). The apatites are typically small (often < 5 µm), rarely reaching diameters of 10 µm. Additional mineral inclusions present within anorthoclase phenocrysts are chevkinite, pyrrhotite, augite, and ilmenite. One apatite microphenocryst, ~8 µm in length, was identified in the highly altered matrix of sample IETG. No other matrix-hosted apatites were identified in samples older than 67 ka.

The sample from the 67 ka eruption (IETR) is also a porphyritic rhyolite, showing few signs of alteration. The matrix is glassy and poor in microlites, with phenocryst phases including plagioclase, augite, ilmenite, apatite, and zircon. Larger phenocrysts are glomeratic, with plagioclase phenocrysts exhibiting sieve textures. Mafic enclaves consisting of olivine-bearing tholeiitic basalt are distributed throughout the sample. Apatites in this sample are dominantly present as isolated phenocrysts within a glassy matrix, but are also present in lower abundances as inclusions in plagioclase phenocrysts and glomerocryst masses. Of all units, apatites in sample IETR are the most abundant and largest (up to 200 µm in length).

The two youngest samples (ITHN, 871 CE; ITN, 1477 CE) are vesicular and porphyritic with glassy rhyolitic, microlite-rich, matrices. As with the 67 ka sample, basaltic enclaves are distributed throughout these young samples, thus, bulk-rock geochemistry reflects a SiO_2 less than that associated with rhyolite (i.e., the bulk-rock analyses classify these samples as dacite). Mineral phases include plagioclase, augite, ilmenite, apatite, and zircon. These phases typically occur as isolated phenocrysts in the matrix, and as glomerocrysts. Plagioclase and augite phenocrysts show thin bright rims in BSE images and exhibit sieve textures and embayments. Within these units apatites are predominantly matrix hosted, with rare plagioclase-hosted apatite inclusions. Apatites are abundant and large (up to 100 µm in length) when compared to those in the oldest samples (IETHB, IETG), but are not as abundant and large compared to those in the 67 ka sample (IETR).

3.4.2. Bulk Rock Compositions

The data set includes five bulk-rock analyses across the five studied units (Table 3.1). Temporal trends in bulk composition feature decreasing SiO_2 and K_2O, consistent Na_2O, and increasing Al_2O_3, FeO, CaO, MgO, TiO_2, and P_2O_5 over time (Figure 3.5).

3.4.3. Glass Compositions

The data set includes 48 matrix glass analyses across the five units studied (Table 3.3; Table SM3). All units are characterized by compositionally heterogeneous matrix glasses (Figure 3.5), which is consistent with macro- and microscopic mixing and mingling features (e.g., Figure 3.3).

3.4.3.1. Major Elements Contents of Matrix Glasses

The matrix glasses (Table 3.3 and Figure 3.5) correspond to trends in bulk-rock data, exhibiting a progressive decrease in SiO_2 content and an increase in CaO, Na_2O, TiO, MgO, and P_2O_5 with decreasing age. Notably, K_2O contents in the matrix glasses are the highest (4.1–5.3 wt%) in the 67 ka (IETR) and 871 CE (ITHN) eruptions, but comparably low (2.6–4.5 wt%) in the 384 ka (IETHB), 278 ka (IETG), and 1477 CE (ITN) eruptions. Concentrations of FeO are low (0.83–2.77 wt%) in the 384 ka (IETHB), 278 ka (IETG), 67 ka (IETR), and 871 CE (ITHN) samples, but high (2.51–5.56 wt%) in the most recent eruption (1477 CE; ITN).

3.4.3.2. Volatile (F, Cl, S) Contents of the Matrix Glasses

Volatile (F, Cl, S) contents in the matrix glasses show significant variability from unit to unit, and within individual units (Table 3.3 and Figure 3.5i–k). For each unit, concentrations of volatiles in the matrix glasses increase with increasing SiO_2 content.

Chlorine values are below the detection limit (~60 µg g^{-1} Cl) in samples older than 67 ka. In the 67 ka unit (IETR), Cl ranges from below the detection limit up to 1560 µg g^{-1} Cl (average ~1010 µg g^{-1} Cl). In the 871 CE unit (ITHN), Cl values are the highest, ranging from 1130 to 3170 µg g^{-1} Cl (average 1950 µg g^{-1} Cl). The 1477 CE unit (ITN) shows Cl values from 860 to 1680 µg g^{-1} Cl (average 1420 µg g^{-1} Cl).

In samples older than 67 ka, F values are lower than detection limit. Fluorine values are highest in the 67 ka unit (IETR), reaching values up to 6000 µg g^{-1} F (average 2340 µg g^{-1} F). In the 871 CE unit (ITHN), F values range from 1230 to 3460 µg g^{-1} F (average 2490 µg g^{-1} F). In the 1477 CE unit (ITN), values range from below detection limit (BDL) to 2730 µg g^{-1} F (average 1080 µg g^{-1} F).

The S contents in the matrix glasses are typically higher in the younger samples (Figure 3.5j). Samples older than 67 ka have S values ranging from BDL to 230 µg g^{-1} S (average 80 µg g^{-1} S). The 67 ka eruption (IETR) shows S values from 240 to 590 µg g^{-1} S (average 390 µg g^{-1} S). Sulfur values are the highest in the 871 CE unit (ITHN), ranging from 240 to 2000 µg g^{-1} S (average 490 µg g^{-1} S). In the 1477 CE sample S values range from 290 to 650 µg g^{-1} S (average 530 µg g^{-1}).

3.4.4. Apatite Hosts and Apatite Compositions

The filtered data set includes a total of 138 analyses collected on 48 apatites from the five units studied (see Table SM2).

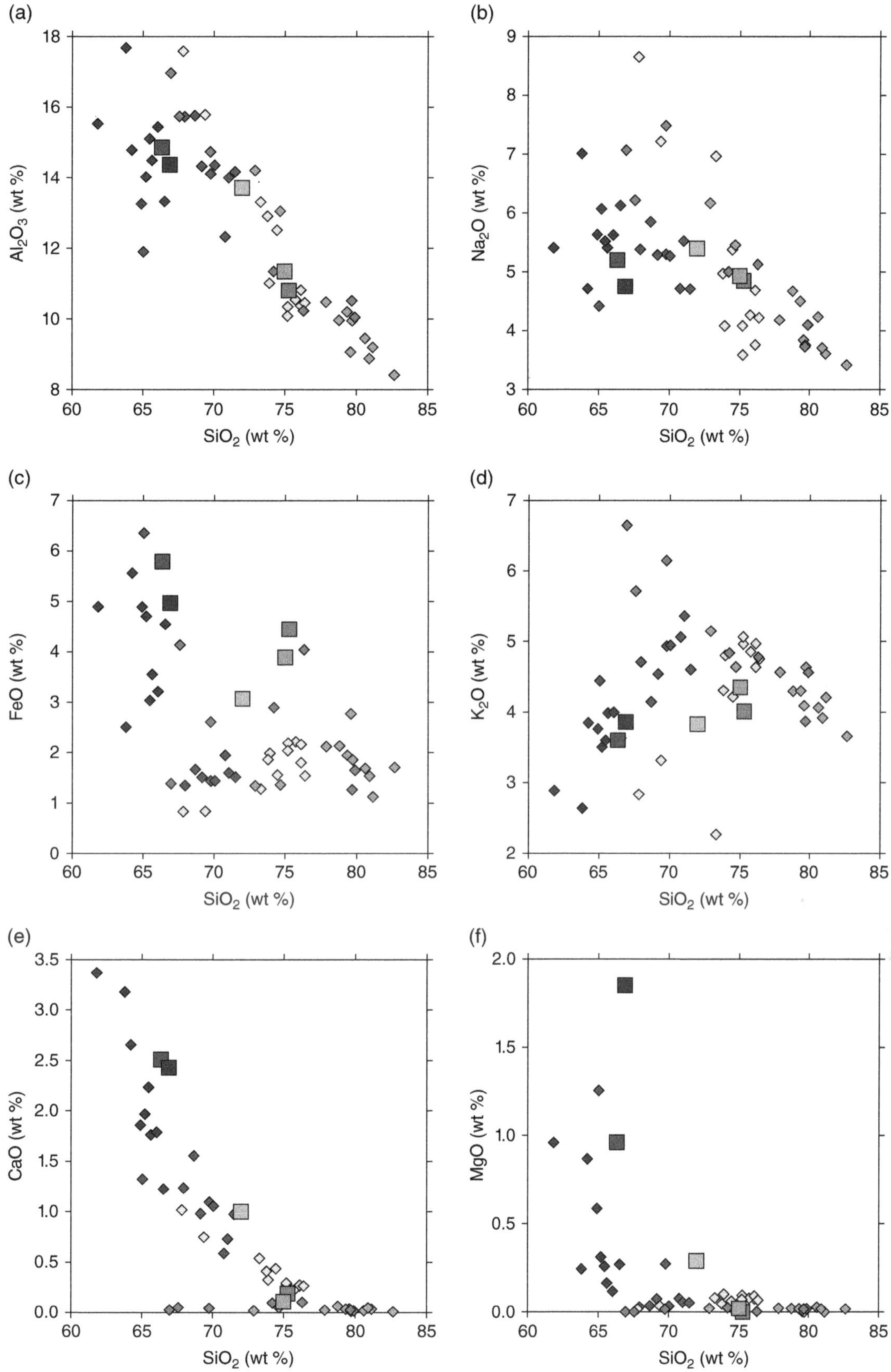

Figure 3.5 Major elements and volatiles in bulk rock samples and matrix glasses. (a)–(h) Harker diagrams showing major element variation (Al_2O_3, Na_2O, FeO, K_2O, CaO, MgO, TiO_2, P_2O_5) in measured volcanic glasses (small diamonds) and in the bulk samples (large squares). (i)–(k)Volatile contents (F, S, and Cl, respectively) are displayed for the matrix glasses. The concentrations of SiO_2 and K_2O generally decrease with time, while the concentrations of all other major elements increase with time. Concentrations of volatile elements are lowest for samples older than 67 ka, and relatively higher for samples 67 ka and younger: F exhibits the highest contents at 67 ka, and Cl exhibits the highest contents at 871 CE. Bulk rock data for ITN and ITHN were previously published by Carley et al. (2011).

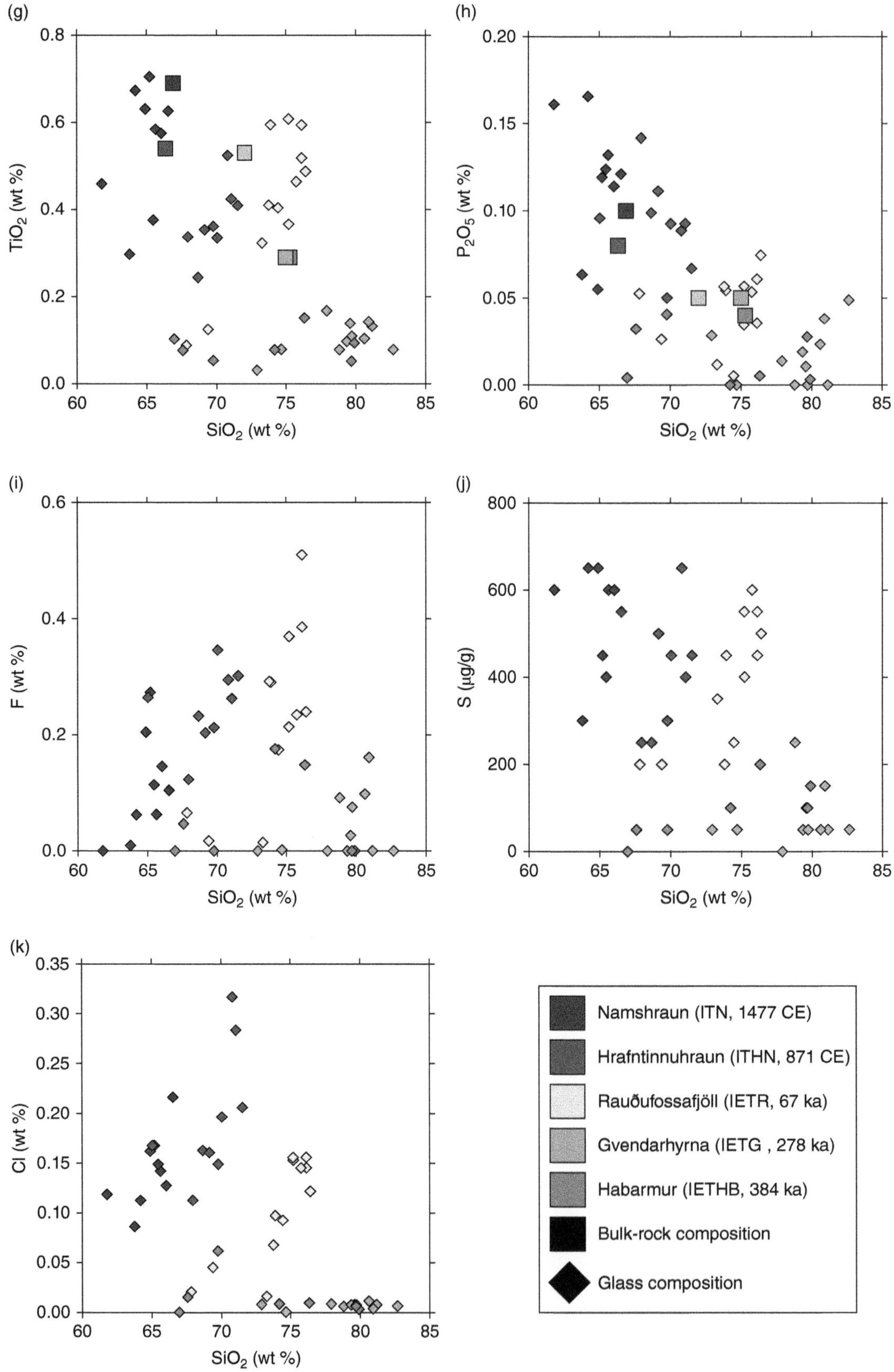

Figure 3.5 (Continued)

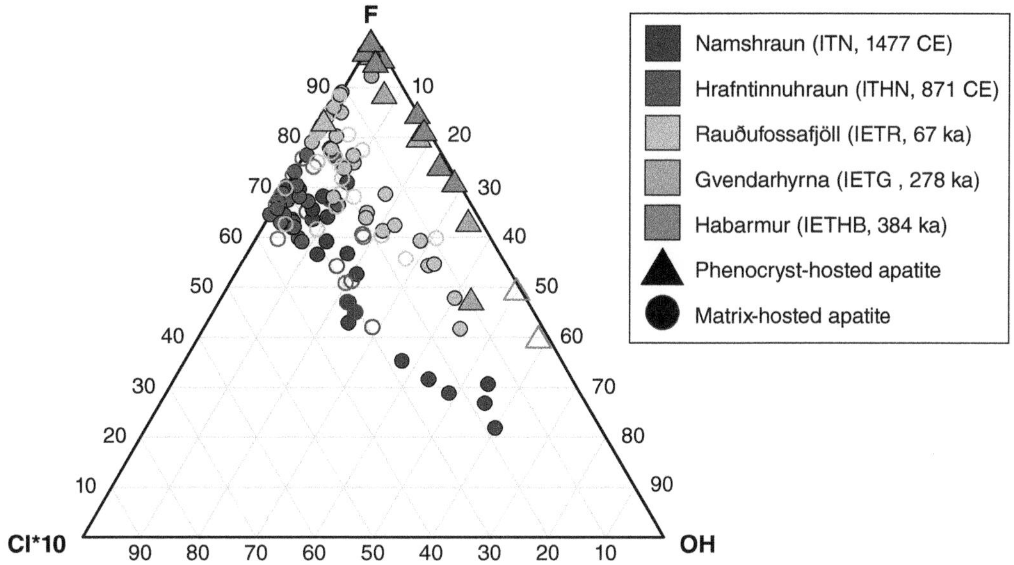

Figure 3.6 Apatite halogen ternary plot showing F, Cl, and OH compositions of Torfajökull apatite hosted in matrix glass (circles) and in feldspar phenocrysts (triangles). Compositions are expressed as atoms per formula unit (apfu). Open icons indicate rim analyses on crystals large enough for multiple analyses. The Cl contents are multiplied by 10 for illustrative purposes.

Except for one matrix-hosted apatite, all apatites from the oldest two units (>67 ka; IETHB and IETG) are anorthoclase hosted, whereas the majority of apatites from units ≤67 ka (IETR, ITHN, ITN) are matrix hosted. Based on the data set available, a correlation between host phase and apatite compositions is not observed (i.e., Figures 3.6–3.8). Dissolution–reprecipitation features, the presence of nearby REE phosphates, REE-depleted (or enriched) areas in apatite, or other features forming as a consequence of metasomatic alteration were not observed (Harlov, 2015; e.g., Figure 3.3). Due to the small size of apatite phenocrysts measured, zoning patterns could not be identified with confidence.

3.4.4.1. Volatile (F, Cl, OH, S) Contents of Apatites

Apatites from Torfajökull are dominantly fluorapatite (Figure 3.6), with F contents ranging from 1.02 to 4.03 wt% F, Cl contents ranging from BDL (~0.013 wt% Cl) to 0.40 wt% Cl, and highly variable calculated OH contents (from 0 to 3.26 wt% OH). Since the OH contents were calculated based on stoichiometry (see section 3.3.5), they vary as a function of F.

Apatites from all samples—regardless of age, magma composition, or host phase (glass or feldspar)—have F contents that range from 1.02 to 4.03 wt% F. In units older than 67 ka (IETHB, IETG), Cl values are close to or below the detection limit. Apatites from the 67 ka eruption (IETR) have higher Cl contents, ranging from 0.07 to 0.35 wt% Cl (average 0.16 wt% Cl). Apatites from samples younger than 67 ka (ITHN, ITN) have the highest Cl contents, ranging from 0.11 to 0.40 wt% Cl (ITHN, average 0.29 wt% Cl) and from 0.13 to 0.34 wt% Cl (ITN, average 0.26 wt% Cl), respectively. Chlorine zonation was identified only in the 67 ka apatites, where core analyses show slightly lower Cl concentrations (average 0.13 wt% Cl) than rim analyses (average 0.18 wt% Cl). This core–rim distinction may be present in apatites from other units that were too small to measure transects. In apatites older than 67 ka, the calculated OH values are highly variable, ranging from 0 to 2.45 wt% OH for unit IETHB (average 0.70 wt% OH) and to 1.75 wt% OH for unit IETG (average 0.13 wt% OH). The OH values in the 67 ka eruption range from 0 to 3.26 wt% OH (average 0.48 wt% OH). Notably, apatites from the 871 CE eruption have the most restricted range of OH values, from 0 to 1.16 wt% OH (ITHN, average 0.24 wt% OH), while apatites from the 1477 CE eruption have a much wider range, from 0 to 3.16 wt% OH (ITN, average 1.16 wt% OH).

Sulfur contents in apatite are typically low, except for apatites in the 67 ka unit (Figure 3.7). In samples older than 67 ka (IETHB, IETG), S contents are typically near the detection limit (20 µg g^{-1} S) and reach maximum values of ~100 µg g^{-1} S. Sulfur in apatites from the 67 ka eruption show the greatest variability, ranging from BDL to 570 µg g^{-1} S. In samples younger than 67 ka (ITHN, ITN), S values typically range from BDL to ~110 µg g^{-1} S. However, each of the <67 ka units contained one measured outlier with significantly (2–4 times) higher S concentrations (ITHN, ~420 µg g^{-1} S; ITN, ~290 µg g^{-1} S).

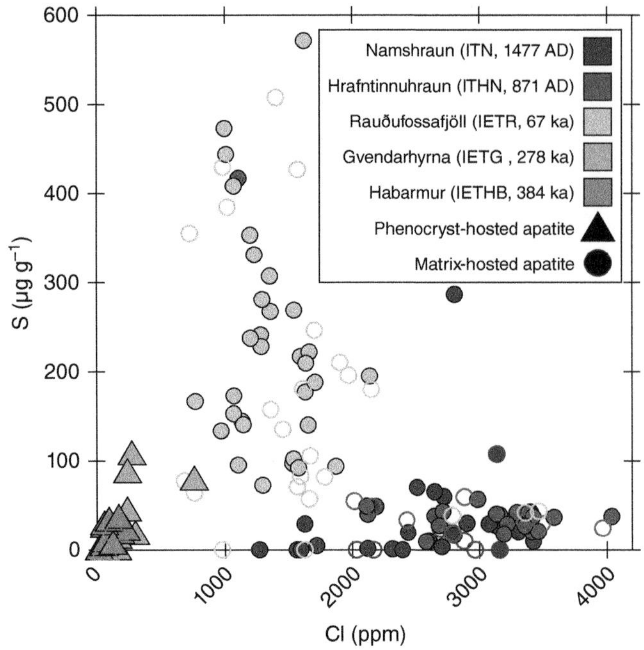

Figure 3.7 Sulfur evolution in Torfajökull apatites as a function of Cl content. Scatter plot showing S and Cl compositions of Torfajökull apatite hosted in matrix glass (circles) and in feldspar phenocrysts (triangles). Open icons indicate rim analyses on crystals large enough for multiple analyses. Chlorine-in-apatite increases with time at Torfajökull as observed with progression from low Cl (<1000 µg g^{-1}) in the 384 ka and 278 ka samples, to high Cl (>1000 µg g^{-1}) in the 871 CE and 1477 CE samples (see also Figure 3.4). Thus, Cl is used as a time proxy in this plot. A significant increase of S-in-apatite is observed for the 67 ka eruption (IETR), whereas the apatites from all other units are characterized by fairly similar S contents.

3.4.4.2. Trace Element Contents of Apatites

Trace elements measured in Torfajökull apatites include SiO_2, Na_2O, FeO, MgO, MnO, light rare earth elements (LREE: Ce_2O_3, La_2O_3, Nd_2O_3), and SO_2.

Apatites from all units are enriched in SiO_2. Apatites from eruptions older than 67 ka have the highest SiO_2 values, with averages of 4.36 wt% (IETHB) and 7.29 wt% (IETG) SiO_2. In units 67 ka and younger, SiO_2 values are generally lower, with averages of 1.25 wt% (IETR), 1.54 wt% (ITHN), and 1.77 wt% SiO_2 (ITN). The apatites from units older than 67 ka show the highest Na_2O concentrations, with values up to 0.63 wt% Na_2O. Units 67 ka and younger have Na_2O contents in apatite BDL (~0.025 wt% Na_2O). Both FeO and MgO values in apatites increase over time. In units older than 67 ka, FeO and MgO values are BDL (~0.045 wt%). In units 67 ka and younger, FeO values range from 0.30 to 1.21 wt% FeO, while MgO contents range from 0.10 to 0.60 wt% MgO. All MnO values are BDL (~0.024 wt% MnO) in units older than 67 ka. In samples 67 ka and younger, MnO values range from BDL to 0.25 wt% MnO.

Apatites from all units show high concentrations of LREEs (Ce_2O_3 + La_2O_3 + Nd_2O_3) (Figure 3.8a and b), up to 18.9 wt% in total $LREE_2O_3$. As Ce, La, and Nd were the only REEs measured in this study, the term LREE will be used to refer to Ce, La, and Nd. Unless stated otherwise, LREE contents will be given as the total of Ce_2O_3 + La_2O_3 + Nd_2O_3. These particular REE were measured because apatite preferentially incorporates LREE over heavy REE within its structure (Pan & Fleet, 2002; Watson & Green, 1981) and Ce, La, and Nd are typically the most abundant (L)REEs in apatite (Fleet & Pan, 1995; Pan & Fleet, 2002; Watson & Green, 1981). Most of the apatite EPMA totals in this study are close to 100% (Table SM1), validating the decision to focus on Ce, La, and Nd as representative of (L)REE in apatite.

A progressive decrease in LREE content and variability is observed at Torfajökull over time (Figure 3.8). The LREE concentration is especially high in anorthoclase-hosted apatite older than 67 ka. The highest REE values are from the 278 ka eruption, with compositions ranging from 8.4 to 18.9 wt% LREE oxides (average 14 wt% LREE). Samples from the 384 ka eruption have similarly high LREEs, ranging from 5.0 to 15.5 wt% (average 10 wt% LREE). While these compositions are certainly higher than most magmatic apatites, P/Ca do confirm these phases are indeed apatite, and not other REE phosphates. Apatites in the 67 ka unit have much lower and restricted LREE contents, ranging from 2.1 to 2.6 wt% (average 2.4 wt% LREE). In the apatites from the 871 CE unit, LREE contents range from 1.9 to 3.3 wt% LREE. In the youngest sample (1477 CE), two distinct populations of apatite with low (0.5–1.3 wt% LREE) and relatively higher (2.0–2.8 wt% LREE) LREEs are present.

3.5. DISCUSSION

The discussion is divided into three subsections in order to facilitate interpretation of the compiled geochemical data and their implications to elucidate magmatic and volcanic processes at Torfajökull.

1. Apatite compositions are placed in a broader global context, identifying features that make Torfajökull apatites specifically (and perhaps Icelandic apatites more generally) compositionally unique (section 3.5.1).
2. Volatile evolution at Torfajökull from the Pleistocene through the Holocene is examined by using new apatite and matrix-glass data (section 3.5.2).
3. Torfajökull's 67 ka eruption, an important inflection point in the system's history, is investigated in greater detail by comparing apatite data with published MI data (section 3.5.3).

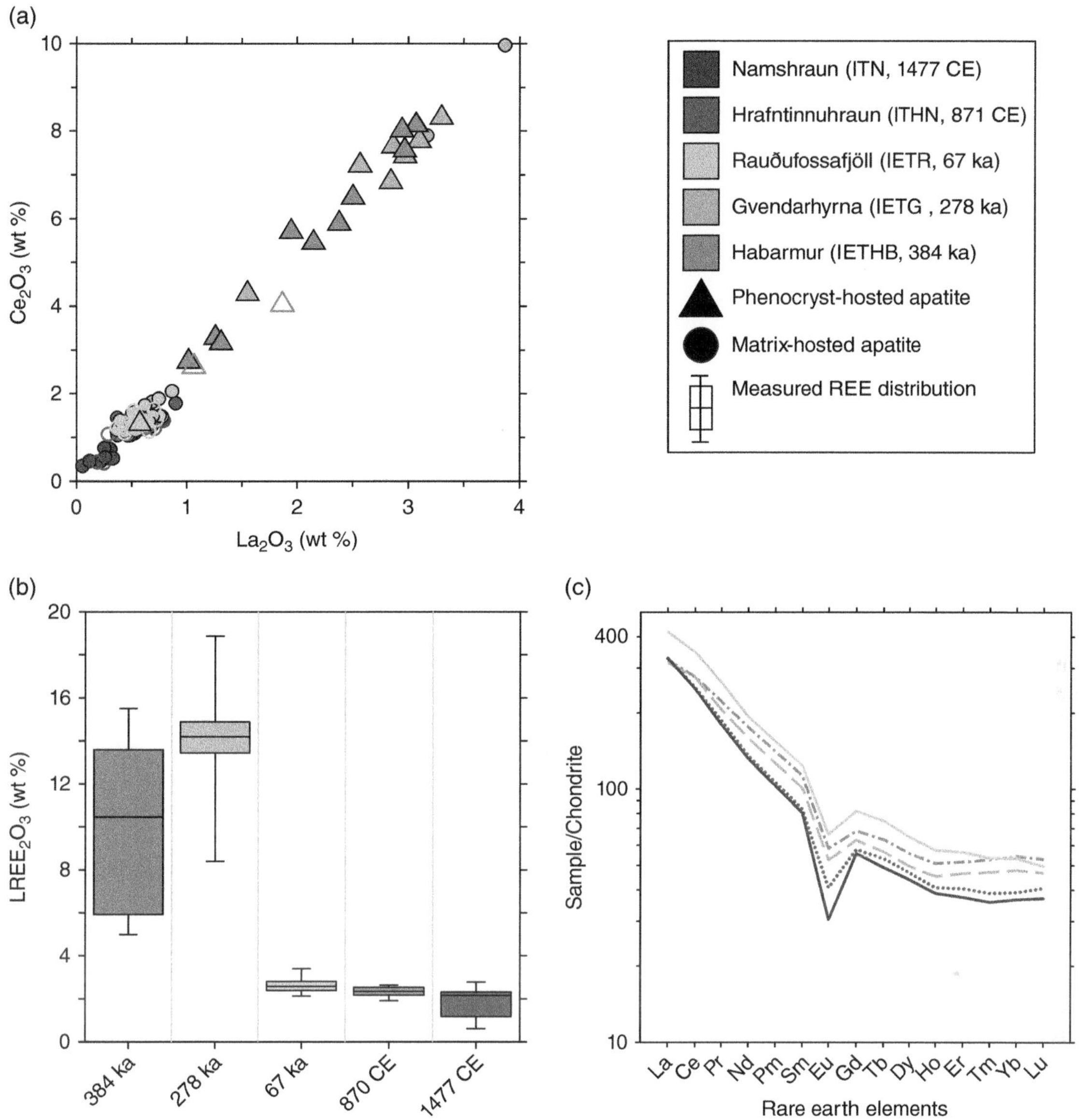

Figure 3.8 Rare-earth-element contents and evolution in apatite. (a) Scatterplot of La_2O_3 versus Ce_2O_3 in apatites, showing a near linear relationship. (b) The LREE content (here LREE = La_2O_3 + Ce_2O_3 + Nd_2O_3; in wt%) are displayed as box and whisker plot as a function of time (or eruptive unit). Whiskers extend to the 5th and 95th percentiles, boxes define the 25th to 75th percentiles, and median values bisect the boxes. (c) Chondrite normalized (McDonough & Sun, 1995) bulk-rock REE compositions for each unit (Supplementary Table SM2). Bulk-rock REE contents generally decrease over time, however the highest REE contents are recorded in the intermediate 67 ka unit (IETR).

Throughout this discussion, we interpret the Torfajökull apatite described in section 3.4 and shown in Figure 3.3 to represent primary magmatic signatures because (a) measured apatites exhibit P/Ca ratios of stoichiometrically ideal apatites and (b) of the absence of dissolution–reprecipitation features that are typically a consequence of metasomatic alteration (Harlov, 2015).

3.5.1. Icelandic Apatites in a Global Context

The Icelandic apatites examined in this study crystallized from magmas formed in an oceanic setting, at the junction of a hotspot and a mid-ocean ridge. In contrast, previous studies of volcanic apatite compositions have focused on highly explosive subduction-related arc systems, such as Mount Pinatubo (Van Hoose et al., 2013),

Santiaguito (Scott et al., 2015), or Augustine (Webster et al., 2017). The composition of volatiles and trace elements in magmas and accessory minerals vary dramatically between tectonic settings (e.g., chlorine (Webster et al., 2017) and zircon (Grimes et al., 2007, 2015)). Thus, Icelandic apatites may also record compositional patterns unique to their magmatic-tectonic setting.

Globally, apatites from most volcanic–plutonic systems share similar compositional characteristics and are typically F-rich (Webster & Piccoli, 2015). Apatites from mafic rocks show the most compositional variability with respect to Cl, whereas apatites from felsic systems are more compositionally restricted (e.g., characteristically low Cl). Fluorine and OH concentrations in apatite typically exceed Cl contents, which has been interpreted as an effect of anion size (Pan & Fleet, 2002; Piccoli & Candela, 2002; Webster & Piccoli, 2015). The REE composition of a typical igneous apatite is, on average, ~0.5 wt% total REE oxides (Pan & Fleet, 2002). Consistent with global apatite trends, Torfajökull apatites are fluorapatite with Cl-poor compositions. However, compared to typical igneous apatites globally, Torfajökull apatites have distinctly higher LREE contents.

3.5.1.1. LREE Evolution

The most distinctive geochemical feature of Torfajökull apatites is their high concentrations of LREEs (Figure 3.8). Apatites from the oldest, peralkaline, samples (384 ka and 278 ka) reach up to 18.9 wt% LREE oxides. Torfajökull apatites from the peralkaline 67 ka unit and the younger, metaluminous, samples (871 CE and 1477 CE) are characterized by LREE contents of ~0.5 to 2.5 wt%, which is still high compared to the global average. All apatite analyses used in this study exhibit a stoichiometric composition for the P (~6 apfu) and Ca (~10 apfu) sites (see section 3.3.5), with P/Ca ratios between 0.49 and 0.68 (see Figure 3.4). This suggests that the measured REE compositions of apatite are real, and not the result of contamination by other phases (e.g., chevkinite) during EPMA analysis.

The elevated LREE compositions observed in Torfajökull apatite crystals raise several questions, including: What substitution mechanism(s) facilitated such extreme LREE compositions in these apatite crystals?; and What magmatic conditions or processes made these apatite compositions possible?

Incorporation of REEs into the apatite structure can be facilitated by several different substitution mechanisms (Pan and Fleet, 2002), including:

$$REE^{3+} + Si^{4+} = Ca^{2+} + P^{5+} \tag{3.1}$$

$$REE^{3+} + Na^{+} = 2Ca^{2+} \tag{3.2}$$

$$2REE^{3+} + \_\_ = 3Ca^{2+} \tag{3.3}$$

where Na^+ can be replaced by other monovalent cations (e.g., K^+) and Si^{4+} can be replaced by other tetravalent cations (e.g., Ti^{4+}). In addition, any combination of equations (3.1)–(3.3) is conceivable, e.g.:

$$REE^{3+} + Si^{4+} + Na^{+} + \_\_ = Ca^{2+} + P^{5+} \tag{3.4}$$

A more encompassing but simplified substitution mechanism may consider all cations (M) measured in apatite:

$$REE^{3+} + M = Ca^{2+} + P^{5+} \tag{3.5}$$

The relevance of these various substitution mechanisms can be tested by applying equations (3.1)–(3.5) to Torfajökull apatite data (Figure 3.9). In each scenario considered, a fitted slope of −1 is ideal, because this represents an empirically ideal substitution mechanism. Apatite compositions from the oldest two units (384 ka and 278 ka) are roughly consistent with equations (3.3) and (3.5): the data follow a slope of approximately -0.9 (equation (3.3); Figure 3.9c; n = 19, 2 excluded) and of approximately −1.5 (equation (3.5); Figure 3.9e; n = 21, 0 excluded), respectively. This suggests that the charge imbalance and structural strain created by the extremely high LREE contents in apatites from these oldest units (up to 20 wt% LREE oxides) is accommodated by a variety of cation substitutions and cation site vacancies, which are partially covered by equations (3.1)–(3.5). The imperfect fit between measured data and the ideal situation may be attributed to unknown contributions of cation vacancy sites. While vacancies are difficult to measure directly, their existence in the apatite structure has been well documented (e.g., Kim et al., 2017; Pan & Fleet, 2002) and provides a plausible explanation for apfu values. This is considered in equation (3.6), which is suggested for apatites containing > 2 apfu REE:

$$REE^{3+} + M \pm n\_\_ = Ca^{2+} + P^{5+} \tag{3.6}$$

Apatite compositions from the youngest three units (67 ka, 871 CE, 1477 CE) are best explained by equations (3.1) and (3.5), with collective slopes of approximately -0.5 (equation 3.1) and -0.7 (equation 3.5), respectively (n = 103; 5 excluded). This suggests that Si^{4+} is mainly responsible for charge balancing the structural incorporation of (L)REEs in the apatites from the 67 ka unit and younger, and the Na^+ plays a minimal role in substitution. Further investigation is required to fully understand the incorporation of REEs into the apatite structure, but substitution mechanisms are likely a complex combination of each of the above-mentioned equations.

In previous studies, the peralkalinity of magmas has been found to encourage REE substitution in apatites (Roeder et al., 1987; Rønsbo, 1989). For example, apatites from the Ilimaussaq intrusion in South Greenland reach total REE concentrations of 16 wt%, SiO_2 values of 7 wt%,

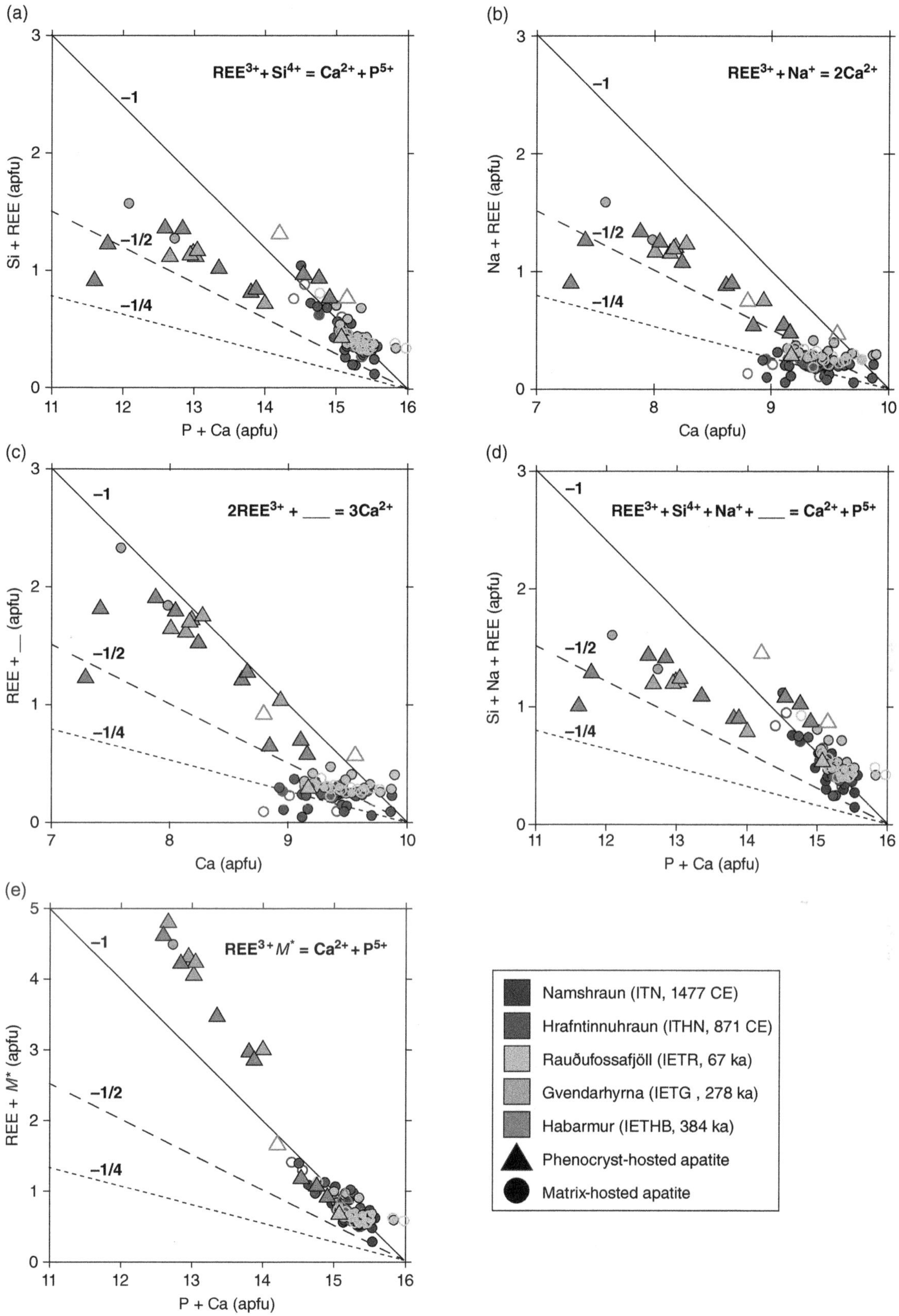

Figure 3.9 Substitution mechanisms for REEs in apatite. Scatterplots displaying the fitness of various substitution mechanisms (equations 3.1–3.5) in Torfajökull apatites, with measured elemental values for cation sites displayed in atoms per formula unit (apfu). Solid slope lines illustrate a perfect fit ($r_2 = 1$) for a given the substitution mechanism, while dashed lines indicate poor fits with the indicated substitution mechanism. The tested substitution mechanisms are displayed on each panel. (a) Sum of REE and Si plotted against sum of P and Ca, corresponding with equation (3.1). (b) Sum of REE and Na plotted against sum of P and Ca, corresponding with equation (3.2). (c) Sum of REE and vacancies plotted against sum of P and Ca (vacancies calculated based on measured REE values), corresponding with equation (3.3). (d) Sum of REE, Si, Na, and vacancy plotted against sum of P and Ca, corresponding with equation (3.4); i.e., the sum of equations (3.1)–(3.3). (e) Sum of all measured cations *M* except P and Ca plotted against sum of P and Ca, corresponding with equation (3.5).

and Na_2O values up to 3 wt% in peralkaline samples (Rønsbo, 1989), consistent with equations (3.1) and (3.2). This relationship between magma composition and apatite's affinity for REEs, Si, and, Na is also observed at Torfajökull, consistent with the compositional "devolution" between the oldest (>67 ka, peralkaline, REE-rich; bulk rock ~75 wt% SiO_2) and the youngest (<67 ka, metaluminous, relatively REE-poor; bulk rock ~66–67 wt% SiO_2) eruptive products (see Figures 3.5 and 3.9; this section and section 3.4.4.2). Such a compositional "devolution" is likely accompanied by increasing magma temperature. Thus, the REE pattern recorded by Torfajökull apatites are entirely consistent with experimental investigations by Watson and Green (1981), which demonstrated that REE apatite–melt partition coefficients decrease with increasing magmatic temperature, and with decreasing silica activity. Considering apatite partition coefficients from Prowatke and Klemme (2006), apatite REE compositions are reasonable for highly enriched Icelandic melts.

A direct relationship between bulk-rock REE contents, apatite REE contents, and coeval mineral phases is expected. However, the clarity of this relationship may be influenced or partially overprinted by various degrees and stages of magma mixing. In the oldest two and the youngest two units at Torfajökull, this relationship seems to apply (Figure 3.8c). The two oldest units show elevated REE contents in bulk-rock data and contain accessory chevkinite, evidence of an REE-enriched magmatic environment. The two youngest units are characterized by comparably low bulk-rock REE contents. These younger units lack chevkinite, but have accessory zircon. Zircon can incorporate a significant amount of REE in its structure (Watson, 1979) and can thus reduce the amount of REE in the melt available for apatite crystallization, even in the absence of chevkinite. The REE evolution of apatites from the oldest two to the youngest two units is consistent with the aforementioned observations. In contrast, the expected relationship between bulk-rock REE, apatite REE, and coeval mineral phases is inconsistent in the 67 ka rhyolite. This "intermediate" unit has the highest REE bulk-rock composition of any unit studied, and has accessory zircon but no observed chevkinite. The apatite in the 67 ka rhyolite has comparably low REE, suggesting that its crystallization likely took place before the injection of REE-depleted mafic melt.

These complex relationships between REE contents (apatite, bulk rock) and the appearance and disappearance of other REE phases (zircon, chevkinite) suggests a yet to be understood compositional (and conditional) evolution at 67 ka. However, the progressive decrease in LREE contents in Torfajökull's apatite over time may be attributed to the increasing influence of the mafic Veiðivötn Fissure Swarm on rhyolitic magmas, both for diluting REE in the host magmas, and for degrading conditions for element partitioning. Further evidence for the increasing importance of magma mixing may be recorded in the youngest unit (1477 CE), which shows two compositionally distinct LREE populations in apatite (Figure 3.8a). Relatively higher LREE contents (2.04–2.78 wt.%) may be indicative of apatites growing from a cooler felsic melt, whereas lower LREE contents (0.46–1.3 wt.%) may signify apatites crystallizing from a hotter mafic melt (for temperature relationship, see Watson & Green, 1981).

3.5.2. Volatile Evolution at Torfajökull (F, Cl, OH)

Volatile components in volcanic apatite can be used to investigate magmatic conditions at the time of apatite crystallization. The controlling factors of halogen distributions in apatite are anion radius (F and OH anions are similar in size, Cl anions are comparably larger; Shannon, 1976), the ratio of halogen fugacities (fHF/fHCl), as well as temperature, pressure, and melt composition (Piccoli & Candela, 1994; Webster et al., 2009, 2017). In the absence of information about halogen fugacity during apatite formation, and assuming constant pressure, elevated Cl (and thus relatively lower F and OH) concentrations in apatite can be correlated to elevated temperature (Piccoli & Candela, 1994, 2002). Thus, volatile contents in apatites can be used as a general proxy for magmatic temperatures. Here, as a first-order approximation, the increasing Cl contents in Torfajökull apatite over time (Figure 3.7) may indicate increasing magma temperature over time.

Chlorine and fluorine contents in the Torfajökull matrix glasses are relatively low in the oldest two units and increase over time. There are notably high concentrations of F in the 67 ka unit glasses and of Cl and S in the 871 CE unit glasses (Figure 3.5). In close correlation with these trends, the ratio of Cl/F and Cl/OH in apatite decreases from the Pleistocene to the Holocene as apatites become more Cl-rich (Figure 3.6). Chlorine solubility in magmas is also dependent on composition, as melts enriched in Ca, Mg, and Fe typically exhibit higher Cl solubility (Mathez & Webster, 2005; Webster et al., 1999). This can help explain why the younger, more mafic units display both higher chlorine contents in glasses and in apatite, indicating that Törfajokull's magmatic system is progressively becoming richer in Cl. Apatite OH contents span similarly wide ranges in all units except for the 871 CE eruption. Apatite in the 871 CE eruption has a very restricted range of OH, perhaps recording drier melt conditions at the time of apatite crystallization.

Apatite saturation temperatures (AST) were calculated for each sample based on average matrix-glass compositions (Table 3.3) by using the equation of Piccoli and

Candela (1994), which is based on felsic systems. Apatites from Torfajökull rhyolites crystallized under a relatively restricted temperature interval of ~780°C to ~890°C. The derived temperatures increase progressively over time (Table 3.3), consistent with our simplified interpretation of the Cl-in-apatite temperature proxy (see above and Figure 3.7). The increase in temperature corresponds directly to the progression of the system towards less evolved compositions, likely due to an increasing influence of high-temperature basalts from the Vedivötn fissure swarm.

3.5.2.1. Sulfur Evolution

Concentrations of S in the matrix glasses increase with decreasing age (Figure 3.5j), likely correlating with an increasing influence of magma mixing processes between Torfajökull rhyolites and Veiðivötn basalts. To better understand S evolution of the Torfajökull system, the S content in the melt at sulfide saturation (hereafter sulfur solubility or SCSS) was modeled for each unit using the method of Fortin et al. (2015) and the measured matrix-glass compositions. The required input parameters for the Fortin et al. (2015) model are the major-element melt composition (Si, Ti, Al, Fe, Mg, Ca, Na, K; see Table 3.3), water content of the melt, pressure, and temperature. We used 4.8 wt% H_2O for the calculations—the maximum H_2O content measured in Torfajökull MIs (Owen et al., 2013)—to represent undegassed, preeruptive conditions. Liquidus temperatures were estimated for each unit using rhyolite-MELTS (Gualda et al., 2012) with an assumed pressure of 100 MPa (Gunnarsson et al., 1998) and an assumed fO_2 at the fayalite–magnetite–quartz (FMQ) buffer, which is consistent with estimations of Torfajökull magmas specifically (Gunnarsson et al., 1998) and Icelandic magmas more generally (Portnyagin et al., 2012).

Sulfur solubility increased in Torfajökull rhyolites with time (Figure 3.10). The model estimates a SCSS of ~150 μg g^{-1} in melts from units older than 67 ka, and ~330 μg g^{-1} S at 1477 CE. The calculations show that the S contents measured in the matrix glasses of the two oldest units are typically below SCSS, whereas the matrix glasses of the 67 ka unit and the two youngest units contain significantly more S than is soluble in these melts at the assumed conditions (on average 1.5–2.5 times more).

Sulfur changes its oxidation state from sulfide-dominated to sulfate-dominated in both melt and apatite above ~FMQ (Jugo et al., 2010; Konecke et al., 2017b). The S contents in silicate melts at anhydrate saturation (SCAS) typically exceed SCSS (e.g., Baker & Moretti, 2011). Thus, the elevated S contents in some matrix glasses of the 67 ka eruption and the two youngest eruptions are best explained by a pre-eruptive or syneruptive oxidation event, resulting in sulfate-bearing melts and higher content S solubilities.

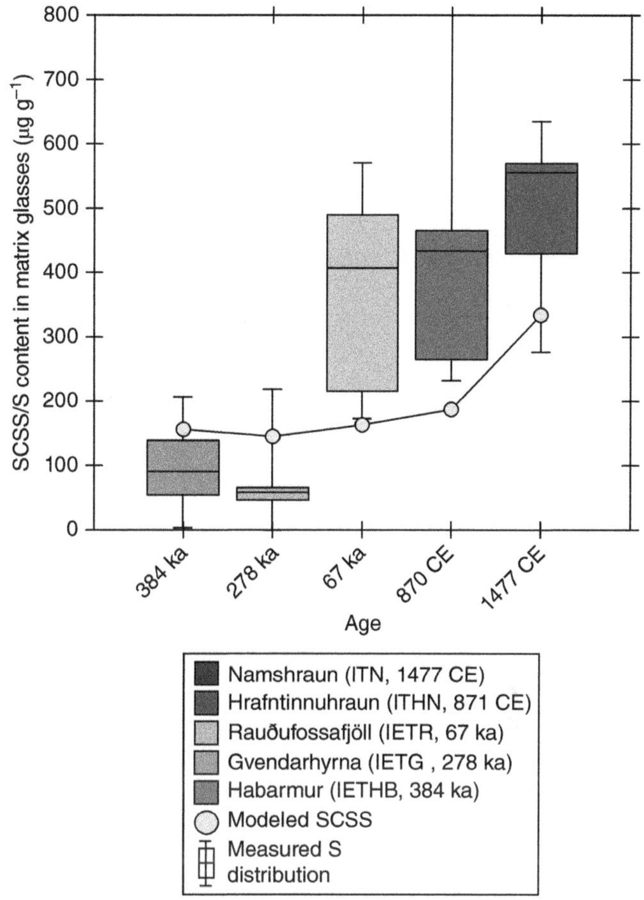

Figure 3.10 Sulfur contents in the matrix glasses compared to sulfide solubility at sulfide saturation (SCSS). Box and whisker plots are shown for the S contents measured in the matrix glasses of the samples from the five studied eruptions of Torfajökull. Circles: The sulfide solubility at sulfide saturation (SCSS) was calculated by using the Fortin et al. (2015) model (see section 3.5.2.1 for details). Black solid lines are to guide the eye; i.e., to illustrate the SCSS evolution through time. Whiskers extend to the 5th and 95th percentiles, boxes define the 25th to 75th percentiles, and median values bisect the boxes.

Torfajökull apatites show consistently low S values (~20–100 μg g^{-1} S) with the exception of the 67 ka unit, where S contents in apatite reach up to 570 μg g^{-1} S (Figure 3.7). To evaluate the S contents in the Torfajökull eruptive products, we applied experimentally derived apatite–melt partition coefficients for S in sulfide-saturated ($D^{S}_{ap/m}$ =0.46) and sulfide-undersaturated ($D^{S}_{ap/m}$ =0.18) reduced rhyolitic systems (Konecke, Fiege, Simon, & Konecke et al., 2017a). The calculations show that the S contents in the apatites of the two oldest units are consistent with average matrix-glass contents when assuming a sulfide saturated scenario (Figure 3.11) and with modeled SCSS values. This is consistent with the presence of anorthite-hosted pyrrhotite (FeS) inclusions in those oldest

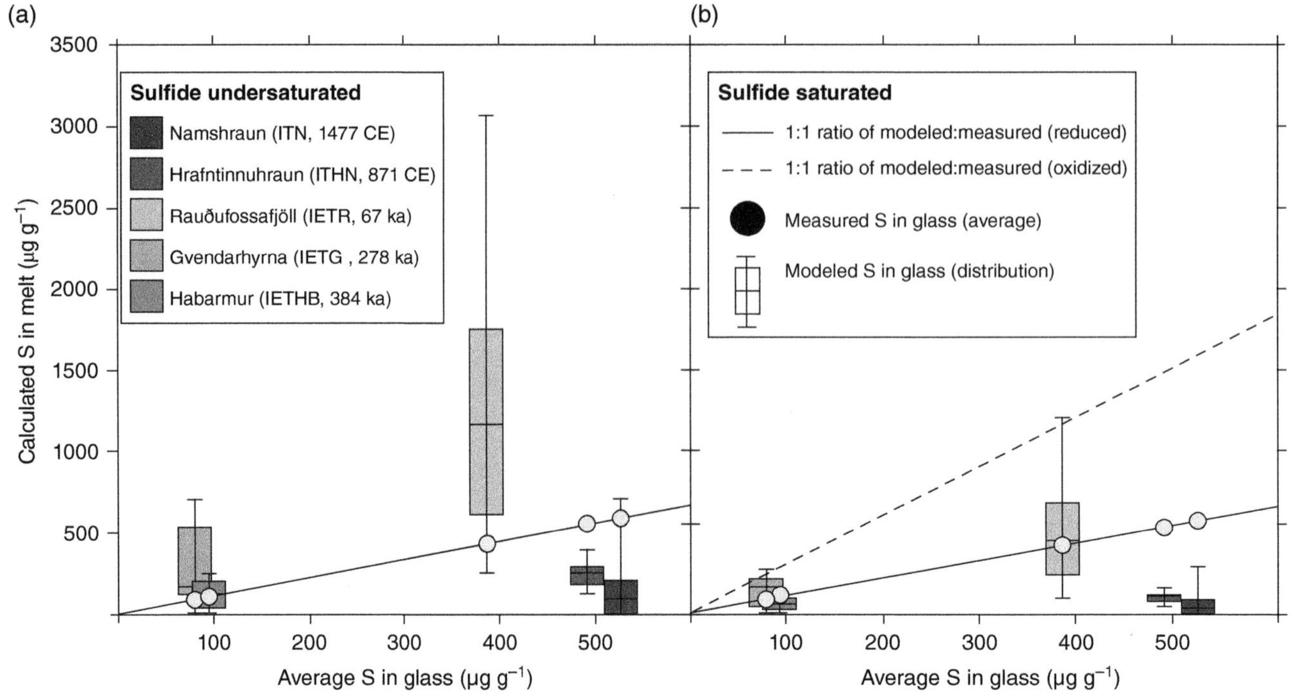

Figure 3.11 Estimation of the S contents in the matrix glasses based on Torfajökull apatite compositions. Apatite–melt partitioning coefficients determined by Konecke et al., 2017b and Konecke et al., 2017a were used to calculate the S contents in the melt at apatite formation (*y*-axis) and plotted against the average S contents measured in the matrix glass of each unit (*x*-axis). (a) Partition coefficient for sulfide-undersaturated conditions was applied ($D^{S}_{ap/m}$ = 0.18). (b) partition coefficient for sulfide-saturated conditions was applied ($D^{S}_{ap/m}$ = 0.46). The solid black lines mark a 1:1 trend. Values that plot near the solid lines are consistent with (a) a sulfide-saturated or (b) a sulfide-undersaturated scenario. The dashed line in (b) is based on ($D^{S}_{ap/m}$ = 1.41, applicable to oxidized rhyolitic systems (see section 3.5.2.1 for details). Whiskers extend to the 5th and 95th percentiles, boxes define the 25th to 75th percentiles, and median values bisect the boxes.

units. The two youngest units are better characterized by a sulfide-undersaturated scenario, based on the S content of the apatites and on the absence of observed sulfide phases. However, the S contents measured in the matrix glasses of the youngest two units required more oxidizing conditions and would favor higher S contents in the apatites, indicating that S may be added to the residual melt (represented by the matrix glass) prior to eruption but after apatite formation. A plausible explanation for the S contents of the youngest units involves syneruptive degassing of an S-bearing fluid. This degassing can cause late-stage oxidation of the residual melt (see Burgisser & Scaillet, 2007) and result in S contents in the matrix glasses that exceed those expected for reduced melts. It is also possible that higher S contents in matrix glasses result from the contribution of relatively more oxidized mafic magmas mixing with the felsic system.

The elevated S contents in many of the 67 ka apatites (47 of 56 analyses S > 80 μg g^{-1}) can not be produced through crystallization from reduced rhyolitic melt (sulfide saturated or sulfide undersaturated), if a reasonable SCSS (~160 μg g^{-1} in the melt) and a maximum $D^{S}_{ap/m}$ value (0.46) for reduced conditions are applied (Figure 3.12). Maximum $D^{S}_{ap/m}$ values are significantly higher for oxidized, sulfur saturated, rhyolitic melts relative to those in reduced environments (e.g., Konecke et al., 2017a; Parat & Holtz, 2005). Notably, experimental $D^{S}_{ap/m}$ values for oxidized rhyolitic systems may be as high as ~14, depending mainly on the S fugacity (Parat & Holtz, 2005).

By applying a $D^{S}_{ap/m}$ value of 1.41 determined by Konecke et al., 2017a for oxidized, sulfate-saturated rhyolites, most (34 out of 56) of the S contents measured in the 67 ka apatites are consistent with the S contents of the matrix glass (dashed line in Figure 3.11). Thus, the S partitioning behavior likely changes at 67 ka, because the substitution of S^{6+} at the P site is more favorable (equation 3.7), in comparison to the substation of S^{2-} at the anion column (equation 3.8) (Kim et al, 2017):

$$S^{6+} + Na^{+} = Ca^{2+} + P^{5+} \tag{3.7}$$

$$S^{2-} = 2F^{-} \tag{3.8}$$

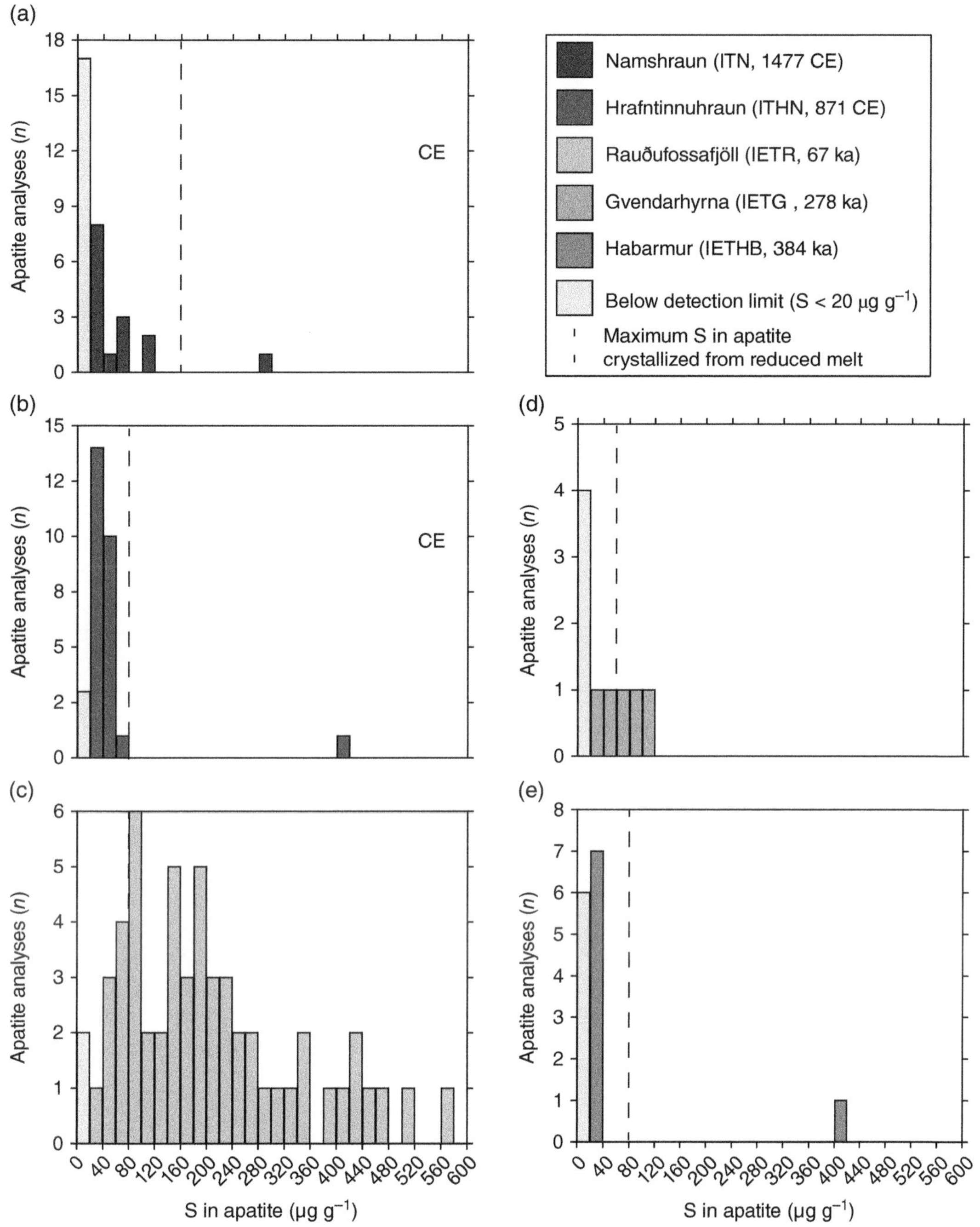

Figure 3.12 Histograms of the S contents in apatite. (a) 1477 CE unit (ITN); (b) 831 CE unit (ITHN); (c) 67 ka unit (IETR); (d) 278 ka unit (IETG); (e) 384 ka unit (IETHB). Measured values at or BDL (20 μg g^{-1} S) are presented in gray. Vertical dashed lines indicate the maximum possible S contents in apatites formed under reducing conditions from the matrix glasses of the different units. These thresholds were estimated using the SCSS values shown in Figure 3.8 and a $D^{S}_{ap/m}$ = value of 0.46 for reduced rhyolites (Konecke et al., 2017a; Konecke, Konecke et al., 2017b). For display purposes, the calculated maximum S contents were rounded to the next bin step (i.e.: ITN ~ 160 μg g^{-1} S; ITHN ~ 80 μg g^{-1} S; IETR ~ 80 μg g^{-1} S; IETG ~ 60 μg g^{-1} S; IETHB ~ 80 μg g^{-1} S). Clearly, S contents in the apatites from the 67 ka unit show the highest variability and many S contents recorded by the apatites of this unit require more oxidizing conditions (i.e., higher S solubility and $D^{S}_{ap/m}$ values; see section 3.5.3).

Sulfur contents in the matrix glass and in the apatites from the two oldest units suggest that S is predominantly present as sulfide in the system, consistent with redox conditions typical for Icelandic magmas. The data indicate that Torfajökull rhyolites were typically sulfide-saturated in the Pleistocene, oxidizing at 67 ka, and dynamic and variable in the Holocene. In the "intermediate" 67 ka unit both matrix glass and apatite S contents record an oxidation. The two youngest, Holocene units are best explained by a scenario in which apatite forms under reducing conditions from a sulfur-depleted melt prior to eruption, while mafic magma injection probably increased sulfur fugacity followed by syneruptive degassing and oxidation, which affected the matrix glass but not the apatite.

3.5.3. Evaluation of the 67 ka Eruption (IETR)

The 67 ka eruption is an inflection point in Torfajökull's history. This eruption took place during the transition from the last (Weichselian) glacial to interglacial climatic state, erupting partially subglacially and partially subaerially (Tuffen et al., 2002). The 67 ka edifices, which erupted in either one singular event or as multiple, closely spaced ring-fracture eruptions, account for the largest volume (26 km^3) of rhyolite exposed at Torfajökull (McGarvie, 1984). Recent geochemical work identified Torfajökull as the source of the Thórsmörk Ignimbrite, a deposit over 30 km away, suggesting that this glaciovolcanic event may have been even larger and more explosive than previously thought (Moles et al., 2019). Petrologically, these samples show the first signs of magmatic interaction with a more mafic source, a characteristic that defines historical Torfajökull eruptions. Geochemically, bulk-rock and glass compositions (Figure 3.5) show a distinct shift towards less-evolved (more mafic) compositions. Apatites crystallized after 67 ka are relatively REE-poor, and are significantly richer in Cl and S when compared to the previous eruptions (Figure 3.7).

To further evaluate the Cl contents in the apatites from the 67 ka eruption, we compared the apatite data from this study with published MI data (Owen et al., 2013) from the same subglacial edifice (Rauðufossafjöll) of the 67 ka eruption. The feldspar hosted MIs measured by Owen et al. (2013) show minimum Cl values of 800 µg g^{-1} Cl, reaching maximum values of 1600 µg g^{-1} Cl. Apatites from the 67 ka eruption have variable Cl concentrations that reach significantly higher values, ranging from 700 to 3500 µg g^{-1} Cl. The relatively higher concentration in apatite versus MI is unsurprising, as Cl exhibits compatible behavior in apatite (Piccoli & Candela, 2002).

A comparison of H–Cl variation in apatites and MIs reveals extraordinary similarities (Figure 3.13). Increasing Cl/H_2O in MIs (from 0.01 to 0.93; Owen et al., 2013) corresponding with decreasing H_2O content is directly related to the relationship between H_2O solubility in the melt, Cl fluid–melt partitioning, and pressure (e.g., Fiege et al., 2014; Witham et al., 2012). This H_2O–Cl degassing trend is confirmed by model calculations performed for a rhyolitic system using the VolatileCalc model (Newman & Lowenstern, 2002), an AST of 865°C, and a fluid–melt partitioning coefficient for Cl of 10 (Fiege et al., 2014); (Figure 3.13). In apatite, increasing Cl/OH (from 0.01 to 1.25) correlates with decreasing OH, following a similar exponential decay trend as the measured MIs and the modeled results (see equations in Figure 3.13b and d). Thus, MIs and apatites of the 67 ka unit seem to record the same degassing event. To our knowledge, this similarity in apatite-versus-MI volatile trends has never before been identified.

Apatites from the 67 ka eruption at Torfajökull have a relatively wide range of S concentrations, from values BDL to 570 µg g^{-1}, reflecting a complicated volatile system at this stage in Torfajökull's history. A degassing event (recorded by both MIs and apatites) in a S-bearing system such as Torfajökull can increase the oxygen fugacity of a magma by up to two log units above its original redox conditions (Burgisser & Scaillet, 2007). Considering that magma storage redox conditions at Torfajökull are estimated to be ~FMQ (Gunnarsson et al., 1998), such a degassing-induced oxidation event can easily change the S oxidation state in the melt from sulfide (S^{2-}) dominated to sulfate (S^{6+}) dominated (e.g., Jugo et al., 2010). As a consequence, sulfides (perhaps originating from mafic Veiðivötn injections) may be oxidized and dissolved. This will increase both the S concentration in the melt (Baker & Moretti, 2011) and $D^{S}_{ap/m}$, the apatite–melt partitioning coefficient of S (Konecke et al., 2017b; Konecke et al., 2017a; Parat et al., 2005). This combination explains the elevated S contents in the 67 ka matrix glass and apatite (section 3.5.2.1; Figures 3.11 and 3.12). In such a scenario, most apatites must form relatively late—after the injection of (eruption triggering?) basaltic magma, and during or after significant magma degassing. Late (syneruptive?) formation of apatite in the 67 ka unit may explain the observed discrepancy between high bulk-rock REE and low apatite REE compositions (see section 3.5.1).

3.6. CONCLUSIONS AND IMPLICATIONS

3.6.1. Magma Evolution at Torfajökull

Torfajökull—a historically active volcano with the greatest exposed volume of rhyolite in modern Iceland—has undergone a compositional "devolution" during its ~400 kyr eruptive history, transitioning from high-SiO_2

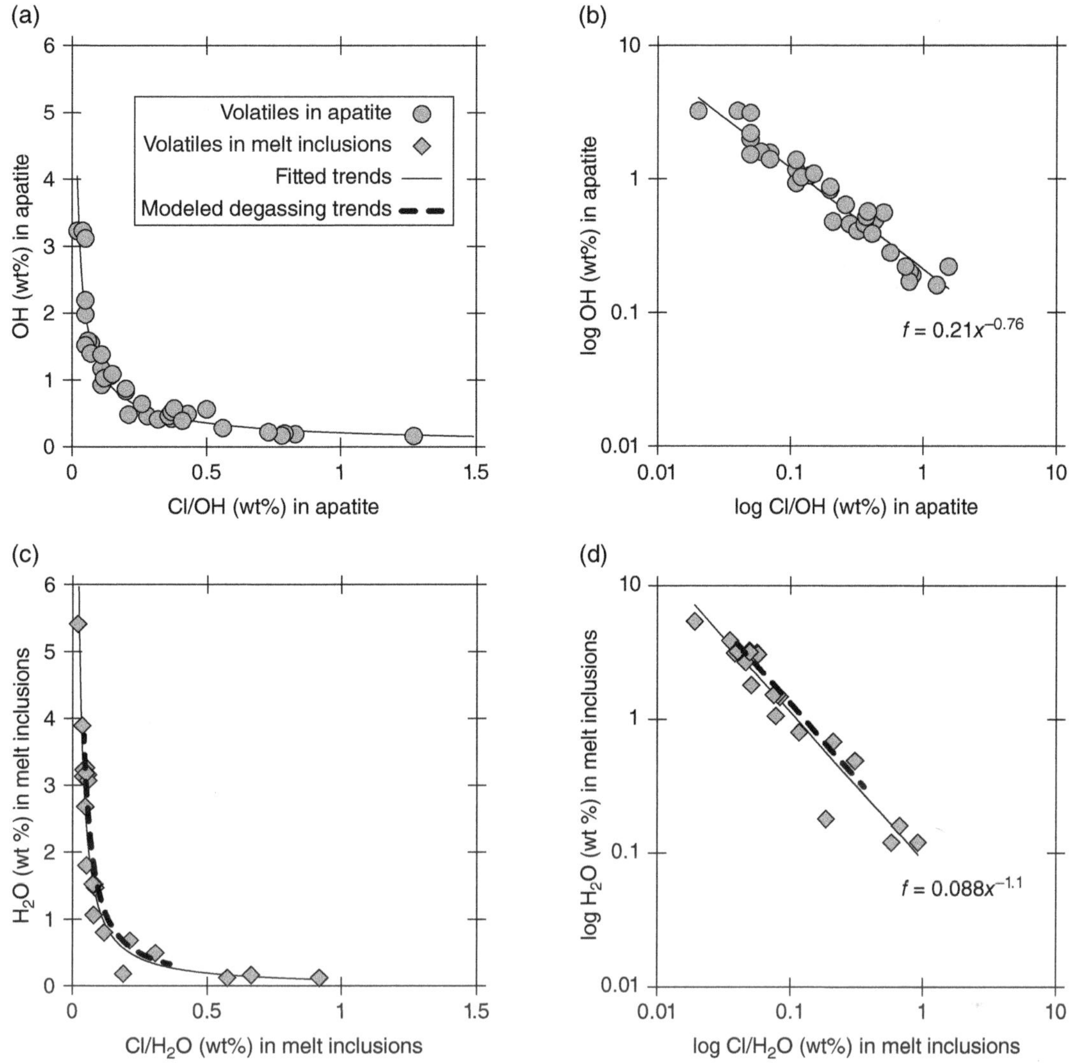

Figure 3.13 Degassing trends recorded by apatite and melt inclusions in the 67 ka eruption. (a) and (b) OH plotted against Cl/OH wt% ratios measured in apatites from sample IETR (67 ka eruption; this study); slope = 0.21*x* – 0.76. (c) and (d) H_2O plotted against Cl/H_2O wt% ratios measured by Owen et al. (2013) in melt inclusions (MIs) in the products of the 67 ka eruption; slope = 0.08 *x* – 1.1. Dashed lines were calculated using the VolatileCalc model (Newman & Lowenstern, 2002) and assuming a fluid–melt partitioning coefficient of 10 for Cl (Fiege et al., 2014); see section 3.5.3 for details.

peralkaline melts to lower SiO_2 metaluminous rhyolitic magmas. Previous studies have attributed this compositional change to increasing influence from the nearby Viedivötn Fissure Swarm as the EVZ propagates southward to Torfajökull's off-rift position (e.g., Gunnarsson et al., 1998; McGarvie et al., 2006; Zellmer et al., 2008). This study—the first to use apatite to conduct an intensive investigation of an Icelandic volcanic system—supports this interpretation and adds valuable insight into volatile and trace element evolution and behavior coinciding with environmental and compositional changes at Torfajökull.

The increase of the apatite saturation temperature over time is consistent with an increasing importance of mafic magma influence, as is the progressive decrease in apatite LREE content beginning around 67 ka. The distribution of other trace elements in apatites also shifts at ~67 ka from being high in SiO_2 and Na_2O, to being rich in FeO and MgO. Both Fe^{2+} and Mg^{2+} can be substituted for Ca^{2+} in the apatite structure (e.g., Pan & Fleet, 2002), and have been identified in significant amounts in apatites from the 67 ka unit and younger units (up to 1 wt% MgO, up to 2 wt% FeO). Their elevated concentrations can be correlated with changing concentrations in the melt. As mafic magma injection increases over time, Na_2O and SiO_2 concentrations decrease and MgO and FeO contents increase in the (hybridized) melts.

The relative variability of volatiles in glasses may be indicative of the heterogeneity of magmas 67 ka and younger due to magma mixing, or variable degrees of degassing within the samples. In particular, the elevated S contents in the matrix glasses and the apatites from the 67 ka eruptive products are best explained by degassing-induced magma oxidation (see Burgisser & Scaillet, 2007), resulting in the transition from a sulfide-dominated to a sulfate-dominated melt and apatite.

3.6.2. Apatite as a Tool for Assessing Magma Evolution

Apatite-focused studies can provide new perspectives of the trace element and volatile evolution of Icelandic rhyolites. Apatites from Torfajökull, and possibly Iceland more broadly, are unique in their ability to incorporate exceptionally high quantities of REE due to the compositions (e.g., high REE) and crystallization conditions (e.g., fO_2, high T) of their host rhyolites. For individual eruptions at Torfajökull, most notably the 67 ka eruption, syneruptively crystallized apatites can provide insights on degassing processes at the scale of individual eruptions, and individual crystals. Comparable H–Cl trends in apatite and MIs highlight the ability of apatite to record relative relationships in pre-eruptive volatile concentrations. Here, Cl concentrations in MIs and apatite show relatively lower rates of degassing of Cl, in comparison to H_2O.

Elevated S contents in apatites from 67 ka are evidence of a dramatic, syneruptive change in S oxidation states in the melt, from sulfide dominated to sulfate dominated. The redox sensitivity of S-in-apatite allows investigation of degassing-induced oxidation processes, i.e., processes that are not recorded in MI data. Hence, apatites have elucidated magmatic evolution and degassing processes at Torfajökull—not previously well understood—and these techniques can now be applied more broadly in Icelandic studies.

Apatite—in contrast to MIs—may contribute a more robust (less prone to alteration) and time-sensitive (core–rim zonation) perspective of pre-eruptive and syneruptive volatile and trace-element evolutions. Future studies may use high-resolution element mapping of apatite (e.g., via nano-SIMS) to better understand temporal volatile and REE evolution based on a single eruptive product. In this future work, experimentally determined trace-element partition coefficients (many of which already exist, some of which will need to be developed) will be an important prerequisite for quantitative assessment and modeling.

ACKNOWLEDGMENTS

This project was supported by US National Science Foundation Research grants to Adam Simon and A.F. (EAR: 1524394) and to Calvin Miller (EAR: 1220523), as well as a Faculty Research Grant to T.L.C. from the Lafayette College Academic Research Committee. We thank Calvin Miller, Lawrence Malinconico, Amanda Leaman, and John Wilson for their support and contributions during the development of this project, and Brian Konecke for his perspective during a preliminary review. To our knowledge, there are no real or perceived financial conflicts of interests for any of the authors.

SUPPLEMENTARY MATERIAL

The full analytical data set (EPMA and bulk XRF results) is provided in Supplementary Material Tables SM1–SM3 available at https://blogs.uoregon.edu/lissieconnors/datasets_iceland/

REFERENCES

Ancey, M., Bastenaire, F., & Tixier, R. (1979) Application of statistical methods in microanalysis. In F. Maurice, L. Meny, R. Tixier (Eds.), *Microanalysis and scanning electron microscopy* (pp. 319–343). Orsay, France: Les Editions de Physique.

Audétat, A., & Lowenstern, J. B. (2014) Melt inclusions. In H. D. Holland, K. K. Turekian (Eds.), *Treatise on geochemistry* (2nd edn, pp. 143–173). Amsterdam: Elsevier.

Baker, D. R., Freda, C., Brooker, R. A., & Scarlato, P. (2005). Volatile diffusion in silicate melts and its effects on melt inclusions. *Annals of Geophysics*, *48*(4–5).

Baker, D. R., & Moretti, R. (2011). Modeling the solubility of sulfur in magmas: a 50-year old geochemical challenge. *Reviews in Mineralogy and Geochemistry*, *73*(1), 167–213.

Belousova, E. A., Griffin, W. L., O'Reilly, S. Y., & Fisher, N. I. (2002). Apatite as an indicator mineral for mineral exploration: trace-element compositions and their relationship to host rock type. *Journal of Geochemical Exploration*, *76*(1), 45–69.

Bergþórsdóttir, I. A. (2018). *The role of apatite in Hekla magmas: trace element partitioning between minerals and melt*. Doctoral dissertation.

Bindeman, I., Gurenko, A., Carley, T., Miller, C., Martin, E., & Sigmarsson, O. (2012). Silicic magma petrogenesis in Iceland by remelting of hydrothermally altered crust based on oxygen isotope diversity and disequilibria between zircon and magma with implications for MORB. *Terra Nova*, *24*(3), 227–232.

Bjarnason, I. (2008), An Iceland hotspot saga. *Jökull*, *58*, 3–16.

Blake, S. (1984). Magma mixing and hybridization processes at the alkalic, silicic, Torfajökull central volcano triggered by tholeiitic Veidivötn fissuring, south Iceland. *Journal of Volcanology and Geothermal Research*, *22*(1–2), 1–31.

Boyce, J. W., & Hervig, R. L. (2008a). Apatite as a monitor of late-stage magmatic processes at Volcán Irazú, Costa Rica. *Contributions to Mineralogy and Petrology*, *157*(2), 135.

Boyce, J. W., & Hervig, R. L. (2008b). Magmatic degassing histories from apatite volatile stratigraphy. *Geology*, *36*(1), 63–66.

Boyce, J. W., Liu, Y., Rossman, G. R., Guan, Y., Eiler, J. M., Stolper, E. M., & Taylor, L. A. (2010). Lunar apatite with terrestrial volatile abundances. *Nature*, *466*(7305), 466.

Boyce, J. W., Tomlinson, S. M., McCubbin, F. M., Greenwood, J. P., & Treiman, A. H. (2014). The lunar apatite paradox. *Science*, *344*(6182), 400–402.

Brenan, J. M. (1993). Partitioning of fluorine and chlorine between apatite and aqueous fluids at high pressure and temperature: implications for the F and Cl content of high P-T fluids. *Earth and Planetary Science Letters*, *117*(1–2), 251–263.

Bucholz C.E., Gaetani G.A., Behn M.D., & Shimizu N. (2013) Post-entrapment modification of volatiles and oxygen fugacity in olivine-hosted melt inclusions. *Earth and Planetary Science Letters*, *374*, 145–155.

Burgisser, A., & Scaillet, B. (2007). Redox evolution of a degassing magma rising to the surface. *Nature*, *445*(7124), 194.

Carley, T.L., Miller, C.F., Fisher, C.M., Hanchar, J.M., Vervoort, J.D., Schmitt, A.K., et al. (2020). Silicic petrogenesis in Iceland through space and time: The isotopic record preserved in zircon and whole rocks. *The Journal of Geology*, *128*(1), 1–28. https://www.journals.uchicago.edu/doi/abs/10.1086/706261

Carley, T. L., Miller, C. F., Wooden, J. L., Bindeman, I. N., & Barth, A. P. (2011). Zircon from historic eruptions in Iceland: Reconstructing storage and evolution of silicic magmas. *Mineralogy and Petrology*, *102*(1–4), 135.

Carley, T. L., Miller, C. F., Wooden, J. L., Padilla, A. J., Schmitt, A. K., Economos, R. C., Bindeman, I. N., & Jordan, B. T. (2014), Iceland is not a magmatic analog for the Hadean: Evidence from the zircon record. *Earth and Planetary Science Letters*, *405*, 85–97.

Carmichael, I. S. E. (1964). The petrology of Thingmuli, a Tertiary volcano in eastern Iceland. *Journal of Petrology*, *5*(3), 435–460.

De Vivo, B, Lima, A., & Webster, J. (2005), Volatiles in magmatic-volcanic systems. *Elements*, *1*, 19–24.

Doherty, A. L., Webster, J. D., Goldoff, B. A., & Piccoli, P. M. (2014). Partitioning behavior of chlorine and fluorine in felsic melt–fluid(s)–apatite systems at 50 MPa and 850–950°C. *Chemical Geology*, *384*, 94–111.

Esposito, R., Hunter, J., Schiffbauer, J. D., Shimizu, N., & Bodnar, R. J. (2014). An assessment of the reliability of melt inclusions as recorders of the pre-eruptive volatile content of magmas. *American Mineralogist*, *99*(5–6), 976–998.

Fiege, A., Behrens, H., Holtz, F., & Adams, F. (2014). Kinetic vs. thermodynamic control of degassing of H_2O–S±Cl-bearing andesitic melts. *Geochimica et Cosmochimica Acta*, *125*, 241–264.

Fleet, M. E., & Pan, Y. (1995). Site preference of rare earth elements in fluorapatite. *American Mineralogist*, *80*(3–4), 329–335.

Fortin, M. A., Riddle, J., Desjardins-Langlais, Y., & Baker, D. R. (2015). The effect of water on the sulfur concentration at sulfide saturation (SCSS) in natural melts. *Geochimica et Cosmochimica Acta*, *160*, 100–116.

Frost, B. R., & Frost, C. D. (2008). A geochemical classification for feldspathic igneous rocks. *Journal of Petrology*, *49*(11), 19551969.

Furman, T., Meyer, P. S., & Frey, F. (1992). Evolution of Icelandic central volcanoes: evidence from the Austurhorn intrusion, southeastern Iceland. *Bulletin of Volcanology*, *55*(1–2), 45–62.

Goldoff, B., Webster, J. D., & Harlov, D. E. (2012). Characterization of fluor-chlorapatites by electron probe microanalysis with a focus on time-dependent intensity variation of halogens. *American Mineralogist*, *97*(7), 1103–1115.

Grimes, C. B., John, B. E., Kelemen, P. B., Mazdab, F. K., Wooden, J. L., Cheadle, M. J., Hanghøj, K., & Schwartz, J. J. (2007). Trace element chemistry of zircons from oceanic crust: A method for distinguishing detrital zircon provenance. *Geology*, *35*(7), 643–646.

Grimes, C. B., Wooden, J. L., Cheadle, M. J., & John, B. E. (2015) "Fingerprinting" tectono-magmatic provenance using trace elements in igneous zircon. *Contributions to Mineralogy and Petrology*, *170*, 1–26.

Gualda, G. A., Ghiorso, M. S., Lemons, R. V., & Carley, T. L. (2012). Rhyolite-MELTS: a modified calibration of MELTS optimized for silica-rich, fluid-bearing magmatic systems. *Journal of Petrology*, *53*(5), 875–890.

Gunnarsson, B., Marsh, B. D., & Taylor Jr, H. P. (1998). Generation of Icelandic rhyolites: silicic lavas from the Torfajökull central volcano. *Journal of Volcanology and Geothermal Research*, *83*(1–2), 1–45. https://www.sciencedirect.com/science/article/pii/S0377027398000171

Hardarson, B. S., & Godfrey Fitton, J. (1997). Mechanisms of crustal accretion in Iceland. *Geology*, *25*(11), 1043–1046.

Harrison, T. M., & Watson, E. B. (1984). The behavior of apatite during crustal anatexis: equilibrium and kinetic considerations. *Geochimica et Cosmochimica Acta*, *48*(7), 1467–1477.

Harlov, D. E. (2015). Apatite: A fingerprint for metasomatic processes. *Elements*, *11*(3), 171–176.

Head, E. M., Shaw, A. M., Wallace, P. J., Sims, K. W., & Carn, S. A. (2011). Insight into volatile behavior at Nyamuragira volcano (DR Congo, Africa) through olivine-hosted melt inclusions. *Geochemistry, Geophysics, Geosystems*, *12*(10). https://doi:10.1029/2011GC00369

Hughes, J. M., & Rakovan, J. (2002). The crystal structure of apatite, $Ca_5(PO_4)_3$ (F, OH, Cl). *Reviews in Mineralogy and Geochemistry*, *48*(1), 1–12.

Jarosewich, E., Nelen, J. A., & Norberg, J. A. (1980). Reference samples for electron microprobe analysis. *Geostandards Newsletter*, *4*(1), 43–47.

Johnson E. R., Kamenetsky V. S., McPhie J., & Wallace, P. J. (2011). Degassing of the H_2O-rich rhyolites of the Okataina Volcanic Center, Taupo Volcanic Zone, New Zealand. *Geology*, *39*(4), 311–314.

Jónasson, K. (1994). Rhyolite volcanism in the Krafla central volcano, north-east Iceland. *Bulletin of Volcanology*, *56*(6–7), 516–528.

Jónasson, K. (2007). Silicic volcanism in Iceland: Composition and distribution within the active volcanic zones. *Journal of Geodynamics*, *43*(1), 101–117.

Jugo, P. J., Wilke, M., & Botcharnikov, R. E. (2010). Sulfur K-edge XANES analysis of natural and synthetic basaltic glasses: Implications for S speciation and S content as function of oxygen fugacity. *Geochimica et Cosmochimica Acta*, *74*(20), 5926–5938.

Ketcham, R. A. (2015). Calculation of stoichiometry from EMP data for apatite and other phases with mixing on monovalent anion sites. *American Mineralogist*, *100*(7), 1620–1623.

Kim, Y. J., Konecke, B., Fiege, A., Simon, A. C., & Becker, U. (2017). An ab-initio study on sulfur-bearing apatite: Energetics and geometry upon incorporation. *American Mineralogist*, *102*, 1646–1656. https://doi:10.2138/am-2017-6044

Konecke, B. A., Fiege, A., Simon, A. C., & Holtz, F. (2017). Cryptic metasomatism during late-stage lunar magmatism implicated by sulfur in apatite. *Geology*, *45*(8), 739–742.

Konecke, B. A., Fiege, A., Simon, A. C., Parat, F., & Stechern, A. (2017). Co-variability of S^{6+}, S^{4+}, and S^{2-} in apatite as a function of oxidation state: Implications for a new oxybarometer. *American Mineralogist*, *102*(3), 548–557.

Larsen, G. (1984). Recent volcanic history of the Veidivötn fissure swarm, southern Iceland—an approach to volcanic risk assessment. *Journal of Volcanology and Geothermal Research*, *22*(1–2), 33–58.

Lloyd, A. S., Plank, T., Ruprecht, P., Hauri, E. H., & Rose, W. (2013). Volatile loss from melt inclusions in pyroclasts of differing sizes. *Contributions to Mineralogy and Petrology*, *165*(1), 129–153.

Lowenstern, J. B. (1995). Applications of silicate-melt inclusions to the study of magmatic volatiles. In J. F. H. Thompson (Ed.), *Magmas, fluids and ore deposits* (Short Course Vol. 23, pp. 71–99). Quebec City, Quebec: Mineralogical Association of Canada.

Lowenstern, J. B. (2003). Melt inclusions come of age: volatiles, volcanoes, and Sorby's legacy. *Developments in Volcanology*, *5*, 1–21.

Macdonald, R., McGarvie, D. W., Pinkerton, H., Smith, R. L., & Palacz, A. (1990). Petrogenetic evolution of the Torfajökull Volcanic Complex, Iceland I. Relationship between the magma types. *Journal of Petrology*, *31*(2), 429–459.

Macdonald, R., Sparks, R. S. J., Sigurdsson, H., Mattey, D. P., McGarvie, D. W., & Smith, R. L. (1987). The 1875 eruption of Askja volcano, Iceland: Combined fractional crystallization and selective contamination in the generation of rhyolitic magma. *Mineralogical Magazine*, *51*(360), 183–202.

Marks, M. A., Wenzel, T., Whitehouse, M. J., Loose, M., Zack, T., Barth, M., et al. (2012). The volatile inventory (F, Cl, Br, S, C) of magmatic apatite: An integrated analytical approach. *Chemical Geology*, *291*, 241–255.

Martin, E., & Sigmarsson, O. (2007). Crustal thermal state and origin of silicic magma in Iceland: the case of Torfajökull, Ljósufjöll and Snæfellsjökull volcanoes. *Contributions to Mineralogy and Petrology*, *153*(5), 593–605.

Martin, E., & Sigmarsson, O. (2010). Thirteen million years of silicic magma production in Iceland: links between petrogenesis and tectonic settings. *Lithos*, *116*(1), 129–144.

Mathez, E. A., & Webster, J. D. (2005). Partitioning behavior of chlorine and fluorine in the system apatite-silicate melt-fluid. *Geochimica et Cosmochimica Acta*, *69*(5), 1275–1286.

McCubbin, F. M., Jolliff, B. L., Nekvasil, H., Carpenter, P. K., Zeigler, R. A., Steele, A., Elardo, S.M., & Lindsley, D. H. (2011). Fluorine and chlorine abundances in lunar apatite: Implications for heterogeneous distributions of magmatic volatiles in the lunar interior. *Geochimica et Cosmochimica Acta*, *75*(17), 5073–5093.

McDonough, W. F., & Sun, S. S. (1995). The composition of the Earth. *Chemical Geology*, *120*(3–4), 223–253.

McGarvie, D. W. (1984). Torfajökull: a volcano dominated by magma mixing. *Geology*, *12*(11), 685–688.

McGarvie, D. W., Burgess, R., Tindle, A. G., Tuffen, H., & Stevenson, J. A. (2006). Pleistocene rhyolitic volcanism at Torfajökull, Iceland; eruption ages, glaciovolcanism, and geochemical evolution, Jökull, *56*, 57–75.

McGarvie, D. W., MacDonald, R., Pinkerton, H., & Smith, R. L. (1990). Petrogenetic evolution of the Torfajökull Volcanic Complex, Iceland II. The role of magma mixing. *Journal of Petrology*, *31*(2), 461–481.

Moles, J. D., McGarvie, D., Stevenson, J. A., Sherlock, S. C., Abbott, P. M., Jenner, F. E., & Halton, A. M. (2019). Widespread tephra dispersal and ignimbrite emplacement from a subglacial volcano (Torfajökull, Iceland). *Geology*, *47*(6), 577–580. https://doi.org/10.1130/G46004.1

Moorbath, S., Sigurdsson, H., & Goodwin, R. (1968). K/Ar ages of the oldest exposed rocks in Iceland. *Earth and Planetary Science Letters*, *4*(3), 197–205.

Moore, L. R., Gazel, E., Tuohy, R., Lloyd, A. S., Esposito, R., Steele-MacInnis, M., et al. (2015). Bubbles matter: An assessment of the contribution of vapor bubbles to melt inclusion volatile budgets. *American Mineralogist*, *100*(4), 806–823.

Mørk, M. B. E. (1984). Magma mixing in the post-glacial Veidivötn fissure eruption, southeast Iceland: a microprobe study of mineral and glass variations. *Lithos*, *17*, 55–75.

Newman, S., & Lowenstern, J. B. (2002). VolatileCalc: a silicate melt–H_2O–CO_2 solution model written in Visual Basic for excel. *Computers and Geosciences*, *28*(5), 597–604.

Nicholson, H., Condomines, M., Fitton, J. G., Fallick, A. E., Grönvold, K., & Rogers, G. (1991). Geochemical and isotopic evidence for crustal assimilation beneath Krafla, Iceland. *Journal of Petrology*, *32*(5), 1005–1020.

Oppenheimer, C., Pyle, D. M., & Barclay, J. (Eds.). (2003). *Volcanic degassing* (Special Publication 213). London: Geological Society.

Owen, J., Tuffen, H., & McGarvie, D. W. (2012). Using dissolved H_2O in rhyolitic glasses to estimate palaeo-ice thickness during a subglacial eruption at Bláhnúkur (Torfa*Jökull*, Iceland). *Bulletin of Volcanology*, *74*(6), 1355–1378.

Owen, J., Tuffen, H., & McGarvie, D. W. (2013). Explosive subglacial rhyolitic eruptions in Iceland are fuelled by high magmatic H_2O and closed-system degassing. *Geology*, *41*(2), 251–254.

Pálmason, G., & Sæmundsson, K. (1974). Iceland in relation to the Mid-Atlantic Ridge. *Annual Review of Earth and Planetary Sciences*, *2*(1), 25–50.

Pan, Y., & Fleet, M. E. (2002). Compositions of the apatite-group minerals: substitution mechanisms and controlling factors. *Reviews in Mineralogy and Geochemistry*, *48*(1), 13–49.

Parat, F., Dungan, M. A., & Streck, M. J. (2002). Anhydrite, pyrrhotite, and sulfur-rich apatite: tracing the sulfur evolution of an Oligocene andesite (Eagle Mountain, CO, USA). *Lithos*, *64*(3–4), 63–75.

Parat, F., & Holtz, F. (2005). Sulfur partition coefficient between apatite and rhyolite: the role of bulk S content. *Contributions to Mineralogy and Petrology*, *150*(6), 643–651.

Parat, F., Holtz, F., & Streck, M. J. (2011). Sulfur-bearing magmatic accessory minerals. *Reviews in Mineralogy and Geochemistry*, *73*(1), 285–314.

Peng, G., Luhr, J. F., & McGee, J. J. (1997). Factors controlling sulfur concentrations in volcanic apatite. *American Mineralogist*, *82*(11–12), 1210–1224.

Piccoli, P., & Candela, P. (1994). Apatite in felsic rocks; a model for the estimation of initial halogen concentrations in the Bishop Tuff (Long Valley) and Tuolumne Intrusive Suite (Sierra Nevada Batholith) magmas. *American Journal of Science*, *294*(1), 92–135.

Piccoli, P. M., & Candela, P. A. (2002). Apatite in igneous systems. *Reviews in Mineralogy and Geochemistry*, *48*(1), 255–292. https://doi.org/10.2138/rmg.2002.48.6

Pichavant, M., Montel, J. M., & Richard, L. R. (1992). Apatite solubility in peraluminous liquids: Experimental data and an extension of the Harrison–Watson model. *Geochimica et Cosmochimica Acta*, *56*(10), 3855–3861.

Portnyagin, M., Hoernle, K., Plechov, P., Mironov, N., & Khubunaya, S. (2007). Constraints on mantle melting and composition and nature of slab components in volcanic arcs from volatiles (H_2O, S, Cl, F) and trace elements in melt inclusions from the Kamchatka Arc. *Earth and Planetary Science Letters*, *255*(1), 53–69.

Portnyagin, M., Hoernle, K., Storm, S., Mironov, N., van den Bogaard, C., & Botcharnikov, R. (2012). H_2O-rich melt inclusions in fayalitic olivine from Hekla volcano: Implications for phase relationships in silicic systems and driving forces of explosive volcanism on Iceland. *Earth and Planetary Science Letters*, *357*, 337–346.

Prestvik, T., Goldberg, S., Karlsson, H., & Grönvold, K. (2001). Anomalous strontium and lead isotope signatures in the off-rift Öræfajökull central volcano in south-east Iceland: evidence for enriched endmember (s) of the Iceland mantle plume? *Earth and Planetary Science Letters*, *190*(3), 211–220.

Prowatke S., & Klemme S. (2006). Trace element partitioning between apatite and silicate melts. *Geochimica et Cosmochimica Acta*, *70*, 4513–4527

Riker, J., Humphreys, M. C., Brooker, R. A., & De Hoog, J. C. (2018). First measurements of OH–C exchange and temperature-dependent partitioning of OH and halogens in the system apatite–silicate melt. *American Mineralogist*, *103*(2), 260–270.

Roeder, P. L., MacArthur, D., Ma, X. P., Palmer, G. R., & Mariano, A. N. (1987). Cathodoluminescence and microprobe study of rare-earth elements in apatite. *American Mineralogist*, *72*(7–8), 801–811.

Rønsbo, J. G. (1989). Coupled substitutions involving REEs and Na and Si in apatites in alkaline rocks from the Ilimaussaq intrusion, South Greenland, and the petrological implications. *American Mineralogist*, *74*(7–8), 896–901.

Sadove, G., Konecke, B., Fiege, A., & Simon, A. C. (2019). Structurally bound S^{2-}, S^{1-}, S^{4+}, S^{6+} in terrestrial apatite: The redox evolution of hydrothermal fluids at the Phillips mine, New York, USA. *Ore Geology Reviews*, *107*, 1084–1096. https://doi.org/10.1016/j.oregeorev.2019.03.033

Sæmundsson, K., & Friðleifsson, G. Ó. (2001). *Geological and geothermal maps of the Torfajökull area* (Report OS-2001, 36, 118). Reykjavik: Orkustofnun.

Sæmundsson, K. (1979). Outline of the geology of Iceland. *Jökull*, *29*, 7–28.

Schattel, N., Portnyagin, M., Golowin, R., Hoernle, K., & Bindeman, I. (2014). Contrasting conditions of rift and off-rift silicic magma origin on Iceland. *Geophysical Research Letters*, *41*(16), 5813–5820.

Schilling, J. G. (1973). Iceland mantle plume: geochemical study of Reykjanes Ridge. *Nature*, *242*(5400), 565.

Scott, J., Humphreys, M., Mather, T., Pyle, D., & Stock, M., (2015) Insights into the behaviour of S, F, and Cl at Santiaguito Volcano, Guatemala, from apatite and glass, *Lithos*, *232*, 375–394.

Shannon, R. D. (1976). Revised effective ionic radii and systematic studies of interatomic distances in halides and chalcogenides. *Acta Crystallographica Section A: Crystal Physics, Diffraction, Theoretical and General Crystallography*, *32*(5), 751–767.

Sigmarsson, O., Hémond, C., Condomines, M., Fourcade, S., & Oskarsson, N. (1991). Origin of silicic magma in Iceland revealed by Th isotopes. *Geology*, *19*(6), 621–624.

Sigurdsson, H. (1977). Generation of Icelandic rhyolites by melting of plagiogranites in the oceanic layer. *Nature*, *269*(5623), 25.

Sigurdsson, H., & Sparks, R. S. J. (1981). Petrology of rhyolitic and mixed magma ejecta from the 1875 eruption of Askja, Iceland. *Journal of Petrology*, *22*(1), 41–84.

Stock, M. J., Humphreys, M., Smith, V. C., Johnson, R. D., & Pyle, D. M. (2015). New constraints on electron-beam induced halogen migration in apatite. *American Mineralogist*, *100*(1), 281–293.

Stock, M. J., Humphreys, M. C., Smith, V. C., Isaia, R., & Pyle, D. M. (2016). Late-stage volatile saturation as a potential trigger for explosive volcanic eruptions. *Nature Geoscience*, *9*(3), 249.

Stoppa, F., & Liu, Y. (1995). Chemical composition and petrogenetic implications of apatites from some ultra-alkaline Italian rocks. *European Journal of Mineralogy*, *7*(2), 391–402.

Stormer, J., Pierson, M. L., & Tacker, R. C. (1993). Variation of F and Cl X-ray intensity due to anisotropic diffusion in apatite. *American Mineralogist*, *78*, 641–648.

Streck, M. J., & Dilles, J. H. (1998). Sulfur evolution of oxidized arc magmas as recorded in apatite from a porphyry copper batholith. *Geology*, *26*(6), 523–526.

Thordarson T., & Hoskuldsson A. (2002) *Iceland. Classic geology in Europe* (Vol. 3). Harpenden, UK: Terra Publishing.

Thordarson, T., & Larsen, G. (2007). Volcanism in Iceland in historical time: Volcano types, eruption styles and eruptive history. *Journal of Geodynamics*, *43*(1), 118–152.

Tuffen, H., McGarvie, D. W., Gilbert, J. S., & Pinkerton, H. (2002). Physical volcanology of a subglacial-to-emergent rhyolitic tuya at Rauðufossafjöll, Torfajökull, Iceland. In J. L. Smellie, M. G. Chapman (Eds.), *Volcano–ice interaction on Earth and Mars* (Special Publication 202, pp. 213–236). London: Geological Society.

Van Hoose, A. E., Streck, M. J., Pallister, J. S., & Wälle, M. (2013). Sulfur evolution of the 1991 Pinatubo magmas based on apatite. *Journal of Volcanology and Geothermal Research*, *257*, 72–89.

Vink, G. E. (1984). A hotspot model for Iceland and the Vøring Plateau. *Journal of Geophysical Research: Solid Earth*, *89*(B12), 9949–9959.

Vogt, P. R. (1971). Asthenosphere motion recorded by the ocean floor south of Iceland. *Earth and Planetary Science Letters*, *13*(1), 153–160.

Walker, G. P. L. (1963). The Breiddalur central volcano, eastern Iceland. *Quarterly Journal of the Geological Society*, *119*(1–4), 29–63.

Wallace, P. J., Kamenetsky, V. S., & Cervantes, P. (2015). Melt inclusion CO_2 contents, pressures of olivine crystallization, and the problem of shrinkage bubbles. *American Mineralogist*, *100*(4), 787–794.

Wallace, P. J., Plank, T., Edmonds, M., & Hauri, E. H. (2015). Volatiles in magmas. In H. Sigurdsson, B. Houghton, S. McNutt, H. Rymer, J. Stix (Eds.), *The Encyclopedia of Volcanoes* (2nd edn; pp. 163–183). Amsterdam: Elsevier.

Watson, E. B. (1979). Zircon saturation in felsic liquids: experimental results and applications to trace element geochemistry. *Contributions to Mineralogy and Petrology*, *70*(4), 407–419.

Watson, E. B., & Green, T. H. (1981). Apatite/liquid partition coefficients for the rare earth elements and strontium. *Earth and Planetary Science Letters*, *56*, 405–421.

Webster, J. D., Goldoff, B. A., Flesch, R. N., Nadeau, P. A., & Silbert, Z. W. (2017). Hydroxyl, Cl, and F partitioning between high-silica rhyolitic melts–apatite–fluid(s) at 50–200 MPa and 700–1000°C. *American Mineralogist*, *102*(1), 61–74.

Webster, J. D., Kinzler, R. J., & Mathez, E. A. (1999). Chloride and water solubility in basalt and andesite melts and implications for magmatic degassing. *Geochimica et Cosmochimica Acta*, *63*(5), 729–738.

Webster, J. D., & Piccoli, P. M. (2015). Magmatic apatite: A powerful, yet deceptive, mineral. *Elements*, *11*(3), 177–182.

Webster, J. D., Tappen, C. M., & Mandeville, C. W. (2009) Partitioning behavior of chlorine and fluorine in the system apatite–melt–fluid. II: Felsic silicate systems at 200 MPa. *Geochimica et Cosmochimica Acta*, *73*, 559–581.

Witham, F., Blundy, J., Kohn, S. C., Lesne, P., Dixon, J., Churakov, S. V., & Botcharnikov, R. (2012). SolEx: A model for mixed COHSCl-volatile solubilities and exsolved gas compositions in basalt. *Computers and Geosciences*, *45*, 87–97.

Wood, D. A. (1978). Major and trace element variations in the Tertiary lavas of eastern Iceland and their significance with respect to the Iceland geochemical anomaly. *Journal of Petrology*, *19*(3), 393–436.

Zellmer, G. F., Rubin, K. H., Grönvold, K., & Jurado-Chichay, Z. (2008). On the recent bimodal magmatic processes and their rates in the Torfajökull–Veidivötn area, Iceland. *Earth and Planetary Science Letters*, *269*(3), 388.

4

Control of Magma Plumbing Systems on Long-Term Eruptive Behavior of Sakurajima Volcano, Japan: Insights from Crystal-Size-Distribution Analysis

Shunsuke Yamashita[1] and Atsushi Toramaru[2]

ABSTRACT

We conducted a crystal-size-distribution (CSD) analysis of plagioclase phenocrysts of four historical lavas after the 15th century from Sakurajima volcano, southern Kyushu, Japan. The CSD analyses of type A (lower An content in core) and type B (higher An content in core) phenocrysts show that the slopes and intercepts of the former were nearly constant over five centuries and that those of the latter substantially increase with time. According to the open-system model of CSD under steady-state conditions, the increases in both the slope and intercept indicate that the effective growth rate of type A phenocrysts was nearly constant in a mixed felsic magma chamber (FMC), whereas that of type B phenocrysts systematically decreased in a mafic magma chamber (MMC) over five centuries. The effective growth rate can be connected to the ascent rate, or the supply rate, of magmas from the mantle. We infer that both the supply rate of mantle-derived mafic magma to MMC and the input rate from MMC to FMC increase with time. We found from CSD that this supply rate of mafic magmas correlates with those from geological data (volumes and intervals between eruptions) for large historical eruptions, suggesting that the supply rate from the mantle controls the triggering of eruption through the excess-volume condition.

4.1. INTRODUCTION

It is important to study the correlation between the long-term evolution of magma plumbing systems inferred from petrological studies and the long-term changes in eruption characteristics such as the dormancy period, volume, and style determined from geological studies. These correlations provide insight into what controls eruption characteristics in terms of long-term behavior. Sakurajima volcano, southern Kyushu, Japan, has produced four historical eruptions, including Plinian eruptions followed by lava flows over the past 550 years (Bunmei eruption, 1471–1476; An-ei eruption, 1779–1782; Taisho eruption, 1914–1915). The latest Showa eruption (1946) consisted mainly of effused lava flows. Sakurajima volcano is one of the most useful volcanoes for examining the above correlation because it exhibits simple repetitive eruptions typical of intermediate stratovolcanoes and is well studied in terms of petrology and geology. In addition, intensive geophysical, petrological, and geochemical monitoring of current eruptions (explosive eruptions occurred more than 700 times in 2015) have been conducted.

Previous studies have clearly shown that magma mixing has occurred and, on the basis of the whole-rock and phenocryst mineral chemistries, that the ratios of two end-member magmas systematically changed during the four historical eruptions (e.g., Nakagawa et al., 2011; Yanagi et al., 1991). More recently, since the 1980s, the basaltic

[1] *Department of Earth and Planetary Sciences, Kyushu University, Fukuoka, Japan. Now at Idemitsu Kosan Co., Ltd., Tokyo, Japan*
[2] *Department of Earth and Planetary Sciences, Kyushu University, Fukuoka, Japan*

Dynamic Magma Evolution, Geophysical Monograph 254, First Edition. Edited by Francesco Vetere.

components have become more significant than for the older historical eruptions. Only a few textural studies have been carried out to in order understand eruption dynamics, such as estimations of decompression rate and microlite crystallization processes in the conduit (Miwa et al., 2013; Miwa & Toramaru, 2013; Nakamura, 2006). However, there have been no textural studies of phenocrysts for the historical eruptions, which can be expected to provide significant information on the evolution of the plumbing systems and dynamics of the magma chambers. This lack of textural studies is probably due to the difficulty in separately analyzing several types of phenocrysts, which record the different origins in terms of their compositions. In turn, we might be able to determine detailed processes in different magma plumbing systems if we analyze phenocrysts by separating their origins. Thus, we attempt to quantitatively study the textural characteristics for phenocrysts, i.e., crystal size distribution (CSD), in order to examine the correlation between the characteristics of magma plumbing systems, as recorded in CSDs for different phenocryst origins, and eruption characteristics.

In addition to Sakurajima volcano, a few other studies have dealt with magma mixing by using the CSD method (e.g., Higgins, 1996; Tomiya & Takahashi, 1995). Since Marsh (1988) introduced the CSD method to petrology it has been recognized as being useful for characterizing various types of igneous rocks (e.g., Cashman & Marsh, 1988; Mangan, 1990; Piochi et al., 2005; Preece et al., 2013; Vinet & Higgins, 2011). In this study we apply CSD analysis of plagioclase phenocrysts to the historical lavas of Sakurajima volcano. The key requirement is that phenocrysts with different chemical and textural characteristics coexist in the same size ranges. Therefore, prior to CSD analysis it is necessary to classify plagioclase phenocrysts in terms of their composition and texture. On the basis of this classification, the CSDs of the individual types of plagioclase phenocrysts are measured, from which a single crystallization model can be determined. First, we present the results of the textural observations and compositional analysis, and then classify plagioclase phenocryst types in lavas from four historical eruptions (Bunmei, An-ei, Taisho, and Showa). Second, we report the results of the CSD analysis for each type. Third, we interpret the CSDs for each type by using an open-system model. Finally, we discuss the correlation between the dynamics of magma plumbing systems and the CSD data, as well as consider implications for the long-term behavior of eruptive activity.

4.2. BACKGROUND

4.2.1. Eruptive History

Sakurajima volcano is a post-caldera volcano that formed 13 ka at the southern rim of the Aira caldera, which formed from an eruption 29 ka (Fukuyama, 1978). Volcanic activity has been very intense since the 1980s, with explosive eruptions occurring more than 1000 times in 2015, approximately 150 times in 2016, 400 times in 2017, and 400 times in 2018, and have mainly been Vulcanian eruptions emitting ash. Sakurajima volcano consists of two active summit craters located parallel to its south–north trend: Kitadake is the northern summit crater and Minamidake is the southern summit crater. Since the 15th century the volcano has had four major eruptions: Bunmei, 1471–1476; An-ei, 1779–1782; Taisho 1914–1915; and Showa, 1946. The Bunmei, An-ei, and Taisho eruptions first produced pumice from Plinian eruptions from newly formed flank vents, and subsequently produced lava flows. The Showa eruption was dominated by lava flows from the crater, following an initial phase producing frequent emissions of small amounts of ash for approximately 3 months (Figure 4.1). Explosive Strombolian and Vulcanian eruptions have frequently occurred from Minamidake crater since 1955,

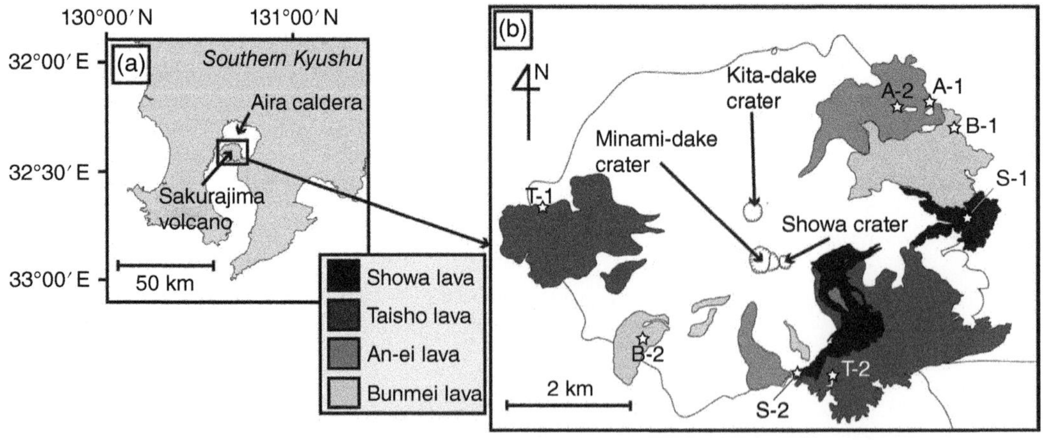

Figure 4.1 (a)Location of Sakurajima Volcano. (b) Geological map of Sakurajima volcano for subaerial lava after the 15th century. [Modified after Kobayashi et al. (2013).] Sampling points are indicated by star symbols.

but since 2006 the dominant explosive activities have shifted to the Showa crater.

Thus, eruptive styles changed from large-volume Plinian eruptions with lava flows (ca. 1 km^3 DRE—dense rock equivalent) in three eruptions (Bunmei, An-ei, Taisho) to small-volume frequent eruptions (one event $< 10^{-3} km^3$) after the Showa eruption in historical time. The total DRE volumes (tephra) are 1.2(0.5) km^3 (Bunmei), 2.1(0.4) km^3 (An-ei), 2.0(0.5) km^3 (Taisho), 0.1(0.02) km^3 (Showa) (e.g., Ishihara et al., 1981; Kobayashi & Tameike, 2002; Kobayashi et al., 2013). The whole-rock chemistry of SiO_2 is 66–68 wt% (Bunmei), 64–66 wt% (An-ei), 59–64 wt% (Taisho), and approximately 61–62 wt% (Showa) (Nakagawa et al., 2011). All lavas include plagioclase, orthopyroxene, clinopyroxene, magnetite, and olivine (only Taisho lavas and Showa lavas) as phenocrysts. The modes of total phenocrysts are approximately 15% (Bunmei), 20% (An-ei), 25–28% (Taisho), and 29% (Showa) (Yanagi et al., 1991). Groundmass minerals are composed of plagioclase, orthopyroxene, clinopyroxene titanomagnetite, and glass.

4.2.2. Plagioclase Phenocrysts

Yanagi et al. (1991) reported that plagioclase phenocrysts in four historical lavas show a bimodal compositional distribution with peaks at around An_{58} and An_{85}, and the frequency and the fraction of An-rich phenocrysts gradually and monotonically increases with eruptive age. Plagioclase phenocrysts are texturally classified into two types in the four historical lavas, a clear type and a dusty type with abundant glass inclusions called a honeycomb texture. However, they do not mention the relationship between the two chemical types (An-poor vs. An-rich) and the two textural types (clear ones vs. dusty ones). Therefore, we conducted both textural observations and chemical analyses to clarify the relationship between the two different classifications. Textural observations were conducted by polarized optical microscope and a field emission scanning electron microscope (FE-SEM; JEOL JSM-7001F), and chemical analysis was carried out by a field emission electron probe microanalyzer (FE-EPMA; JXA-8530F) (both at the Department of Earth and Planetary Sciences, Kyushu University). Most plagioclase phenocrysts are euhedral and are approximately 3.5 mm in maximum length for all samples. The modes of plagioclase phenocrysts determined by measurement of the area fraction on back-scatted electron (BSE) images taken by FE-SEM are 8.4%, 13.5% 19.5%, and 20.3%, respectively, for the Bunmei, An-ei, Taisho, and Showa lavas. In the four historical lavas, two textural types of plagioclase phenocrysts, clear and dusty with glass or mineral inclusions (honeycomb texture), coexist uniformly in space. At the later phase in the Taisho lava, dusty plagioclase phenocrysts with many fine glass inclusions (approximately 10 μm) rarely exist.

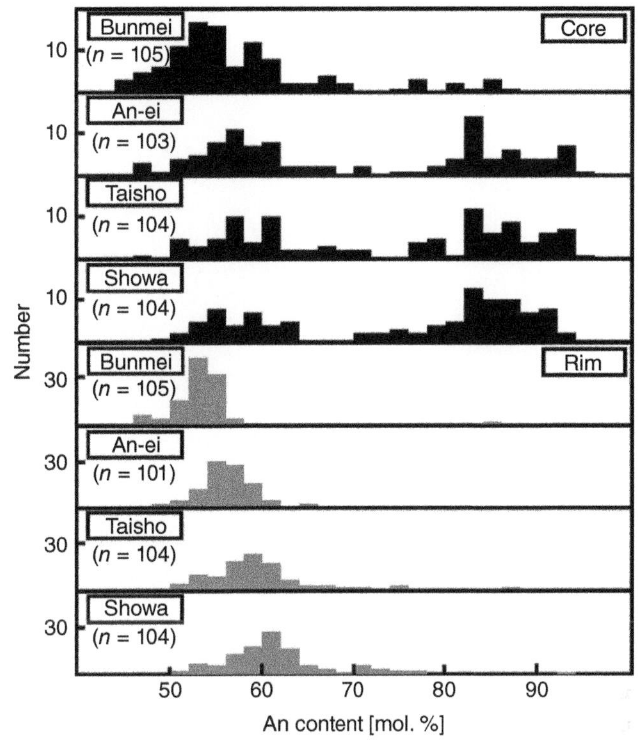

Figure 4.2 Compositional distributions of clear plagioclase at both cores and rims for four historical lavas. The rim compositions of the two types of plagioclase phenocrysts are not distinguishable because they nearly completely overlapped for An content except higher An content with small fractions in the Showa lava.

Figure 4.2 shows the frequency distributions of the compositions in the cores and rims of clear plagioclase from our analyses. Analyzed points are determined as the approximate centers of crystals for cores judging from the zoning pattern, and as positions at 10 μm inside the phenocryst–groundmass boundary. The analysis conditions were an acceleration voltage of 15 keV, a beam current of 5.0 nA, and a beam size of 3.0 μm for FE-EPMA. The core compositions are characterized by bimodal distributions with peaks around An_{55-60} (An-poor) and around An_{84} (An-rich). The An contents at peak frequencies for the An-poor plagioclases increase slightly with time from An_{55} (Bunmei) to An_{60} (Showa) (Figure 4.2).

From Figure 4.2, it is found that clear plagioclases are classified into two types according to the chemical compositions: one with An-poor (An_{50-75}) and the other with An-rich (An_{75-95}) cores. In their textural characteristics, the former (type A) commonly shows oscillatory zoning within a range of An_{50-70} in their core compositions for the four historical eruptions. The latter (type B) is relatively homogeneous in its core compositions (around

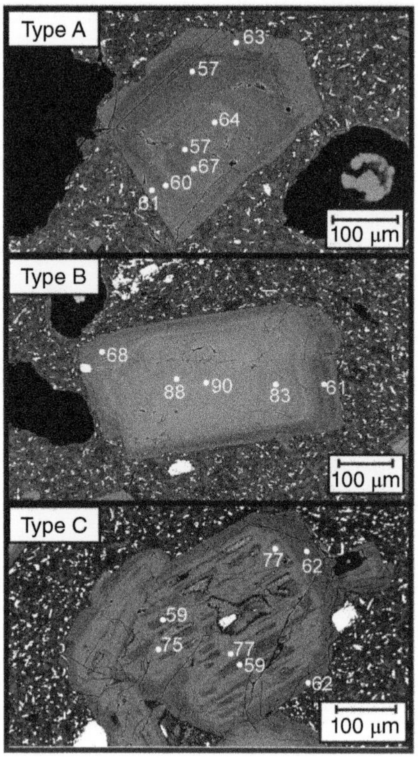

Figure 4.3 Representative backscattered electron images of type A–C plagioclase phenocrysts.

An_{84}) and rim compositions (around An_{60}), and this shows that the compositional contrast between the core and the overgrowth-rim is sharp.

The compositional distributions of the rims are characteristically unimodal, showing that the An contents with single peaks in the frequencies change from An_{55} to An_{62} with time, similar to the behavior of the An-poor cores. By combining the textural observations and compositional analyses of individual phenocrysts with their core compositions, plagioclase phenocrysts can be classified into three types (Figure 4.3) as follows:

• A, clear texture and lower An content (around $An_{50–70}$) in the core;

• B, clear texture and higher An content (around $An_{75–95}$) in the core;

• C, dusty texture and heterogeneous compositions (patchy distributions of An poor and An rich) in the core.

We preliminarily measured the rim widths for type B phenocrysts with clearly recognized rims and obtained the average values for euhedral plagioclase phenocrysts. As a result, the rim widths vary from 10 to 250 μm in all lavas, while the younger lavas have wider ranges of variation. The rim widths of type A phenocrysts were not measured because there was no clear compositional difference to identify the rim zones.

4.3. CSD ANALYSIS

4.3.1. Method

We conducted CSD measurements over the whole area of two thin sections for each eruption (one thin section (5 cm × 3 cm) per sampling point for different lava flow units; see Figure 4.1). The CSD analysis was conducted on BSE images taken separately by a field emission scanning electron microscope (FE-SEM; JEOL JSM-7001F) for different types of plagioclase phenocrysts (Figure 4.4). We used 35–70 BSE images (magnification of ×30) per thin section to cover the area of the whole thin-section. The magnification was constant and was sufficient to identify the smallest phenocrysts. The individual types can be easily distinguished by the brightness of the core in the BSE images once correlation between the brightness and the core An content was established for several plagioclase phenocrysts for each image. We used the additional factor of the contrasting transition from core to rim: in type A, the brightness shifts from dark to light, corresponding to the core and rim An contents, whereas in type B, the brightness shifts from light to dark. We manually enclosed plagioclase phenocrysts on a PC using Illustrator (Adobe) software, binarized the images and brought them into Image J software for two-dimensional textural measurements. We converted two-dimensional CSDs into three-dimensional CSDs by CSD correction software (Higgins, (2000, version 1.51) using stereological methods.

4.3.2. Results

The CSD data obtained are shown in Figure 4.5. The CSD plots of both type A and type B can be approximated by log-linear CSDs. Detailed data are listed in Table 4.1 with other textural data.

We characterize CSD data by a log-linear CSD expressed by equations:

$$n(L) = n_0 \exp\left(-\frac{L}{G_0\tau_{ex}}\right) \tag{4.1}$$

$$n_0 = \frac{J_0}{G_0} \tag{4.2}$$

where $n(L)$ and L are population density (number mm^{-4}) and crystal size (mm), respectively. n_0, G_0, τ_{ex}, and J_0 are nucleation density (number mm^{-4}), effective growth rate (mm s^{-1}), timescale defining crystal extraction rate (s^{-1}), and nucleation rate (number mm^{-3} s^{-1}), respectively. In a CSD plot of $\ln n(L)$ versus size L plot, the absolute value of the slope equals $|1/G\tau_{ex}|$, and the intercept equals $\ln n_0$. The above formulation is based on the assumptions

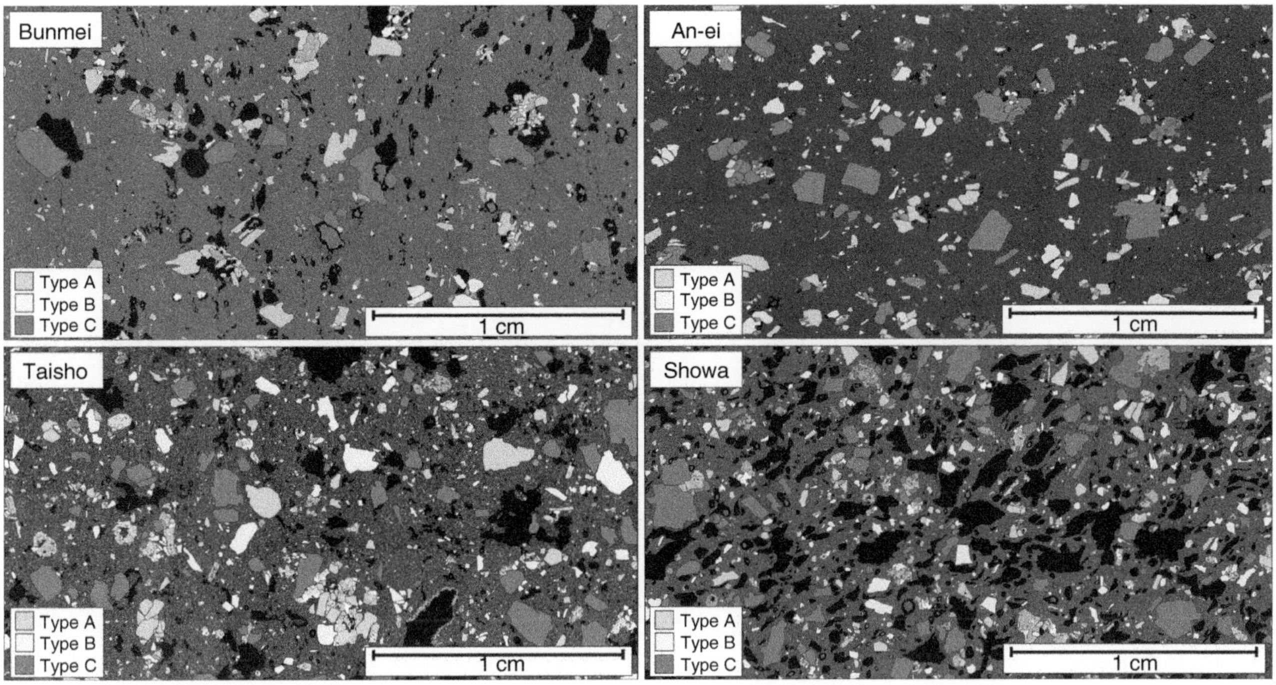

Figure 4.4 Example of color maps for the crystal size distribution analyses of separate types of plagioclase phenocrysts: green, type A; yellow, type B; red, type C.

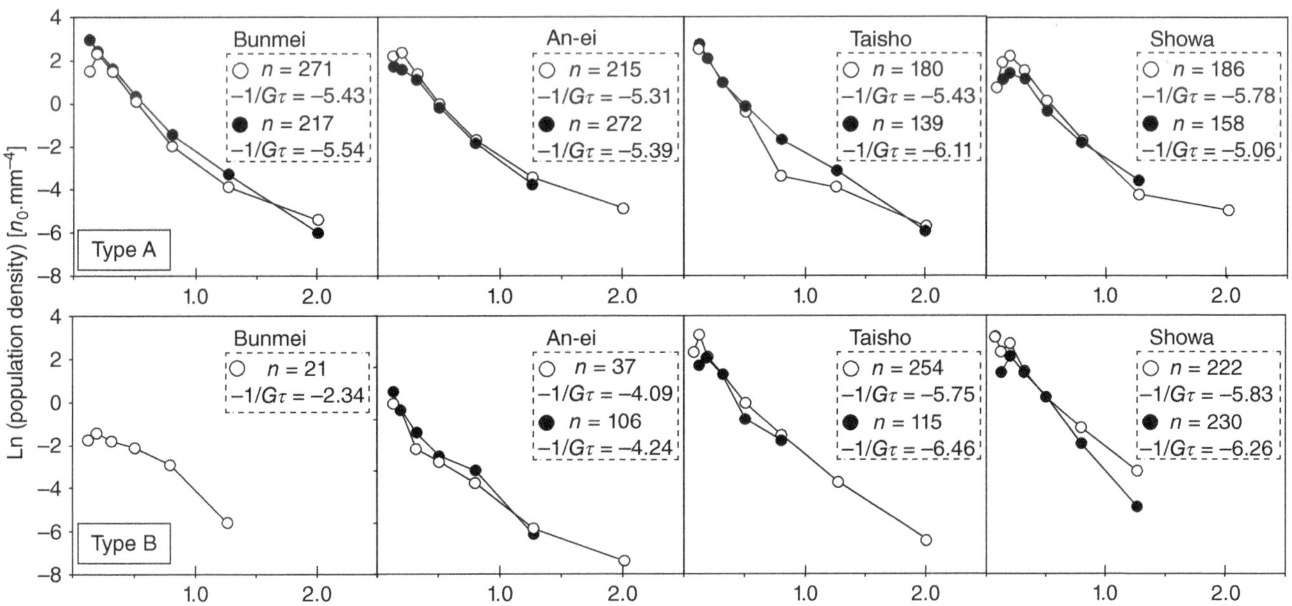

Figure 4.5 Crystal size distribution (CSD) for (upper panel) type A and (lower panel) type B plagioclase phenocrysts. Solid and open symbols correspond respectively to samples suffixed by –1 and –2 from the lava units of different eruptions (sample locations shown in Figure 4.1). Liner regression and regression slope were obtained using CSD corrections (Higgins, 2000).

Table 4.1 Summary of Plagioclase Phenocryst CSD Analyses and Textural Data for Four Historical Lavas from Sakurajima Volcano

Eruption			Input factors				Results	
Type	Sample	Age (CE)	Vesicularity (vol%)	Measured area (mm^2)	Shape	Number	Intercept (mm^{-4})	|Slope| (mm^{-1})
A	Bunmei-1	1471–1476	1.77	230.031	1:2.22:2.22	217	3.41	5.54
	Bunmei-2	1471–1476	10.9	424.218	1:2.22:2.22	271	2.85	5.43
	An-ei-1	1779	ND	559.471	1:2.86:2.86	272	2.67	5.39
	An-ei-2	1779	23.12	350.983	1:1.54:1.54	215	2.99	5.31
	Taisho-1	1914–1915	14.3	252.853	1:2.86:2.86	139	2.87	5.08
	Taisho-2	1914–1915	11.98	382.890	1:2.00:2.00	180	2.92	6.11
	Showa-1	1946	23.9	451.686	1:2.00:2.00	158	2.47	5.06
	Showa-2	1946	14.72	285.670	1:2.00:2.00	186	3.1	5.78
B	Bunmei-2	1471–1476	10.9	419.000	1:1.54:1.54	21	−0.38	3.6
	An-ei-1	1779	ND	562.011	1:1.54:1.54	106	1.15	4.24
	An-ei-2	1779	23.12	355.177	1:3.33:3.33	37	0.78	4.09
	Taisho-1	1914–1915	14.3	253.234	1:1.54:1.54	115	3.17	6.46
	Taisho-2	1914–1915	11.98	380.586	1:2.5:2.5	254	3.3	5.75
	Showa-1	1946	23.9	453.746	1:1.54:1.54	230	3.29	6.26
	Showa-2	1946	14.72	288.267	1:2.22:2.22	222	3.49	5.83

Note. ND, not detected.

proposed by Marsh (1988): (a) open system; (b) influx <<outflux for crystals; (c) the system volume, V, is constant; (d) the growth rate, G_0, and the nucleation rate, J_0, are constant on average during the individual intervals among successive eruptions, while the values may change between successive individual eruptions through historical time.

The above equations are derived as a solution of the population balance equation with the boundary condition of a constant nucleation rate, J_0, a constant growth rate, G_0, and a constant rate of crystal extraction rate, $1/\tau_{ex}$:

$$\frac{\partial n(L)}{\partial t} + G_0 \frac{\partial n(L)}{\partial L} = -\frac{n(L)}{\tau_{ex}} \tag{4.3}$$

where the right-hand term indicates the extraction or the outflux of phenocrysts, in which the timescale defining the extraction rate is represented by τ_{ex}. Thus, we have three independent parameters, J_0, G_0, and τ_{ex}. Marsh (1988) interprets that the timescale, τ_{ex}, is identical to the residence time. However, we discriminate the time for crystal growth and the time for crystal extraction. The former is a residence time, τ_{re}, and the latter is a timescale defining the extraction rate τ_{ex}. The interpretation of the CSD will be considered according this model.

Under these assumptions, we quantitatively interpret CSD data for types A and B. As type C are thought to be formed by partial melting or skeletal growth (e.g., Kawamoto, 1992; Nakamura & Shimakita, 1998), they do not match the above conditions so are excluded for measurement and discussion.

The slopes and the intercepts of type A are constant regardless of eruptive age, and those of type B increase with time, i.e., the Showa lavas show the steepest slope. Considering equations 4.1 and 4.2, these results indicate that the values of $G_0\tau$ and J_0/G_0 for type A are constant, while the values for type B increase with time.

4.4. DISCUSSION

4.4.1. The Origins of Types A and B

The core An contents of clear plagioclase show a bimodal compositional distribution (Figure 4.2), indicating the mixing of binary magmas as proposed by Yanagi et al. (1991). The relative frequency of type B (high An) increases with time. The peak positions of An content for cores of type A (low An) increase slightly from An_{60} (Bunmei) to An_{65} (Showa). The increase in An content with time is more noticeable in the rim compositions for both types (depicted by the same gray color in Figure 4.2). In addition, it is found that the relative abundances of type A decrease with time. These results suggest that the compositions of felsic magmas containing type A plagioclase gradually became slightly enriched in their mafic components over hundreds of years by repeated injection of mafic magmas conveying type B plagioclases. The same rim compositions for both type A and B indicate that the rims crystallized from the same mixed magma.

Type A and B rim compositions are similar to, or slightly higher than, those of the type A cores, indicating

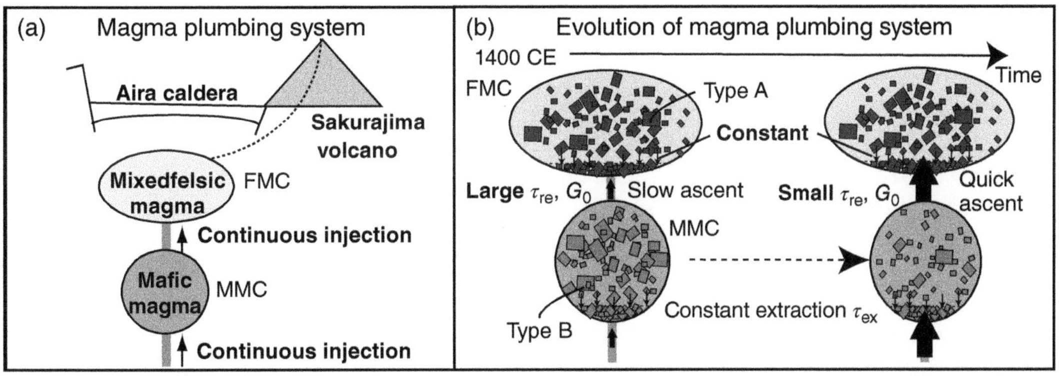

Figure 4.6 A model of the magma plumbing system of Sakurajima volcano since the 15th century. (a) Traditional model of a magma plumbing system. Type A phenocrysts crystallized in the residual mixed felsic magma chamber (FMC), and type B phenocrysts crystallized in the mafic magma chamber (MMC). (b) Evolution of the magma plumbing system to account for the CSDs of the different eruptions. A slow ascent rate from MMC to FMC leads to a long residence time, t_{re}, resulting in a large effective (average) growth rate corresponding to domination by relatively large crystal sizes. The historical increase in ascent rate generates a monotonic increase in the slopes and the intercepts of the CSDs of type B phenocrysts. The ascent rate is equivalent to the supply rate from the mantle if the plumbing system experiences no large change in geometry.

that the rim formed during or/and after the successive mixing events. Relatively wide variations in the overgrowth-rim widths of type B (10–200 µm) indicate greater variation in crystallization time after the injection of mafic magmas into felsic magmas, suggesting continuous injection of mafic magma into the felsic magma chamber (mixed magmas), where rim crystallization proceeded.

4.4.2. Location of Magma Chambers

We utilized geophysical data to constrain the locations of the two magma chambers: the upper "felsic" and the lower "mafic" magma chambers. Mogi (1958) reported that the pressure source of ground deformation for the Taisho eruption was located at a depth of 10 km beneath the Aira caldera. After the Taisho eruption, continuation of ground uplift suggests that the magmas continuously accumulated beneath the Aira caldera (Iguchi et al., 2008).

Sekine et al. (1979) reported other constraints by experimentally reproducing the mineral assemblage at 79–106 MPa (a saturated H_2O content of 2.9–3.5 wt%) and temperature of 950°C. This indicates depths 3–4 km shallower than for the geophysically estimated depth. The experimental petrological constraints represent the final equilibration pressure and temperature, and the corresponding water content may be undersaturated. Triggering the eruption, however, does not require saturation or oversaturation of water in the magmas if the eruption is triggered by the "excess volume" condition in our model.

More detailed geodetic modeling of recent ground uplift indicates continuous injection of mafic magma into the source beneath the Aira caldera (Hickey et al., 2016). Thus, we interpret that the FMC, which is recorded in mineral assemblages just prior to the eruption, corresponds to the magma chamber beneath the Aira caldera. The existence of a MMC has not been detected by geophysical observations, however, at least, the "mafic magma chamber" as the field of crystallizing type B must exist deeper than the felsic magma chamber (Figure 4.6).

4.4.3. Interpretation of CSD and Control Processes in the Magma Plumbing System

In general, in an open system, CSD characteristics such as slope and intercept are controlled by the outflux, i.e., the rate at which crystals are extracted from the system, and the influx, i.e., the rate at which crystals are introduced in addition to the continuous crystallization within the system. The outflux comprises two processes: crystal settling and continuous magma discharge from the chamber that conveys crystals throughout the plumbing system. The influx comprises one process: continuous magma supply transporting crystals from the other magma sources. We interpret that the Sakurajima volcano has two magma chambers, the FMC and the MMC, in which type A and type B CSDs are formed, respectively.

We assume that magmas supplied into the respective chambers are free of crystal types characteristic of the other (i.e., no input of type A into FMC and no input of type B into MMC), because the core compositions for each type have distinctive signatures indicating crystallization in situ in the corresponding magma chambers. Therefore, continuous crystallization in each magma chamber is the only factor that explains the respective characteristic CSD, a factor that is already represented in the steady-state model by the constant nucleation, J_0, and growth, G_0, assumptions.

With regard to the type A phenocrysts, with constant slopes and intercepts over 500 years, the values of the parameter J_0, G_0, and τ_{ex} are thought to be constant. In other words, the crystallization conditions such as nucleation rate, average growth rate, and extraction rate have been constant for 500 years in the FMC. The slopes and intercepts of type B phenocrysts, however, increase with time over 500 years, thus the parameters of J_0/G_0 and $1/(G_0\tau_{ex})$ increase with time. As both the slope and intercept are observed to increase, the simplest explanation is that a common parameter, G_0, decreases over time in the MMC. If the timescale, τ_{ex}, characterizing the extraction rate decreases with time, then the increase in slope can be explained but increase in the intercept cannot.

As G_0 is the average or effective growth rate of high An plagioclase phenocrysts in the MMC, we can assume the unimodal distribution of An content indicates unique conditions of the MMC under which type B phenocrysts crystallize. The average growth rate may be an integrated growth rate represented by the growth length. So, under steady-state conditions with a constant geometry of the magma plumbing system, the integrated growth length in the MMC equals the average residence time controlled by the influx from the mantle, which is balanced with the outflux to the FMC. For instance, if the influx from the mantle is high, then the residence time is short. Thus, it can be expected that the average growth rate is proportional to the residence time, t_{re}, and is inversely proportional to the influx from the mantle (supply rate), Q:

$$G_0 \propto t_{re} \propto Q^{-1} \tag{4.4}$$

This equation means that the slope and intercept of CSD are proportional to the supply rate:

$$|\text{slope}| \propto Q_{CSD} \tag{4.5}$$

$$|\text{intercept}| \propto Q_{CSD} \tag{4.6}$$

where Q_{CSD} is the supply rate estimated by CSD method.

4.4.4. Average Growth and Mantle-Supply Rates: Factors Controlling the CSD Properties of Plagioclase Phenocrysts

To examine the hypothesis that the slope and intercept represent the supply rate, we consider the correlation between the slope and the discharge rate. Discharge rates are calculated by the erupted mass divided by the repose time before the eruption. Figure 4.7 indicates that for large eruptions with a DRE volume on the order of 1 km^3 the slopes are positively correlated with discharge, except for the Showa eruption. This correlation supports our hypothesis that magma supply controls the triggering of eruptions. Correlation is not linear, however, and does not show proportionality between the slope and supply rate, so other factors with secular variations may influence the triggering of large-scale eruptions.

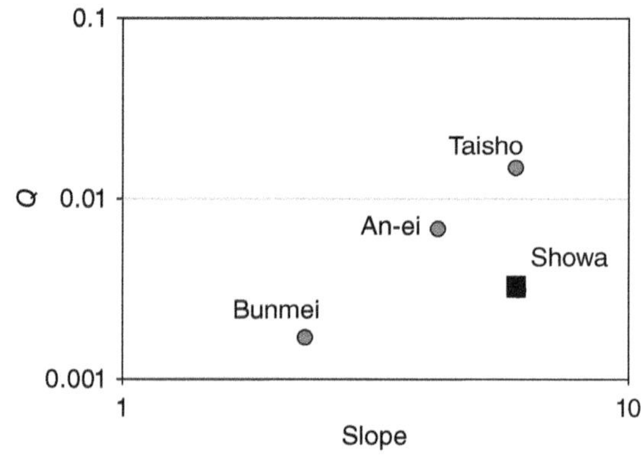

Figure 4.7 Correlation between discharge rates and slopes of CSD of high An plagioclase phenocrysts.

The correlation is represented by

$$|\text{slope}| \propto Q_{geol}^{1/2} \tag{4.7}$$

where Q_{geol} is the value calculated by the geological data.

Compared with the results from the CSD method (equation 4.5), there is a discrepancy with the geological method, as follows:

$$Q_{geol} = Q_{CSD}^2 \tag{4.8}$$

This indicates that the discharge rate at the surface is apparently exaggerated compared with the supply rate estimated by the CSD method. Nevertheless, it is an interesting suggestion that the supply rate from the mantle controls the discharge rate at the surface.

Finally, we provide comments on the Showa eruption, which is different in terms of volume and scale from typical large eruptions consisting of Plinian ash fall followed by lava flows from a stratovolcano. It has been observed that eruptive styles changed after the Showa eruption to small-volume, frequent Vulcanian eruptions. The magmatic behavior, such as the crystallization processes in the FMC, remains at a constant pace, so processes in the FMC have no influence on eruptive phenomena. On the other hand, the MMC is located at a deeper level and controls the behavior of eruptive phenomena, such as the frequency of eruption events and dominant eruption styles of the Vulcanian and Strombolian types, through increasing the supply rate of mantle-derived mafic magmas. Thus, we suggest the possibility that the plumbing system drastically changed after the Showa eruption.

4.5. CONCLUSIONS

We present the results of CSD analysis forming the basis of a new classification of plagioclase phenocrysts from Sakurajima lavas of four major historical eruptions after the 15th century. The magma plumbing system consists of two magma chambers in which the crystallization of two types of plagioclase phenocrysts proceeds: in the mixed felsic magma chamber (FMC), type A low-An phenocrysts crystallize, and in the mafic magma chamber (MMC), type B high-An phenocrysts crystallize. In the MMC, both the mantle-derived mafic magma supply rate and supply rate to the FMC increase over time. The long-term changes in the magma supply rate in the deeper levels and in the production rate in the mantle may control the repose times and the plumbing system over time for the Sakurajima volcano.

ACKNOWLEDGMENTS

We thank T. Ikeda and T. Miyamoto for their helpful discussions, suggestions, and encouragement. The FE-SEM and FE-EPMA analyses were supported by Mr. K. Shimada. This work was funded by a Takachiho Scholarship, the Late Professor Tatsuro Matsumoto Scholarship.

REFERENCES

Cashman, K. V., & Marsh, B. D. (1988). Crystal size distribution (CSD) in rocks and the kinetics and dynamics of crystallization II: Makaopuhi lava lake *Contributions to Mineralogy and Petrology*, *99*, 292–305.

Fukuyama, H. (1978). Geology of Sakurajima volcano, southern Kyushu. *Journal of the Geological Society of Japan*, *84*, 309–316.

Hickey, J., Gottsmann, J., Nakamichi, H., & Iguchi, M. (2016). Thermomechanical controls on magma supply and volcanic deformation: application to Aira caldera, Japan. *Nature Scientific Reports*, *6*, 32691.

Higgins, M. D. (1996). Magma dynamics beneath Kameni volcano, Thera, Greece, as revealed by crystal size and shape measurements. *Journal of Volcanology and Geothermal Research*, *70*, 37–48.

Higgins, M. D. (2000). Measurement of crystal size distributions. *American Mineralogist*, *85*, 1105–1116.

Iguchi, M., Takayama, T., Yamazaki, T., Tada, M., Suzuki, A., Ueki, S., Ohta, Y., & Nakao, S. (2008). Movement of magma at Sakurajima Volcano revealed by GPS observation. *Annals of Disaster Prevention Research Institute, Kyoto University*, *51*(51), 241–246.

Ishihara, K., Takayama, T., Tanaka, Y., & Hirabayashi, J. (1981). Lava flows at Sakurajima volcano (1)—volume of the historical lava flows. *Annals of Disaster Prevention Research Institute, Kyoto University*, *24*(B-1), 1–10.

Kawamoto, T. (1992). Dusty and honeycomb plagioclase: indicators of processes in the Uchino stratified magma chamber, Izu Peninsula, Japan. *Journal of Volcanology and Geothermal Research*, *49*, 191–208. doi: 10.1016/0377-0273(92)90014-5

Kobayashi, T. Miki, D., Sasaki, H., Iguchi, M., Yamamoto, T., & Uto, K. (2013). *Geological map of Sakurajima volcano* (2nd edition; 1:25000). Geological Survey of Japan.

Kobayashi, T., & Tameike, T. (2002). History of eruptions and volcanic damage from Sakurajima Volcano, southern Kyushu, Japan. *The Quaternary Research (Daiyonki-Kenkyu)*, *41*(4), 269–278.

Mangan, M. T. (1990). Crystal size distribution systematics and the determination of magma storage times: The 1959 eruption of Kilauea volcano, Hawaii. *Journal of Volcanology and Geothermal Research*, *44*, 295–302. doi: 10.1016/0377-0273(90)90023-9

Marsh, B. D. (1988) Crystal size distribution (CSD) in rocks and the kinetics and dynamics of crystallization I. *Theory. Contributions to Mineralogy and Petrology*, *99*, 277–291.

Miwa, T., Geshi, N., & Shinohara, H. (2013). Temporal variation in volcanic ash texture during a Vulcanian eruption at the Sakurajima Volcano, Japan. *Journal of Volcanology and Geothermal Research*, *260*, 80–89. doi: 10.1016/j.jvolgeores.2013.05.010

Miwa, T., & Toramaru, A. (2013). Conduit process in vulcanian eruptions at Sakurajima volcano, Japan: Inference from comparison of volcanic ash with pressure wave and seismic data. *Bulletin of Volcanology*, *75*, 1–13. doi: 10.1007/s00445-012-0685-y

Mogi, K. (1958). Relations between the eruptions of various volcanoes and the deformations of the ground surfaces around them. *Bulletin of the Earthquake Research Institute*, *36*, 99–134. doi: 10.1016/j.epsl.2004.04.016

Nakagawa, M., Matsumoto, A., Amma-Miyasaka, M., Togashi, Y., & Iguchi, M. (2011). Change of mode of eruptive activity and the magma plumbing system of Sakurajima Volcano since the 20th century. In *Study on preparation process of volcanic eruption based on integrated volcano observation 2010* (pp. 85–94). Sakurajima Volcano Research Center.

Nakamura, K. (2006). Textures of plagioclase microlite and vesicles within volcanic products of the 1914–1915 eruption of Sakurajima Volcano, Kyushu, Japan. *Journal of Mineralogical and Petrological Sciences*, *101*, 178–198. doi: 10.2465/jmps.101.178

Nakamura, M., & Shimakita, S. (1998) Dissolution origin and syn-entrapment compositional change of melt inclusion in plagioclase. *Earth and Planetary Science Letters*, *161*, 119–133. doi: 10.1016/S0012-821X(98)00144-7

Piochi, M., Mastrolorenzo, G., & Pappalardo, L. (2005). Magma ascent and eruptive processes from textural and compositional features of Monte Nuovo pyroclastic products, Campi Flegrei, Italy. *Bulletin of Volcanology*, *67*, 663–678. doi: 10.1007/s00445-005-0410-1

Preece, K., Barclay, J., Gertisser, R., & Herd, R. A. (2013) Textural and micro-petrological variations in the eruptive products of the 2006 dome-forming eruption of Merapi volcano, Indonesia: Implications for sub-surface processes. *Journal of Volcanology and Geothermal Research*, *261*, 98–120. doi: 10.1016/j.jvolgeores.2013.02.006

Sekine, T., Katsura, T., & Aramaki, S. (1979). Water saturated phase relations of some andesites with application to the estimation of the initial temperature and water pressure at the time of eruption. *Geochimica et Cosmochimica Acta*, *43*(8), 1367–1376.

Tomiya, A., & Takahashi, E. (1995). Reconstruction of an evolving magma chamber beneath Usu volcano since the 1663 eruption. *Journal of Petrology*, *36*(3), 617–636. doi: 10.1093/petrology/36.3.617

Vinet, N., & Higgins, M. D. (2011) What can crystal size distributions and olivine compositions tell us about magma solidification processes inside Kilauea Iki lava lake, Hawaii? *Journal of Volcanology and Geothermal Research*, *208*(3–4), 136–162. doi: 10.1016/j.jvolgeores.2011.09.006

Yanagi, T., Ichimaru, Y., & Hirahara, S. (1991). Petrochemical evidence for coupled magma chambers beneath the Sakurajima volcano, Kyushu, Japan. *Geochemical Journal*, *25*, 17–30.

Part II
Physical Properties in Magma

5

Dynamics of Volcanic Systems: Physical and Chemical Models Applied to Equilibrium Versus Disequilibrium Solidification of Magmas

Letizia Giuliani[1], Gianluca Iezzi[2], and Silvio Mollo[3]

ABSTRACT

The solidification of magmas can occur by cooling ($\Delta T/\Delta t$) and/or degassing-induced decompression ($\Delta P/\Delta t$), as a function of *solidus*, glass transition, and melting temperatures, respectively T_s, T_g, and T_m. These three parameters strongly depend on the bulk composition (X) of the system and vary with T, P, fO_2, and H_2O. In recent decades, physical and chemical models of magmas have been profoundly refined, such as thermal, rheological and density behaviors, the formation conditions of mineral/melt phases, and volatile solubility. However, variations of physical and chemical features of magmas through time are still poorly constrained. This means that mechanisms binding the formation of different phases under *equilibrium* and *disequilibrium* conditions are still poorly quantified, although they are of paramount importance for reconstructing magmatic dynamics encrypted in volcanic rocks. Here, the most relevant solidification conditions of magmas leading to possible crystallization paths are tested in physical models. The textural and, especially, chemical attributes of minerals are reviewed in order to summarize the salient features able to discriminate between formation of *equilibrium* and *disequilibrium* phases. Finally, the reconstruction of magmatic intensive variables deduced from composition of minerals is discussed.

5.1. INTRODUCTION

The transformation of a magma (melt ± crystals ± dissolved/exsolved volatiles) into a solid rock (glass ± crystals ± bubbles) is one of the most important and voluminous processes in the lithosphere. Magma solidifies in plutonic or volcanic environments by cooling and/or as a result of decompression. General reappraisal of magmatic solidification behaviour has been reviewed in many studies in recent decades (Armienti, 2008; Blundy & Cashman, 2008; Cashman, 1990; Hammer, 2008; Iezzi et al., 2009; Kirkpatrick, 1981; Lasaga, 1997; Marsh, 1996, 2006; Mollo & Hammer, 2017; Muncill & Lasaga, 1988; Putirka, 2017; Zhang, 2008). These investigations were mostly based on experimental studies *ex-situ* as a function of $\Delta T/\Delta t$ (Baker & Grove, 1985; Brandeis et al., 1984; Conte et al., 2006; Corrigan, 1982; Del Gaudio et al, 2010; Gibb, 1974; Grove & Raudsepp, 1978; Iezzi et al., 2008, 2011, 2014; Kirkpatrick, 1981, 1983; Kolzenburg, Di Genova, et al., 2018; Kolzenburg, Giordano, et al., 2018; Lesher et al., 1999; Mollo et al., 2010; Mollo, Lanzafame, et al., 2011; Orlando et al., 2008; Pupier et al., 2007; Uhlmann et al., 1979, 1981; Vetere et al., 2013, 2015) and of $\Delta P/\Delta t$ (Applegarth et al., 2013; Arzilli & Carroll, 2013; Arzilli et al., 2015; Brugger & Hammer, 2010; Fiege et al., 2015;

[1] *Department of Engineering and Geology (InGeo), University G. d'Annunzio of Chieti-Pescara, Italy*

[2] *Department of Engineering and Geology (InGeo), University G. d'Annunzio of Chieti-Pescara, Italy and National Institute of Geophysics and Volcanology (INGV) Rome, Italy*

[3] *National Institute of Geophysics and Volcanology (INGV), Rome, Italy and Department of Earth Sciences, Sapienza University of Rome, Rome, Italy*

Dynamic Magma Evolution, Geophysical Monograph 254, First Edition. Edited by Francesco Vetere.

Hammer & Rutherford, 2002; La Spina et al., 2016; Martel & Schmidt, 2003; Métrich & Rutherford, 1998; Shea & Hammer, 2013). Additional studies have been conducted using techniques *in-situ* (Masotta & Keppler, 2017; Masotta et al, 2014; Ni et al., 2014; Polacci et al., 2018; Schiavi et al., 2009;). Crystal chemistry and textural data determined for magmatic rocks, coupled with experimental constraints, have significantly and further strengthened the solidification models (Cooper, 2017; Lanzafame et al., 2013, 2017; Masotta et al., 2018; Mollo, Lanzafame, et al., 2011; Mollo et al., 2015).

In contrast to simplified synthetic systems (Fokin et al., 2006; Lasaga, 1997; Zhang, 2008), natural magmas include a great number of chemical constituents. Due to crystal fractionation phenomena, the solidified phases (crystals ± glass ± bubbles) exhibit chemistries slightly to greatly different to the bulk magma composition. As a function of P, T, fO_2, and H_2O conditions, some chemical components may be more (compatible elements) or less (incompatible elements) easily incorporated in minerals, melts, and fluids (Lasaga, 1997; Richet & Ottonello, 2010; Zhang, 2008). As a consequence, cation exchange reactions between minerals and melts are frequently used to reconstruct the intensive variables of the solidifying system, which are referred to as thermometers and barometers (Masotta et al., 2013; Mollo et al., 2010; Mollo, Putirka, et al., 2011; Nimis & Ulmer, 1998; Putirka, 2008). Attaining *equilibrium* conditions requires that all chemical species have enough time to reach the low-energy states when chemical gradients cease at the interface between crystal, melt, and/or bubble (Dingwell, 2006; Lasaga, 1997; Pichavant et al., 2007; Zhang, 2008). In volcanic environments, however, this situation is frequently not attained and *disequilibrium* processes take place when the cation diffusion phenomena are too sluggish with respect to crystal and bubble growth conditions (Pichavant et al., 2007; Zhang & Cherniak, 2010).

The solidification of magma starts with the initial crystallization of one or more mineral phases, so that chemically different (residual) melts are derived as a function of the amount and composition of the phases formed (i.e., crystals + bubbles). Alternatively, magmas or residual melts may not nucleate crystals, thus producing glassy rocks (Vetere et al., 2015). When early formed crystals and/or bubbles favor successive nucleation events, the process is heterogeneous, rather than homogeneous (Mollo et al., 2012; Shea & Hammer, 2013; Vetere et al., 2013). The *solidus* temperature (T_s) is the virtual temperature at which the solidification process ends, and the resulting rock is holocrystalline. Alternatively, the glass transition temperature T_g (T of melt viscosity is 10^{12} Pa·s) is the expression of vitrophiric natural products (Dingwell & Webb, 1990; Giordano et al., 2005, 2008). Hence, with respect to the *liquidus* or melting temperature (T_m), natural magmas can solidify between $T_m - T_s$ or $T_m - T_g$, even though the rate of solidification and the bulk chemical system (X) greatly influence the solidification process (Vetere et al., 2015).

The variation of T_m, T_s, and T_g depends on the melt composition, pressure, and amount of volatiles in the system. For instance, the presence of H_2O drastically reduces T_m, T_s, and T_g. As a consequence, H_2O exsolution induced by decompression imposes strong kinetic effects that do not follow *equilibrium* principles (Asimow & Ghiorso, 1998; Gualda et al., 2012). When the degree of cooling or decompression is fast, chemical species in melts with high viscosities have little chance to attain their low-energy states, especially cations with high charges and low ionic radii, such as Si, the most abundant constituent in magmas (Iezzi et al., 2014; Roskoz et al., 2006a, 2006b; Zhang & Cherniak, 2010).

All these very complex processes are recorded principally by the textures and compositions of both phenocrysts and matrix crystals. A first, nontrivial consideration, is thus to discriminate between *equilibrium* and *disequilibrium* compositions and textures in volcanic rocks (Iezzi et al., 2014; Lanzafame et al., 2013, 2017; Mollo & Hammer, 2017; Putirka, 2008, 2017).

In this chapter we mostly focus on the possible timing of P–T changes and their influence on the solidification conditions of magmas, especially in complex and vertically developed magma storage regions (Cashman et al., 2017). Subsequently, we will describe and discuss some tools to decipher the textural and chemical *(dis)equilibrium* aspects of volcanic rocks.

5.2. MAGMAS AT CRUSTAL DEPTHS

Heat dissipation from magmas is the primary cause of melt differentiation and solidification. Magmas form principally in the upper mantle and, during their ascent, may undergo polybaric crystallization at different crustal levels, or they may be erupted directly to the surface (Annen et al., 2015; Bachmann & Huber, 2016; Cashman et al., 2017; Gonnermann & Manga, 2007; Marsh, 1996; Putirka, 2017). In the past two decades, the common view of large, shallow, and prevalently melted magma chambers has been significantly reinterpreted in favor of vertically rising magmas that, at various crustal depths, deflect and expand horizontally to form sills. Repeated injections produce lenses of magmas or even plutons (Annen et al., 2015; Gudmundsson, 2011; Menand, 2008; Menand et al., 2015; Taisne & Jaupart, 2009; Tait & Taisne, 2013). The final result is a complex plumbing system that may involve the entire crust overlying active mantle melting regions, the so-called trans-crustal magmatic system (TCMS; Cashman et al., 2017). Geochemical and geophysical data suggest that these magma batches may originate crystal mushes,

Table 5.1 Parameters used for the Two-Dimensional Model of Heat Transfer

	ρ (kg m^{-3})	C_p (J Kg^{-1} K^{-1})	K (W m^{-1} K^{-1})	α (mm^2 s^{-1})
Rock 1[a]	2700	2200	2.4	0.4
Magma[b]	2550	1285	1.6 4.1 6.6	0.5 1.3 2

Note. [a] Values of ρ, $C_{p,}$ and K were computed following Nabelek et al. (2012) and Whittington et al. (2009). The value of α was mathematically deduced by $\alpha = K/C_p \times \rho$. [b]The input composition is the average between the basaltic (B_{100}) and rhyolitic (R_{100}) data investigated by Vetere et al. (2015). The temperature is 1250 °C. The value of K was computed by using α from Annen (2017).

consisting mainly of cold and uneruptible crystal-rich regions for the most part of their lifetime (Bachmann & Bergantz, 2004; Cashman et al., 2017; Cooper, 2017; Dufek & Bachmann, 2010; Ganne et al., 2018; Marsh, 2015; Masotta et al., 2016; Putirka, 2017). Episodically, crystal mushes and possibly wall rocks are partially to totally (re)melted by the heat supplied from an active mantle source. These melts may then rise to shallow crustal levels, as a function of density contrast, gas-pressure, and tectonic events (Cooper, 2017; Marsh, 1996, 2015; Rivalta et al., 2015). Moreover, the rejuvenation of crystal mushes can (re)connect magma batches distributed throughout the TCMS (Bachman & Bergantz, 2004; Caricchi & Blundy, 2015; Cashman et al., 2017; Cooper, 2017; Marsh, 2015), thus explaining the complex chemical and textural features observed in many volcanic rocks (Putirka, 2017). An implicit consequence of TCMS and possible (re) activation processes is that crystals and melts solidifying under variable P–T–fO_2–H_2O conditions and timescales can mix over relatively short times before eruptions (Burgisser & Bergantz, 2011; Cooper & Kent, 2014; Davidson et al., 2007; Druitt et al., 2012; Pamukcu et al., 2015; Turner & Costa, 2007). Hence, the volcanic rocks correspond to eruptible (unlocked) magma portions that are (re)mobilized in relatively short times and subjected to variable degrees of mixing, mingling, and/or contamination phenomena (Sparks & Cashman, 2017).

5.3. *P-T-t* PATHS DURING MAGMA ASCENT AND EMPLACEMENT

Cooling and decompression numerical models represent valuable tools to better constrain the possible $\Delta T/\Delta t$ and $\Delta P/\Delta t$ paths of magmas as a function of melt volume, temperature, bulk composition, and volatile exsolution (Blundy & Annen, 2016; Dufek et al., 2013; Griffiths, 2000; Harris et al., 2005; Nabelek et al., 2012; Neri, 1998; Whittington et al., 2009; Xu & Zhang, 2002). Following the approach described in Vetere et al. (2015), two-dimensional numerical simulations based on conductive heat loss have been performed with the Heat3D code (Wohletz et al., 1999), in order to constrain the $\Delta T/\Delta t$ changes of a hypothetical magma rising through different crustal levels (Blundy & Annen, 2016). The physicochemical parameters of the magma are reported in Table 5.1. The wall-rock parameters are maintained constant in the numerical simulations, whereas the magma thermal conductivities (K) are 1.6, 4.1, and 6.6 W m^{-1} K^{-1} (Annen, 2017; Nabelek et al., 2012; Whittington et al., 2009). The low and intermediate values of K simulate stationary and convective melts, respectively, that are in contact with less conductive wall rocks. Conversely, the higher value of K indicates melts migrating through highly conductive wall rocks (Annen, 2017; Nabelek et al., 2012).

In Figure 5.1, a silicate melt layer with a thickness of 1 km is instantly intruded between a 3-km-thick roof rock under a geothermal gradient of 100 °C km^{-1} (a high heat flux typical in active volcanic settings) and an underlying 2-km-thick floor rock at 500 °C. All the three numerical simulations show that the innermost and outermost portions of the virtual magmatic layers cool down on the order of 10^{-6} and 10^{-5} °C h^{-1}, respectively. Cooling rates higher than 10^{-5} °C h^{-1} can be found only in the proximity of the magma–wall-rock interface. Therefore, more than 90–95 vol% of the silicate layer is cooled at $\Delta T/\Delta t < 10^{-5}$ °C h^{-1} (Figure 5.1). Similar silicate layers with thickness > 1 km or sandwiched between hotter wall rocks (low crustal levels) are cooled at $\Delta T/\Delta t < 10^{-6}$ °C h^{-1}.

At shallower levels, a 100-m-thick sill is emplaced below a 300-m-thick rock and above a 200-m-thick floor, experiencing $\Delta T/\Delta t$ on the order of 10^{-3} °C h^{-1} (Figure 5.2). Only ~5–10 vol% of the initial silicate liquid cools at $\Delta T/\Delta t > 10^{-3}$ °C h^{-1}. Consequently, any sill with thickness ≥ 0.1 km ponding at crustal levels below 0.3 km (Figures 5.1 and 5.2) would prevalently cool at $\Delta T/\Delta t \leq 10^{-3}$ °C h^{-1}.

A further numerical simulation has been performed for a magma feeding a 20-m-width dike intruding in a host rock with thickness and height of 0.03 and 3 km, respectively (Table 5.1 and Figure 5.3). The geothermal gradient has been maintained at 100 °C km^{-1} and the heat transfer has been modeled on three horizontal profiles at depths of 2.5, 1.5, and 0.5 km (Figure 5.3). The resulting

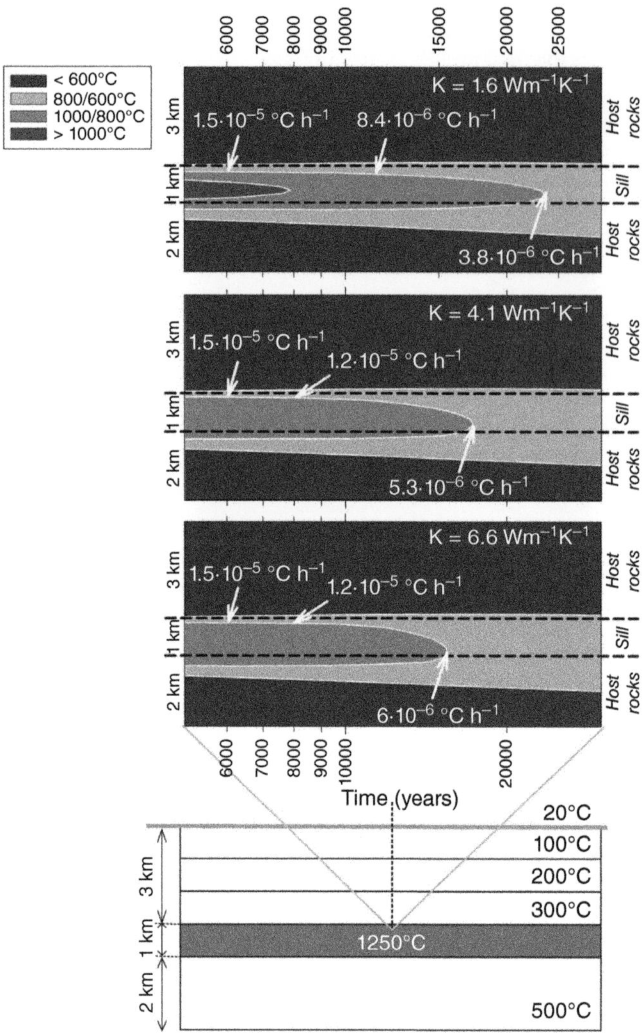

Figure 5.1 Simulations of cooling conditions of a sill 1 km thick, located at a depth of 3 km. The conductive heat transfer has been modeled on the vertical profile, in the center of the sill. The thermodynamic parameters are listed in Table 5.1. A vertical thermal gradient of 100 °C km^{-1} has been applied to the rock layers. The thermodynamic parameters of the deepest rock have been computed following Nabelek et al. (2012) and Whittington et al. (2009): $T = 500$ °C; $C_p = 2100$ J Kg^{-1} K^{-1}, $K = 1.7$ Wm^{-1} K^{-1} and $\alpha = 0.3$ mm^2 s^{-1}.

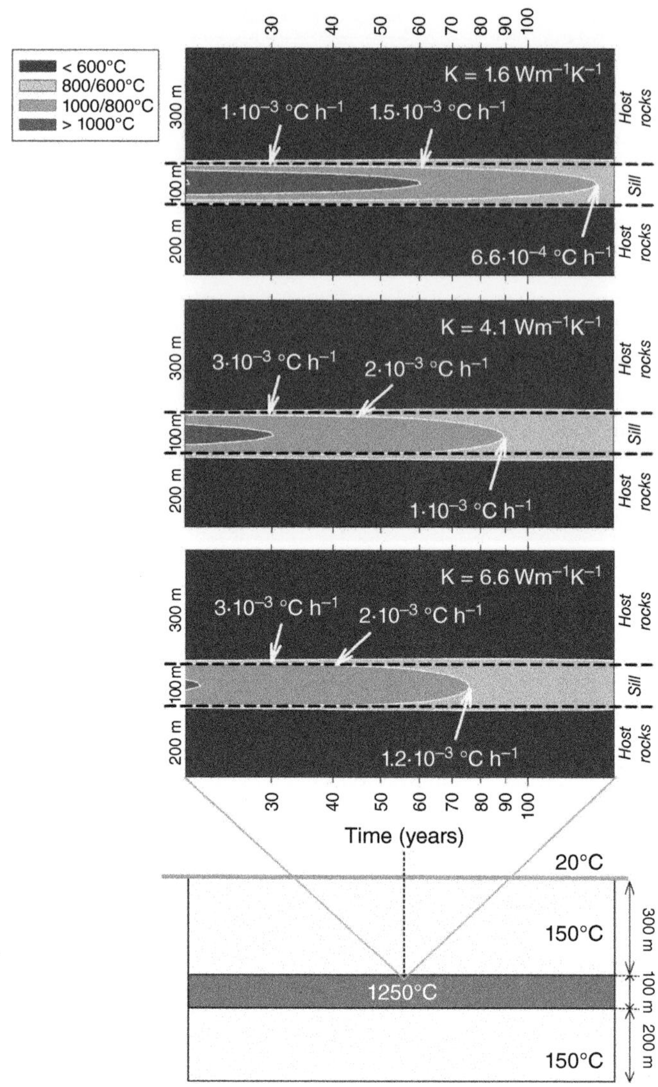

Figure 5.2 Simulations of cooling conditions of a sill 100 m thick, located at depth of 300 m. The conductive heat transfer has been modeled on the vertical profile, in the center of the sill. The thermodynamic parameters are listed in Table 5.1. A constant temperature of 150 °C has been considered for the rock layers.

values of $\Delta T/\Delta t$ are in the order of 10^{-3}–10^{-1} °C h^{-1} (Figure 5.3) and partially overlap with the fastest cooling rate determined for the 0.1-km-thick sill (Figure 5.2).

Considering a 1-m-thick lava flowing at subaerial conditions (Figure 5.4), the numerical simulations yield $\Delta T/\Delta t$ values of 10^1 and 10^2 °C h^{-1} for the innermost and outermost parts, respectively. Compared to the lava flow, only the thinner dike portions cool at a rate higher than 10^2 °C h^{-1} (Figure 5.4). A similar simulation holds also for volcanic bombs with metric dimensions, whereas pyroclastic lapilli and ashes can cool at rates significantly faster than 10^2 °C h^{-1}, owing to their centimeter to millimeter dimensions (Vetere et al., 2013, 2015; Xu & Zhang, 2002). Lavas with thickness greater than 1 m experience cooling rates in the order of 10^0 to 10^{-1} °C h^{-1} in their intermediate and central portions (Vetere et al., 2015). Overall, most of the solidifying portions of basaltic to rhyolitic magmas erupted to the surface are exposed to $\Delta T/\Delta t$ that varies from 10^{-6} to 10^2 °C h^{-1}.

During ascent in the upper crust, the physicochemical state of H_2O-bearing magmas is greatly affected by pressure changes. The solubility of 1, 3 and 5 wt% H_2O has been modeled by the thermodynamic equation of Papale et al. (2006), using as input data the basaltic (B_{100})

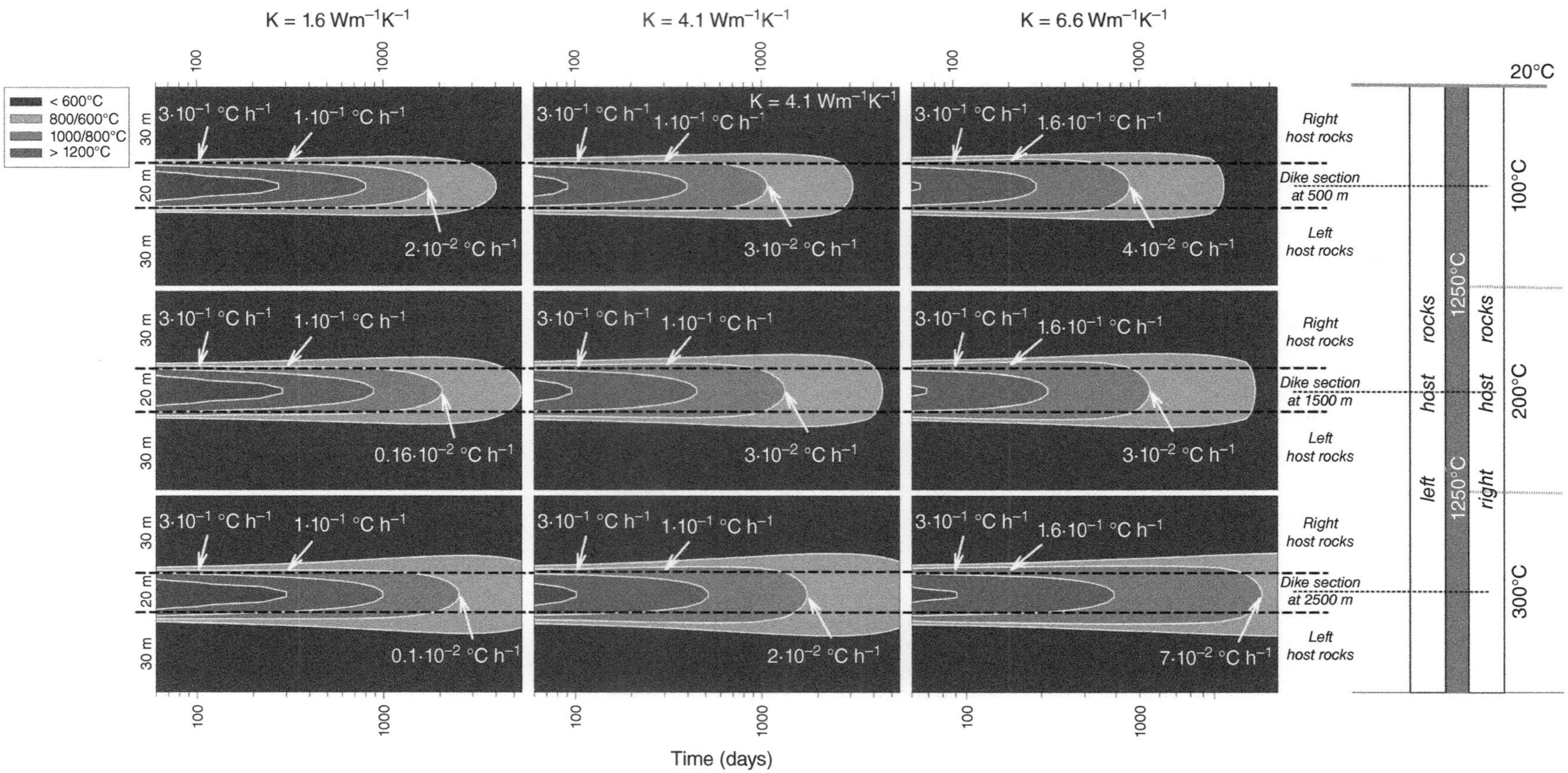

Figure 5.3 Simulations of cooling conditions of a dike 20 m wide intruded from a depth of 3 km. The conductive heat transfer has been modeled on three horizontal profiles, at depths of 2.5, 1.5, and 0.5 km. A vertical thermal gradient of 100°C km⁻¹ has been applied to the host rock. The thermodynamic parameters are listed in Table 5.1.

and rhyolitic (R_{100}) compositions previously used by Vetere et al. (2015; see Tables 5.2 and 5.3 for a complete list of the input parameters). The dissolved H_2O concentrations in the magma (or exsolved from it) are shown in Figure 5.5. Assuming high (2900 kg m^{-3}) and low (2600 kg m^{-3}) density values for the rock piles, representing oceanic (Carlson & Raskin, 1984) and continental (Hacker et al., 2015) crusts respectively, it is found that the exsolution of 5, 3, and 1 wt% of H_2O initiates at about 11–10, 5–4 and <2–1 km below the surface (Figure 5.5). At lower depths, the magma is bubble-bearing and its density progressively decreases with increasing exolution (Gonnermann & Manga, 2007; Lanzafame et al., 2017). At 2000, 1000 and 200/300 bar, H_2O does not exsolve from the magma in the absence of crystallization (see below). Notably, 5 wt% H_2O corresponds to a threshold value commonly documented for several magmatic systems; indeed, H_2O concentrations higher than 5 wt% are rarely reported for terrestrial magmas (Edmunds & Wallace, 2017; Gonnermann & Manga, 2007; Gonnermann & Taisne, 2015; Lanzafame et al., 2017, and reference therein). The exsolution/solubility paths shown in Figure 5.5 well approximate most of the *P*-driven processes described by the TCMS model. Assuming a total *P* difference of 4000 bar, travel times of 1 h, 1 day, 1 week, and 1 month, and rock densities of 2600 and 2900 kg m^{-3}, the $\Delta P/\Delta t$ rate ranges from 10^0 to 10^3 bar h^{-1}, translating to virtual magma ascent velocities of 10^{-2} to 10^1 km h^{-1} (see Table 5.4 for further details). H_2O-bearing minerals (amphibole and mica) may also crystallize at relative low temperatures (700–900°C), storing up to 4 wt% OH groups (Deer et al., 2013). Taking an amount of 10% by mass or volume of amphibole + mica, it is found that only 0.25–0.4 wt% H_2O dissolved in a magma can be eventually sequestrated. Consequently, the prominent crystallization of anhydrous mineral phases leads to an inevitable enrichment of dissolved/exsolved H_2O species (Burgisser & Degruyter, 2015; Edmunds & Wallace, 2017, and references therein).

In a first approximation, cooling-induced solidification ($\Delta T/\Delta t$) is progressively less effective moving from the

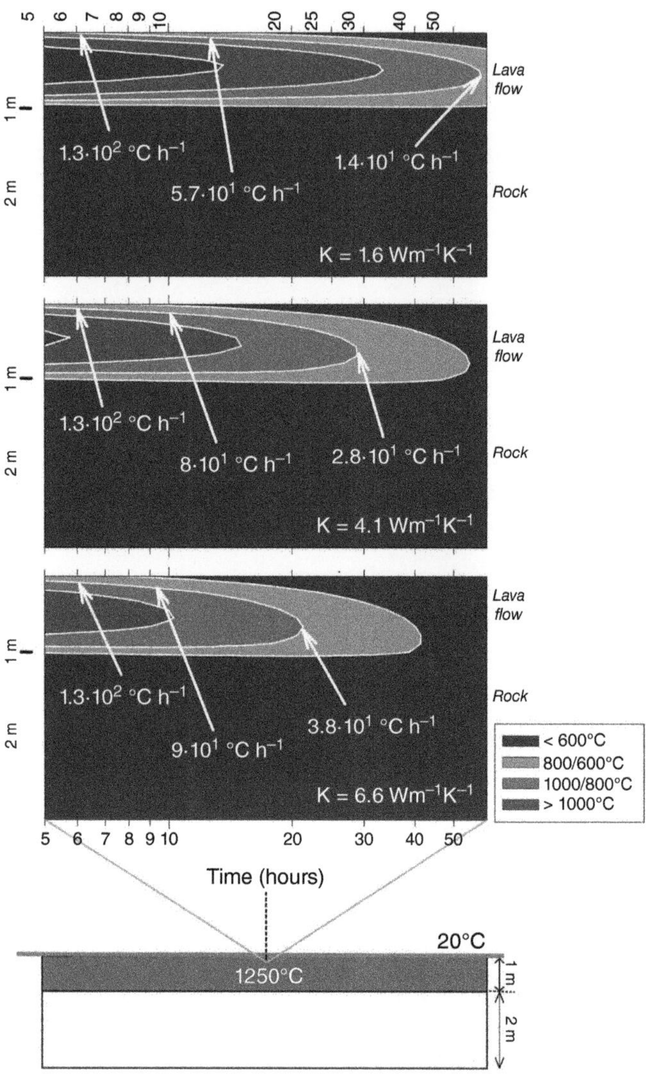

Figure 5.4 Simulations of cooling conditions of a lava flow 1 m thick. Conductive heat transfer has been modeled on the vertical profile, in the center of the flow. Thermodynamic parameters are listed in Table 5.1.

Table 5.3 Input Data for the H_2O Solubility Model of Papale et al. (2006)

T (°C)	1300–800
P (bar)	4000–50
H_2O (wt%)	1–3–5
CO_2 (wt%)	0.1

Table 5.2 Chemical Compositions (wt%) of Basalt and Rhyolite used for all the Simulations, with Standard Deviation in Parentheses

Magma	SiO_2	TiO_2	Al_2O_3	Fe_2O_3	MnO	MgO	CaO	Na_2O	K_2O	P_2O_5	Total	Fe^{2+}/Fe_{tot}
B_{100}	48.02	0.98	15.59	11.37	0.18	9.42	13.20	1.79	0.04	0.06	100.65	0.386
	(0.40)	(0.08)	(0.19)	(0.25)	(0.04)	(0.11)	(0.14)	(0.05)	(0.01)	(0.02)	(0.67)	(0.02)
R_{100}	73.97	0.12	13.48	2.29	0.08	0.44	1.36	3.75	4.89	0.03	100.41	0.342
	(0.67)	(0.06)	(0.17)	(0.16)	(0.05)	(0.05)	(0.08)	(0.17)	(0.08)	(0.02)	(0.84)	(0.02)

Note. Data from Vetere et al. (2013, 2015).

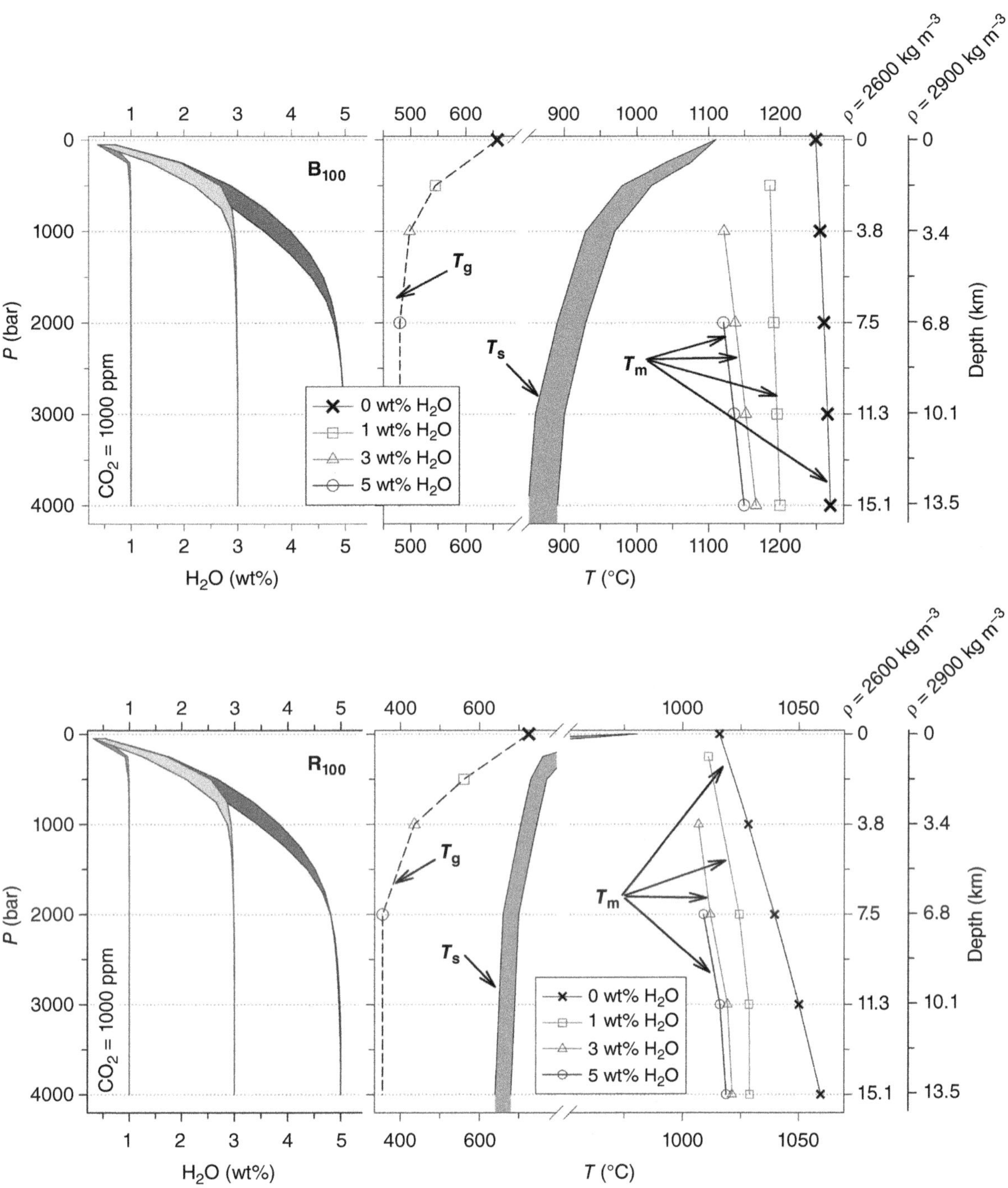

Figure 5.5 Modeling data for the solubility of H_2O in B_{100} and R_{100} magmas as a function of P in the left panels. The thermal ranges are 1300–800 °C and 1100–700 °C for B_{100} and R_{100}, respectively. The oxygen fugacity is buffered to NNO. The H_2O concentrations are 0, 1, 3, and 5 wt% (Table 5.2). Data were obtained using the thermodynamic code of Papale et al. (2006). On the right, the variations of T_m, T_s and T_g are shown as a function of P and H_2O. The values of T_m were estimated using the Pele code (Boudreau, 1999). The values of T_s for B_{100} and R_{100} were deduced from Latypov (2003) and Pistone et al. (2017), respectively. The values of T_g were obtained at a melt viscosity of 10^{12} Pa·s, in agreement with Giordano et al. (2008).

Table 5.4 Simulated ΔP/Δt Rates and Related Velocity Values

	ΔP	ΔP/Δt	P = 2600 kg m⁻³			P = 2900 kg m⁻³		
t (h)	(bar)	(bar h⁻¹)	Depth (km)	v (km h⁻¹)	v (m s⁻¹)	Depth (km)	v (km h⁻¹)	v (m s⁻¹)
1	4000	4×10^3	~15	1.5×10^1	4.2×10^0	~13	1.5×10^1	4.2×10^0
24 (1 day)		1.7×10^{-2}		6.3×10^{-1}	1.7×10^1		5.4×10^{-1}	1.5×10^1
168 (1 week)		2.4×10^{-1}		8.9×10^{-2}	2.5×10^{-2}		7.7×10^{-2}	2.1×10^{-2}
720 (1 month)		5.6×10^0		2.1×10^{-2}	5.8×10^{-3}		1.8×10^{-2}	5.0×10^{-3}

Earth's surface down to mantle depth and degassing-induced solidification ($\Delta P/\Delta t$) progressively shifts towards shallow depths (from 10/11 to a few kilometers) under decreasing amounts of H_2O (Figures 5.1–5.5).

5.4. PHYSICAL PROPERTIES OF MAGMAS

The *liquidus* or melting temperature (T_m), the *solidus* temperature (T_s), and the glass transition temperature (T_g) depend on the X–P–fO_2–H_2O conditions of the system and constrain the solidification path of magmas (Lasaga, 1997; Zhang, 2008). In this study, the variation of T_m has been computed through thermodynamically derived codes from the literature (Ghiorso & Sack, 1995; Gualda et al., 2012), using as input data B_{100} and R_{100} equilibrated at 100–4000 bar, 0–5 wt% H_2O, 1000 ppm CO_2, and fO_2 fixed at the NNO buffer. Figure 5.5 shows that the values of T_m for B_{100} and R_{100} compositions decrease with increasing H_2O concentration in the melt. At $P < 2000$ bar, the H_2O exsolution determines a substantial increase of T_m that is accompanied by a degassing-induced crystallization path. At $P > 2000$ bar, T_m increases only as a function of P (Brugger & Hammer, 2010; Fiege et al., 2015; Gonnerman & Manga 2007, 2012; Lipman et al., 1985). Likewise, the increase of H_2O dissolved in B_{100} and R_{100} significantly lowers T_s (Figure 5.5), but this temperature change is less effective with increasing P (Latypov, 2003; Pistone et al., 2017). T_g has been calculated for B_{100} and R_{100} using the viscosity model of Giordano et al. (2008). With both P and H_2O decreasing, it is found that T_g progressively decreases to depict a characteristic parabolic trend (Figure 5.5). An anhydrous or poorly hydrated magma will instead only solidify under the effect of cooling dynamics (Figures 5.1–5.5) down to T_g of the residual melt (Dingwell, 2006; Vetere et al. 2013, 2015; Zhang, 2008).

The rheological paths of lavas and magmas are strongly controlled by the composition of the silicate melt, as well as the type and amount of crystals (Applegarth et al., 2013; Burgisser & Degruyter, 2015; Gonnerman & Taisne, 2015; Harris, 2013; Hon et al., 2003; Hoover et al., 2001; Kilburn, 2004; Robert et al., 2014; Saar et al., 2001; Szramek et al., 2010). Following the procedure described in Lanzafame et al. (2013), the viscosities of B_{100} and R_{100} melts have been calculated using the model of Giordano et al. (2008) for melt-H_2O concentrations of 0, 1, 3 and 5 wt% (see data in Table 5.5). The viscosity values have then been used as input data for the model of Vona et al. (2011), in order to determine the apparent viscosity of the magma (melt + crystals). This approach does not account for the progressive change in the residual melt composition induced by the increasing crystal content. Therefore, results from calculations should be considered as conservative values for the viscosity of the system. In Figure 5.6, the viscosity of both B_{100} (T = 800–1200°C) and R_{100} (T = 700–1100°C) increases slightly, moderately, and abruptly at crystal contents of <30, 30–40 and 45 vol%, respectively (Figure 5.6). As the amount of H_2O increases from 0 to 5 wt%, the viscosity difference between B_{100} and R_{100} decreases (Figure 5.6). At 700°C (*solidus*), R_{100} is invariably above or close to T_g when either under anhydrous or hydrous conditions (at crystal contents >40 vol%). At 1100°C (*liquidus*), the viscosity of anhydrous R_{100} is 10^6–10^9 Pa·s, then decreases by three orders of magnitude with increasing H_2O (Figure 5.6). Close to the *liquidus* temperature, R_{100} is moderately to highly viscous (10^3–10^6 Pa·s) under hydrous conditions (Figure 5.6). On the other hand, B_{100} is extremely fluid (<10^3 Pa·s) close to its *liquidus*, whereas it is poorly viscous (10^3–10^6 Pa·s) at the *solidus* region, for crystal content <40 vol% and H_2O >1 wt%.

Table 5.5 Data Input for Viscosity and Density Modeling of B_{100} and R_{100} Magmas

Variable	B_{100}	R_{100}
T (°C)	1200–800	1100–700
H_2O (wt%)	0–1–3–5	
Crystals (vol%)	10–20–30–40–50	

An important rheological threshold is also provided by the carapace of lava flows, especially during emplacement of basaltic lavas, the most abundant on Earth's surface. Indeed, Hawaiian basalt effusions can develop *pahoehoe* crusts only at low to moderate strain rates, before reaching a yield strength corresponding to crystal contents <10–20 vol%. Transitional carapaces form when the crystal content is <20 vol%, whereas *aa* morphologies require degrees of crystallization of 30–40 vol% (Hon

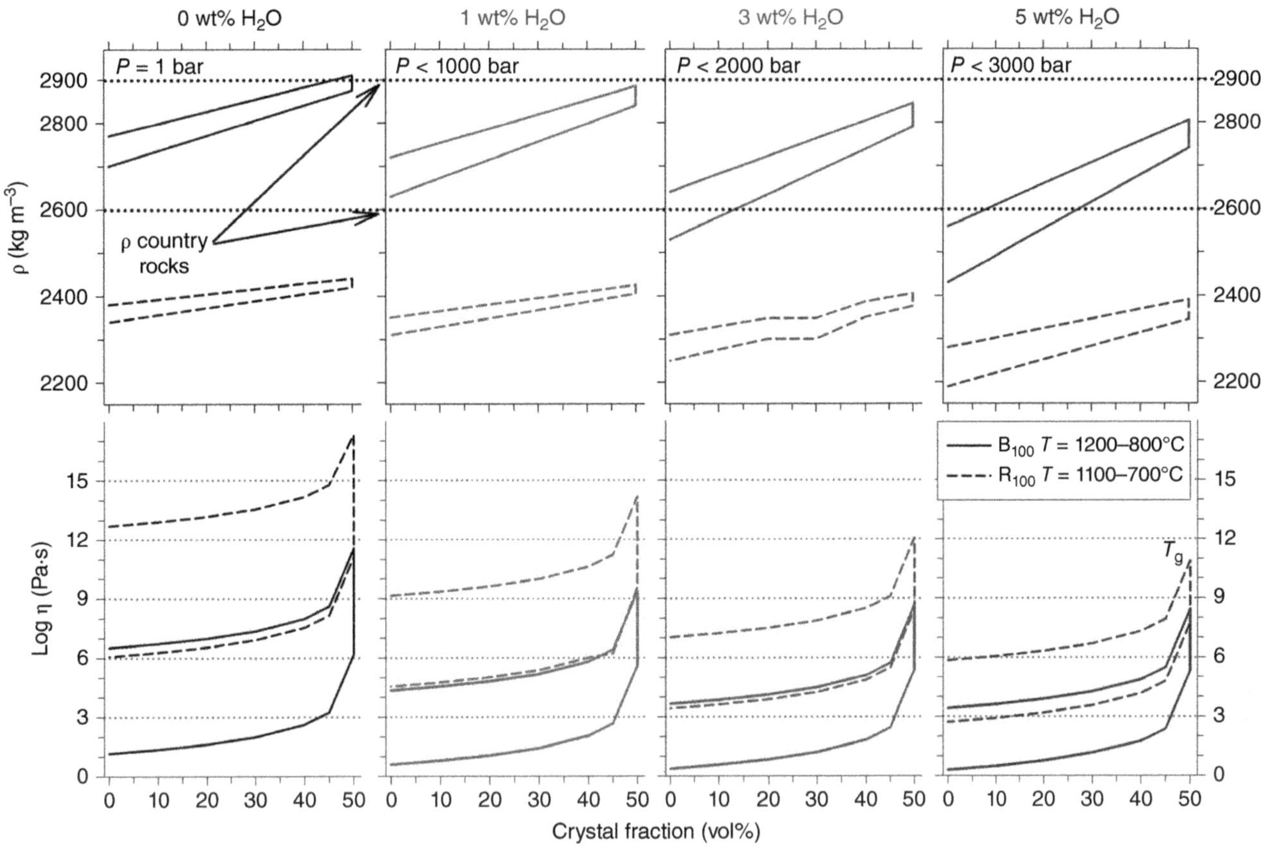

Figure 5.6 Viscosity (top panels) and density (bottom panels) modeling data of B_{100} and R_{100} melts as a function of crystal content (vol%). The parameters were calculated in the thermal ranges 1200–800 °C and 1100–700 °C for B_{100} and R_{100}, respectively. The viscosity was estimated following Giordano et al. (2008), using a melt strain rate of 0.1 s^{-1} and a crystal aspect ratio of 3:1. The density of suspension (magma + crystals) was calculated assuming 2700, 3400, and 2500 kg m^{-3} for plagioclase, clinopyroxene, and k-feldspar respectively, and using the Glass Density Calc v.3.1 code to determine the density of the magma (Lange & Carmichael, 1990).

et al., 2003; Hoover et al., 2001; Saar et al., 2001; Sehlke et al., 2014). Notably, these threshold values shift towards lower crystal contents for eruptive products with compositions more evolved than primitive basaltic melts (i.e., SiO_2 >46 wt%), such as lava flows emplaced at Mount Etna volcano (Kilburn, 2004; Lanzafame et al., 2013).

Another non-negligible parameter controlling magma dynamics is density, which may change greatly as a function of the degree of chemical differentiation and H_2O concentration. Using the values reported in Table 5.5 as input data for the model of Lange and Carmichael (1990), the melt-density change has been calculated assuming the same crystal proportions of clinopyroxene (3400 kg m^{-3}) and plagioclase (2700 kg m^{-3}) in B_{100}, as well as alkali-feldspar (2500 kg m^{-3}) in R_{100}. Results from numerical simulations show that B_{100} is invariably denser than R_{100} (Figure 5.6). Overall, the density progressively decreases with increasing H_2O and T. However, as the crystal content increases, more marked density changes are observed for B_{100} relative to R_{100} (Figure 5.6).

The ascent rate of magmas is controlled by the density contrast between the magma and wall rocks, as well as the magmatic overpressure (Gonnerman & Taisne, 2015; Lanzafame et al., 2017; Rivalta et al., 2015; Taisne & Jaupart, 2009). Assuming that the density of the continental crust is 2600 kg m^{-3}, it is found that R_{100} and only hydrated and crystal-poor B_{100} melts can ascend towards the surface. In contrast, poorly hydrated and crystal-rich B_{100} melts stall at depths (Figure 5.6). On the other hand, with respect to a typical oceanic crust with density of 2900 kg m^{-3}, the buoyant ascent of R_{100} and B_{100} magmas invariably takes place within the rock piles (Figure 5.6).

5.5. MAGMA ASCENT AND ERUPTION

Magmas with densities greater than those of the wall rocks can rise up through low to intermediate crustal depths (Figure 5.6). However, this is a necessary but insufficient condition. In a laminar regime within a volcanic dike or conduit, the ascent velocity (v) of a magma can be approximated by considering the conduit width (w), the gravitational acceleration ($g = 9.81$ m s^{-2}), the viscosity of the magma (η_{eff}), and density contrast ($\Delta\rho$) between magma and wall rocks (Del Gaudio et al., 2010; Lister & Kerr, 1991). The equation relating these variables is expressed in the form:

$$v = -[(w/2)^2/(3\eta_{eff})] \times \Delta\rho g$$

The ascent velocities of B_{100} and R_{100} magmas have been calculated as a function of the crystal content, assuming conduit widths of 5, 20, and 50 m, wall-rock densities of 2600 and 2900 kg m^{-3}, and magma viscosities and densities as discussed above (Figure 5.6). A conservative limit of $v < 1$ m s^{-1} is considered as indicative of absence of magma uprising and excluding the effect of magma overpressure and bubble formation (i.e., $P > P_{H2O}$; Figure 5.5). Results indicate that the upward migration of B_{100} magma is always possible above T_s, when the wall-rock density is 2900 kg m^{-3} (Figure 5.7a). Conversely, at the lower density of 2600 kg m^{-3}, the ascent of B_{100} magma may take place only under hydrous and crystal-poor conditions (close to *liquidus*). The same finding applies to the R_{100} magma (Figure 5.7b), thus excluding any possibility of upward migration under anhydrous and crystal-rich conditions.

At the base of the crust (i.e., at depth > 10–11 km), the crystallization of magma is mostly controlled by *equilibrium* processes where the intensive variables of the system remain almost constant over long periods, characterized by low $\Delta T/\Delta t$ (Figure 5.1) and negligible $\Delta P/\Delta t$ (Figure 5.5) changes (Lasaga, 1997; Scaillet & MacDonald, 2001; Zhang, 2008); as corroborated by the numerical simulations that form part of this study. Considering the Eyring equation, Si and Al are the most abundant cations in magmas and their diffusion length is on the order of 10^1 and 10^4 μm year^{-1} (cf. Vetere et al., 2013). Therefore, the *disequilibrium* characteristics of minerals formed at the base of the crust (depth of several kilometers) cannot be ascribed to rapid P–T changes. These *disequilibrium* features can be determined only by magma dynamics related to magma mixing and/or rejuvenation events (Cashman et al., 2017). Data plotted in Figures 5.5–5.7 delineate two reliable scenarios:

1. In the high-density (2900 kg m^{-3}) oceanic crust, B_{100} magma can attain high crustal levels (< 10–11 km) with moderate to low amounts of crystals.
2. In the low-density (2600 kg m^{-3}) continental crust, only high-T, hydrous, and crystal-poor B_{100} and R_{100} magmas can migrate towards the shallow crustal levels (Figures 5.6 and 5.7).

Indeed, the eruption of these magmas is possible only when the crystal content is lower than 50 vol% (Figure 5.6). Both B_{100} and R_{100} magmas can potentially solidify up to 50 vol% of minerals via degassing-induced crystallization within the continental crust (in the last 10–11 km), and B_{100} can also crystallize abundant mineral phases within the oceanic crust (Figures 5.6 and 5.7).

In oceanic divergent margins, a hypothetical B_{100} magma is prevalently degassed and effects related to $\Delta P/\Delta t$ are thus unimportant. By contrast, both B_{100} and R_{100} magmas in a convergent margin or stalling in shallow continental crust are easily enriched in H_2O. Consequently, solidification phenomena induced by $\Delta P/\Delta t$ effects are of primary importance during magma ascent and crystallization (Cassidy et al., 2018; Gonnermann & Manga, 2007; Lanzafame et al., 2017).

Except for convergent plate margins, the upper continental crust may extend from 10 to 14 km up to the surface, whereas the upper oceanic crust may be a few kilometers thick (Hacker et al., 2015). For H_2O contents ≤ 5 wt% and $P < 2500$ bar, T_m and T_s change drastically as a function of depth (Figure 5.5). Moreover, $\Delta P/\Delta t$ effects exert a primary control on the nucleation and growth of new crystal phases, as well as further overgrowth events on early-formed phenocrysts. These phenomena are determined by H_2O exsolution and the consequent expansion of the Tm region (Figure 5.5), while the chemical diffusivity of Si and Al cations remains the most important effect for the solidification kinetics (Vetere et al., 2015). These considerations are also strongly supported by results from decompression experiments conducted on hydrated systems (Arzilli & Carroll, 2013; Brugger & Hammer, 2010; Fiege et al, 2015; Hammer & Rutherford, 2002; Mètrich & Rutherford, 1998).

5.6. TEXTURES

The driving force for the crystallization is the degree of undercooling (ΔT), expressed as the difference between T_m and the actual T of the magma, as well the time required to attain a certain T (Lasaga, 1998; Zhang, 2008). Recently, Vetere et al. (2015) performed experiments using the same $\Delta T/\Delta t$ rates for six anhydrous melt compositions comprised between B_{100} and R_{100}. The general outcomes can be sumamrized as follows:

1. Melts with low viscosity (SiO_2-poor) rapidly nucleate crystals, whereas high viscous ones (SiO_2-rich) tend to vitrify or only slightly nucleate.
2. Fast kinetics and high SiO_2 content cause the suppression of plagioclase and favor the growth of pyroxene and spinel.
3. At a constant amount of SiO_2, fast kinetics ($\Delta T/\Delta t$ and possibly $\Delta P/\Delta t$) lead to dendritic textures and tiny crystal sizes.
4. At constant kinetic effects ($\Delta T/\Delta t$ and possibly $\Delta P/\Delta t$), the melt become progressively enriched in SiO_2 and the crystal texture changes from large, equant to tiny, dendritic morphologies (Figure 5.8).

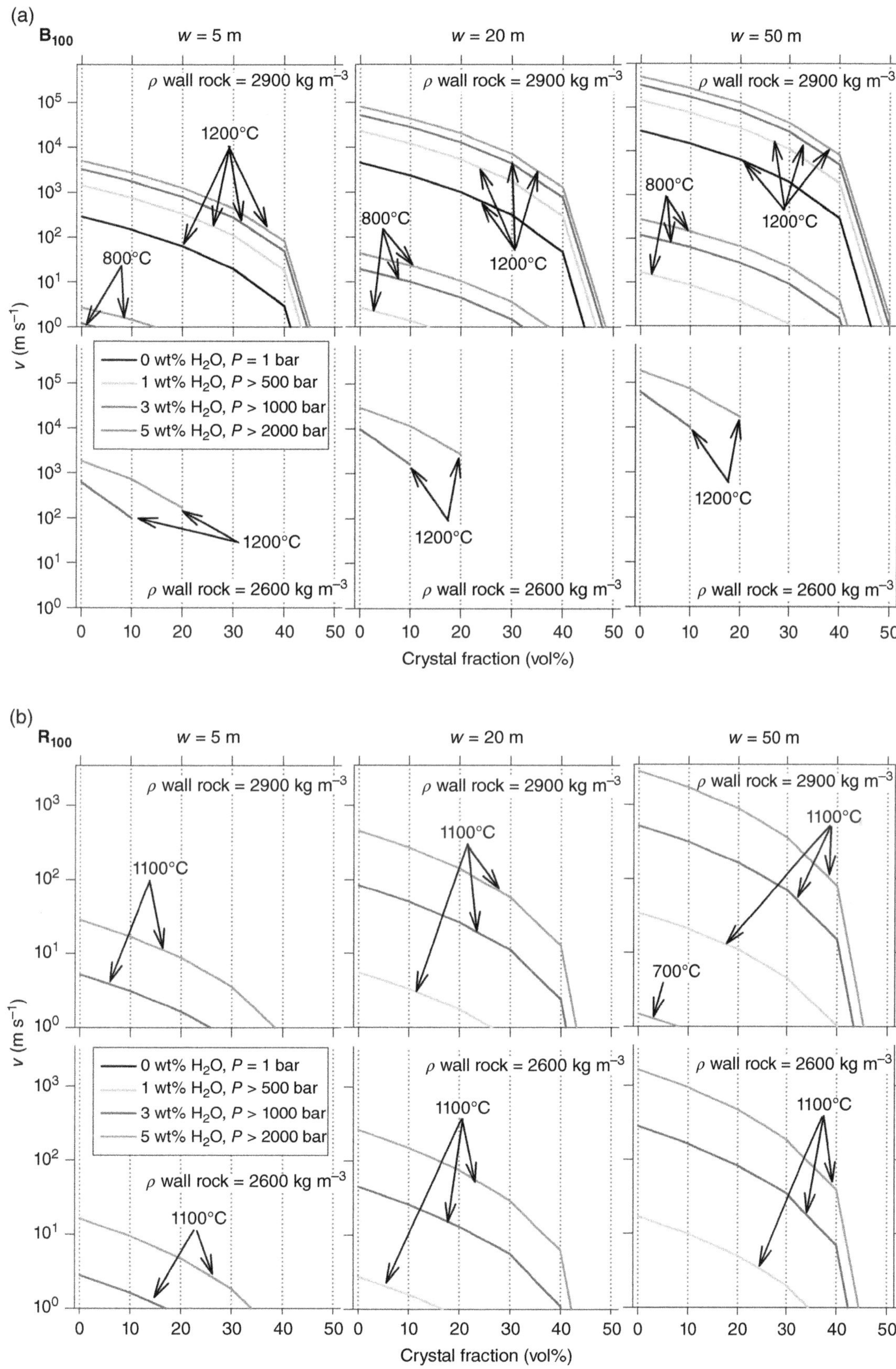

Figure 5.7 Estimates of magma ascent velocities for (a) B_{100} and (b) R_{100} as a function of the crystal content (vol%). The velocity values were obtained following Lister and Kerr (1991) and Del Gaudio et al. (2010), as well as considering different values of the conduit diameter (*w*) and the density of the country rock (*ρ*).

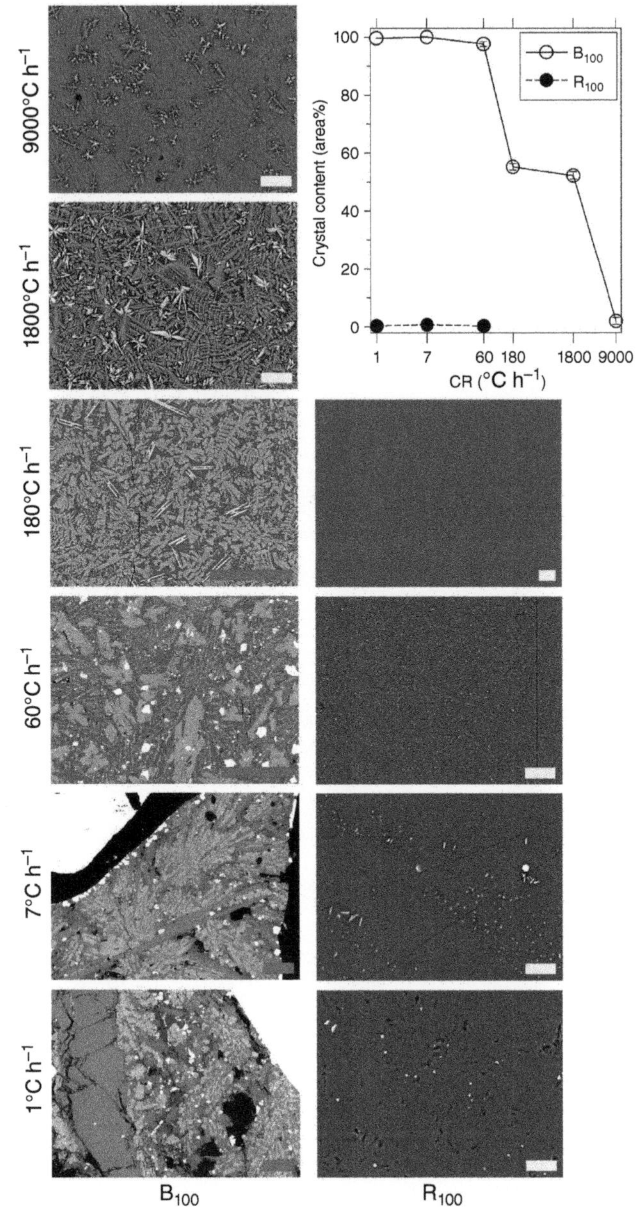

Figure 5.8 BSE-SEM microphotographs showing the textural features of (left) B_{100} and (right) R_{100} experimental products from Vetere et al. (2013, 2015). The red and yellow bars correspond to 100 and 10 μm, respectively. Crystal content is in the top right of the figure. Clinopyroxene, plagioclase, and spinel crystallize from the B_{100} melt, whereas only spinel nucleates from the R_{100} melt.

The critical cooling rate (R_c) is the minimum rate required to solidify ≤2 area% of crystals. The experimental data from Vetere et al. (2013, 2015) indicate that R_c changes from 10^4 to $<10^0$ °C h^{-1}, shifting from B_{100} to R_{100}. These two melts also show a very different glass-forming ability (GFA): B_{100} still nucleates at 9000°C h^{-1}, whereas R_{100} only slightly nucleates at 1°C h^{-1} (Figure 5.8). Therefore, crystal nucleation and growth rates are significantly delayed or even hampered with increasing the SiO_2 content in the melt (Iezzi et al., 2009; Vetere et al., 2015). This general picture also can be expanded to the hydrated systems, albeit more investigations are required to quantify the relationship between $\Delta P/\Delta t$ and X. However, it remains the fact that both $\Delta T/\Delta t$ and $\Delta P/\Delta t$ can exert a profound influence on nucleation and crystal-growth kinetics, as well as melt degassing phenomena.

5.7. KINETICALLY CONTROLLED CATION EXCHANGES IN CLINOPYROXENE AND PLAGIOCLASE

At slight undercooling ($<<\Delta T$) and low to moderate solidification rates, the advancing crystal surface maintains partitioning *equilibrium* with the surrounding melt and the crystal growth is considered interface controlled. Both crystal and melt show homogeneous compositions because chemical diffusion in the melt may persistently supply fresh cations to the advancing crystal surface, accounting for the slow growth rate of the crystal. In contrast, large undercooling ($>>\Delta T$) and high solidification rates, favor the development of a diffusive boundary layer in the melt next to the growing crystal face and the crystal growth is considered diffusion controlled. In the latter case, the deviation from *equilibrium* increases, as the nucleation and crystal-growth rates greatly exceed the diffusion rates of chemical species in the melt (Kirkpatrick, 1981; Lasaga, 1997; Zhang, 2008). At this condition, compositional gradients develop in the melt feeding the crystal growth due to the need for transport of chemical species to and from the crystal surface (Zhang, 2010). The components compatible with the crystal are depleted in the melt near the crystal surface, whereas the incompatible components are rejected from the advancing crystal surface and enriched in the adjacent melt. The disparity in growth *versus* diffusion rate causes the diffusive boundary layer forming at the crystal–melt interface to be preferentially enriched in elements incompatible with the crystal lattice, so that the advancing crystal surface incorporates cations at *nonequilibrium* proportions (Kirkpatrick, 1981). This latter situation enhances the development of dendritic crystal shapes (Iezzi et al., 2014).

5.8. CLINOPYROXENE DISEQUILIBRIUM GROWTH

Figure 5.9a shows the composition of the melt next to the surface of clinopyroxene crystals synthesized in the laboratory at both isothermal (1100 °C = 0 °C h^{-1}) and cooling rate (2.5, 10, and 50 °C h^{-1} from 1250 down to 1000 °C) conditions (Mollo, Blundy, et al., 2013; Mollo, Putirka, et al., 2013). The experiments were carried out

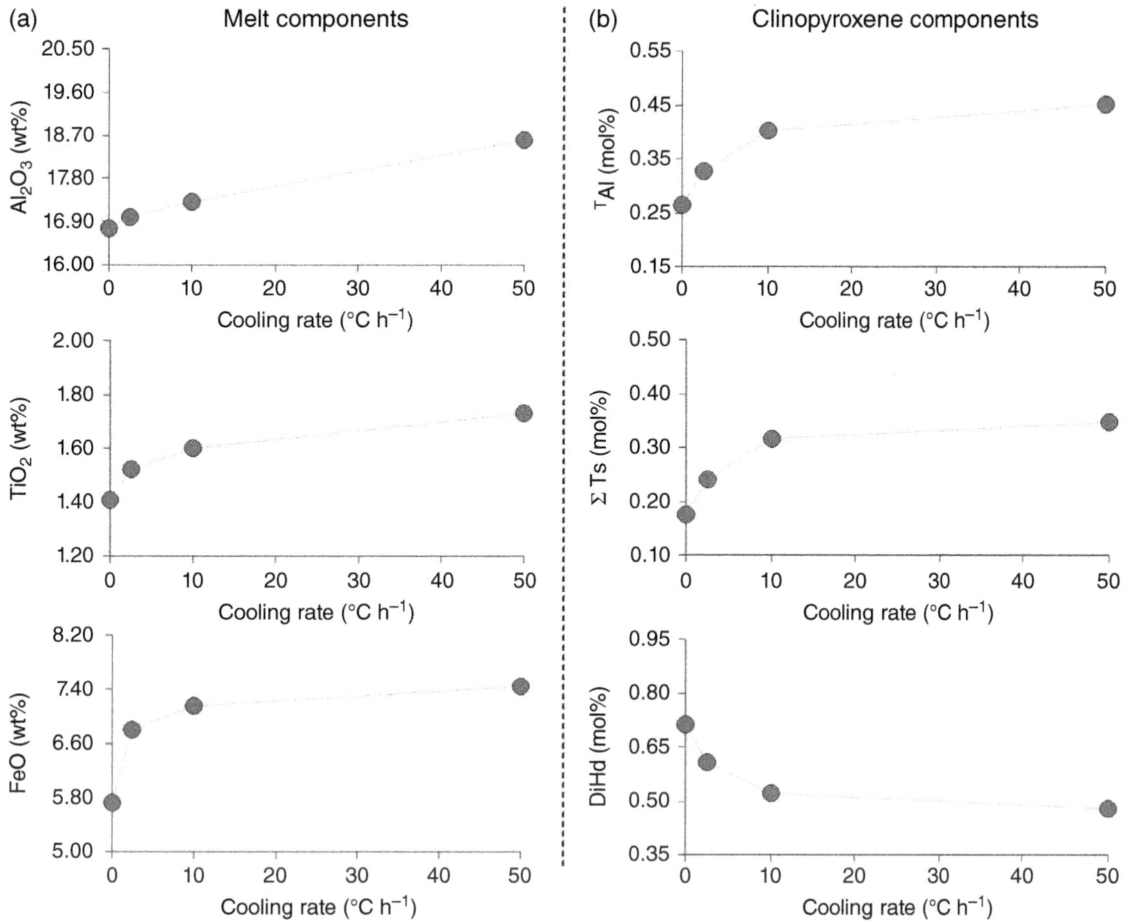

Figure 5.9 (a) Al_2O_3, TiO_2, and FeO in the melt next to the advancing crystal surface of clinopyroxene are plotted *versus* the cooling rate condition. (b) ^{T}Al, ΣTs, and DiHd in *(dis)equilibrium* clinopyroxene compositions are plotted *versus* the cooling rate condition. Data are from the experiments of Mollo, Blundy, et al. (2013) and Mollo, Putirka, et al. (2013) carried out at 1 atm and NNO+1.5 buffer, using as starting melt composition a primitive trachybasalt from Mount Etna volcano (Sicily). The *equilibrium* crystals were obtained at the isothermal temperature of 1100 °C (0 °C h^{-1}), whereas the (dis)equilibrium clinopyroxenes were grown under cooling rates of 2.5, 10, and 50 °C h^{-1} from 1250 down to 1100 °C. $\Sigma Ts = CaTs + CaFeTs + CaTiTs = Ca^{T}Al^{M2}AlSiO_6 + CaFeSiAlO_6 + CaTiAl_2O_6$. $DiHd = Di + Hd = CaMgSi_2O_6 + CaFeSi_2O_6$.

at 1 atm and NNO+1.5 buffer, using as starting melt composition a primitive trachybasalt from Mount Etna volcano (Sicily, Italy). The melt feeding the crystal growth is enriched in Al_2O_3, TiO_2, and FeO (Figure 5.9a), showing a clear dependence with the cooling rate condition due to the fact that cations incompatible with the crystal lattice build up at the clinopyroxene–melt interface. As the cooling rate increases, the crystal growth rate increases but the diffusion coefficients of chemical elements in the melt decrease with decreasing temperature (Kirkpatrick, 1981). Thus, the thickness of the diffusive boundary layer at the crystal–melt interface increases with increasing cooling rate, as the cations are less efficiently rejected away from the crystal interface, having less time to reequilibrate with the original bulk composition of the far-field melt (or relaxed melt in which all chemical gradients cease; Dowty, 1980; Hammer, 2008; Kirkpatrick, 1981; Mollo & Hammer, 2017).

The composition of clinopyroxene is sympathetic with the enrichment or depletion of cations in the diffusive boundary layer supplying nutrients to the advancing crystal surface. Figure 5.9b shows that, with respect to the isothermal temperature of 1100 °C (0 °C h^{-1}), clinopyroxenes are progressively enriched in tetrahedrally coordinated aluminium (^{T}Al), so that the amount of Tschermak (Ts) components ($\Sigma Ts = CaTs + CaFeTs + CaTiTs = Ca^{T}Al^{M2}AlSiO_6 + CaFeSiAlO_6 + CaTiAl_2O_6$)

increases at the expense of diopside and hedenbergite (DiHd = Di + Hd = $CaMgSi_2O_6$ + $CaFeSi_2O_6$). Mollo, Blundy, et al. (2013) and Mollo, Putirka, et al. (2013) have illustrated that, under the effect of undercooling, the chemistry of clinopyroxene is controlled by the exchange of ^{M2}[Mg, Fe^{2+}] with ^{M1}[Al, Fe^{3+}] coupled with the substitution of Si with ^{T}Al. The substitution of Na for Ca may also contribute to maintain charge balance requirements (Dobosi & Jenner, 1999). Moreover, highly charged cations, such as Ti, are also accommodated in the M1 site of clinopyroxene to balance the charge deficiency caused by the increasing concentration of aluminium (Grove & Raudsepp, 1978; Mollo, Blundy, et al., 2013; Mollo, Putirka, et al., 2013; Scarlato et al., 2014). As the cooling rate and the crystal growth rate increase, jadeite (Jd = $NaAlSi_2O_6$), enstatite (En = $Mg_2Si_2O_6$) and ferrosilite (Fs = $Fe_2Si_2O_6$) also increase (Mollo, Blundy, et al., 2013; Mollo, Putirka, et al., 2013). Note that large undercooling delays the crystal nucleation at lower temperatures and the depression of clinopyroxene saturation surface favors formation of Fe-rich crystals (Grove & Bence, 1977). On this basis, the kinetically controlled cation exchange in clinopyroxene can be expressed as $[Di + Hd]_{ss}$ + melt ↔ $[Ts + Jd + En + Fs]_{ss}$, thus reappraising the marked and continuous chemical variation of clinopyroxene with the increment of cooling rate observed either in previous dynamic experimental studies conducted on terrestrial and extraterrestrial materials (Baker & Grove, 1985; Gamble & Taylor, 1980; Grove & Bence, 1977; Grove & Raudsepp, 1978; Lofgren et al., 2006; Mollo et al., 2010, 2012; Mollo, Blundy, et al., 2013; Mollo, Putirka, et al., 2013) or in naturally cooled igneous rocks, such as lava flows and margins of magmatic intrusions (Baginski et al., 2009; Coish & Taylor, 1979; Faraone et al., 1988; Mevel & Velde, 1976; Mollo, Lanzafame, et al., 2011; Mollo et al., 2015; Scarlato et al., 2014; Smith & Lindsley, 1971; Ujike, 1982).

An important consequence of crystal formation at fast growth rates and the development of a thick diffusive boundary layer in the melt is that the value of the partition coefficient ($K_i = {}^{xls}C_i/{}^{melt}C_i$, where ${}^{xls}C_i$ and ${}^{melt}C_i$ are the concentrations of a chemical element i in the crystal and in the melt, respectively) and exchange partition coefficient ($K_{i\text{-}j} = K_i/K_j = {}^{xls}(C_i/C_j)/{}^{melt}(C_i/C_j)$ where i and j are two distinct chemical elements) may greatly change. The partitioning of chemical elements departs from *equilibrium* with the tendency for $K >> 1$ to decrease with increasing cooling rate, and to progressively increase for $K << 1$ (Grove & Bence, 1979; Grove & Raudsepp, 1978; Lofgren et al., 1979). For example, K_{Fe}, K_{Ti}, and K_{Na} increase under the effect of cooling rate (Figure 5.10a), responding to the inability of short-range diffusion in the melt to supply fresh cations to the rapidly growing surface of clinopyroxene (Gamble & Taylor, 1980; Lofgren et al., 1979). The chemistry of clinopyroxene becomes more similar to the melt supersaturated with incompatible cations, thus the partition coefficient approaches hypothetically to unity (Grove & Bence, 1979).

One of the most powerful tools used to verify the attainment of *equilibrium* crystallization between clinopyroxene and melt is the Fe–Mg exchange reaction of $K_{Fe\text{-}Mg} = 0.27 \pm 0.03$ (Putirka et al., 2003), despite the value derived from more recent experimental observations being $K_{Fe\text{–}Mg} = 0.28 \pm 0.08$ (Putirka, 2008). The Fe–Mg exchange is not perfectly invariant, accounting for a slightly temperature dependency that leads to the calibration of a *T*-sensitive model for $K_{Fe\text{–}Mg}$ (Putirka, 2008). Figure 5.10b shows that values of $K_{Fe\text{–}Mg}$ measured from *equilibrium* and kinetic crystallization experiments slightly increase from 0.24 to 0.29 with increasing cooling rate. Indeed, the FeO/MgO ratio in clinopyroxene increases with Ts, as the rapid crystal growth causes enrichments in *nonequilibrium* components. This compositional variation also has been documented for natural clinopyroxene phenocrysts from an ankaramite lava flow near the summit of Haleakala volcano at Hawaii (Hammer et al., 2016). Importantly, irrespective of the *disequilibrium* uptake of cations in rapidly growing crystals, the Fe–Mg exchange substantially departs from the *equilibrium* ranges of 0.27 ± 0.03 (Putirka et al., 2003) or 0.28 ± 0.08 (Putirka, 2008). This is much more evident when clinopyroxene–melt pairs are tested for *equilibrium* using the *T*-sensitive model for $K_{Fe\text{–}Mg}$ (Putirka, 2008), yielding an invariant value of 0.26 (Figure 5.10b). Evidently, both K_{Fe} and K_{Mg} tend towards an almost constant K_{Fe}/K_{Mg} ratio when the crystal growth rate is much greater than the mobility of both Fe and Mg in the melt (Mollo & Hammer, 2017).

The constancy of $K_{Fe\text{–}Mg}$ under rapid crystal growth conditions has important ramifications in the modeling of magmatic processes, leading to erroneous identification of *equilibrium* data. On the other hand, the Jd–DiHd exchange *equilibrium* formulated by Putirka et al. (1996) is highly sensitive to the thermal path of the system (Figure 5.10b), responding to the greater incorporation of Na cations in clinopyroxene and the lower solubility of DiHd caused by [Si] ↔ [^{T}Al] cation substitutions (Figure 5.9b). The Jd–DiHd exchange is remarkably sensitive to temperature and weakly sensitive to pressure because the volume change of this *equilibrium* is much smaller than for Jd formation. Therefore, clinopyroxene forming from rapidly cooled magmas are more inclined to record higher saturation temperatures with respect to slow cooling conditions (see below).

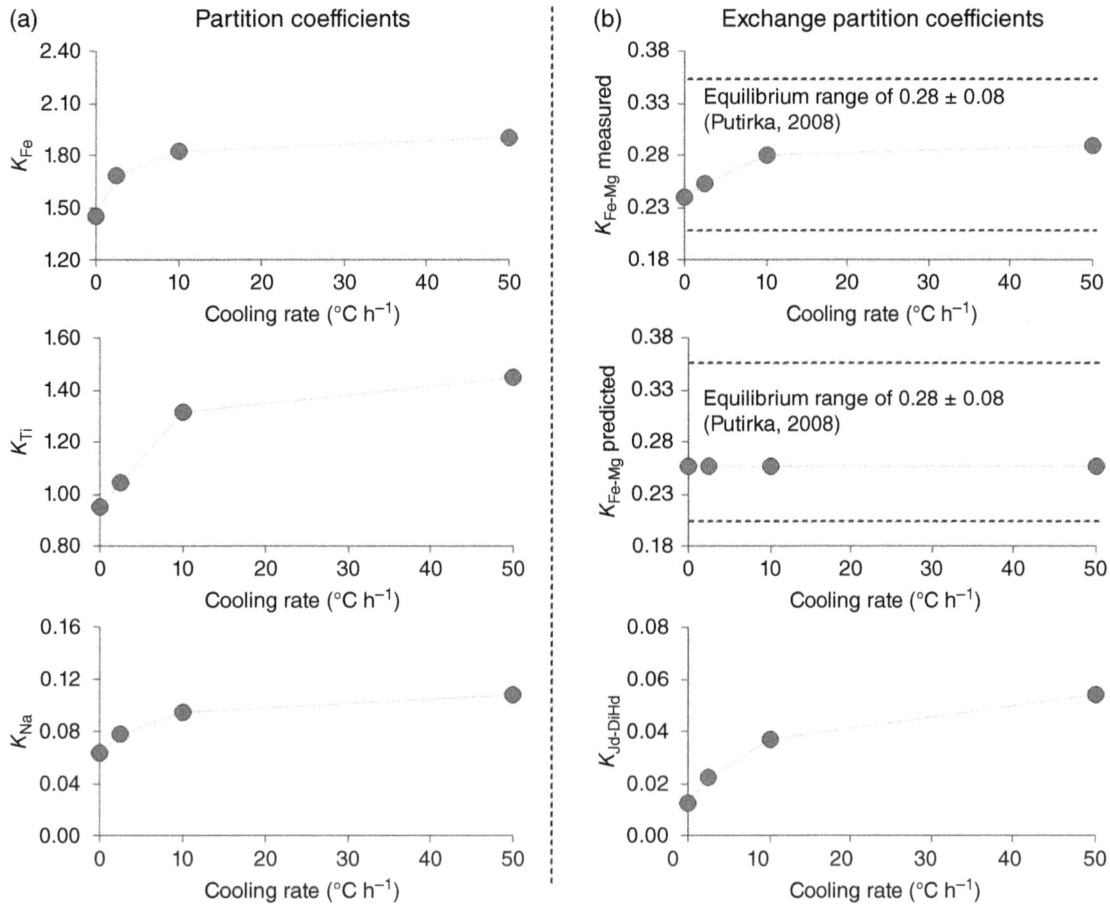

Figure 5.10 (a) K_{Fe}, K_{Ti}, and K_{Na} measured between clinopyroxene and melt are plotted *versus* the cooling rate condition. (b) $K_{Fe\text{–}Mg}$ measured, $K_{Fe\text{–}Mg}$ predicted, and $K_{Jd\text{-}DiHd}$ are plotted *versus* the cooling rate condition. $K_{Fe\text{–}Mg}$ predicted has been obtained through the *T*-sensitive model of Putirka (2008), while $K_{Jd\text{-}DiHd}$ has been formulated using that of Putirka et al. (1996).

5.9. PLAGIOCLASE DISEQUILIBRIUM GROWTH

Figure 5.11a shows the different composition of the melt feeding the growth of plagioclase crystals obtained in the laboratory at isothermal (1000 °C = 0 °C h^{-1}) and cooling rate (0.5, 2.1, 3, 15, and 9.4 °C min^{-1} from 1250 down to 1000 °C) conditions (Mollo, Putirka, et al., 2011). The starting composition used for the experiments is a trachybasalt from Mount Etna volcano solidified at 500 MPa and the redox state of NNO+1.5 buffer (Mollo, Putirka, et al., 2011). The effect of cooling rate on the melt layer is related directly to the rapid growth of plagioclase, whereby nominal incompatible elements are rejected less efficiently into the melt next to the advancing crystal surface. Therefore, the amounts of CaO, FeO, and MgO progressively increase in the diffusive boundary layer that develops at the crystal–melt interface (Figure 5.11a), so that plagioclase crystals incorporate higher proportions of anorthite (An = $CaAl_2Si_2O_8$), FeO, and MgO with increasing cooling rate (Figure 5.11b). Overall, the *disequilibrium* growth of plagioclase is accompanied by the cation exchange [Si + Na + K] ↔ [Al + Ca + Fe + Mg], reproducing most of the natural plagioclases crystallized at large undercooling during magma ascent in volcanic conduits (Singer et al., 1995; Scarlato et al., 2017) or at the rapidly chilled margin of magmatic intrusions (Loomis, 1981; Loomis & Welber, 1982; Mollo, Lanzafame, et al., 2011; Pietranik et al., 2006; Ujike, 1982). The replacement of [K + Na] with [Fe + Mg] is ascribed to the inability of plagioclase to reequilibrate with the rapidly cooled melt, thus producing high-temperature, An-rich compositions (Drake, 1975). This results from the very slowly coupled CaAl–NaSi interdiffusion, which inhibits the intracrystalline cation redistribution and favors the retention of early cation concentrations (Baschek & Johannes, 1995; Liu & Yund, 1992). Indeed, plagioclase crystals sensibly enriched in An are typically observed in either decompression

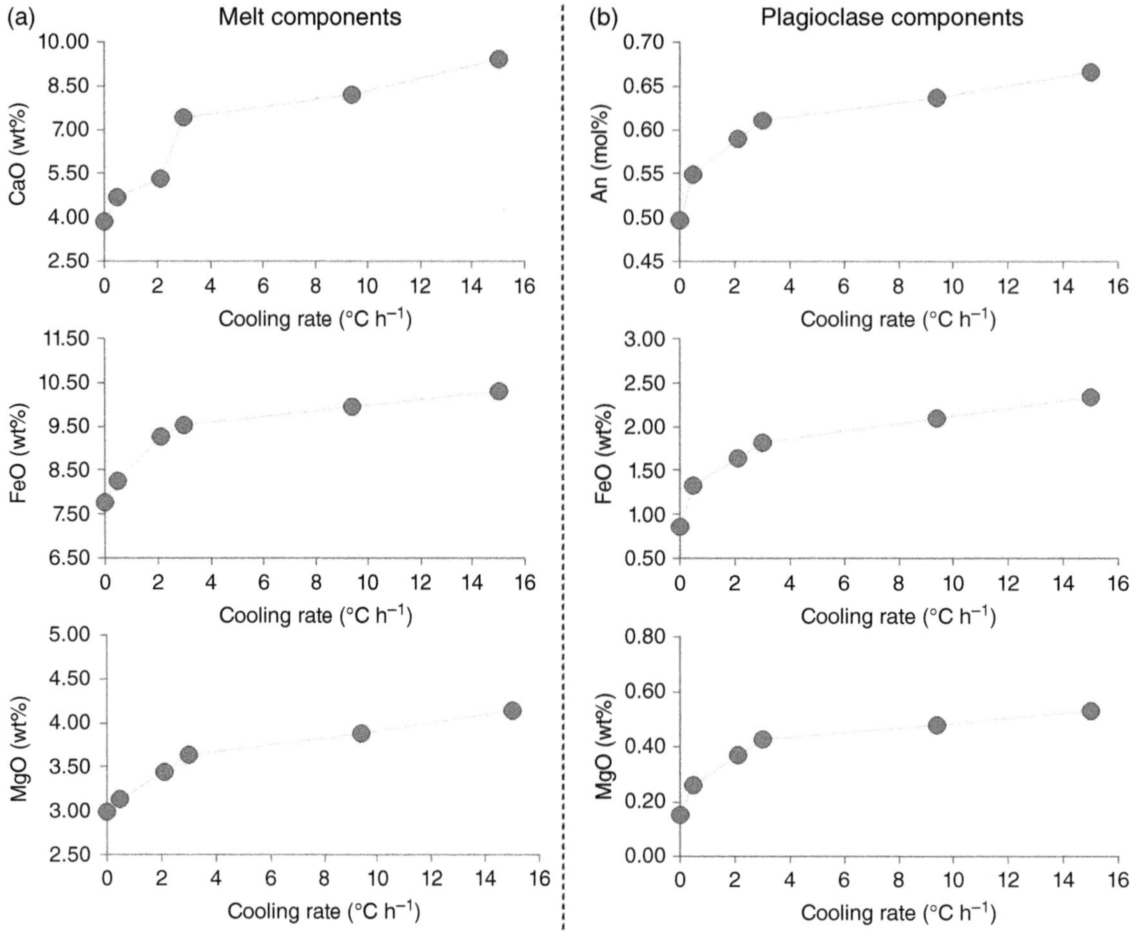

Figure 5.11 (a) CaO, FeO, and MgO in the melt next to the advancing crystal surface of plagioclase are plotted *versus* the cooling rate condition. (b) An, FeO, and MgO in *(dis)equilibrium* plagioclase compositions are plotted *versus* the cooling rate condition. Data are from the experiments of Mollo, Putirka, et al. (2011), carried out at 500 MPa and NNO+1.5 buffer using as starting melt composition a primitive trachybasalt from Mount Etna volcano (Sicily). The *equilibrium* crystals were obtained at the isothermal temperature of 1000 °C (0 °C h^{-1}), whereas the *disequilibrium* plagioclases were grown under cooling rates of 0.5, 2.1, 3, 9.4, and 15 °C h^{-1} from 1250 down to 1000 °C. An = $CaAl_2Si_2O_8$.

(Brugger & Hammer, 2010) and cooling experiments (Iezzi et al., 2008, 2011; Mollo, Putirka, et al., 2011; Mollo et al., 2012). Since the strength of the ^{T}Si–O bonds in the melt is greater than that of ^{T}Al–O bonds (Kirkpatrick, 1983; Iezzi et al., 2014), the transfer rate of Al cations from the melt to the surface of rapidly growing crystals is facilitated more than that of Si cations. As a consequence, the [Si + Na] ↔ [Al + Ca] exchange parallels the higher mobility of Al in the melt relative to that of Si under kinetically controlled conditions (Zhang, 2010). In this context, Fe increases in the crystal lattice due to the fact that decreasing amounts of Si cations leaves a larger number of available T and M sites for Fe substitution with increasing An (Smith, 1983; Smith & Brown, 1988; Tegner, 1997).

Figure 5.12a shows that the partition coefficients of Al, Fe, and Mg increase monotonically with cooling rate, responding to the *disequilibrium* uptake of cations incompatible with plagioclase crystal structure (cf. Scarlato et al., 2017). On the other hand, the Ca–Na exchange sensibly decreases with increasing cooling rate (Figure 5.12b), despite the effects of temperature and pressure being negligible for the value of $K_{Ca\text{-}Na}$, which, in turn, is largely controlled by the amount of H_2O in *equilibrium* with plagioclase (Sisson & Grove, 1993; Ushioda et al., 2014). Therefore, the remarkable variations of Ca–Na exchange along a temperature-dependent path (Figure 5.12b) are ascribable to the rapid growth of plagioclase and the development of a diffusive boundary layer in the melt adjacent to the crystal face. The Ab–An exchange also has

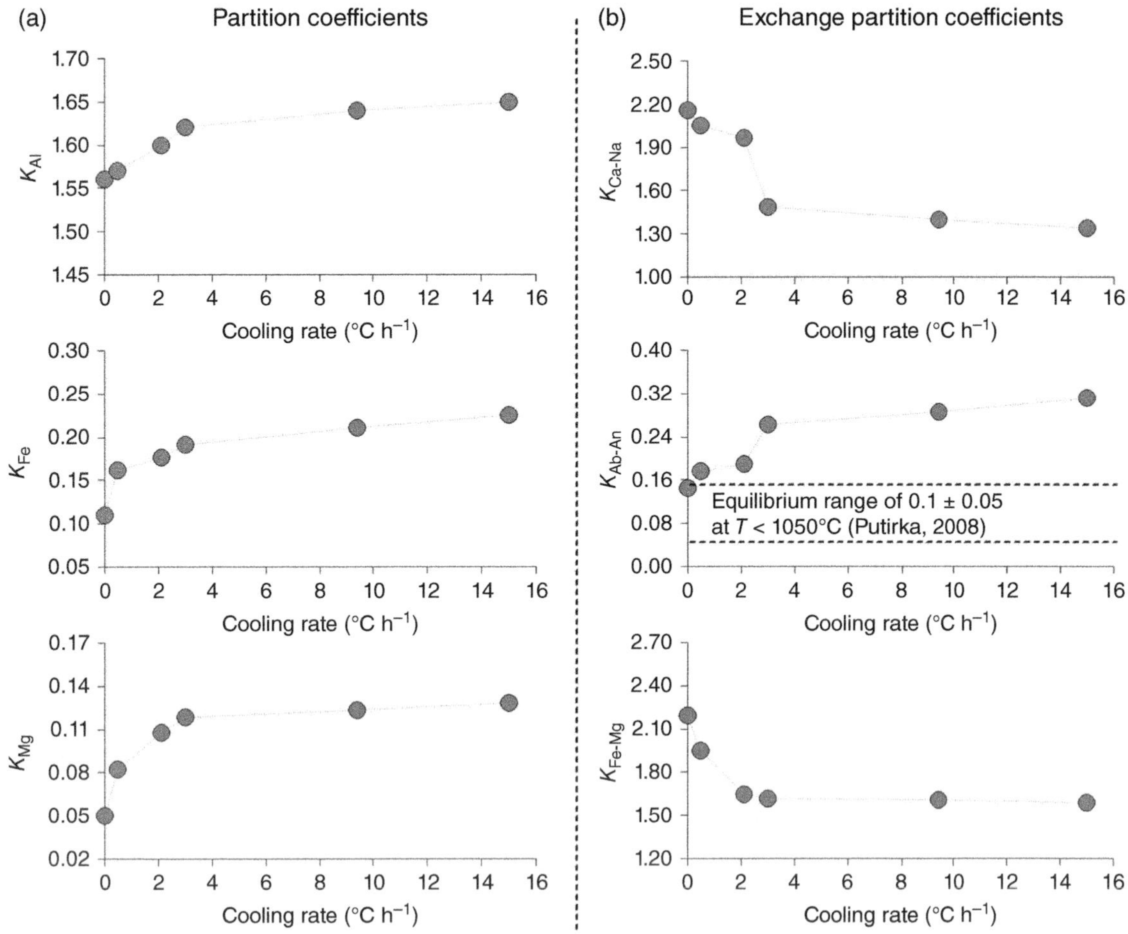

Figure 5.12 (a) K_{Al}, K_{Fe}, and K_{Mg} measured between plagioclase and melt are plotted *versus* the cooling rate condition. (b) $K_{Ca\text{-}Na}$, $K_{Ab\text{-}An}$, and $K_{Fe\text{-}Mg}$ are plotted *versus* the cooling rate condition. The Ab–An exchange also has been proposed by Putirka (2008) as a possible test for *equilibrium* due to its constant values of $K_{Ab\text{-}An} = 0.10 \pm 0.05$ and 0.27 ± 0.11 at $T < 1050\,°C$ and $T \geq 1050\,°C$, respectively.

been proposed by Putirka (2008) as a possible test for *equilibrium* due to its constant values of $K_{Ab\text{-}An} = 0.10 \pm 0.05$ and 0.27 ± 0.11 at $T < 1050\,°C$ and $T \geq 1050\,°C$, respectively. $K_{Ab\text{-}An}$ increases from 0.14 (i.e., isothermal temperature of 1000 °C = 0 °C h^{-1}) to 0.31 (i.e., maximum cooling rate of 16 °C h^{-1}), denoting strong departure from *equilibrium* relative to the resting temperature of the experiments (Figure 5.12b). This finding is not surprising because the *disequilibrium* chemistry of plagioclase attempts to mimic the An-rich compositions (Figure 5.11b) typical of high-temperature (i.e., $T \geq 1050\,°C$ and $K_{Ab\text{-}An} = 0.27 \pm 0.11$) crystallizing systems (Mollo, Putirka, et al., 2011; Mollo et al., 2012; Iezzi et al., 2014). At the same time, the Fe–Mg exchange decreases by ~27% under the influence of kinetic effects (Figure 5.12b), in apparent contrast with the weak variation observed for clinopyroxene (Figure 5.10b). Since Fe and Mg cations are incompatible with plagioclase, they are incorporated in the crystal lattice at higher *nonequilibrium* proportions with increasing crystal growth rate (Figure 5.11a). This leads to the conclusion that, different to that observed for clinopyroxene, the Fe–Mg exchange reaction has the potential to reveal strong departure from *equilibrium* for plagioclase-bearing cooling magmas (Mollo, Lanzafame, et al., 2011; Mollo, Putirka, et al., 2011).

5.10. ASSESSMENT OF EQUILIBRIUM AND ESTIMATE OF *P*–*T*–H_2O CRYSTALLIZATION CONDITIONS

Putirka (1999) proposed for the first time a model to test for *equilibrium* between clinopyroxene and host magma based on the difference (Δ) between DiHd predicted for clinopyroxene via regression analyses of clinopyroxene–melt pairs from *equilibrium* experiments, and DiHd measured in natural phenocrysts under investigation. Furthermore, Mollo, Putirka, et al. (2013) refined this test for *equilibrium* extending its application to a

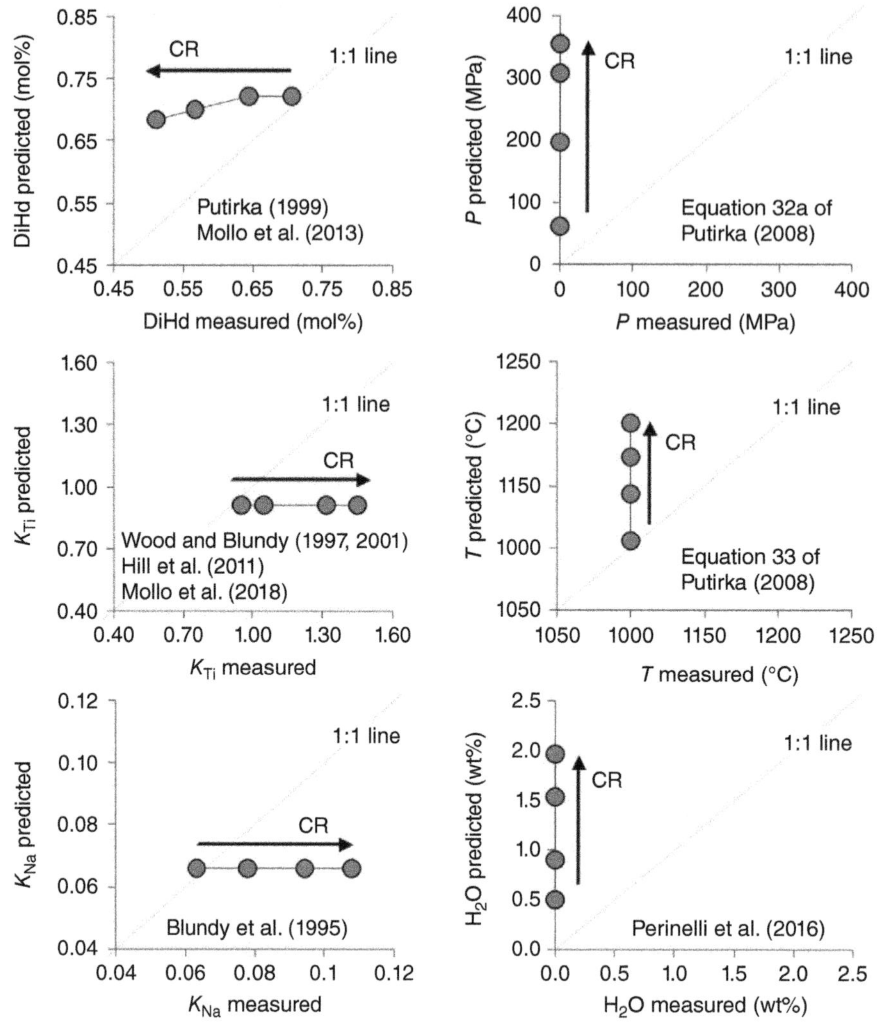

Figure 5.13 The departure from *equilibrium* of clinopyroxene compositions obtained under the effects of increasing cooling rate (CR) conditions has been modeled through the *equilibrium* expressions of DiHd (Putirka, 1999; Mollo, Putirka, et al., 2013), K_{Ti} (Hill et al., 2011; Mollo et al., 2018; Wood & Blundy, 1997, 2001), and K_{Na} (Blundy et al., 1995), as well as barometric, thermometric (Putirka, 2008), and hygrometric equations (Perinelli et al., 2016).

broader range of compositions and providing evidence that, under variable undercooling conditions, ΔDiHd is a more robust test for *equilibrium* than $K_{Fe\text{-}Mg}$ (Figure 5.12b). Accordingly, Figure 5.13 shows that not all DiHd values measured for *equilibrium* (isothermal temperature of 1100 °C = 0 °C h^{-1}) and *disequilibrium* (cooling rates of 2.5, 10, and 50 °C h^{-1}) experiments align along the one-to-one *equilibrium* line. As the degree of cooling increases, the rapid growth of clinopyroxene is accompanied by the exchange [DiHd] ↔ [Ts] and, consequently, measured *versus* predicted DiHd components progressively depart from the *equilibrium* (i.e., ΔDiHd increases as the measured DiHd component decreases). The *disequilibrium* incorporation of cations incompatible with the clinopyroxene crystal lattice (i.e., Ti and Na) is also indicated by the thermodynamically derived models for K_{Ti} and K_{Na} (Figure 5.13). More specifically, the partitioning behavior of Ti between clinopyroxene and melt has been estimated through the model of Mollo et al. (2018), in which the formalisms of Wood & Blundy (1997, 2001) and Hill et al. (2011) have been integrated. This leads to a formulation of the lattice-strain theory that accounts for the increasing probability of cations entering a locally charge-balanced site as ^{T}Al increases and the electrostatic work done on transferring one mole of cations from silicate melt to the clinopyroxene crystallographic site. Figure 5.13 shows that K_{Ti} measured for rapidly growing clinopyroxenes differs from the value expected at the thermodynamic *equilibrium*. As the [DiHd] ↔ [Ts] exchange reaction proceeds, the proportion of M1 sites charge-balanced by Ti increases with decreasing Mg. K_{Ti} is positively correlated with ^{T}Al because Ti enters the M1 site of $CaAl_2SiO_6$ (3+ charge on M1) much more readily than $CaMgSi_2O_6$ (2+ charge on M1). Additionally, the increase in proportions of Na and ^{T}Al cations serves to increase K_{Ti} by increasing the charge-balanced configurations available to accommodate the highly charged cations in the M1 sites (Hill et al., 2000; Lundstrom et al., 1994; Schosnig & Hoffer, 1998;). This is attested by the concurrent increase of $K_{Jd\text{-}DiHd}$ (Figure 5.10b) with both K_{Ti} and K_{Na} (Figure 5.13), where the *equilibrium* partitioning of Na is predicted by the thermodynamic expression of Blundy et al. (1995) calibrated through thermochemical data on the jadeite melting *equilibrium*.

The immediate effect of crystal-growth kinetics is that clinopyroxene-based barometers, thermometers, and hygrometers respond to the apparent partitioning of chemical elements between the rapidly growing crystal face and the diffusive boundary layer in the melt. As a consequence, these models tend to overestimate the pressure (from equation 32a of Putirka, 2008), temperature (from equation 33 of Putirka, 2008), and melt-H_2O content (from the expression of Perinelli et al., 2016) in *equilibrium* with clinopyroxene (Figure 5.13). Importantly, there is a positive correlation between the degree of undercooling, the value of ΔDiHd, and the magnitude of overestimation (Figure 5.13). The errors of *P*, *T*, and H_2O estimates caused by the mistaken use of *disequilibrium* compositions progressively decrease as ΔDiHd goes to zero (Mollo & Masotta, 2014). Thus, the value of ΔDiHd is a proxy of the degree of *disequilibrium* of a cooling system, allowing the estimate error to be minimized. The monotonic increase of *P*, *T*, and H_2O estimates with increasing undercooling can be intuitively explained through thermometric considerations. Within the thermal range of ~100 °C estimated by the thermometer, the clinopyroxene crystals grow over an everchanging temperature whose saturation surface is bounded between the liquidus and the final resting temperature of the crystal growth (Mollo & Masotta, 2014; Mollo, Putirka, et al., 2013). During cooling, there is no single *equilibrium* temperature but a range of relevant temperatures in which the crystal growth is rate controlled by the diffusion of chemical species in the melt. In fact, according to the Arrhenius expression, the diffusion coefficient decreases with decreasing temperature, causing a convex relationship between the apparent partition coefficient and the cooling rate (Figure 5.10). Thus, with increasing cooling rate, the thermometer predicts increasing crystallization temperatures (i.e., the Jd–DiHd exchange increases; Figure 5.13) somewhere between the saturation temperature of clinopyroxene and the closure temperature of crystal-growth kinetics (i.e., the temperature of quenching or the temperature at which cation exchanges are frozen in).

With respect to the most common igneous minerals, plagioclase is characterized by a marked sensitivity to pressure, temperature, and melt-H_2O content variations during magmatic processes. Therefore, either a decrease in the saturation temperature due to magma cooling or the increase in the *liquidus* temperature of the melt due to decompression under H_2O-saturated conditions may cause substantial deviation from *equilibrium* for plagioclase composition. Figure 5.14 shows results from calculations performed using compositions from *equilibrium* and *disequilibrium* plagioclase-bearing experiments reported in Mollo, Putirka, et al. (2011). These data have been tested with the *equilibrium* model of Namur et al. (2012), which predicts the An content in plagioclase as a function of melt composition. The model tends to overestimate the An content in plagioclase at cooling rates higher than 2.1 °C h^{-1} due to the CaO-rich composition of the melt feeding the rapid growth of plagioclase (Figure 5.14). At the same time, the complementary model of Hamada and Fuji (2007) predicts that the value of $K_{Ca\text{-}Na}$ is higher than that measured for *disequilibrium* plagioclase–melt pairs (Figure 5.14).

The most important consequence of kinetically controlled cation redistributions at the crystal–melt interface is that plagioclase-based thermometers, barometers, hygrometers, and oxygen barometers are affected by large uncertainties. For example, the oxygen barometer of Lundgaard and Tegner (2004) accounts for the partitioning of ferric and ferrous iron between plagioclase and melt at *equilibrium* proportions. The Fe_2O_3–FeO exchange is positively correlated with An in plagioclase and, thus, the oxygen barometer predicts higher redox states for the rapidly cooled plagioclase crystals (Figure 5.14). Systematic overestimates are also observed for the barometer, thermometer, and hygrometer (equations C, A, and H in Putirka, 2005), which are calibrated through the Ca–Na exchange. The models yield maximum errors of 400 MPa, 130 °C, and 2 wt% H_2O at the fastest cooling rate of 15 °C h^{-1} (Figure 5.14). In general, cooling rates higher than 2.1 °C h^{-1} are fast enough to hamper the partitioning of cations at *equilibrium* proportions, responding to the development of a diffusive boundary layer at the crystal–melt interface (Figure 5.14). This results in the plagioclase composition recording the apparent temperature of the early stage of the crystal growth rather than the effective temperature of the attainment of *equilibrium* with the surrounding melt. The magnitude of this effect increases with increasing disparity between crystal growth rate and diffusion of chemical components in the melt. Consequently, the precision of thermometers, barometers, and hygrometers decreases (Figure 5.14), producing the effect of an increasing degree of *disequilibrium* due

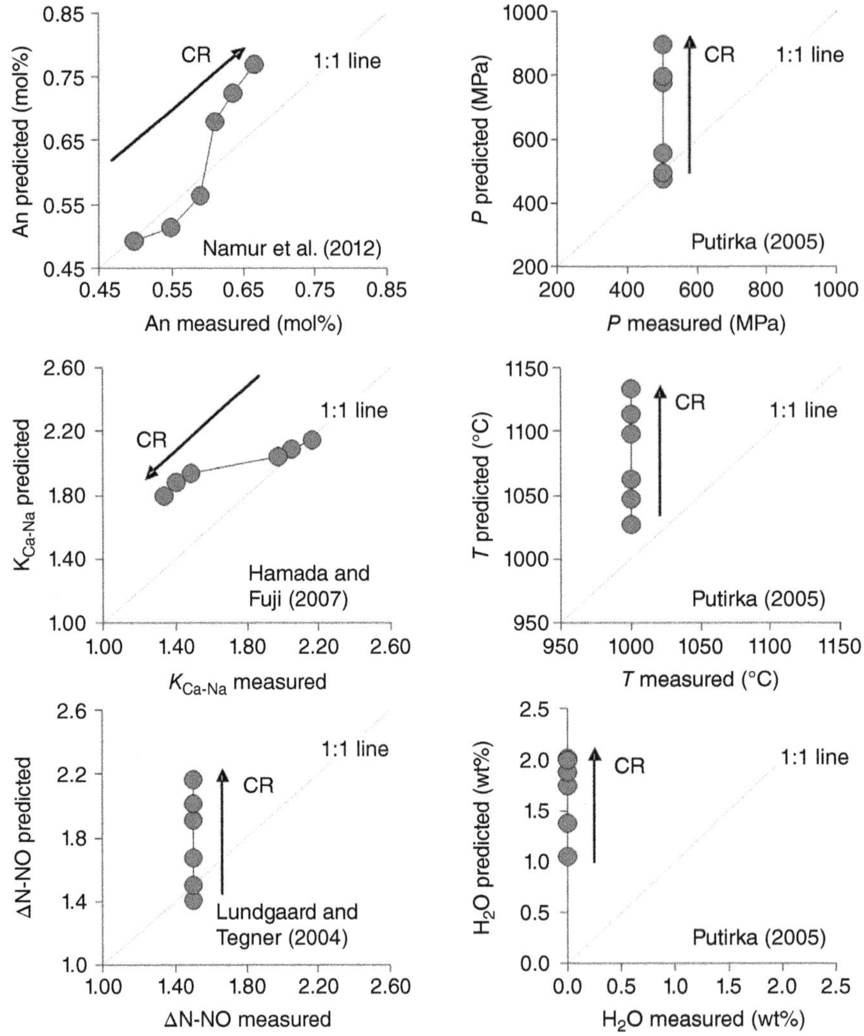

Figure 5.14 The departure from *equilibrium* of plagioclase compositions obtained under the effects of increasing cooling rate (CR) conditions (Mollo, Putirka, et al., 2011) has been modeled through the *equilibrium* expressions of An (Namur et al., 2012) and $K_{Ca\text{-}Na}$ (Hamada & Fuji, 2007), as well as oxygen barometric (Lundgaard & Tegner, 2004), barometric, thermometric, and hygrometric (Putirka, 2005) equations.

to kinetic variations in element partitioning at the crystal–melt interface.

5.11. THE IMPORTANCE OF MODEL CALIBRATION TO RETRIEVE THE INTENSIVE VARIABLES OF MAGMA

An intriguing feature of magma modeling based on clinopyroxene chemistry is that a large number of barometers and thermometers has been proposed in literature over recent decades (Lindsley, 1983; Masotta et al., 2013; Mollo et al., 2018; Neave & Putirka, 2017; Nimis, 1999; Nimis & Taylor, 2000; Nimis & Ulmer, 1998; Putirka, 1999, 2008; Putirka et al., 1996, 2003). As anticipated above, these models are affected by an increasing uncertainty as the cooling rate of the system increases (Figure 5.15). Barometers and thermometers are based on the assumption of *equilibrium* and, therefore, any form of *disequilibrium* is source of error. The ability prediction of barometers and thermometers is controlled by a number of factors that may or may not change simultaneously along the decompression and cooling path of magmas. Moreover, these models are calibrated using different crystal and melt compositional parameters that are more or less sensitive to the [DiHd] ↔ [Ts] exchange. If we inadvertently use *disequilibrium* mineral–melt pairs to predict the crystallization conditions of magmas, the magnitude of the systematic error of estimate may change significantly as a function of the model calibration. Consequently, some barometers and thermometers are

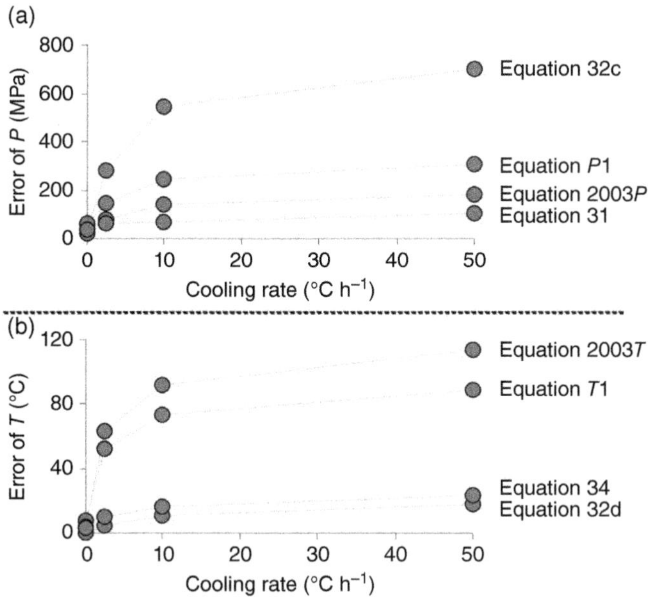

Figure 5.15 Errors of (a) pressure and (b) temperature estimates caused by the *disequilibrium* growth of clinopyroxene with increasing cooling rate have been plotted for equations P1 (Putirka et al., 1996), 2003P (Putirka et al., 2003), 31, and 32c (Putirka, 2008) selected as barometers, and equations T1 (Putirka et al., 1996), 2003T (Putirka et al., 2003), 32d, and 34 (Putirka, 2008) selected as thermometers.

more suitable for minimizing the error of estimate caused by a diffusion-controlled crystal growth regime. To quantitatively understand this phenomenon, equations P1 (Putirka et al., 1996), 2003P (Putirka et al., 2003), 31, and 32c (Putirka, 2008) have been selected as barometers, whereas equations T1 (Putirka et al., 1996), 2003T (Putirka et al., 2003), 32d, and 34 (Putirka, 2008) have been selected as thermometers.

The selection procedure accounts for the different calibration parameters used to derive the predictive equations, as well as the different response of the equations to the use of *disequilibrium* compositions. In the case of barometers, equations 2003P and 31 are weakly sensitive to the rapid growth of clinopyroxene, showing maximum errors of 100 and 180 MPa, respectively (Figure 5.15a). These models depend prevalently on melt parameters (i.e., SiO_2, MgO, CaO, Na_2O, and K_2O) that, however, explore limited compositional changes in the diffusive boundary layer next to the crystal–melt interface. The compositional variations are less than 30%, lowering as a function of cation diffusivity in the melt (i.e., $D_{Na} \approx D_K > D_{Mg} \approx D_{Ca} > D_{Si}$). In contrast, equations P1 and 32c are largely controlled by clinopyroxene components, showing maximum errors of 300 and 700 MPa, respectively (Figure 5.15a). While equation P1 is related to the Jd–melt exchange reaction, equation 32c is highly dependent on the amount of CaTs, which increases remarkably under the effect of crystallization kinetics (Figure 5.10b).

In the case of thermometers, maximum errors of 18 and 23 °C are measured for equatios 34 (mostly related to En in clinopyroxene) and 32d (mostly related to SiO_2, MgO, and CaO in the melt), respectively (Figure 5.15b). These low uncertainties are the result of small variations for the crystal and melt parameters used to calibrate the models and their scarce semblance to cooling rate conditions. Conversely, the Jd–DiHd exchange is greatly influenced by crystallization kinetics (Figure 5.10b) and is the most controlling parameter for equations T1 and 2003T, causing maximum errors of estimates of 90 and 110 °C, respectively (Figure 5.15b). Overall, calculations conducted to test the accuracy of barometers and thermometers illustrate that models including both crystal and melt compositional parameters are affected by a greater error of estimate relative to those in which the melt and the crystal compositions are considered separately. Moreover, clinopyroxene compositional parameters that more effectively reflect the kinetic exchange $[Si + Mg \pm Fe^{2+} \pm Ca] \leftrightarrow [Al + Ti \pm Fe^{3+} \pm Na]$ reduce significantly the accuracy of barometers and thermometers.

This is an important outcome, if we consider that (a) a magma begins to rise to the surface with a relatively low crystal content, (b) during ascent, the magma can be superheated, remaining so over much of its ascent, (c) superheating precludes changing of the magma composition by crystal fractionation, and (d) phenocrysts initiate to growth at high temperature and typically form very near the surface, experiencing variable undercooling conditions due to volatile exsolution or temperature changes (e.g., Blundy & Cashman, 2008; Blundyet al., 2006; Cashman, 1993; Hammer, 2008).

5.12. THE COMPLEXITY OF CLINOPYROXENE SECTOR ZONING

As anticipated above, the effect of undercooling on the crystallization of clinopyroxene can be categorized roughly into:

1. *Equilibrium* crystallization driven by an interface-controlled regime, characterized by crystals with euhedral and prismatic faces with homogeneous compositions, as well as growth rates of the same order of chemical exchanges between the surrounding melts and the forming minerals.

2. *Disequilibrium* crystallization driven by a diffusion-controlled regime, where the growing faces of an euhedral to anhedral (hopper to dendritic) crystal exhibit highly variable compositions, the rate of crystal growth is faster than the rate of diffusion of chemical elements in the melt, and a thin to thick diffusive boundary layer forms at the crystal-melt interface (Mollo & Hammer, 2017).

However, a wide spectrum of clinopyroxene morphologies and compositions may develop between these two scenarios as a function of the degree of undercooling. Through a comparative approach based on cooling-rate experiments (Kouchi et al., 1983; Mollo et al., 2010; Mollo, Blundy, et al., 2013; Mollo, Putirka, et al., 2013) and theoretical models (Watson & Liang, 1995; Watson & Muller, 2009), the crystal-growth kinetics of clinopyroxene have been applied to the effects of high ($\Delta T >> 45\,°C$), moderate ($\Delta T = 45\,°C$), low ($\Delta T = 13\,°C$), and very low ($\Delta T << 13\,°C$) degrees of undercooling that, in turn, translate to different textural evolutions and chemical changes of the mineral.

At $\Delta T >> 45\,°C$, the interface kinetics are extremely rapid and only micrometer to submicrometer crystal faces develop, resulting in dendritic and skeletal clinopyroxene morphologies (Mollo et al., 2010; Mollo, Blundy, et al., 2013; Mollo, Putirka, et al., 2013). The crystal growth rate greatly exceeds the ability of chemical species to diffuse towards or away from the melt adjacent to the growing crystal face. Thus, cations (i.e., Al, Fe^{3+}, Ti, and Na) incompatible with the clinopyroxene crystal lattice are rejected by the advancing crystal surface and concentrated in the diffusive boundary layer that develops in the adjacent melt. This causes the successive and rapid growth of crystal faces to incorporate impurity cations at *nonequilibrium* proportions, leading to the formation of *disequilibrium* clinopyroxene crystals (Mollo, Blundy, et al., 2013; Mollo, Putirka, et al., 2013). Additionally, the combination of a thick diffusive boundary layer in the melt and a fast crystal growth rate can overwhelm the crystallographic controls on the compositions of dendritic and skeletal faces, so that sector zoning does not develop (Lofgren et al., 2006; Watson & Muller, 2009).

At $\Delta T = 45\,°C$, the cation exchange reaction proceeds in a direction opposite to that documented at $\Delta T = 13\,°C$, as reported below (Kouchi et al., 1983). The incorporation of $Al + Ti \pm Fe^{3+} \pm Na$ is better facilitated in the fast growing (-111) sector, leading to reverse compositional zoning of the advancing crystal faces (Hammer et al., 2016; Welsch et al., 2016). These crystal-growth kinetics can be interpreted as intermediate between the rapid formation of dendritic and skeletal crystals, with all faces indistinctively enriched in $Al + Ti \pm Fe^{3+} \pm Na$ (i.e., $\Delta T >> 45\,°C$; Mollo et al., 2010, 2012; Mollo, Blundy, et al., 2013; Mollo, Putirka, et al., 2013) and the growth of euhedral crystals with sector zoning $[Si + Mg + Fe^{2+} + Ca]_{(-111)} = [Al + Ti + Fe^{3+} + Na]_{(100)}$ (i.e., $\Delta T = 13\,°C$; Dowty, 1976; Hollister & Gancarz, 1971; Leung, 1974; Nakamura, 1973; Skulski et al., 1994). Rationally, at $\Delta T = 45\,°C$ (see below), kinetic effects due to diffusion-limited reactions may overprint the control exercised by the structural sites exposed by the advancing crystal face (Downes, 1974; Dowty, 1976; Kouchi et al., 1983; Shimizu, 1981). In this context, chemical enrichments and depletions in the melt adjacent to the crystal surface are directly related to the different cation diffusivities, in the order $MgO > CaO > FeO > TiO_2 > SiO_2 \approx Al_2O_3$ (Zhang, 2010). Over the timescale of the slow growing (100) sector, the diffusive boundary layer results in weak enrichment with Al_2O_3, as the bulk melt far from the crystal sector more efficiently supplies fresh chemical nutrients at the crystal–melt interface. In contrast, the diffusive boundary layer next to the fast growing (-111) sector is rapidly supersaturated with Al_2O_3, leading to the growth of crystal compositions greatly controlled by diffusion-limited cation exchange reactions (Mollo & Hammer, 2017, and references therein).

Conversely, at $\Delta T < 45\,°C$, the advancing faces of clinopyroxene have enough time to grow at the micrometer to millimeter scale, forming euhedral morphologies characterized by compositional sector zoning (Kouchi et al., 1983). Several types of sector zoning can be found due to the concurrent effects of different processes related to the growth rate of the crystal face, the spatial distribution of structural sites in the advancing crystal surface, the length scale of lattice diffusion, and the diffusion of chemical elements in the melt (Leung, 1974; Kouchi et al., 1983; Watson & Liang, 1995). Overall, a sector-zoned clinopyroxene can be disassembled into distinct polyhedra, with the crystal faces representing bases of pyramids, designated as (100), (010), and (110) for the prism faces, and (-111) for the nonprism face, with all apices converging at the crystal center.

At $\Delta T = 13\,°C$, the relative order of enrichment of incompatible cations in the crystal structure is related to the faces $(100) \approx (110) > (010) > (-111)$, as the result of an opposite growth rate for the sectors $(-111) > (010) > (110) \approx (100)$ (Brophy et al., 1999; Dowty, 1976; Duncan & Preston, 1980; Dunworth et al., 2001; Ferguson, 1973; Harkins & Hollister, 1977; Hollister & Garcanz, 1971; Kouchi et al., 1983; Leung, 1974; Nakamura, 1973; Shimizu, 1981; Skulski et al., 1994; Ubide et al., 2019; Wass, 1973; Watson & Liang, 1995). The replacement of $[Si]_{(-111)}$ for $[Al]_{(100)}$ in a tetrahedral site is balanced by substitution of divalent cations $[Mg \pm Fe^{2+} \pm Ca]_{(-111)}$ for trivalent and tetravalent cations $[Ti \pm Fe^{3+}]_{(100)}$ in octahedral sites, as well as by the coupled substitution $[Mg + Ca]_{(-111)} = [Na + Al]_{(100)}$. These cation-exchange mechanisms lead to preferential formation of $CaAl_2SiO_4$ (CaTs), $CaFeAlSiO_6$ (CaFeTs), $CaTiAl_2O_6$ (CaTiTs), and $NaAlSi_2O_6$ (Jd) components at the slow growing (100) sector relative to the fast growing (-111) one (e.g., Dowty, 1976; Hollister & Gancarz, 1971; Kouchi et al., 1983; Skulski et al., 1994).

Therefore, each clinopyroxene face preserves a distinct composition due to enrichment of elements in one sector relative to the other, whilst the different types of chemical substitutions in the sectors depend on the structures of the growing layers (i.e., the spatial structural distributions of M and T sites) and their slower or faster mechanisms of growth (Dowty, 1976; Ferguson, 1973; Hollister & Gancarz, 1971; Leung, 1974; Nakamura, 1973; Skulski et al., 1994). If all sectors are characterized by an identical three-dimensional atomic configuration, it is also true that the structure of one single growing crystal sector consists of two-dimensional atomic arrangements that may be markedly different. Single layers of M2 sites alternating with double layers of M1 sites form the atomic arrangement of the fast growing (-111) sector. This structure requires concomitant addition of divalent cations and silicate tetrahedra, as M and T sites are exposed simultaneously (Leung, 1974; Nakamura, 1973). Consequently, the melt adjacent to the fast growing (-111) sector is prevalently depleted in Si + Mg ± Fe^{2+} ± Ca (more suitable for the clinopyroxene structure) and enriched in Al + Ti ± Fe^{3+} ± Na (incompatible with the crystal lattice). In contrast, the geometry of the structural sites on the slow growing (100) sector includes alternating layers of M and T sites that are not exposed simultaneously. This implies that large amounts of divalent cations are required to fill layers composed entirely of M1 and M2 sites. As a consequence, the melt feeding the crystal growth is rapidly depleted in Mg ± Fe^{2+} ± Ca, so that Ti ± Fe^{3+} ± Na may enter the M site at higher proportions. Similarly, the accretion of the T layer requires the formation of silicate chains, thus the number of Si atoms available in the melt next to the crystal surface rapidly decreases, favoring incorporation of Al in tetrahedral coordination (Dowty, 1976; Ferguson, 1973; Hollister & Gancarz, 1971; Leung, 1974; Nakamura, 1973; Skulski et al., 1994). The interface kinetics of clinopyroxene growth are also ascribed to an inverse relationship between the lateral advancing rate of the adsorption layer and the normal growth rate of the crystal face (Nishizawa et al., 1972). A faster lateral growth rate of the adsorption layer provides an increased number of structural sites for element incorporation, so that cations with a high charge and small ionic radius are more firmly bonded to the structural sites of the slow growing (100) sector relative to the fast growing (-111) one (Kouchi et al., 1983).

At $\Delta T << 13\,°C$, the surfaces of the slow-growing clinopyroxene crystals can be continuously in *equilibrium* with the melt feeding the crystal growth (Watson & Liang, 1995; Watson & Muller, 2009). The length of the advancing crystal–melt interface corresponds to a distance many orders of magnitude smaller than the effective diffusive transport distance of chemical species in the melt. Under such circumstances, sector zoning is not entirely attributable to a diffusion-controlled crystal growth mechanism related to a diffusive boundary layer in the melt or a *disequilibrium* uptake of impurity cations at the crystal–melt interface. In turn, the occurrence of sector zoning depends on the competition between the crystal growth rate and the diffusivity of a surface-enriched element in the crystal lattice (Hollister, 1970; Tiller & Ahn, 1980). During each crystal growth step, cations incompatible with the crystal lattice are invariably enriched in the thin surface layer that develops at the crystal–melt interface. This surface layer becomes part of the crystal volume and any retention of impurity cations within the newly created crystal volume causes an excess of concentration that is eliminated by lattice diffusion (Watson & Liang, 1995). When the rate of lattice diffusion is much greater than the rate of crystal growth, the interior of the crystal has enough time to exchange cations with the surrounding melt through the surface layer. Therefore, any concentration gradient in the interior of the crystal ceases and *equilibrium* cation distribution prevails. However, as the rate of crystal growth increases, lattice diffusion through the surface layer becomes progressively less efficient and impurity cations in the surface layer are trapped in the interior of the newly grown crystal, leading to a sector of concentration enrichment behind the surface layer (Watson & Liang, 1995). It is apparent that rapidly growing clinopyroxene crystals do not have sufficient time to eliminate compositional zoning by solid-state diffusion, showing *disequilibrium* compositions due to the more pronounced effects of chemical gradients in the melt.

To better understand the importance of sector zoning for the interpretation of magma dynamics, Figure 5.16 shows the compositions measured in sector-zoned clinopyroxenes reported in Hollister and Gancarz (1971), as well as the estimates of pressure (equation P1 of Putirka et al., 1996), temperature (equation T1 of Putirka et al., 1996), and melt-H_2O content (hygrometric expression of Perinelli et al., 2016). Hollister & Gancarz (1971) have investigated the chemical variations of clinopyroxenes from a leucite phonolite belonging to the alkaline igneous province designated the Roman Province (central Italy) (e.g., Mollo & Vona, 2014, and references therein). Downes (1974) documented the sector zoning in clinopyroxene phenocrysts found in a trachybasaltic cone developed during the eruption of 1669 on the southern flank of Mount Etna volcano (Sicily). The fast growing (-111) sector is enriched in SiO_2, MgO, and CaO, and depleted in Al_2O_3, TiO_2, and Na_2O, responding to the control of the spatial distribution of structural sites (Figure 5.16). On the other hand, the slow growing (100) sector shows an opposite behavior (Figure 5.16). According to the early suggestion of Hollister & Gancarz (1971), it is apparent from Figure 5.16 that the inverse relationship between fast growing (-111) and slow growing (100) sectors leads to different *P*, *T*, and melt-H_2O estimates. The fast growing (-111) sector may overestimate the storage

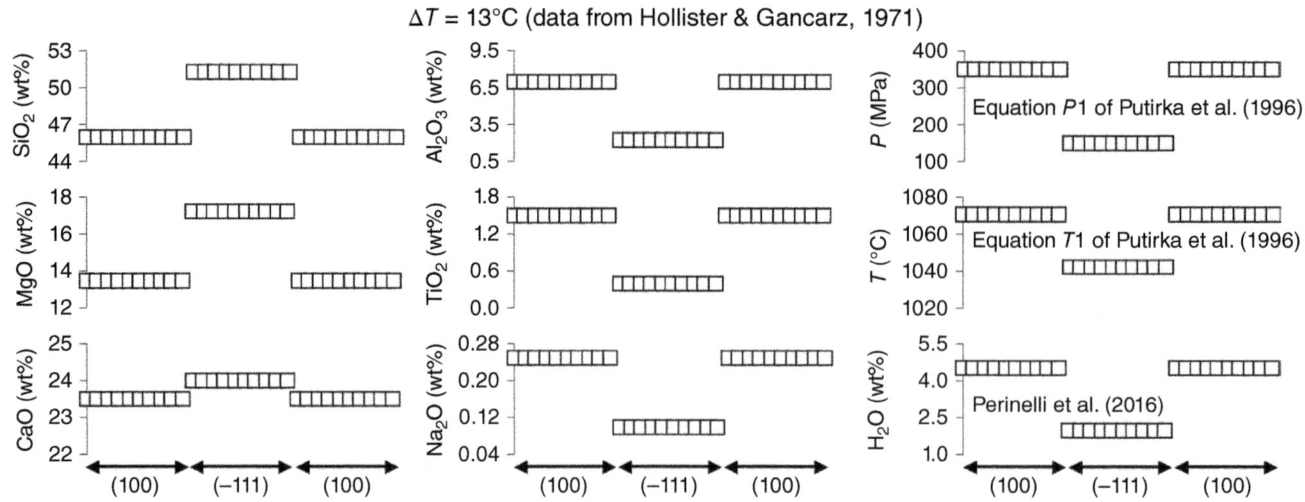

Figure 5.16 Reverse compositional changes of (100) and (-111) sectors in clinopyroxene, as well as barometric, thermometric, and hygrometric estimates of two different natural crystals characterized by opposite cation exchanges of (a) $[Si + Mg \pm Fe^{2+} \pm Ca]_{(-111)} = [Al + Ti \pm Fe^{3+} \pm Na]_{(100)}$ and (b) $[Si + Mg \pm Fe^{2+} \pm Ca]_{(100)} = [Al + Ti \pm Fe^{3+} \pm Na]_{(-111)}$. The two different natural crystals have been described in detail by Hollister and Gancarz (1971) and by Downes (1974). Estimates of pressure, temperature, and melt-H_2O content have been performed through equation P1 of Putirka et al. (1996), equation T1 of Putirka et al. (1996), and the hygrometric expression of Perinelli et al. (2016).

conditions of magma (Figure 5.16), so that the *P*–*T*–H_2O estimates a shift from $[350\text{ MPa}; 1070\,°C; 4.5\text{ wt\% } H_2O]_{(100)}$ to $[150\text{ MPa}; 1040\,°C; 2.0\text{ wt\% } H_2O]_{(-111)}$. This poses questions on the interpretation of thermobarometric data under the effects of kinetic conditions.

Clinopyroxene-based thermobarometric equations rely on thermodynamic expressions describing the relationship between the *equilibrium* constant of a specific clinopyroxene–melt exchange reaction and the intensive variables of the system. In the Jd–melt exchange, the formation of Jd is accompanied by a large change in molar volume, resulting in a *P*-dependent reaction (e.g., Blundy et al., 1995). Conversely, in the Jd–DiHd and CaTsDiHd equilibria, the volume changes for the solution of Jd and CaTs into DiHd are small, leading to *T*-dependent reactions where the amount of H_2O dissolved in the melt may depress the saturation surface of clinopyroxene (e.g., Putirka et al., 1996). The spatial atomic arrangements of the fast growing (-111) sector favors the incorporation of DiHd components, causing *P*–*T*–H_2O underestimates (Figure 5.16). Similarly, the geometry of the structural sites (rather than the thickness of the diffusive boundary layer in the melt) determines the enrichment of incompatible elements in the slow growing (-111) sectors, promoting *P*–*T*–H_2O overestimates (Figure 5.16).

Therefore, caution should be exercised in interpreting the sector zoning of clinopyroxene phenocrysts. The cation exchange $[Si + Mg \pm Fe^{2+} \pm Ca]_{(-111)} = [Al + Ti \pm Fe^{3+} \pm Na]_{(100)}$ cannot be resolved unambiguously in terms of enrichments in CaTs and Jd at the expense of DiHd, due to the substantial disparity between the crystal growth and the mobility of components in the melt (Hammer et al., 2016; Mollo & Hammer, 2017; Mollo et al., 2010). This is true especially at large degrees of undercooling where strong departure from *equilibrium* arises mainly at the fast growing (-111) sector that is characterized by a CaTs–Jd-rich composition due to the great difference in the rate of crystal–melt interface advance and the rate of chemical diffusion in the melt. In contrast, the slow growing (100) sector has more time to reequilibrate with the melt, showing a DiHd-rich composition that minimizes the error of estimate caused by crystal growth kinetics, thus providing more reliable and accurate predictions of the crystallization conditions of magma (Hammer et al., 2016; Welsch et al., 2016). When a thin diffusive boundary layer develops next to the slow growing (100) sector, the surrounding melt more easily supplies fresh cations to the advancing crystal surface (Watson & Liang, 1995) and, consequently, the sector composition is weakly influenced by *disequilibrium* cation partitioning. However, it is worth stressing that, at small degrees of undercooling, the interface kinetics of clinopyroxene growth are much more difficult to interpret because deviations from *equilibrium* are not so effective. When sector zoning develops in response to sluggish crystallization kinetics, such as during slow magma ascent, cations are partitioned at near-*equilibrium* proportions between the slow-growing crystal and the feeding melt. Under such circumstances, sector-zoned clinopyroxene compositions can be used as imput data for thermobarometric models, thus providing reliable P–T–H_2O constraints on magma decompression and cooling histories (Ubide et al., 2019).

5.13. CONCLUSIONS

Magma solidification can be strongly influenced by $\Delta T/\Delta t$ and $\Delta P/\Delta t$ rates. However, numerical simulations of conductive heat dissipation show that cooling rates higher than 10^{-1}–10^{-2} °C h^{-1} are not attained before eruptions or may affect only limited portions of erupted products. Even at shallow crustal levels (i.e., depths of a few kilometers) magmas experience cooling rates on the order of 10^{-5}–10^{-6} °C h^{-1}. In contrast, cooling rates higher than 10^{-1} °C h^{-1} can be easily established after eruption. On the other hand, hydrated magmas may be influenced by variable $\Delta P/\Delta t$ rates, especially at depths ≤ 10 km for the case of highly hydrated magmas (5 wt% H_2O), but less than 5 km or even less for mildly and poorly hydrated conditions (< 3 and < 1 wt.% H_2O). In volcanic systems, magmas forming lavas or pyroclats can be erupted from their vents only if the crystal content is ≤ 50 vol%. In turn, $\Delta T/\Delta t$ equal to or higher than 10^0 °C h^{-1} influences and still controls the solidification of molten portions. Generally, minerals solidifying prior to eruptions are able to record the magnitude of $\Delta T/\Delta t$ experienced by magmas. This can be retrieved through thermometers, barometers, and/or hygrometers, if kinetics of solidification were low or moderate as predicted by physicochemical models (Figures 5.1–5.3). At shallow crustal depths either *equilibrium* or *disequilibrium* processes can occur as a function of the magnitude of $\Delta P/\Delta t$ rates, especially for hydrated magmas (Figure 5.5). Similarly, mineral phases solidified after eruptions easily experience *disequilibrium* processes as a function of the inevitably high $\Delta T/\Delta t$, at least for thin lava flows (Figure 5.4).

Also important is the role of bulk composition on kinetics, as demonstrated by experimental results conducted on B_{100} (basaltic) and R_{100} (rhyolitic) compositions (Vetere et al., 2015). In addition, density and viscosity numerical models impose further constraints on magma ascent conditions (Figures 5.6 and 5.7). Anhydrous or hydrous basaltic melts can rise in oceanic domains only if close to their *liquidus* (i.e., crystal-poor magmas), whereas only highly hydrated basaltic melts can be extracted from the continental crust. Rhyolitic melts require to be almost fully melted and with significant amount of water, in order to ascend towards the surface in continental crustal domains.

The textural and chemical features of mineral phases are valuable tools to decipher the solidification paths of magmas when the physicochemical conditions of the system change from *equilibrium* to *disequilibrium* crystal-growth conditions. The immediate effect of crystal-growth kinetics is the attainment of apparent cation partitioning between the rapidly growing crystal face and the diffusive boundary layer in the melt. This causes kinetically controlled cation exchanges where the crystal compositions changes with increasing degree of undercooling, such as during magma ascent in volcanic conduits or cooling at the rapidly chilled margin of magmatic intrusions. The application of barometers, thermometers, and hygrometers to *disequilibrium* crystal compositions leads invariably to error of P–T–H_2O estimates that, in some cases, may be very large. However, the magnitude of departure from *equilibrium* remains poorly investigated and, for example, under sluggish crystallization kinetics, cations can be partitioned at near-*equilibrium* proportions between the slow-growing crystal and the feeding melt. Such a complex spectrum of mineral and melt compositional changes should provide a natural guide towards future research focusing on the calibration and application of more effective *equilibrium* models to correctly interpret and quantify the crystallization conditions of natural magmas.

ACKNOWLEDGMENTS

The authors are grateful to Dr. F. Vetere and Dr. A. Fiege for their editorial work. The comments of Professor M. Masotta and one anonymous reviewer have greatly improved the quality of this work.

REFERENCES

Annen, C. (2017). Factors affecting the thickness of thermal aureoles. *Frontiers in Earth Science*, *5*, 1–13.

Annen, C. Blundy, J. D., Leuthold, J., Stephen, R., & Sparks, J. (2015). Construction and evolution of igneous bodies: Towards and integrated perspective of crustal magmatism. *Lithos*, *230*, 206–221.

Applegarth, L. J., Tuffen, H., James, M. R., & Pinkerton, H. (2013). Degassing-driven crystallisation in basalts. *Earth Science Reviews*, *116*, 1–16.

Armienti, P. (2008). Decryption of igneous rock textures: Crystal size distribution tools. *Reviews in Mineralogy and Geochemistry*, *69*, 623–649.

Arzilli, F., Agostini, C., Landi, P., Fortunati, A., Mancini, L., & Carroll, M. R. (2015). Plagioclase nucleation and growth kinetics in a hydrous basaltic melt by decompression experiments. *Contributions to Mineralogy and Petrology*, *170*(5–6), 1–16.

Arzilli, F., & Carroll, M. R. (2013). Crystallization kinetics of alkali feldspars in cooling and decompression-induced crystallization experiments in trachytic melt. *Contributions to Mineralogy and Petrology*, *166*, 1011–1027.

Asimow, P. D., & Ghiorso, M. S. (1998). Algorithmic modifications extending MELTS to calcolate subsolidus phase relations. *American Mineralogist*, *83*, 1127–1131.

Bachmann, O., & Bergantz, G. W. (2004). On the origin of crystal-poor rhyolites: extracted from batholithic crystal mushes. *Journal of Petrology*, *45*, 1565–1582

Bachmann, O., & Huber, C. (2016). Silicic magma reservoirs in the Earth's crust. *American Mineralogist*, *101*, 2377–2404.

Baginski, B., Dzierzanowski, P., Macdonald, R., & Upton, B. G. J. (2009). Complex relationships among coexisting pyroxenes: the Paleogene Eskdalemuir dyke, Scotland. *Mineralogical Magazine*, *73*, 929–942.

Baker, M. B. & Grove, T. L. (1985). Kinetic controls on pyroxene nucleation and metastable liquid lines of descent in a basaltic andesite. *American Mineralogist*, *70*, 279–287.

Baschek, G. & Johannes, W. (1995). The estimation of NaSi–CaAl interdiffusion rates in peristerite by homogenization experiments. *European Journal of Mineralogy*, *7*, 295–307.

Blundy, J. D., & Annen, C. J. (2016). Crustal magmatic systems from the perspective of heat transfer. *Elements*, *12*, 115–120.

Blundy, J., & Cashman, K. (2008). Petrologic reconstruction of magmatic system variables and processes. *Reviews in Mineralogy and Geochemistry*, *69*, 179–239. https://doi.org/10. 2138/rmg.2008.69.6

Blundy, J. D., Cashman, K. V., & Humphreys, M. C. S. (2006). Magma heating by decompression-driven crystallisation beneath andesite volcanoes. *Nature*, *443*, 76–80.

Blundy, J. D., Falloon, T. J., Wood, B. J., & Dalton, J. A. (1995). Sodium partitioning between clinopyroxene and silicate melts. *Journal of Geophysical Research Solid Earth*, *100*, 15501–15515. http://dx.doi.org/10.1029/95JB00954

Boudreau, A. E. (1999). PELE—a version of the MELTS software program for the PC platform. *Computers and Geosciences*, *25*, 201–203.

Brandeis, G., Jaupart, C., & Allegre, C. J. (1984). Nucleation, crystal growth and the thermal regime of cooling magmas. *Journal of Geophysical Research*, *89*, 10161–10177.

Brugger, C. R., & Hammer, J. E. (2010). Crystallization kinetics in continuous decompression experiments: implications for interpreting natural magma ascent processes. *Journal of Petrology*, *51*, 1941–1965.

Brophy, J. G., Whittington, C. S., & Park, Y. K. (1999). Sector-zoned augite megacrysts in Aleutian high alumina basalts: implications for the conditions of basalt crystallization and the generation of calc-alkaline magmas. *Contributions to Mineralogy and Petrology*, *135*, 277–290.

Burgisser, A., & Bergantz, G. W. (2011). A rapid mechanism to remobilize and homogenize highly crystalline magma bodies. *Nature*, *471*, 212–215.

Burgisser, A., & Degruyter W. (2015). Magma ascent and degassing at shallow levels. In H. Sigurdsson, B. Houghton, S. McNutt, H. Rymer, J. Stix (Eds.), *The Encyclopedia of Volcanoes* (2nd edn, pp. 225–236). Academic Press.

Caricchi, L. & Blundy, J. (2015). The temporal evolution of chemical and physical properties of magmatic systems. In L. Caricchi, J. Blundy (Eds.), *Chemical, physical and temporal evolution of magmatic systems* (Special Publication 422, pp. 1–15). London: Geological Society.

Carlson, R. L., & Raskin, G. S. (1984). Density of the ocean crust. *Nature*, *311*, 555–558.

Cashman, K. V. (1990). Textural constraints on the kinetics of crystallization on igneous rocks. Reviews in *Mineralogy*, *24*, 259–309.

Cashman, K. V. (1993). Relationship between plagioclase crystallization and cooling rate in basaltic melts. *Contributions to Mineralogy and Petrology*, *113*, 126–142.

Cashman, K. V., Steohem, R., Sparks, J., & Blundy, J. D. (2017). Vertically extensive and unstable magmatic systems: A unified view of igneous processes. *Science*, *355*. doi:10.1126/science.aag3055

Cassidy, M., Manga M., Cashman K., & Bachmann O. (2018). Controls on explosive-effusive volcanic eruption styles. *Nature Communications*, *9*, 2839. doi: 10.1038/s41467-018-05293-3

Coish, R. A. & Taylor, L. A. (1979). The effect of cooling rate on texture and pyroxene chemistry in DSDP Leg34 basalt: a microprobe study. *Earth and Planetary Science Letters*, *42*, 389–398.

Conte, A., Perinelli, C., & Triglia, R. (2006). Cooling kinetics experiments on different Stromboli lavas: effects on crystal morphologies and phase composition. *Journal of Volcanology and Geothermal Research*, *155*. 179–200.

Cooper, K. M. (2017). What does a magma reservoir look like? The "crystal's-eye" view. *Elements*, *13*, 23–28.

Cooper, K. M., & Kent, A. J. R. (2014). Rapid remobilization of magmatic crystals kept in cold storage. *Nature*, *506*, 480–483.

Corrigan, G. M. (1982). Supercooling and the crystallization of plagioclase, olivine and clinopyroxene from basaltic magmas. *Mineralogical Magazine*, *46*, 31–42.

Davidson, J. P., Morgan, D. J., Charlier, B. L. A., Harlou, R., & Hora, J. M. (2007). Microsampling and isotopic analysis of igneous rocks: Implications for the study of magmatic systems. *Annual Reviews in Earth and Planetary Sciences*, *35*, 273–311.

Deer, W. A., Howie, R. A., & Zussmanm, J. (2013). *An introduction to the rock-forming minerals*. London: Mineralogical Society of Great Britain and Ireland.

Del Gaudio, P., Mollo S., Ventura G., Iezzi G., Taddeucci J., & Cavallo A. (2010). Cooling rate-induced differentiation in anhydrous and hydrous basalts at 500 MPa: implications for the storage and transport of magmas in dykes. *Chemical Geology*, *270*, 164–178.

Dingwell, D. B. (2006). Transport properties of magmas: diffusion and rheology. *Elements*, *2*, 281–286.

Dingwell D. B & Webb, S. L. (1990). Relaxation in silicate melts. *European Journal of Mineralogy*, *2*, 427–449.

Dobosi, G. & Jenner, G. A. (1999). Petrological implications of trace element variation in clinopyroxene megacrysts from the Norgrad volcanic province, north Hungary: a study by laser ablation microprobe-inductively coupled mass spectrometry. *Lithos*, *46*, 731−749.

Downes, M. J. (1974). Sector and oscillatory zoning in calcic augites from M. Etna, Sicily. *Contributions to Mineralogy and Petrology*, *47*(3),187–196. doi:10.1007/bf00371538

Dowty, E. (1976). Crystal structure and crystal growth 2: sector zoning in minerals. *American Mineralogist*, *61*, 460–469

Dowty, E. (1980). Crystal growth and nucleation theory and the numerical simulation of igneous crystallization. In R. B. Hargraves (Ed.), *The physics of magmatic processes* (pp. 419–485). Princeton, NJ: Princeton University Press.

Drake, M. J. (1975). The oxidation state of europium as an indicator of oxygen fugacity. *Geochimica et Cosmochimica Acta*, *39*, 55–64.

Druitt, T. H., Costa, F., Deloule, E., Dungan, M., & Scaillet B. (2012). Decadal to monthly timescales of magma transfer and reservoir growth at a caldera volcano. *Nature*, *482*, 77–80. doi: 10.1038/nature10706; pmid: 22297973

Dufek, J. & Bachmann, O. (2010). Quantum magmatism: Magmatic compositional gaps generated by melt–crystal dynamics. *Geology*, *38*, 687–690.

Dufek, J., Huber, C., & Karlstrom, L. (2013). Magma chamber dynamics and thermodynamics. In S. Fagents, T. Gregg, R. Lopes (Eds.), *Modelling volcanic processes: the physics and mathematics of volcanism* (pp. 5–31). Cambridge, UK: Cambridge Press.

Duncan, A. M., & Preston, R. M. P. (1980). Chemical variation of clinopyroxene phenocrysts from the trachybasaltic lavas of Mount Etna, Sicily. *Mineral Magazine*, *43*, 765–770.

Dunworth, E. A., Neumann, E. R., & Rosenbaum, J.M. (2001). The Skien lavas, the Oslo Rift: Petrological disequilibrium and geochemical evolution. *Contributions to Mineralogy and Petrology*, *140*, 701–719.

Edmunds, M., & Wallace, P. J. (2017). Volatiles and exsolved vapor in volcanic systems. *Elements*, *13*, 35–40.

Faraone, D., Molin, G., & Zanazzi, P. F. (1988). Clinopyroxene from Vulcano (Aeolian Islands, Italy): crystal chemistry and cooling history. *Lithos*, *22*, 113–126.

Ferguson, A. (1973). On hour-glass sector zoning in clinopyroxene. *Mineralogical Magazine*, *29*, 321–325.

Fiege A., Vetere F., Iezzi G., Simon A., & Holtz F. (2015). The roles of decompression rate and volatiles (H_2O + Cl ± CO_2 ± S) on crystallization in (trachy-)basaltic magma. *Chemical Geology*, *411*, 310–322.

Fokin, V. M., Zanotto, E. D., Yuritsyn, N. S., & Schmelzer, J. W. P. (2006). Homogeneous crystal nucleation in silicate glasses: A 40 years perspective. *Journal of Non-Crystalline Solids. 27*, 2681–2714.

Gamble, R. P. & Taylor, L. A. (1980). Crystal/liquid partitioning in augite: effects of cooling rate. *Earth and Planetary Science Letters*, *47*, 21–33.

Ganne, J., Bachmann, O., & Feng, X. (2018). Deep into magma plumbing systems: Interrogating the crystal cargo of volcanic deposits. *Geology*, *46*, 415–418.

Ghiorso, M. S., & Sack, R. O. (1995). Chemical mass transfer in magmatic processes IV. A revised and internally consistent thermodynamic model for the interpolation and extrapolation of liquid–solid equilibria in magmatic system at elevated temperatures and pressures. *Contributions to Mineralogy and Petrology*, *119*, 197–212.

Gibb, F. G. F. (1974). Supercooling and crystallization of plagioclase from a basaltic magma. *Contributions to Mineralogy and Petrology*, *119*, 197–212.

Giordano, D., Nichols A. R. L., & Dingwell, D. B. (2005). Glass transition temperatures of natural hydrous melts: a relationship with shear viscosity and implications for the welding process. *Journal of Volcanology and Geothermal Research*, *142*, 105–118.

Giordano, D., Russell J. K., & Dingwell, D. B. (2008). Viscosity of magmatic liquids: a model. *Earth and Planetary Science Letters*, *271*, 123–134

Gonnermann, H. M., & Manga, M. (2007). The fluid mechanics inside a volcano. *Annual Reviews of Fluid Mechanics*, *39*, 321–356.

Gonnermann, H. M., & Manga, M. (2012). Dynamics of magma ascent in the volcanic conduit. In S. Fagents, T. Gregg, R. Lopes (Eds.), *Modelling volcanic processes: the physics and mathematics of volcanism* (pp. 55–84). Cambridge, UK: Cambridge Press.

Gonnermann, H. M., & Taisne, B. (2015). Magma transport in dykes. In H. Sigurdsson, B. Houghton, S. McNutt, H. Rymer, J. Stix (Eds.), *The Encyclopedia of Volcanoes* (2nd edn, pp. 215–224). Academic Press.

Griffiths, R. W. (2000). The dynamic of lava flows. *Annual Review of Fluid Mechanics*, *32*, 477–518.

Grove, T. L., & Bence, A. E. (1977). *Experimental study of pyroxene–liquid interaction in quartz-normative basalt 15597.* Paper presented at 8th Lunar and Planetary Science Conference (Proceedings, pp. 1549 1579), Houston, TX.

Grove, T. L. & Raudsepp, M. (1978). *Effects of kinetics on the crystallization of quartz-normative basalt 15597: an experimental study*. Paper presented at 10th Lunar and Planetary Science Conference (Proceedings, pp. 585–599), Houston, TX.

Gualda, G. A. R., Ghiorso, M. S., Lemons, R. V., & Carley, T. L. (2012). Rhyolite-MELTS: A modified calibration of MELTS optimized for silica-rich, fluid-bearing magmatic systems. *Journal of Petrology*, *53*, 875–890.

Gudmundsson, A. (2011). Deflection of dykes into sills at discontinuities and magma chamber formation. Tectonophysics, *500*, 50–64.

Hacker, B. R., Kelemen, P. B., & Behn, M. D. (2015). Continental Lower Crust. *Annual Reviews in Earth and Planetary Sciences*, *43*, 167–205

Hamada, M., & Fuji, T. (2007). H_2O-rich island arc low-K tholeiite magma inferred from Ca-rich plagioclase-melt inclusion equilibria. *Geochemical Journal*, *41*, 437–461

Hammer, J. E. (2008). Experimental studies of the kinetics and energetics of magma crystallization. In K. D. Putirka, F.J. Tepley (Eds.), *Minerals, inclusions and volcanic processes. Reviews in Mineralogy and Geochemistry*, *69*. 9–59.

Hammer, J. E., Jacob, S., Welsch, B., Hellebrand, E., & Sinton, J. M. (2016). Clinopyroxene in post shield Haleakala ankaramite 1. Efficacy of thermobarometry. *Contributions to Mineralogy and Petrology*. *171*, 7. https://doi.org/10.1007/s00410-015-1212-x

Hammer, J. E., & Rutherford, M. J. (2002). An experimental study of decompression-induced crystallization in silicic melt. *Journal of Geophysical Research*, *107*(B1),8-1–8-24.

Harkins, E. & Hollister, L. S. (1977). Sector zoning of clinopyroxene from a weakly metamorphosed diabase. *American Mineralogist*, *62*, 390–394.

Harris, A. J. L. (2013). Lava flows. In S. Fagents, T. Gregg, R. Lopes (Eds.), *Modelling volcanic processes: the physics and mathematics of volcanism* (pp. 85–106). Cambridge, UK: Cambridge Press.

Harris, A. J. L., Bailey, J., Calvari, S., & Dehn, J. (2005). Heat loss measured at a lava channel and its implications for down-channel cooling and rheology. In M. Manga, G. Ventura (Eds.), *Kinematics and dynamics of lava flows* (Special Paper 396, pp. 125–146). Boulder, CO: Geological Society of America.

Hill, E., Blundy, J. D., & Wood, B. J. (2011). Clinopyroxene–melt trace element partitioning and the development of a predictive model for HFSE and Sc. *Contributions to Mineralogy and Petrology*, *161*, 423–438. http://dx.doi.org/10.1007/s00410-010-0540-0

Hill, E., Wood, B. J., & Blundy, J. D. (2000). The effect of Ca-Tschermaks component on trace element partitioning between clinopyroxene and silicate melt. *Lithos*, *53*, 203–215. http://dx.doi.org/10.1016/S0024-4937(00)00025-6

Hollister, L. S. (1970). Origin, mechanism, and consequences of compositional sector-zoning in staurolite. *American Mineralogist*, *55*, 742–766

Hollister, L. S., & Gancarz, A. J. (1971). Compositional sector-zoning in clinopyroxene from the Narce area, Italy. *American Mineralogist*, *56*, 959–979.

Hon, K., Gansecki, C., & Kauahikaua, J. (2003). The transition from a′a′ to pahoheoe crust on flows emplaced during the Pu'u O'o-Kupaianaha eruption. *US Geological Survey Professional Paper*, *1676*, 89–104.

Hoover, S. R., Cashman, K. V., & Manga, M. (2001). The yield strength of subliquidus basalts: experimental results. *Journal of Volcanology and Geothermal Research*, *107*, 1–18.

Iezzi, G., Mollo, S., Shaini, E., Cavallo, A., & Scarlato, P. (2014). The cooling kinetics of plagioclase revealed by electron microprobe mapping, *American Mineralogist*, *99*, 898–907. http://dx.doi.org/10.2138/am.2014.4626

Iezzi, G., Mollo, S., Torresi, G., Ventura, G., Cavallo, A., & Scarlato, P. (2011). Experimental solidification of an andesitic melt by cooling. *Chemical Geology*, *283*, 261–273. https://doi.org/10.1016/j.chemgeo.2011.01.024

Iezzi, G., Mollo, S., & Ventura, G. (2009). Solidification behavior of natural silicate melts and volcanological implications. In N. Lewis, A. Moretti (Eds.), *Volcanoes: formation, eruptions and modelling* (pp.127–151). Hauppauge, NY: Nova Science Publishers.

Iezzi, G., Mollo S., Ventura G., Cavallo A., & Romano C. (2008). Experimental solidification of anhydrous latitic and trachytic melts at different cooling rates: the role of nucleation kinetics. *Chemical Geology*, *253*, 91–101. doi: 10.1016/j.chemgeo.2008.04.008

Kilburn, C. R. J. (2004). Fracturing as a quantitative indicator of lava flow dynamics. *Journal of Volcanology and Geothermal Research*, *132*, 209–224.

Kirkpatrick, R. J. (1981). Kinetics of crystallization of igneous rocks. In A. C. Lasaga, R. J. Kirkpatrick (Eds.), *Kinetics of geological processes*. (Reviews in Mineralogy, Vol. 8, pp. 321–397). Washington, DC: Mineralogical Society of America.

Kirkpatrick, R. J. (1983). Theory of nucleation in silicate melts. *American Mineralogist*, *68*, 66–77.

Kolzenburg, S., Di Genova, D., Giordano, D., Hess, K. U., & Dingwell, D. B. (2018). The effect of oxygen fugacity on the rheological evolution of crystallizing basaltic melts. *Earth and Planetary Science Letters*, *487*, 21–32.

Kolzenburg, S., Giordano, D., Hess, K. U., & Dingwell, D. B. (2018). Shear rate-dependent disequilibrium rheology and dynamics of basalt solidification. *Geophysical Research Letters*, *45*, 6466–6475.

Kouchi, A., Sugawara, Y., Kashima, K. & Sunagawa, I. (1983). Laboratory growth of sector zoned clinopyroxenes in the system $CaMgSi_2O_2$–$CaTiAl_2O_6$. *Contributions to Mineralogy and Petrology*, *83*, 986–990.

Lange, R. L. & Carmichael, S. E. (1990). Thermodynamic properties od silicate liquids with emphasis on density, thermal expansion and compressibility. *Reviews in Mineralogy and Geochemistry*, *24*, 25–64.

Lanzafame, G., Iezzi, G., Mancini, L., Lezzi, F., Mollo, S., & Ferlito, C. (2017). Solidification and turbulence (non-laminar) during magma ascent: insights from 2D and 3D analyses of bubbles and minerals in an Etnean dyke. *Journal of Petrology*, *58*, 1511–1534.

Lanzafame, G., Mollo, S., Iezzi, G., Ferlito C., & Ventura, G. (2013). Unravelling the solidification path of a pahoehoe "cicirara" lava from Mount Etna volcano. *Bulletin of Volcanology*, *75*(4,) 1–16.

Lasaga, A. C. (1997). *Kinetic theory in the earth sciences*. Princeton, NJ: Princeton University Press.

La Spina, G., Burton, M., De'Michieli Vitturi, M., & Arzilli, F. (2016). Role of syn-eruptive plagioclase disequilibrium crystallization in basaltic magma ascent dynamics. *Nature Communications*, *7*, 13402.

Latypov, R. M. (2003). The origin of marginal compositional reversals in basic±ultrabasic sills and layered intrusions by Soret fractionation. *Journal of Petrology*, *44*, 1579–1618.

Lesher, C. E., Cashman, K. V., & Mayfield, J. D. (1999). Kinetic controls on crystallization of Tertiary North Atlantic basalt and implications for the emplacement and cooling history of lava at Site 989, Southeast Greenland rifted margin. *Proceeding of the ODP Scientific Results*, *163*, 135–148.

Leung, I. S. (1974). Sector-zoned titanaugites: morphology, crystal chemistry, and growth. *American Mineralogist*, *59*, 127–138.

Lindsley, D. H. (1983). Pyroxene thermometry. *American Mineralogist*, *68*, 477–493.

Lipman P. W., Banks N. G., & Rhodes J. M. (1985). Degassing-induced crystallization of basaltic magma and effects on lava rheology. *Nature*, *317*, 604–607.

Lister, J. R. & Kerr, R. C. (1991). Fluid–mechanical models of crack propagation and their application to magma transport and dyke. *Journal of Geophysical Research*, *96*, 10049–10077.

Liu, M., & Yund, R. A. (1992). NaSi–CaAl interdiffusion in plagioclase. *American Mineralogist*, *77*, 275–283.

Lofgren, G. E., Grove, T. L., Brown, R. W., & Smith, D. P. (1979). *Comparison of dynamic crystallization techniques on Apollo 15 quartz normative basalts*. Paper presented at 10th Lunar and Planetary Science Conference (Proceedings Vol. 1, pp. 423–438), Houston, TX.

Lofgren, G. E., Huss, G. R., & Wasserburg, G. J. (2006). An experimental study of trace-element partitioning between Ti-Al-clinopyroxene and melt: Equilibrium and kinetic effects including sector zoning. *American Mineralogist*, *91*, 1596–1606.

Loomis, T. P. (1981). An investigation of disequilibrium growth processes of plagioclase in the system anorthite–albite–water by methods of numerical simulation. *Contributions to Mineralogy and Petrology*, *81*, 219–229.

Loomis, T. P. & Welber, P. W. (1982). Crystallization processes in the rocky hill granodiorite pluton, California: an interpretation based on compositional zoning of plagioclase. *Contributions to Mineralogy and Petrology*, *81*, 230–239.

Lundgaard, K. L., & Tegner, C. (2004). Partitioning of ferric and ferrous iron between plagioclase and silicate melt. *Contributions to Mineralogy and Petrology*, *147*, 470–483.

Lundstrom, C. C., Shaw, H. F., Ryerson, F. J., Phinney, D. L., Gill, J. B., & Williams, Q. (1994). Compositional controls on the partitioning of U, Th, Ba, Pb, Sr and Zr between clinopyroxene and haplobasaltic melts: implications for uranium series disequilibria in basalts. *Earth and Planetary Science Letters*, *128*, 407–423. http://dx.doi.org/10.1016/0012-821X(94)90159-7

Marsh, B. D. (1996). Solidification fronts and magmatic evolution. *Mineralogical Magazine*, *60*, 5–40.

Marsh, B. D. (2006). Dynamics of magmatic systems. *Elements*, *2*, 287–292.

Marsh, B. D. (2015). Magmatism, magma and magma chambers. In G. Schubert (Ed.), *Treatise on geophysics* (2nd edn, Vol. 6, pp. 273–323). Amsterdam: Elsevier.

Martel, C. & Schmidt, B. C. (2003). Decompression experiments as an insight into ascent rates of silicic magmas. *Contributions to Mineralogy and Petrology*, *144*, 397–415.

Masotta, M., & Keppler, H. (2017). A new hydrothermal moissanite cell apparatus for optical in-situ observations at high pressure and high temperature, with applications to bubble nucleation in silicate melts. *American Mineralogist*, *102*, 2022–2031.

Masotta, M., Mollo, S., Freda, C., Gaeta, M., & Moore, G. (2013). Clinopyroxene–liquid thermometers and barometers specific to alkaline differentiated magmas. *Contributions to Mineralogy and Petrology*, *166*, 1545–1561. https://doi.org/10.1007/s00410-013-0927-9

Masotta, M., Mollo, S., Nazzari, M., Trecchiato, V., Scarlato, P., Papale, P., & Bachmann, O. (2018). Crystallization and partial melting of rhyolite and felsite rocks at Krafla volcano: a comparative approach based on mineral and glass chemistry of natural and experimental products. *Chemical Geology*, *483*, 603–618.

Masotta, M., Mollo, S., Gaeta, M., & Freda, C. (2016). Melt extraction in mush zones: The case of crystal-rich enclaves at the Sabatini Volcanic District (central Italy). *Lithos*, *248–251*, 288–292.

Masotta, M., Ni, H., & Keppler, H. (2014). In situ observations of bubble growth in basaltic, andesitic and rhyodacitic metls. *Contributions to Mineralogy and Petrology*, *167*, 976.

Menand, T. (2008). The mechanics and dynamics of sills in layered elastic rocks and their implications for the growth of laccoliths and other igneous complexes. *Earth and Planetary Science Letters*, *267*, 93–99.

Menand, T., Annen, C., & De Saint Blanquat, M. (2015). Rates of magma transfer in the crust: insights into magma reservoir recharge and pluton growth. *Geology*, *43*, 199–202.

Métrich, N. & Rutherford, M. (1998). Low pressure crystallization paths of H_2O-saturated basaltic-hawaiitic melts from Mt. Etna: Implications for open-system degassing of basaltic volcanoes. *Geochimica et Cosmochimica Acta*, *62*(7), 1195–1205.

Mevel, C., & Velde, D. (1976). Clinopyroxenes in Mesozoic pillow lavas from the French Alps: influence of cooling rate on compositional trends. *Earth and Planetary Science Letters*, *32*, 158–164.

Mollo, S., Blundy, J., Scarlato, P., De Cristofaro, S. P., Tecchiato, V., Di Stefano, F., et al. (2018). An integrated P–T–H_2O-lattice strain model to quantify the role of clinopyroxene fractionation on REE + Y and HFSE patterns of mafic alkaline magmas: Application to eruptions at Mt. Etna. *Earth-Science Reviews*, *185*, 32–56. https://doi.org/10.1016/j.earscirev.2018.05.014

Mollo, S., Blundy, J., Scarlato, P., Iezzi, G., & Langone, A. (2013). The partitioning of trace elements between clinopyroxene and trachybasaltic melt during rapid cooling and crystal growth, *Contributions to Mineralogy and Petrology*, *166*, 1633–1654. doi: 10.1007/s00410-013-0946-6

Mollo, S., Del Gaudio, P., Ventura, G., Iezzi, G., & Scarlato P. (2010). Dependence of clinopyroxene composition on cooling rate in basaltic magmas: Implications for thermobarometry. *Lithos*, *118*, 302–312. doi: 10.1016/j.lithos.2010.05.006

Mollo, S., Giacomoni, P. P., Andronico, D., & Scarlato, P. (2015). Clinopyroxene and titanomagnetite cation redistributions at Mt. Etna volcano (Sicily, Italy): Footprints of the final solidification history of lava fountains and lava flows. *Chemical Geology*, *406*, 45–54.

Mollo, S., & Hammer, J. E. (2017). Dynamic crystallization in magmas. In W. Heinrich, R. Abart (Eds.), *Mineral reaction kinetics: microstructures, textures, chemical and isotopic signatures* (EMU Notes in Mineralogy, Vol. 16, pp. 1–46), European Mineralogical Union and the Mineralogical Society of Great Britain and Ireland. doi: 10.1180/EMU-notes.16.12

Mollo, S., Lanzafame, G., Masotta, M., Iezzi, G., Ferlito, C., & Scarlato, P. (2011). Cooling history of a dike as revealed by mineral chemistry: A case study from Mt. Etna volcano, *Chemical Geology*, *283*, 261–273. doi: 10.1016/j.chemgeo.2011.06.016

Mollo, S., & Masotta, M. (2014). Optimizing pre-eruptive temperature estimates in thermally and chemically zoned magma chambers. *Chemical Geology*, *368*, 97–103. http://dx.doi.org/10.1016/j.chemgeo.2014.01.007

Mollo, S., Misiti, V., Scarlato, P., & Soligo, M. (2012). The role of cooling rate in the origin of high temperature phases at the chilled margin of magmatic intrusions, *Chemical Geology*, *322–323*, 28–46. doi: 10.1016/j.chemgeo.2012.05.029

Mollo, S., Putirka, K., Iezzi, G., Del Gaudio, P., & Scarlato P. (2011). Plagioclase–melt (dis)equilibrium due to cooling dynamics: implications for thermometry, barometry and hygrometry. *Lithos*, *125*, 221–235. doi: 10.1016/j.lithos.2011.02.008

Mollo, S., Putirka, K., Misiti, V., Soligo, M., & Scarlato, P. (2013). A new test for equilibrium based on clinopyroxene-melt pairs: Clues on the solidification temperatures of Etnean alkaline melts at post-eruptive conditions. *Chemical Geology*, *352*, 92–100. doi:10.1016/j.chemgeo.2013.05.026

Mollo, S., & Vona, A. (2014). The geochemical evolution of clinopyroxene in the Roman Province: A window on decarbonation from wall-rocks to magma. *Lithos*, *192–195*, 1–7. https://doi.org/10.1016/j.lithos.2014.01.009

Muncill, G. E., & Lasaga, A. C. (1988). Crystal-growth kinetics of plagioclase in igneous system: Isotherml H_2O-saturated experiments and extension of a growth model to complex silicate melts. *American Mineralogist*, *73*, 982–992.

Nabelek, P. I., Hofmeister, A. M., & Whittington, A. G. (2012). The influence of temperature-dependent thermal diffusivity on the conductive cooling rates of plutons and temperature–time paths in contact aureoles. *Earth and Planetary Science Letters*, *317*, 157–164.

Nakamura, Y. (1973). Origin of sector-zoning of igneous clinopyroxenes. *American Mineralogist*, *58*, 986–990.

Namur, O, Charlier, B, Toplis, M. J., & Vander Auwera, J. (2012). Prediction of plagioclase–melt equilibria in anhydrous silicate melts at 1-atm. *Contributions to Mineralogy and Petrology*, *163*, 133–150.

Neave, D. A., & Putirka, K. D. (2017). A new clinopyroxene–liquid barometer, and implications for magma storage pressures under Icelandic rift zones. *American Mineralogist*, *102*, 777–794.

Neri, A. (1998). A local heat transfer analysis of lava cooling in the atmosphere: application to thermal diffusion-dominated lava flows. *Journal of Volcanology and Geothermal Research*, *81*, 215–243.

Ni, H., Keppler, H., Walte, N., Schiavi, F., Chen, Y., Masotta, M., & Li, Z. (2014). In situ observation of crystal growth in a basalt melt and development of crystal size distribution in igneous rocks. *Contributions to Mineralogy and Petrology*, *167*, 1003.

Nimis, P. (1999). Clinopyroxene geobarometry of magmatic rocks. Part 2. Structural geobarometers for basic to acid, tholeiitic and mildly alkaline magmatic systems. *Contributions to Mineralogy and Petrology*, *135*, 62–74.

Nimis, P., & Taylor, W. R. (2000). Single clinopyroxene thermobarometry for garnet peridotites. Part 1 Calibration and testing of a Cr-in-cpx barometer and an enstatite-in-cpx thermometer. *Contributions to Mineralogy and Petrology*, *139*, 541–554.

Nimis, P., & Ulmer, P. (1998). Clinopyroxene geobarometry of magmatic rocks part 1: An expanded structural geobarometer for anhydrous and hydrous, basic and ultrabasic systems. *Contributions to Mineralogy and Petrology*, *133*, 122–135.

Nishizawa, J. I., Terasaki, T., & Shimbo, M. (1972). Layer growth in silicon epitaxy. *Journal of Crystal Growth*, *13*(14), 297–301.

Orlando, A., D'Orazio, M., Armienti, P., & Borrini, D. (2008). Experimental determination of plagioclase and clinopyroxene crystal growth rates in an anhydrous trachybasalt from Mt. Etna (Italy). *European Journal of Mineralogy*, *20*, 653–664.

Pamukcu, A. S., Gualda, G. A. R., Bégué, F., & Gravley, D. M. (2015). Melt inclusion shapes: timekeepers of short lived giant magma bodies. *Geology*, *43*, 947–950.

Papale, P., Moretti, R., & Barbato, D. (2006). The compositional dependence of the saturation surface of H_2O + CO_2 fluids in silicate melts. *Chemical Geology*, *229*, 78–95.

Perinelli, C., Mollo, S., Gaeta, M., De Cristofaro, S. P., Palladino, D. M., Armienti, P., et al. (2016). An improved clinopyroxene-based hygrometer for Etnean magmas and implications for eruption triggering mechanisms. *American Mineralogist*, *101*, 2774–2777. http://dx.doi.org/10.2138/am-2016-5916

Pichavant, M., Costa, F., Burgisser, A., Scaillet, B., Martel, C., & Poussineau, S. (2007). Equilibration scales in silicic to intermediate magmas: Implications for experimental studies. *Journal of Petrology*, *48*, 1955–1972.

Pietranik, A., Koepke, J., & Puziewicza, J. (2006). Crystallization and resorption in plutonic plagioclase: implications on the evolution of granodiorite magma (Gęsiniec granodiorite, Strzelin Crystalline Massif, SW Poland). *Lithos*, *86*, 260–280.

Pistone, M., Blundy, J., & Brooker, A. M. (2017). Water transfer during magma mixing events: Insights into crystal mush rejuvenation and melt extraction processes. *American Mineralogist*, *102*, 766–776.

Polacci, M., Arzilli, F., La Spina, G., Le Gall, N., Cai, B., Hartley, M. E., et al. (2018). Crystallisation in basaltic magmas revealed via in situ 4D synchrotron X-ray microtomography. *Nature Scientific Reports*, *8*(1),8377. doi: 10.1038/s41598-018-26644-6

Pupier, E., Duchene, S., & Toplis, M.J. (2007). Experimental quantification of plagioclase crystal size distribution during cooling of a basaltic liquid. *Contributions to Mineralogy and Petrology*, *155*, 555–570.

Putirka, K. D. (1999). Clinopyroxene + liquid equilibria. *Contributions to Mineralogy and Petrology*, *135*, 151–163.

Putirka, K. D. (2005). Igneous thermometers and barometers based on plagioclase + liquid equilibria: test of some existing models and new calibrations. *American Mineralogist*, *90*, 336–346.

Putirka, K. D. (2008). Thermometers and barometers for volcanic systems. In: K. D. Putirka, F. Tepley (Eds.), *Minerals, inclusions, and volcanic processes. Reviews in Mineralogy and Geochemistry*, *69*, 61–120. https://doi.org/10.2138/rmg.2008.69.3

Putirka, K. D. (2017). Down the crater: where magmas are stored and why they erupt. *Elements*, *13*, 11–16.

Putirka, K., Johnson, M., Kinzler, R., & Walker, D. (1996). Thermobarometry of mafic igneous rocks based on clinopyroxene–liquid equilibria, 0–30 kbar. *Contributions to Mineralogy and Petrology*, *123*, 92–108.

Putirka, K., Mikaelian, H., Ryerson, F., & Shaw, H. (2003). New clinopyroxene-liquid thermobarometers for mafic, evolved, and volatile-bearing lava compositions, with applications to lavas from Tibet and the Snake River Plain, Idaho. *American Mineralogist*, *88*, 1542–1554.

Rivalta, E., Taisne, B., Bunger, A. P. & Katz, R. F. (2015). A review of mechanical models of dyke propagation: Schools of thought, results and future directions. *Tectonophysics*, *638*, 1–42.

Richet P., & Ottonello, G. (2010). Thermodynamics of phase equilibria in magma. *Elements*, *6*, 315–320.

Robert, B., Harris, A., Gurioli, L., Métard, E., Sehlke, A., & Whittington, A. (2014). Textural and rheological evolution of basalt flowing down lava channel. *Bulletin of Volcanology*, *76*, 824.

Roskoz, M., Toplis, M. J., & Richet, P. (2006a). Kinetic vs thermodynamic control of crystal nucleation and growth in molten silicates. *Journal of Non-Crystalline Solids*, *352*, 180–184.

Roskoz, M., Toplis, M. J., & Richet, P. (2006b). Crystallization of highly supercooled silicate melts. *Advanced Engineering Materials*, *8*, 1224–1228.

Saar, M. O., Manga, M., Cashman, K. V., & Fremouw, S. (2001). Numerical models of the onset of yield strength in crystal-melt suspensions. *Earth and Planetary Science Letters*,*187*, 367–379.

Scaillet, B., & MacDonald, R. (2001). Phase relations of peralkaline silicic magmas and petrogenetic implications. *Journal of Petrology*, *42*, 825–845.

Scarlato, P., Mollo, S., Blundy, J., Iezzi, G. & Tiepolo, M. (2014). The role of natural solidification paths on REE partitioning between clinopyroxene and melt. *Bulletin of Volcanology*, *76*, 810.

Scarlato, P., Mollo, S., Del Bello, E., Von Quadt, A., Richard, J. B., Gutierrez, E., et al. (2017). The 2013 eruption of Chaparrastique volcano (El Salvador): Effects of magma storage, mixing, and decompression. *Chemical Geology*, *448*, 110–122. http://dx.doi.org/10.1016/j.chemgeo.2016.11.015

Schiavi, F., Walte, N., & Keppler, H. (2009). First in situ observation of crystallization processes in a basaltic–andesitic melt with the moissanite cell. *Geology*, *37*(11), 963–966.

Schosnig, M., & Hoffer, E. (1998). Compositional dependence of REE partitioning between diopside and melt at 1 atmosphere. *Contributions to Mineralogy and Petrology*, *133*, 205–216. http://dx.doi.org/10.1007/s004100050448

Sehlke, A., Whittington, A., Robert, A., Harris, A., Gurioli, L., & Médard E. (2014). Pahoehoe to `a`a transition of Hawaiian lavas: an experimental study. *Bulletin of Volcanology*, *76*, 876.

Shea, T., & Hammer, J. E. (2013). Kinetics of cooling- and decompression-induced crystallization in hydrous mafic-intermediate magmas. *Journal of Volcanology and Geothermal Research*, *260*, 127–145.

Shimizu, N. (1981). Trace element incorporation into growing augite phenocryst. *Nature (London)*, *289*, 575–577.

Singer, B. S., Dungan, M. A. & Layne, G. D. (1995). Textures and Sr, Ba, Mg, Fe, K and Ti compositional profiles in volcanic plagioclase: clues to the dynamics of calcalkaline magma chambers. *American Mineralogist*, *80*, 776–798.

Sisson, T. W., & Grove, T. L. (1993). Experimental investigations of the role of water in calcalkaline differentiation and subduction zone magmatism. *Contributions to Mineralogy and Petrology*, *113*, 143–166.

Skulski, T., Minarik, W., & Watson, E. B. (1994). High-pressure experimental trace-element partitioning between clinopyroxene and basaltic melts. *Chemical Geology*, *117*, 127–147. http://dx.doi.org/10.1016/0009-2541(94)90125-2

Smith, J. V. (1983). Phase equilibria of plagioclase. In Ribbe, P. H. (Ed.), *Feldspar mineralogy* (2nd edn, Reviews in Mineralogy, Vol. 2, pp. 223–239). Washington, DC: Mineralogical Society of America.

Smith, J. V., & Brown, W. L. (1988). *Feldspar Minerals, Volume I*. Berlin: Springer-Verlag, 828 pp.

Smith, D., & Lindsley, D.H. (1971). Stable and metastable augite crystallization trends in a single basalt flow. *American Mineralogist*, *56*, 225–233.

Sparks, R. S. J., & Cashman, K. V. (2017). Dynamic magma systems: implications for forecasting volcanic activity. *Elements*, *13*, 35–40.

Szramek, L., Gardner, J. E., & Hort, M. (2010). Cooling-induced crystallization of microlite crystals in two basaltic pumice clasts. *American Mineralogist*, *95*, 503–509.

Taisne, B., & Jaupart, C. (2009). Dyke propagation through layered rocks. *Journal of Geophysical Research Solid Earth*, *114*, B09203.

Tait, S., & Taisne, B. (2013). The dynamics of dyke propagation. In S. Fagents, T. Gregg, R. Lopes (Eds.), *Modelling volcanic processes: the physics and mathematics of volcanism* (pp. 32–54). Cambridge, UK: Cambridge Press.

Tegner, C. (1997). Iron in plagioclase as a monitor of the differentiation of the Skaergaard intrusion. *Contributions to Mineralogy and Petrology*, *128*, 45–51.

Tiller, W. A., & Ahn, K. S. (1980) Interface field effects on solute redistribution during crystallization. *Journal of Crystal Growth*, *49*, 483–501.

Turner, S. & Costa, F. (2007). Measuring timescales of magmatic evolution. *Elements*, *3*, 267–272.

Ubide, T., Mollo, S., Zhao, J., Nazzari, M., & Scarlato, P. (2019). Sector-zoned clinopyroxene as a recorder of magma history eruption triggers, and ascent rates. *Geochimica et Cosmochimica Acta*, *251*, 265–283.

Uhlmann, D. R., Onorato, P. I. K., & Scherer, G. W. (1979). *A simplified model for glass formation*. Paper presented at 10th Lunar and Planetary Science Conference (Proceedings Vol. 1, pp. 375–381), Houston, TX.

Uhlmann, D.R., Yinnon, H., Fang, C.H. (1981). *Simplified model evaluation of cooling rates for glass-containing lunar composition*. Paper presented at 12th Lunar and Planetary Science Conference (Proceedings, pp. 281–288), Houston, TX.

Ujike, O. (1982). Microprobe mineralogy of plagioclase, clinopyroxene and amphibole as records of cooling rate in the Shirotori-Hiketa dike swarm, northeastern Shikoku, Japan. *Lithos*, *15*, 281–293.

Ushioda, M., Takahashi, E., Hamada, M., & Suzuki, T. (2014). Water content in arc basaltic magma in the northeast Japan and Izu arcs: an estimate from Ca/Na partitioning between plagioclase and melt. *Earth, Planets, Space*, *66*, 1–10. http://dx.doi.org/10. 1186/1880-5981-66-127

Vetere, F., Iezzi, G., Behrens, H., Cavallo, A., Misiti, V., Dietrich, M., et al. (2013). Intrinsic solidification behavior of basaltic to rhyolitic melts: A cooling rate experimental study. *Chemical Geology*, *354*, 233–242.

Vetere, F., Iezzi, G., Behrens, H., Holtz, F., Ventura, G., Misiti, V., et al. (2015). Glass forming ability and crystallization behavior of sub-alkaline silicate melts. *Earth-Science Reviews*, *150*, 25–44.

Vona, A., Romano, C., Dingwell, D. B., & Giordano, D. (2011). The rheology of crystal-bearing basaltic magmas from Stromboli and Etna. *Geochimica et Cosmochimica Acta*, *75*, 3214–3236.

Wass, S. Y. (1973). The origin and petrogenetic significance of hour-glass zoning in titaniferous clinopyroxenes. *Mineralogical Magazine*, *39*, 133–144

Watson, E. B., & Liang, Y. (1995). A simple model for sector zoning in slowly grown crystals: implications for growth rate and lattice diffusion, with emphasis on accessory minerals in crustal rocks. *American Mineralogist*, *80*(11–12), 1179–1187.

Watson, E. B., & Müller, T. (2009). Non-equilibrium isotopic and elemental fractionation during diffusion-controlled crystal growth under static and dynamic conditions. *Chemical Geology*, *267*(3–4), 111–124. doi: 10.1016/j.chemgeo.2008.10.036

Welsch, B., Hammer, J.E., Baronnet, A., Jacob, S., Hellebrand, E., & Sinton, J. (2016). Clinopyrozene in postshield Haleakala ankaramite: textures, compositional zoning and supersaturation in the magma. *Contributions to Mineralogy and Petrology*, *171*, 6.

Whittington, A. G., Hofmeister, A. M., & Nabelek, P. I. (2009). Temperature-dependent thermal diffusivity of Earth's crust and implications for magmatism. *Nature*, *458*, 319–321.

Wohletz, K., Civetta, L., & Orsi, G. (1999). Thermal evolution of the Phlegraeanmagmatic system. *Journal of Volcanology and Geothermal Research*, *91*, 381–414.

Wood, B. J., & Blundy, J. D. (1997). A predictive model for rare earth element partitioning between clinopyroxene and anhydrous silicate melt. *Contributions to Mineralogy and Petrology*, *129*, 166–181. http://dx.doi.org/10.1007/s004100050330

Wood, B. J., & Blundy, J. D. (2001). The effect of cation charge on crystal–melt partitioning of trace elements. *Earth and Planetary Science Letters*, *188*, 59–71. http://dx.doi.org/10.1016/S0012-821X(01)00294-1

Xu, Z., & Zhang, Y. (2002). Quench rates in water, air and liquid nitrogen and inference of temperature in volcanic eruption columns. *Earth and Planetary Science Letters*, *200*, 315–330.

Zhang, Y. (2008). *Geochemical kinetics*. Princeton, NY: Princeton University Press.

Zhang, Y. (2010) Diffusion in minerals and melts: theoretical background. In Y. Zhang, D. Cherniak (Eds.), *Diffusion in minerals and melts. Reviews in Mineralogy and Geochemistry*, *72*, 5–60.

Zhang Y., & Cherniak D. J. (Eds.) (2010). *Diffusion in minerals and melts. Reviews in Mineralogy and Geochemistry*, *72*, 1038 pp.

6

Architecture of the Magmatic System in the Main Ethiopian Rift

Sabrina Nazzareni[1], Stefano Rossi[1], Maurizio Petrelli[1], and Luca Caricchi[2]

ABSTRACT

The Main Ethiopian Rift represents a young continental rift with an associated large volume of magmatism that forms one of the major large igneous provinces. Plio-Quaternary magmatic activity is located in the Main Ethiopian Rift (MER). Transitional to peralkaline mafic and felsic magmatic products dominate the volcanism, which is related to complex variation in the crustal structure across the rift. Clinopyroxene, olivine, and plagioclase are liquidus phases in the transitional and alkaline mafic magmas. Here we used the ability of clinopyroxene to record variations of P, T and fO_2 to reconstruct the host-rock's geological history. Transitional and alkaline basalts and basaltic trachyandesite (with MgO > 7 wt%) from across the MER have augitic phenocrysts as liquidus phase. Euhedral centrimetric megacrysts of similar composition are sometime associated with alkaline rift basalts. Clinopyroxene geobarometry performed combining X-ray diffraction with mineral chemistry points to a complex polybaric plumbing system active since 7.5–3.7 Ma. The MER clinopyroxene crystallize in a wide pressure range from the lower crust (ca. 10 kbar) up to middle crust (ca. 4 kbar) in the plumbing system beneath Debre Zeyit Fault Zone and the Wonji Fault Belt. The continuous polybaric MER clinopyroxene crystallization from the lower to middle crust can be explained by a plumbing system composed of a complex of dykes and sills where magmas rise, stall, and crystallize.

6.1. INTRODUCTION

The Ethiopian rift valley (Figure 6.1) is home to one of the major large igneous provinces (LIPs) representing a site where intense continental break-up has been accompanied by widespread magmatic activity (flood basalts and silicic volcanics), which gave rise to a large Oligocene to Miocene plateau and extensive Plio-Quaternary volcanism in the Afar depression and along the Main Ethiopian Rift (MER; Mohr & Zanettin, 1988; Peccerillo et al., 2003). The East African Rift (MER and Afar) is a recent intracontinental rift where the upwelling of two distinct plumes from the deep mantle has impinged on the overlying lithosphere to produce an incomplete continental breakup (e.g., Furman, 2007; George et al., 1998; Macdonald et al., 2001; Marty et al., 1996; Pik et al., 1999; White & McKenzie, 1989). The mantle upwelling generated a large volume of magmatism producing rocks that cover an area of about $6 \times 10^5\,km^2$ with thickness that locally reaches 1500 m (Boccaletti et al., 1999).

Most studies on Ethiopian rift magmatism have focused on the regional scale, particularly to explain the origin of the bimodal distribution of mafic and felsic products associated with the scarcity of intermediate rock compositions (i.e., the Daly Gap; e.g., Ayalew, 2000; Ayalew et al., 2002; Baker et al., 1977, 1996; Boccaletti et al., 1999; Macdonald et al., 1995; Trua et al., 1999). In particular, silicic rocks dominate the rift-floor sequences and mainly consist of peralkaline rhyolites with minor trachytes, emplaced mostly as pyroclastic flow and fall deposits. Mafic rocks, in contrast, mostly comprise transitional basalts, often forming cinder cones and lava flows, with minor tuff rings and cones (Peccerillo et al., 2007).

The nature of the Daly Gap cannot be simply explained by fractional crystallization of mafic magma or by partial melting, although crustal assimilation (i.e., assimilation

[1] *Department of Physics and Geology, University of Perugia, Perugia, Italy*

[2] *Department of Earth Sciences, University of Geneva, Geneva, Switzerland*

Dynamic Magma Evolution, Geophysical Monograph 254, First Edition. Edited by Francesco Vetere.

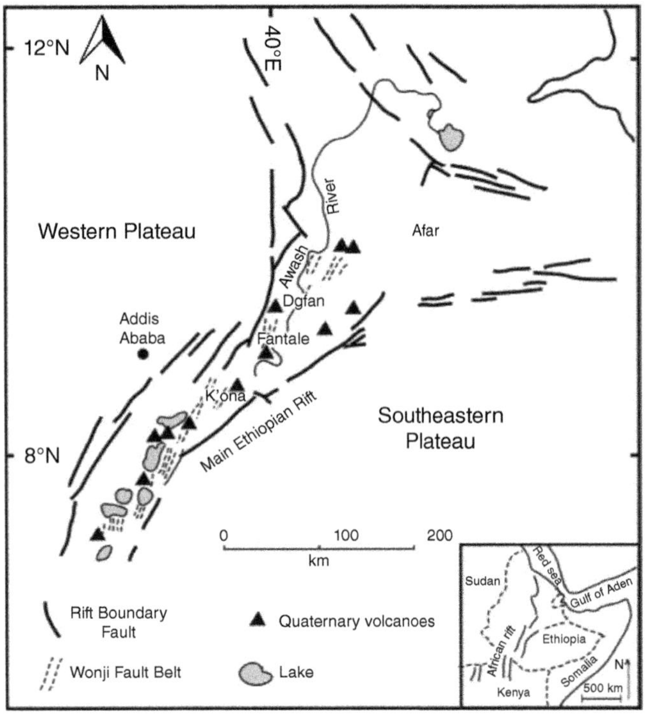

Figure 6.1 Location map of the area studied. [Modified from Williams et al. (2004).]

and fractional crystallization—AFC) processes may have played a role in the rhyolite generation (e.g., Ayalew, 2000; Ayalew et al., 2002; Baker et al., 2000; Peccerillo et al., 2003; Trua et al., 1999; Yirgu, 1997). The study of peralkaline silicic magmatism not only served to constrain the mechanisms of magma emplacement and evolution (polybaric or isobaric differentiation processes), but also to reconstruct the architecture of the plumbing system (Caricchi et al., 2006; Peccerillo et al., 2003; Rooney et al., 2005, 2007, 2011).

The geodynamic setting of the MER is complex as it connects the East African Rift system with the Afar triple junction (Figure 6.1). The MER is characterized by a sequence of half-graben with changes in direction and overlapping zones between rift segments with a mean NE–SW direction (Acocella & Kome, 2002). The Wonji Fault Belt is the youngest part of the MER and is characterized by active NNE–SSW trending extension (Acocella & Kome, 2002). The recent volcanic and tectonic activity is located in the central MER (Wonji Fault Belt and Silti-Debre Zeyit Fault), with fissure and central eruptions along faults and fractures suggesting a close link between magmatism and tectonics (Acocella & Kome, 2002).

Even though the link between magmatism and tectonics has been widely described, the conditions of magma storage still remain poorly constrained (e.g., Peccerillo et al., 2003; Rooney et al., 2007). Two main models of the plumbing system involve the ponding of melt in large magma chambers or in a complex dyke zone distributed throughout the crust (e.g., Acocella & Kome 2002; Gleeson et al., 2017; Maccaferri et al., 2014; Mahatsente et al., 1999; Peccerillo et al., 2007; Rooney et al., 2007, 2011).

In order to provide further constraints on the architecture of the volcanic plumbing system of the MER, we first review geochemical data available in the literature, and then focus on clinopyroxene crystals that are a common phenocryst phase in the mafic volcanics of the Main Ethiopian Rift. Together with olivine, pyroxene is a liquidus phase (Caricchi et al., 2006) and therefore can be used to investigate the variation of intensive parameters such as P, T, and fO_2 within the magmatic system in the period preceding eruption (e.g., Brizi et al., 2003; Nazzareni et al., 2001, 2011; Neave & Putirka, 2017; Perugini et al., 2003; Ubide et al., 2019). Pyroxenes may also record subsolidus processes, since the intracrystalline kinetics of cations is related directly to the cooling rate of the host rocks. The pioneering work by Dal Negro et al. (1982, 1989) demonstrated that the chemical signature and the structural parameters of clinopyroxene could be used to reconstruct the geological history of the magmatic system, particularly for defining the pressures of crystallization.

In particular, Dal Negro et al. (1982) studied pyroxenes from fissural basalts (Aiba, Alaji, and Ashangi formations) of the Ethiopian Trap series, and the central cone of the Termaber Formation aiming to use the crystallographic parameters for the classification and characterization of magmatic rocks. Their work identified a subcalcic trend defined by low-Ca clinopyroxene ($Ca < 0.8$ atoms per formula unit—apfu) and a calcic trend defined by high-Ca clinopyroxene ($Ca > 0.8$ apfu), with the former group related to the tholeiitic lavas and the latter to the alkaline-peralkaline rhyolites. Low-Ca clinopyroxene is characterized by a decrease of Ca in the M2 site, substituted by Fe^{2+} and Mg, and the substitution by ions ($Ti^{4+} > Fe^{3+}$) smaller than Mg and Fe^{2+} in the octahedral M1 site. In high-Ca clinopyroxene the variation of Ca content is smaller than in the low-Ca clinopyroxene and the tetrahedral Al substitution is mainly compensated by Fe^{3+} and Ti^{4+} in the M1 site. The crystallographic geometrical parameters (bond lengths, monoclinic angle β, polyhedral volumes) of these pyroxenes define two groups as well and were used as a petrological tool to discriminate the parental magma type.

We used a similar approach to Dal Negro et al. (1982) and measured crystallographic geometrical parameters on selected clinopyroxenes phenocrysts from the MER to estimate their crystallization pressure. Combining clinopyroxene geobarometric investigation, volcanic bulk geochemistry, experimental constraints, and geophysical data, we aimed to better constrain the magma storage conditions and the magmatic evolution leading to the formation of the Ethiopian volcanic province.

6.1.1. Literature Review

Detailed study of the dynamics and structure of the MER magmatic system requires knowledge of petrological, geochemical and isotopic characteristics of the magmatism in the area. Thus, we have reviewed the geochemical, petrogenetic, and geodynamic evolution of the studied area (e.g., Ayalew et al., 2002, 2016; Barberio et al., 1999; Brotzu et al., 1980, 1981; Caricchi et al., 2006; Dal Negro et al., 1982; Peccerillo et al., 2003, 2007; Pik et al., 1998, 1999; Rooney, 2010; Rooney et al., 2007; Trua et al., 1999; Williams et al., 2004; Yirgu, 1997). Our data are complemented with both whole-rock and clinopyroxene data from the GEOROC database (http://georoc.mpch-mainz.gwdg.de).

6.2. METHODOLOGY

6.2.1. Sample Selection

Mafic rocks from the Ethiopian Rift include lavas, strombolian scoriae, and hydrovolcanic lapilli and ashes. Felsic products (i.e., trachytes and rhyolites) include lavas, pumices, and welded ignimbrites. Intermediate rocks are rare (Peccerillo et al., 2007). We focus on mafic samples from the plateau, the transition zone, and the rift zone that represent the three main phases of MER evolution. The volcanic rocks used in this work were collected across the rift in the Wonji Fault Belt from the western margin (Ethiopian Central Plateau region) to the axial portion of the MER in the area between Debre Zeit and the National Park of the Fantale volcano. Samples from Blue Nile Basalts (BN; age ca. 50–35 Ma), Addis Abeba Basalts (AA; age ca. 8–7.5 Ma), and Rift Series Basalts (RFT; age ca. 2–1.6 Ma) were selected from among the porphyritic volcanics having pyroxene as phenocrysts. Most of the samples studied are lavas, apart from RFT19 and RFT33, which were collected from a scoria cone. We also selected three centimeter-scale megacrysts (Mcpx, Cpx2, Cpx3) that are associated with the Rift Series Basalts.

A total-alkali versus silica (TAS) plot of the selected rocks is reported in Figure 6.2a (BN28, AA, and RFT). The composition of the AA1, the starting material utilized by Caricchi et al. (2006) for the experimental investigations on MER magmas, is also reported in Figure 6.2a (white triangle) as a reference. Major and trace-element composition of bulk rock is reported in Table 6.1.

Mafic products have a porphyritic texture with phenocrysts of olivine, plagioclase, and clinopyroxene surrounded by a microcrystalline to hypocrystalline groundmass. Total phenocryst abundance varies significantly between 5 and 20% by volume (with peaks of 40–50%) and centimeter-scale plagioclase and clinopyroxene megacrysts were also found in lavas and scoria products (Peccerillo et al., 2007; Figure 6.3).

Forsteritic olivine is the dominant phase and it occurs prevalently as euhedral crystals, sometimes zoned and altered in iddingsite. Plagioclase phenocrysts, often zoned, show a bytownitic–labradoritic composition. Clinopyroxene crystals, augitic in composition, are less frequent as phenocrysts but ubiquitous in the groundmass. Identical mineralogical phases also occur in the groundmass, along with minor Fe–Ti oxides.

Rare intermediate rocks are characterized by aphyric to porphyritic textures with phenocrysts and megacrysts of plagioclase, olivine, and clinopyroxene. Zoned oligoclasic–bytownitic plagioclase represent the most abundant phase. Mg-olivine (forsterite) is often present as corroded crystals altered in iddingsite, while clinopyroxene is alkaline, green in color, and diopsitic in composition (Barberio et al., 1999; Peccerillo et al., 2003). Figure 6.3 reports examples of representative textures observed in some selected samples.

6.2.2. Clinopyroxene Selection

For X-ray diffraction investigation, single-crystal fragments of clinopyroxene phenocrysts and megacrysts have been extracted under a microscope from a ca. 100-μm-thick rock section. Using this approach, we identified crystal fragments from the core or the rim of phenocrysts and megacrysts, selecting optically a homogeneous portion of the crystal.

Megacrysts from the Rift Series Basalts show a centimeter-scale euhedral habit: $1.5 \times 2.5 \times 2$ cm for Mcpx, $2 \times 2 \times 1.8$ cm for Cpx2, and $3 \times 2 \times 2$ cm for Cpx3, respectively. Megacrysts were embedded in resin and cut parallel to the crystallographic *c* axis to obtain 100-μm-thick oriented sections. At least two fragments representing different portions of the megacrysts were extracted using a scalpel.

Selected clinopyroxene phenocrysts showed no evidence of disequilibrium with the surrounding matrix, Electron microprobe and LA-ICP-MS investigations were also performed on megacrysts.

6.2.3. Single-Crystal X-ray Diffraction

Crystallographic geometrical parameters were obtained by single-crystal X-ray diffraction data using both a Philips Pw1100 and an Xcalibur (Rigaku-Oxford diffraction) diffractometers. Diffraction data were collected up to $\theta = 32°$ by using graphite monochromatized MoKα radiation, and after correction for Lorentz, polarization, and absorption, intensity data were merged to obtain a set of independent reflections (Table SM1—see Supplementary Material (SM) section).

Anisotropic structural refinements have been carried out in the *C2/c* space group, by using the SHELXl

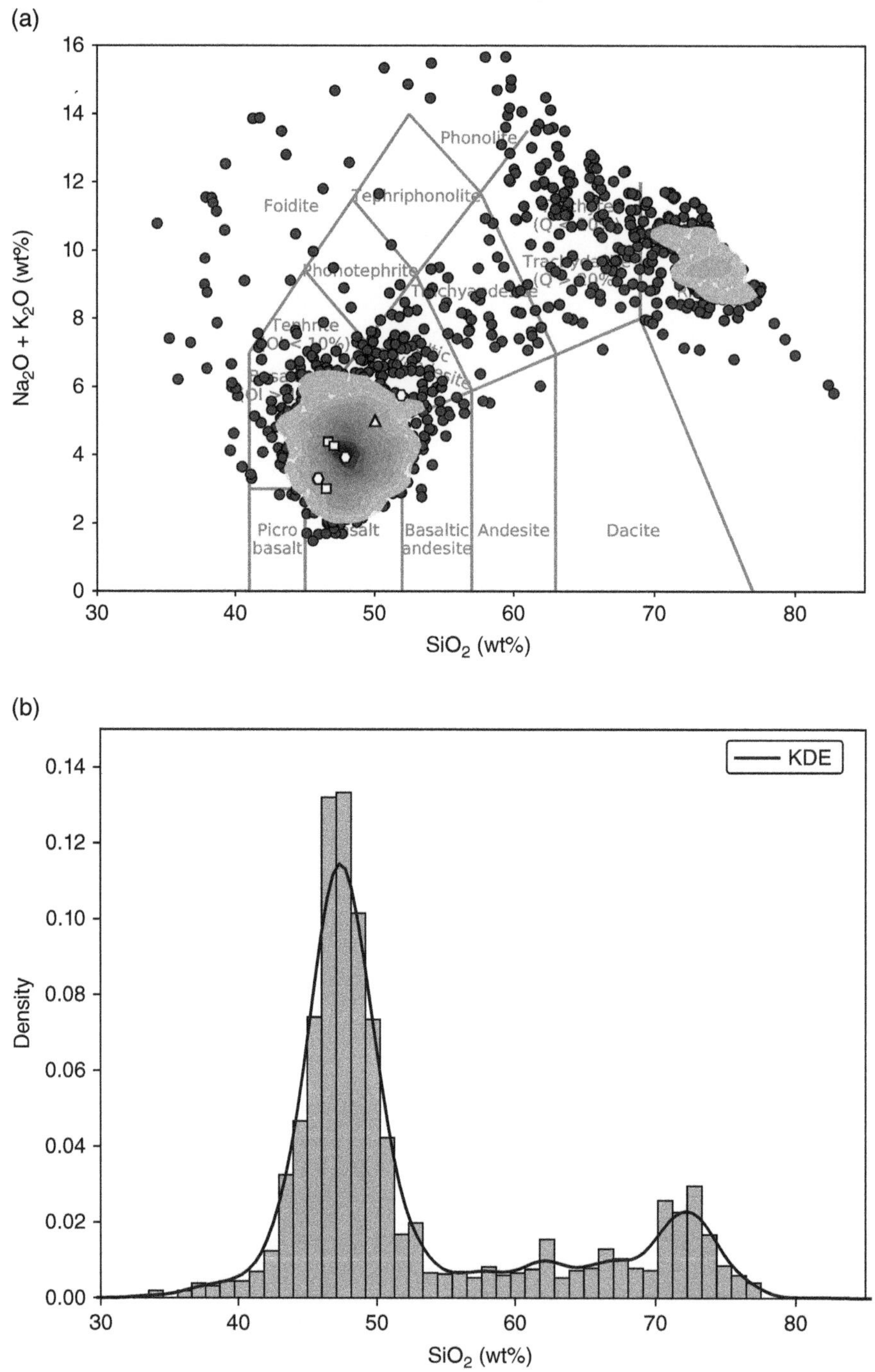

Figure 6.2 (a) Total-alkali versus silica classification diagram. Literature data from GEOROC database; white circles, selected rocks for this study. (b) Density histogram of SiO_2: KDE, kernel density estimation. See text for discussion.

(Sheldrick, 2008) software package starting from the diopside coordinate (Nazzareni et al., 1998). Details of the refinements are reported in Table SM1. In the refinements, we used partly ionized scattering factor, ($Si^{2.5+}$, $O^{1.5-}$, a mixed $95Mg^{2+}/5Al^{3+}$ and mixed $85Ca^{2+}/15Na^{+}$), and occupancies of Mg/Al and Fe^{2+} in M1 and Ca and Na in M2 have been refined. The refinements converged after few cycles using isotropic thermal factors first and then anisotropic thermal factors to final R1 discrepancy factors between 1.37 and 3.09% (Table SM1). In all samples a residue peak in the electron density map occurred early in the refinement cycles at around 0.6 Å from the

Table 6.1 Bulk Composition for Selected Rock Samples

	Sample									
Element	RFT19	RFT33	RFT34	BN11	BN27	BN28	AA1	AA3	AA5	AA6
K	1	1	1	5	1	1	7	7	7	7
SiO_2	46.53	47.06	46.66	39.83	45.82	45.29	50.01	45.9	51.91	47.83
TiO_2	1.62	2.04	2.17	3.97	1.35	1	1.91	1.47	2.01	1.76
Al_2O_3	16.48	16.78	16.32	10.95	16.87	17.15	16.1	15.17	17.37	16.29
Fe_2O_3	9.16	7.42	3.89	9.7	3.33	5.98	5.19	5.48	6.79	3.04
FeO	2.04	4.38	7.88	5.29	7.4	5.4	5.26	5.19	2.97	7.21
MnO	0.16	0.18	0.19	0.17	0.17	0.18	0.17	0.17	0.15	0.16
MgO	9.38	7.36	7.44	10.18	8.93	10.15	7.31	12.1	4	8.73
CaO	9.51	9.65	9.73	13.11	9.42	9.73	8.16	10.26	7.58	9.38
Na_2O	2.42	3.35	3.33	0.17	0.81	0.52	1.5	0.6	1.74	0.72
K_2O	0.59	0.89	0.99	2.43	1.91	1.96	3.44	2.63	3.91	3.13
P_2O_5	0.34	0.36	0.38	0.75	0.24	0.24	0.33	0.25	0.39	0.4
H_2O^a	1.76	0.52	1.01	3.44	3.75	2.41	0.61	0.78	1.19	1.34
Nb	19.00	21.00	23.00	23.00	19.00	19.00	20.00	18.00	25.00	19.00
Y	21.00	25.00	26.00	27.00	21.00	21.00	25.00	20.00	34.00	26.00
Rb	2.00	17.00	21.00	0.00	8.00	10.00	31.00	8.00	40.00	6.00
Zr	89.00	137.00	155.00	216.00	86.00	107.00	146.00	93.00	222.00	121.00
Sr	598.00	540.00	574.00	1039.00	375.00	290.00	551.00	441.00	642.00	630.00
Ni	165.00	102.00	90.00	73.00	120.00	222.00	142.00	271.00	33.00	100.00
xCr	714.00	460.00	427.00	263.00	419.00	513.00	401.00	1003.00	58.00	497.00
V	252.00	274.00	281.00	484.00	219.00	201.00	242.00	230.00	244.00	232.00
Ba	295.00	363.00	446.00	598.00	231.00	126.00	472.00	259.00	643.00	338.00
xLa	21.00	26.00	29.00	54.00	12.00	15.00	30.00	16.00	41.00	24.00
xCe	51.00	64.00	53.00	120.00	29.00	36.00	61.00	44.00	84.00	48.00
Cr	0.00	397.00	0.00	228.00	341.00	0.00	0.00	842.00	0.00	0.00
La	0.00	26.20	0.00	48.50	12.50	0.00	0.00	17.10	0.00	0.00
Ce	0.00	58.50	0.00	95.40	25.40	0.00	0.00	33.20	0.00	0.00
Nd	0.00	29.90	0.00	43.40	15.00	0.00	0.00	16.80	0.00	0.00
Sm	0.00	6.20	0.00	9.91	3.91	0.00	0.00	4.15	0.00	0.00
Eu	0.00	1.93	0.00	2.72	1.24	0.00	0.00	1.28	0.00	0.00
Tb	0.00	0.68	0.00	0.89	0.52	0.00	0.00	0.52	0.00	0.00
Yb	0.00	2.07	0.00	1.46	1.95	0.00	0.00	1.70	0.00	0.00
Lu	0.00	0.30	0.00	0.20	0.28	0.00	0.00	0.24	0.00	0.00
Sc	0.00	29.00	0.00	41.00	30.00	0.00	0.00	27.00	0.00	0.00
Cs	0.00	0.00	0.00	0.00	0.18	0.00	0.00	0.00	0.00	0.00
Hf	0.00	3.19	0.00	5.50	2.30	0.00	0.00	2.30	0.00	0.00
Ta	0.00	2.14	0.00	3.90	1.20	0.00	0.00	1.40	0.00	0.00
Th	0.00	2.50	0.00	4.80	1.40	0.00	0.00	1.70	0.00	0.00
U	0.00	0.50	0.50	0.80	0.00	0.00	0.00	0.30	0.30	0.30

Note. [a] H_2O values are determined by the loss-on-ignition method.

M2 site, in agreement with the M2' site of Rossi et al. (1978). We thus refined the scattering factor of Fe^{2+} at the M2' position in all the samples. Bond lengths, polyhedral volumes, and occupancies resulting from the refinements are reported in Table SM1.

6.2.4. Electron Microprobe

Major elements were analyzed by wavelength-dispersive spectroscopy (WDS) using a Cameca/Camebax SX50 (at the CNR-Istituto di Geoscienze e Georisorse (IGG), Section of Padova, Italy) and a JEOL JXA8600 (at the IGG, Section of Florence, Italy) electron microprobe. Working conditions were 15 kV and 15 nA, and X-ray counts were converted into oxide weight percentages. Up to four points were measured on each sample and an average chemical value was obtained. The analytical data obtained are reported in Table 6.2 and Table SM2. The accuracy of the analytical data is better than 2% for all elements analyzed.

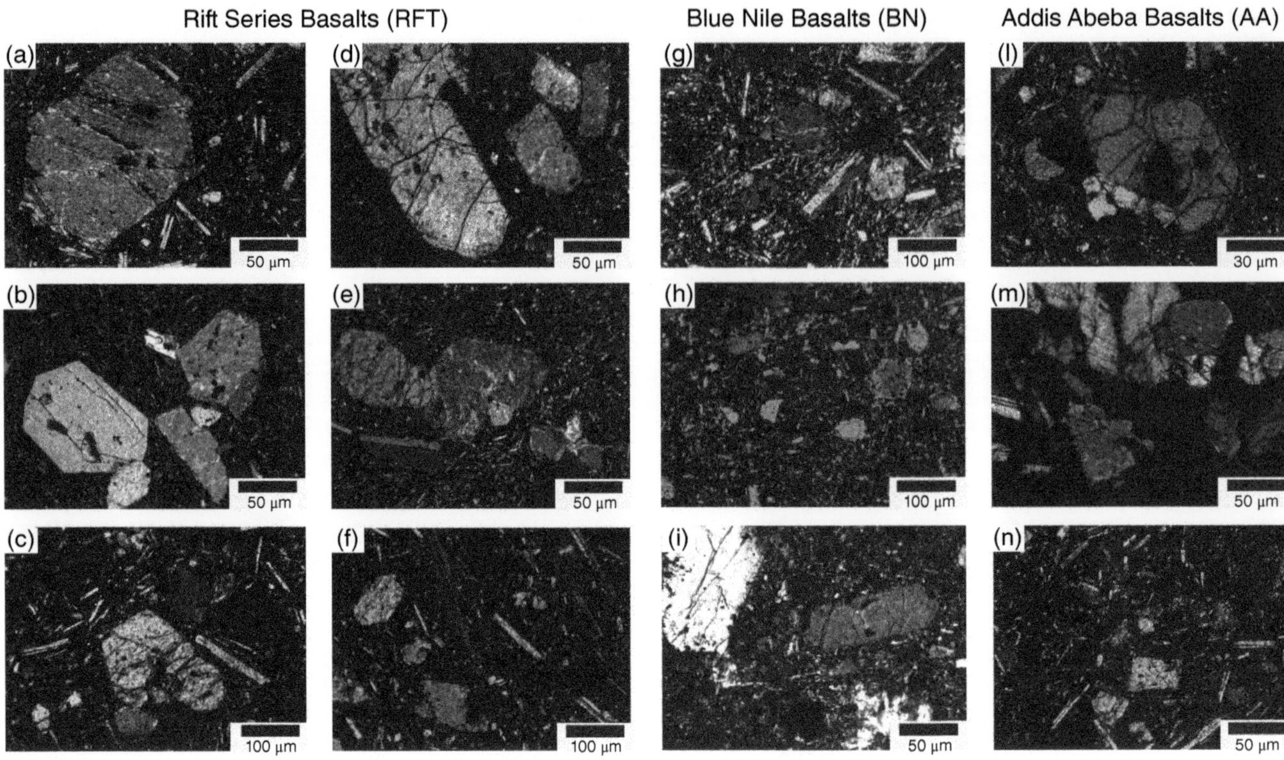

Figure 6.3 Microphotographs of selected MER samples at different magnifications.

Table 6.2 Chemical Data of Selected Clinopyroxenes

Site	Element	BN28	AA5	Mcpx	RFT19
T	Si	1.862	1.854	1.84	1.798
	Al	0.138	0.146	0.16	0.202
M1	Al	0.031	0.042	0.042	0.065
	Ti	0.04	0.043	0.048	0.039
	Fe^{3+}	0.047	0.059	0.061	0.084
	Fe^{2+}	0.125	0.112	0.107	0.076
	Cr	0.01	0.001	0	0.008
	Mn	0.003	0.001	0.003	0.002
	Mg	0.744	0.739	0.738	0.726
M2	Ca	0.882	0.82	0.816	0.859
	Na	0.029	0.043	0.039	0.032
	Fe^{2+}	0.046	0.093	0.088	0.046
	Mg	0.041	0.043	0.054	0.061
	Mn	0.001	0.001	0.003	0.002
T_c (°C)		1063	797	870	1033

Note. T_c closure temperature. Cation partitioning is based on six oxygens (expressed in apfu). Fe–Mg partitioning between M1 and M2 polyhedral sites was obtained combining chemistry and SC-XRD. Uncertainty is less than 10% (1σ).

6.2.5. LA-ICP-MS

Trace-element determinations were performed by laser ablation inductively coupled plasma mass spectrometry (LA-ICP-MS) at the Department of Physics and Geology (University of Perugia), using a Teledyne Photon Machine G2 laser system coupled to a Thermo iCAP-Q ICP-MS. Helium was used as carrier gas with Ar and N_2 added after the ablation cell to avoid plasma destabilization and enhance the instrumental sensitivity. Analyses were performed on the sector zoning of the Mcpx megacrysts, along the exact coordinates of previously analyzed electron probe microanalysis (EPMA) profiles. Data were collected for Sc, Ti, V, Cr, Mn, Ni, Rb, Sr, Zr, Nb, Cs, Ba, Hf, plus rare earth elements (REEs), using spot sizes of 20 or 30 μm, depending on the width of the zone being analyzed, with a repetition rate and fluence of 8 Hz and 3.5 J cm^{-2} respectively. Dwell time was 10 ms per analyte. The NIST 610 reference material, Si concentrations determined by EPMA, and USGS BCR2G glass were used as the calibration standard, internal standard, and quality control, respectively. Data reduction was carried out using the Iolite v.3 software package (Paton et al., 2011). Under the reported analytical conditions, precision and accuracy are better than 10% (Petrelli et al., 2008, 2016). Data are reported in Table SM2.

6.2.6. Thermobarometric Analysis

Magmatic phase equilibria are influenced by the competition of composition, crystallinity, oxygen fugacity, temperature, and pressure (Neave & Putirka, 2017). The quantification of these parameters is essential for estimating magma storage depths and to integrate petrological, geochemical, seismological, and geodynamic data in volcanically active regions. Pyroxenes represent one of the most sensitive mineral to pressure changes in magmatic systems (Neave & Putirka, 2017).

Here, we combine the application of the chemical-based thermometer of Putirka (2008) and the barometer of Neave and Putirka (2017) to estimate the crystallization pressure for our clinopyroxene megacrysts' composition. The barometer has been calibrated from experimental data in the range from 1 atm to 20 kbar on hydrous and anhydrous ultramafic to intermediate rock samples (Neave & Putirka, 2017). Table SM2 reports the composition of clinopyroxene megacrysts along the investigated chemical profiles and the relative pressure estimates. We obtained a *P–T* estimate that is in good agreement with those reported by Caricchi et al. (2006; reported in the Figure SM1).

6.2.7. Viscosity and Magma Ascent Velocities Calculations

We calculated the expected magma viscosities at the *P–T* conditions relevant for our system. Melt viscosity, η_{mt}, has been calculated according to the model of Giordano et al. (2008) in the temperature range 1125–1140°C, considering a melt with an initial H_2O content between 2.0 and 4.0 wt%, which is within the range used by Caricchi et al. (2006). The effect of crystals, depending on both their shape (i.e. aspect ratio, *r*) and abundance (i.e. crystal content, ϕ_s), has been evaluated using equation (6.1) in agreement with Laumonier et al. (2014).

$$\log \eta_{rel} = \frac{\alpha}{1+\beta \exp^{\varphi\phi_s}} + \frac{\delta}{1+\varepsilon \exp^{\xi\phi_s}} \qquad (6.1)$$

where α, β, δ, ε, φ, and ξ are fitting parameters and ϕ_s represents the crystal content (ranging from 0 to 40 vol% in our case).

The average flow velocity (V) for an ascending dike characterized by a height $2a$, have been estimated in agreement with Rubin (1993, 1995) using the following equation:

$$V = \frac{1}{3\eta} \frac{\Delta P_{excess}}{[\mu/(1-\nu)]^2} a \qquad (6.2)$$

where η is the magma viscosity, ΔP_{excess} is the excess pressure and $\mu/(1-\nu)$ is the rock stiffness. Note that for a given ΔP_{excess}, the average flow velocity is inversely proportional to magma viscosity. The values of the parameters assumed to fit equation (6.2) are reported in Table SM3.

6.3. RESULTS AND DISCUSSION

6.3.1. Whole-Rock Geochemical Data from the Literature

The Ethiopian Rift Valley is one of the youngest LIPs, covering an area of more of 6×105 km², dominated by fissure-fed basaltic lavas (Figure 6.2a; up to 3×10^5 km³ in volume) along with rhyolitic and trachytic pyroclastic products associated with minor lavas (Boccaletti et al., 1999; Peccerillo et al., 2007). Magmatic activity in the area started during the Oligocene (ca. 50 Ma) forming a large plateau (up to the Miocene), shifting to Afar and along the rift during the Plio-Quaternary (e.g. Merla et al., 1979; Mohr & Zanettin, 1988).

The MER magmatic system had a peak of volcanic activity around 30–31 Ma when tholeiitic to transitional–basaltic lava flows were erupted over a very short period (1–2 Myr; Hofmann et al., 1997) to form the Ethiopian Plateau (e.g., Ashange and Aiba Basaltic Formations; Mohr & Zanettin, 1988). The latest episodes were characterized by alternating basaltic and ignimbrite eruptions, which formed the so-called Amba Alaji Rhyolitic Formation (e.g., Merla et al., 1979). At around 10–5 Ma the formation of the rift valley started, the transition between plateau formation and the rift phase was characterized by the formation of several shield volcanoes (Termaber Basalt Formation). Transitional to Na-alkaline basalts with minor trachytes characterize this evolution phase (Piccirillo et al., 1979; Peccerillo et al., 2003, 2007). In the Plio-Quaternary (ca. 5 Ma to present) magmatic activity was directly linked to the main opening of the MER and Afar.

In the MER, recent volcanism consists of trachytic and peralkaline rhyolitic ignimbrites (e.g., pumiceous and ash-fall deposits) and minor lavas. Volumetrically subordinate basalts, mostly forming cinder cones, tuff rings, and cones (Peccerillo et al., 2003), are associated with silicic rocks erupted from large central volcanoes with summit calderas (e.g. Fantale, Gedemsa, Kone, Bede Gebabe), mostly aligned along extensional faults of the Wonji Fault Belt (e.g., Wonji Basalts; Barberio et al., 1999; Mohr, 1971; Peccerillo et al., 2007). Rocks of intermediate compositions are sporadic, imprinting on the volcanism of the area a bimodal character (Figure 6.2b).

The water-free SiO_2 histogram and the two-dimensional kernel density estimations (KDE) reported in Figures 6.2, 6.4, and 6.5, allowed us to better define the proportions between the erupted chemical compositions, and to clearly define a modal composition for the most frequently sampled basalts (i.e., the largest abundance in terms of specific chemical composition).

In detail, Figure 6.2a and b highlights the well-known bimodal distribution (i.e., Daly Gap) that characterizes MER volcanic products. Applying simple statistical considerations, we can provide further constraints. As an example, integrating the area below the KDE reported in Figure 6.2b, it emerges that SiO_2 compositions between 43 and 52 wt% and between 68 and 77 wt%, represent 64% and 14% of the total erupted SiO_2 compositions respectively. Globally, they represent 77% of all erupted compositions in terms of SiO_2 contents, with a ratio of about 5/1 between the mafic ($43 < SiO_2 < 52$ wt%) and felsic ($68 < SiO_2 < 77$ wt%) end-members respectively. The

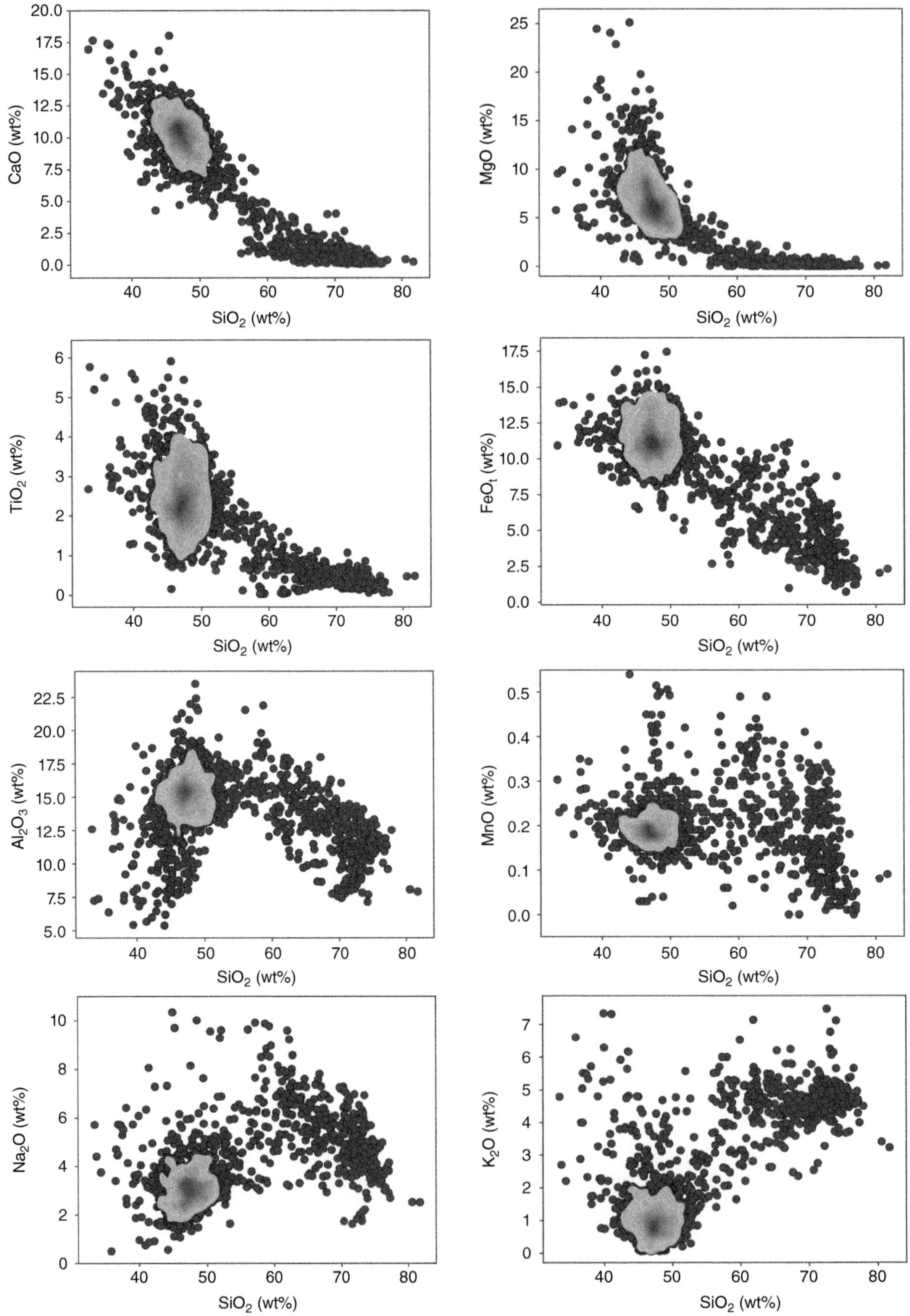

Figure 6.4 Variation diagrams of major elements against silica.

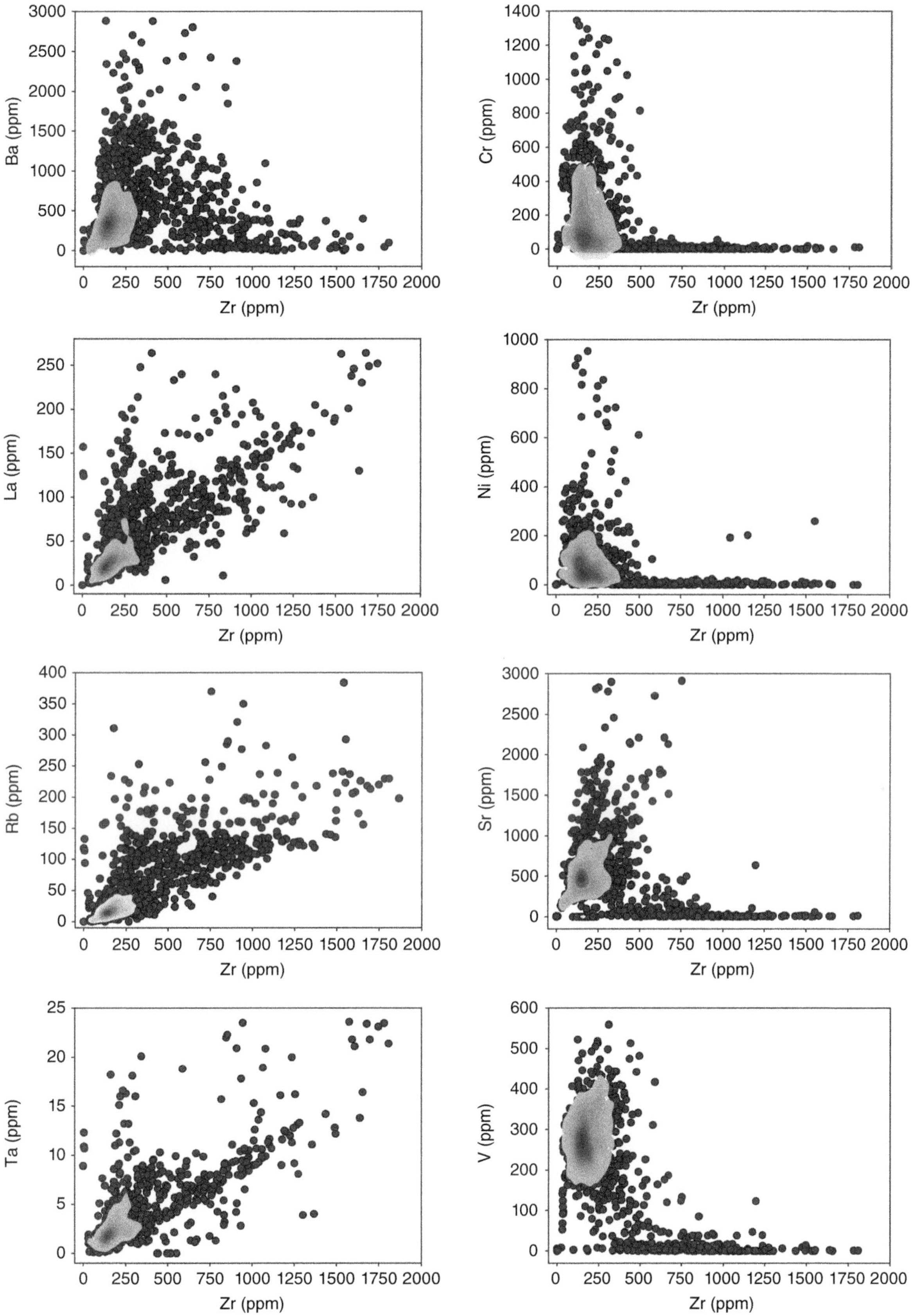

Figure 6.5 Variation diagrams of trace elements against Zr.

Table 6.3 Modal composition for both analyzed major and trace elements

Major elements		Trace elements	
SiO_2	48.34	Cr	53
TiO_2	2.24	V	270
Al_2O_3	15.81	Ni	50
FeO_t	11.73	Rb	14
MnO	0.15	Sr	470
MgO	6.83	Ba	330
CaO	10.61	La	20
Na_2O	3.06	Ta	1.8
K_2O	0.71	Zr	140
Cr_2O_3	0.00		
P_2O_5	0.51		
Total	100.00		

Note. This composition has been utilized as a reference liquid for thermobarometric estimates based on clinopyroxene analyses.

calculated SiO_2 modal compositions fall at ~47.5 and ~72.5 wt% for the mafic and the felsic members respectively.

Focusing now on the most represented SiO_2 contents (i.e., the mafic magmas: $43 < SiO_2 < 52$ wt%), Figure 6.4 shows that a well-characterized modal composition can be calculated through the KDE analysis, for all analyzed major elements (i.e., CaO, MgO, TiO_2, FeO, Al^2O_3, MnO, Na_2O, K_2O and P_2O_5). The same information also can be observed in Figure 6.5, where the modal compositions are reported for the analyzed trace elements (e.g., Ba, La, Ni, Ta, and V).

The resulting modal composition (Table 6.3) is used as a reference liquid for thermobarometric estimations based on clinopyroxene analyses.

6.3.2. Thermobarometry Based on the Structural Geometry of the Analyzed Clinopyroxenes

Clinopyroxene crystals have a diopside/augitic composition with a chemical variability from BN28 ($Wo_{48}En_{43}Fs_9$) to AA5 ($Wo_{45}En_{43}Fs_{12}$) to RFT19 ($Wo_{50-46}En_{47-42}Fs_{10-6}$). Associated with the chemical variation a change in pyroxene color has been reported by Dal Negro et al. (1982) from pale/light green (low-Ca pyroxene Ca < 0.8 apfu) in tholeiitic basalts to deep green or greenish brown or brownish/violet (high-Ca pyroxene Ca > 0.8 apfu) in alkaline basalts. Our samples are high-Ca pyroxene deep green in color, only rarely having greenish brown color.

Clinopyroxene minerals are a solid solution characterized by complex heterovalent cation substitutions that can involve tetrahedral and M2 sites, $(R^{3+}, R^{4+})T = (R^{2+}, R^{+})M2$, octahedral M1 and M2 sites $(R^{3+}, R^{2+})M1 = (R^{+}, R^{2+})M2$, or all the crystallographic sites in the structure $(R^{4+}, R^{2+})M1 = (R^{3+}, R^{4+})T, (R^{+}, R^{2+})M2$. The cation substitution in pyroxene is the result of the *P–T–X* conditions of the melt from which they crystallize, and strongly influence the crystallographic structure of the pyroxene (bond lengths and polyhedral volumes). The MER pyroxenes presented in this study have the same scheme of cation substitution as the low-Ca and high-Ca pyroxenes of Dal Negro et al. (1982):

$$\left(Al^{3+}, Si^{4+}\right)T = \left(Ca^{2+}, Na^{+}\right)M2,$$
$$\left(Fe^{3+}, Ti^{4+}, Mg^{2+}\right)M1 = \left(Al^{3+}, Si^{4+}\right)T$$
$$\left(Ca^{2+}\right)M2 = \left(Mg, Fe^{2+}\right)M2$$

acting at different extents in the two different groups.

The pyroxenes studied are characterized by coupled substitution of Al^{IV} for Si in T, Al^{VI}, Ti^{4+}, and Fe^{3+} for Mg in M1 sites, and Na for Ca in M2 sites to charge balance the structure. All samples have Ca > 0.8 apfu, with Ca content varying more than Na content. To balance the variable Ca content Mg and Fe^{2+} enter into M2 site.

The chemistry of the clinopyroxene megacrysts associated with Rift Series Basalts is characterized by a zonation in color from dark green to light green. Samples Cpx2 and Cpx3 show only a thin rim zone, but Mcpx is characterized by sector zoning. The chemical variation of Mcpx was investigated along three different profiles of nearly 150m in length across the zoning in the section containing the *c* axis (Figure 6.6). Major-element variation across the zoning is small and mainly involves Si and tetrahedral Al, as well as Mg and octahedral Al and Ti^{4+}, together with a variation in Ca and Na in M2 site. The enrichment of trace elements from the dark green zone to the light green zone (Figure 6.6) suggests that the sector zoning is related to a different magma chemistry during the crystal growth.

The MER clinopyroxenes are characterized by a similar geometry of the tetrahedral site (VT 2.236–2.269 Å^3, mean T-O 1.637–1.646 Å) reflecting a similar Si and Al content. The largest variation occurs in the M1 and M2 sites by the substitution of Na for Ca and the entering of Fe^{2+}, with Mg in M2 sites strongly influencing the polyhedral geometry; the polyhedral volume increases from the AA (plus RFT19) to BN and RFT sample series, in particular both the shortest M2–O2 and the longest M2–O3C2 bond distances increase (Table SM1). The crystallographic β angle reflects the entering of Fe^{2+} and Mg in the Cpx structure, i.e., this angle decreases at the entering of Fe^{2+} in the octahedral site M1. Furthermore, the entering of Ca and Na in M2 makes this site more regular, with the parameter DM2 (M3–O2 - (M2O3 + M2–O2 + M2–O1)/3) expressing the polyhedral M2 distortion.

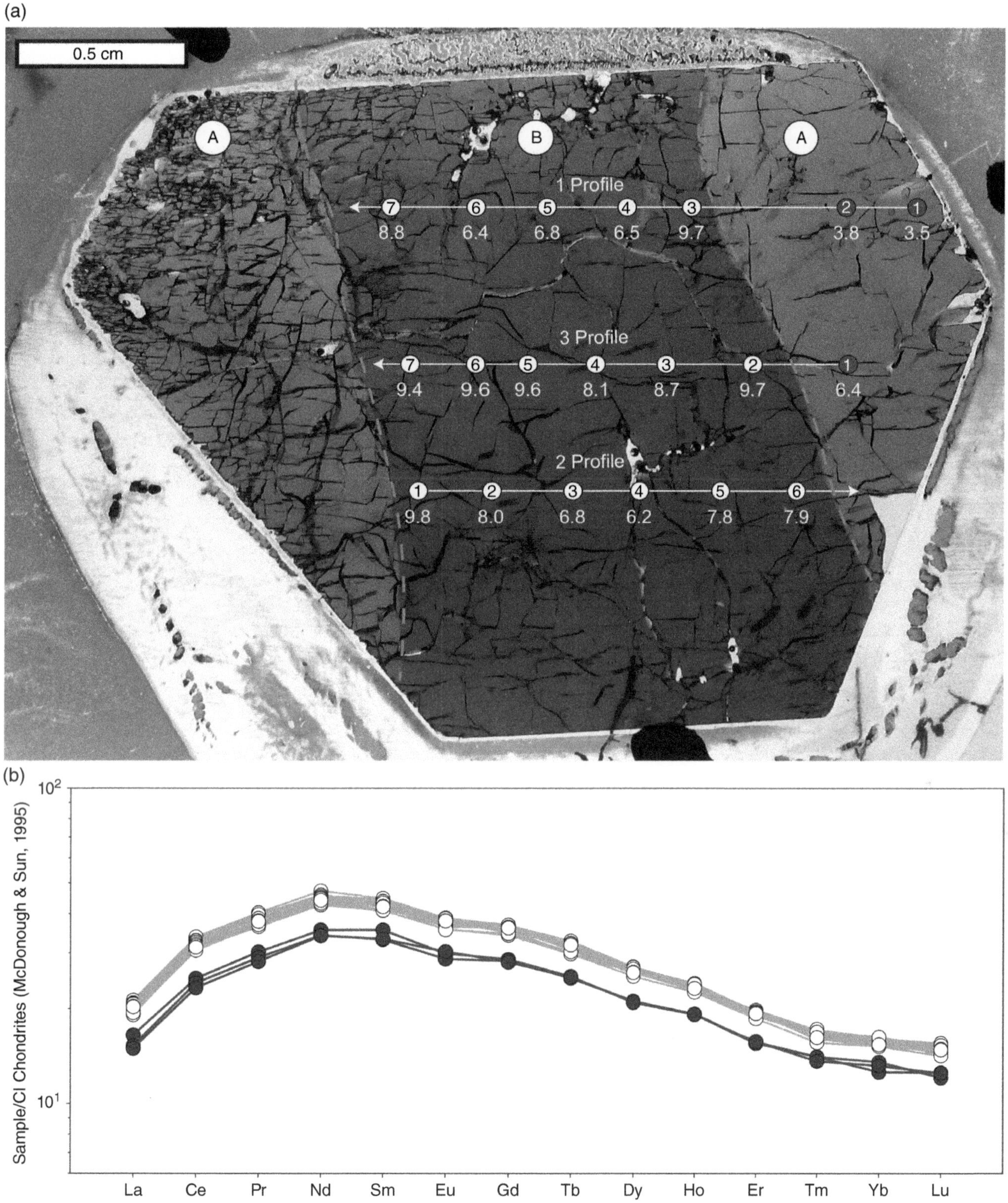

Figure 6.6 Megacryst Mcpx: (a) photograph with the reported profile lines used for the EMPA and LA-ICP-MS measurements across the sector zoning; (b) corresponding spider diagram.

Pyroxene is very sensitive to changes in the magma chemistry and responds by changing its major (and trace) element composition according to the cation exchange vector described above. Thus, some crystallographic geometrical parameters and their relationships, like that between the β angle and DM2, record the magma chemical characteristics, such as the degree of magma evolution or the tholeiitic to alkaline degree of the parental magma (Dal Negro et al., 1989). The MER pyroxenes plot in the alkaline/transitional field defined by Dal Negro et al. (1982) in agreement with the bulk-rock composition (Table 6.1).

Octahedral M1 bond distances are shorter in the AA (plus RFT19) than in BN and RFT sample series and consequently the octahedral M1 decrease in the same samples (Table SM1). Megacrysts have an intermediate behavior: M2–O2 and M2–O3C2 bond distances and M1-O bond distances are longer than AA (plus RFT19) and shorter than BN and RFT sample series.

6.3.2.1. Closure Temperatures

Subsolidus processes can be recorded by pyroxene because the Mg and Fe^{2+} cations intracrystalline exchange between the M1 and M2 crystallographic sites is dependent on temperature and time (Dal Negro et al., 1982). The closure temperature (T_c) for this exchange reaction is related to the cooling rate of the magma (Alvaro et al., 2015; Murri et al., 2018). Thus, we can use this parameter to reconstruct the subsolidus history of the host rock, with high T_c corresponding to fast cooling (see Malgarotto et al., 1993).

Dal Negro et al. (1982) calculated the Mg–Fe^{2+} intracrystalline distribution for the Ethiopian pyroxenes, obtaining generally low closure temperature values for all samples. We recalculated these T_c values with the new formulation of Murri et al. (2018), obtaining temperatures that vary from 496°C to 876°C for the low-Ca pyroxenes and from 420°C to 732°C for the high-Ca pyroxenes.

We selected samples BN28, AA5, and RFT19 and the megacryst sample to calculate the T_c (Table 6.2) and obtained closure temperatures between 1063°C and 797°C, which are significantly higher than those calculated from the Dal Negro et al. (1982) data. The megacryst sample has a closure temperature of 870°C, suggesting a fast cooling rate in agreement with the other samples. The Dal Negro et al. (1982) pyroxenes were selected from a well-crystallized groundmass that had a slower cooling rate than phenocrysts (like our samples), thus explaining the observed difference of closing temperatures. Moreover, closure temperature has been proven to record the difference in stratigraphic cooling rate across a dike (Malgarotto et al., 1993), with a fast cooling rate at the margins (T_c = 798°C, cooling rate = 34° h^{-1}) and slower cooling rate in the middle (T_c = 627°C, cooling rate = 1°C h^{-1}). The same could apply to lava flows, obtaining the highest T_c near the rapidly cooled surface.

6.3.2.2. Geobarometric Estimations

Pyroxene is an early liquidus mineral phase in mafic magmas. Although its crystallization is function of *P–T–X* variables, clinopyroxene is especially sensitive to pressure. Such a characteristic has been used to define a geobarometer based on clinopyroxene crystal chemistry (e.g., Neave & Putirka, 2017; Nimis, 1995; Nimis & Ulmer, 1998; Putirka, 2008).

Dal Negro et al. (1989) observed that in clinopyroxene a linear relationship exists between cell and octahedral M1 volumes with pressure of crystallization. At high pressure both volumes have small values that increase with decreasing pressure irrespective of magma bulk composition. The V_{cell}/V_{M1} relationship is related to the cation substitutions affecting the structure, (Al^{3+}, Mg^{2+})M1 = (Ca^{2+}, Na^{+})M2, (Al^{3+}, Mg^{2+})M1 = (Al^{3+}, Si^{4+})T, and (Ca^{2+})M2 = (Mg,Fe^{2+})M2, which at a given *P* is related to the crystallization conditions (aCaO, $aSiO_2$). The V_{cell}/V_{M1} relationship is very sensitive to pressure crystallization conditions and even small variations are recorded by the pyroxene structure, for example as observed in core–rim pairs (Nazzareni et al., 2003).

The cell volume of MER clinopyroxenes varies from 436.0(8) to 442.18(2) Å^3, octahedral M1 volume varies from 11.593(2) to 12.058(5) Å^3, and polyhedral M2 volume varies from 25.306(4) Å^3 to 25.714(9) Å^3 (Table SM1). Cation substitutions are responsible for the geometrical variation of the polyhedral volumes: the octahedral M1 shrinks as Al or Fe^{3+} substitute Mg or Fe^{2+} and the polyhedral M2 expands as Na substitutes Ca. Cell volume, however, reflects the whole structure geometry.

The pyroxenes belonging to AA rocks show the smallest volumes (Table SM1), with limited variation among selected core fragments. The largest values are recorded by the rim fragment of sample AA5 and from the core of a green/brown phenocryst of sample AA3. Sample RFT19 pyroxenes have the smallest volume values of all the MER samples, with values comparable to pyroxene from spinel lherzolite rock (Figure 6.7).

Most of the MER clinopyroxenes have similar ranges of cell and M1 volume values (pyroxene fragments from core phenocrysts of BN and RFT samples) plotting as a cluster (cluster 1) in the V_{M1} versus V_{cell} plot (Figure 6.7). Sample RFT109 has the largest cell and M1 volumes (442.18(2) Å^3, 12.058(5) Å^3) suggesting a crystallization at very shallow conditions (i.e., close to the Earth's surface or after eruption).

Megacrysts have intermediate volume values between the BN plus RFT sample series pyroxenes (cluster 1) and AA (plus RFT19) samples, with partial overlapping with the low-Ca pyroxenes (Figure 6.7). In particular, sample

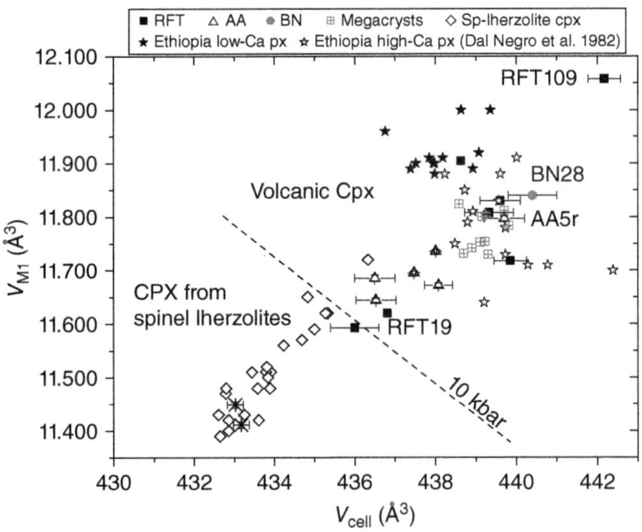

Figure 6.7 Volume of M1 octahedron plotted against the cell volume. High-pressure boundary taken from Nazzareni et al. (2001). Asterisks in the spinel-lherzolite field are clinopyroxene from pyroxenite nodules enclosed in Oahu alkaline lava (Hawaii) crystallized at 17 kbar (Nazzareni et al., 2003).

Mcpx is characterized by a green sector zoning, darker in the central part and lighter towards the edges (Figure 6.6). Under an optical microscope we extracted fragments in the (100) plane from the outermost rim zone of the dark green and light green parts and from the dark green center of the crystal. The volume values measured for the dark-green and light-green parts of sample Mcpx are very similar, reflecting slight chemical variation in the sector zoning and also suggesting a similar crystallization pressure. A partial agreement is evident when comparing our data with those of Dal Negro et al. (1982) on MER pyroxenes.

Most of our samples have similar geometrical parameters (bond distances and volumes) compared to the high-Ca pyroxene of Dal Negro et al. (1982). Only AA and RFT19 samples greatly differ from the preexisting database of Dal Negro et al. (1982).

The relationship between cell and octahedral M1 volumes, as discussed in Dal Negro et al. (1989), can thus be used to constrain the crystallization pressure of pyroxenes and the magma storage conditions. The chemistry (and structure) of clinopyroxene is greatly influenced by the activity of magma components (i.e., $aSiO_2$, aAl^2O_3, aNa_2O), thus to minimize the effect of magma chemical composition on the clinopyroxene we selected mafic basalts and a basaltic trachyandesite (Figure 6.2a), with MgO > 7 wt% apart from sample AA5, which has a MgO content of 4 wt%.

In the cell versus M1 volume plot (Figure 6.7) we can observe that MER pyroxenes are distributed over a wide space in the volcanic field, from high-pressure, defined by clinopyroxene from spinel lherzolites (Dal Negro et al., 1989), to lower pressures. The MER pyroxenes from AA basalts crystallized at the highest pressures close to the mantle field (as defined by the spinel lherzolite pyroxene). RFT19 phenocryst pyroxene cores partially overlap the high-pressure (mantle) field in the V_{cell}–V_{M1} plot, recording the highest crystallization pressure among all the samples studied (Figure 6.7). Nevertheless most of our samples (BN and RFT series) plot as a cluster (overlapping with the literature data) suggesting lower crystallization pressures for the plateau and rift host-rocks.

Crystallographic data collected for the megacrystals suggest that they crystallized at intermediate pressures between the high-pressure AA and RFT19 pyroxenes and the BN and RFT pyroxenes. The AA and RFT19 samples plot at the limit of the spinel-lherzolite clinopyroxene stability field (P = 20–10 kbar) suggesting a maximum pressure of crystallization around 9–10 kbar. Two samples from pyroxenite mantle nodules enclosed in an alkaline lava from Oahu (Hawaii), with crystallization pressure estimated to be 17 kbar (Nazzareni et al., 2003), are also reported in Figure 6.7.

Clinopyroxenes from the Blue Nile basalts (BN) and Wonji Fault Belt alkaline basalts have volume values comparable to those of clinopyroxenes from the ancient alkaline centers of Mount Etna, for which a crystallization pressure around 5–6 kbar has been inferred (Nazzareni et al., 2003).

There is general agreement between our data and the low-Ca pyroxene of Dal Negro et al. (1982). Most of the tholeiitic pyroxenes of Dal Negro et al. (1982), although shifted to higher V_{M1} values in the plot, suggest similar P values (Figure 6.7). Sample RFT109, which has the largest cell and octahedral M1 volumes, can be considered as a "surface or near-surface" end-member of the MER magmatic system.

6.3.3. Thermobarometric Estimations Based on Clinopyroxene's Crystal Chemistry

The combined application of the chemical-concentration-based Neave and Putirka (2017) barometer with the Putirka (2008) thermometer is reported in Figure 6.8. In detail, we applied the P–T estimations based on the chemistry of the clinopyroxene in equilibrium with the modal chemical composition of the erupted basalts to (a) the analyzed crystals (Figure 6.8a) and (b) data from Pik et al. (1998), Brotzu et al. (1981), and Del Negro et al. (1982) for both the plateau and rift areas. The equilibrium between melt composition and clinopyroxene was assessed following Putirka (1999).

The analyzed clinopyroxene provided temperature and pressure estimations in the range from ~1140 to ~1200°C and from ~3.5 to ~10 kbar (Figure 6.8a). These estimates are in agreement with those reported for clinopyroxene

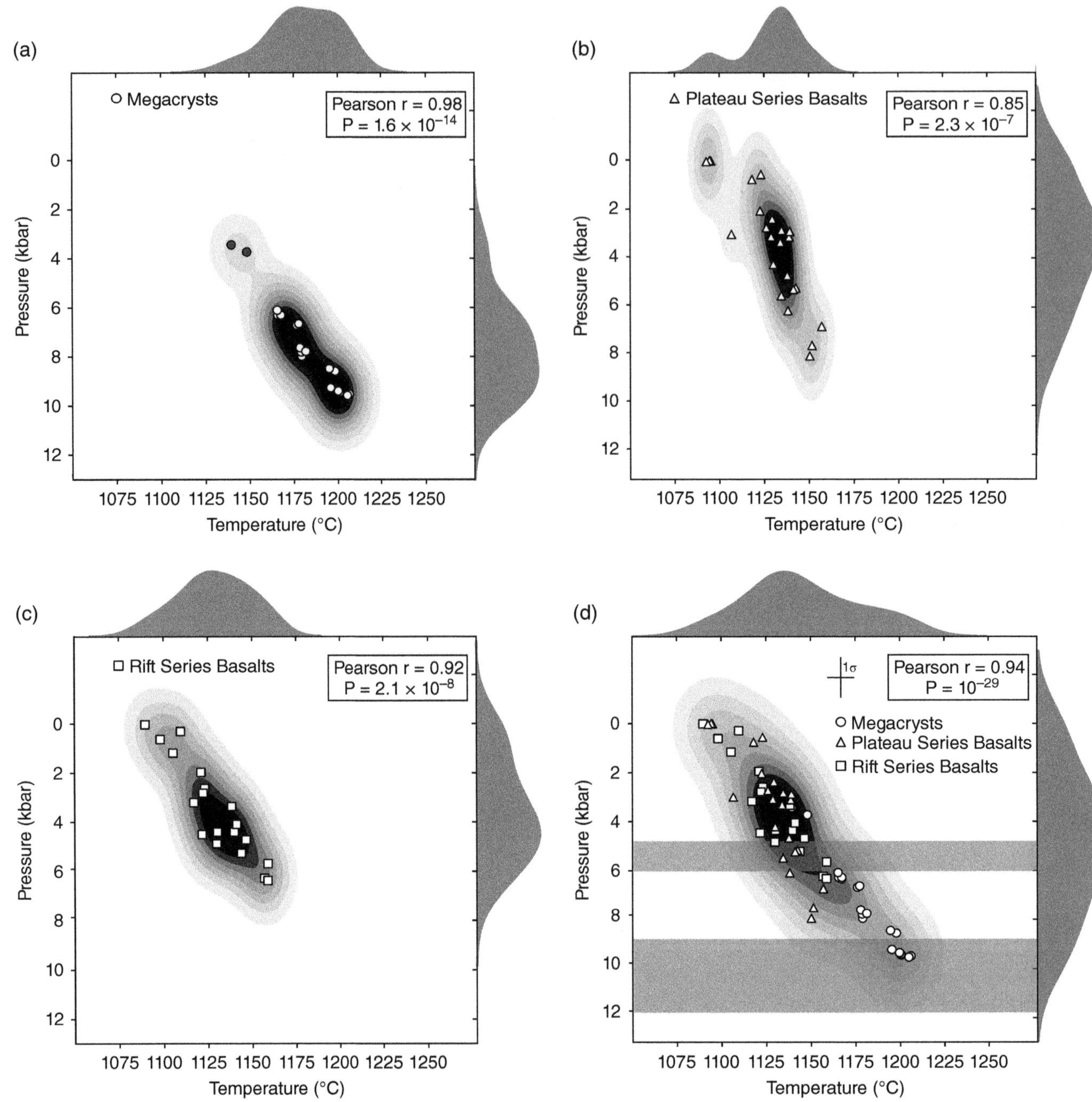

Figure 6.8 Pressure versus temperature diagram. Data obtained combining Neave and Putirka (2017) barometer with the Putirka (2008) thermometer.

in the literature for the plateau and the rift zones, respectively (Figure 6.8b–d). We obtained a P–T estimation in good agreement with those reported by Caricchi et al. (2006; Figure SM1). Combining all the available data, we obtained temperature and pressure between ~1080 and ~1200°C and 0 to ~10 kbar (Figure 6.8d). The most likely T and P estimates are ~1130°C and 4 kbar (Figure 6.8d). Pressure values obtained on the basis of structural geometries of the analyzed clinopyroxenes are also reported in Figure 6.8d for comparison.

6.4. TOWARDS A VOLCANOLOGICAL MODEL

Following early formation of the basaltic plateau by fissure eruptions, continental rifting proceeded by ductile crustal stretching to form a rift valley through normal faulting (McKenzie, 1978). The geophysical models for the Ethiopian Rift Valley suggest that crustal thickness decreases from 35–50 km beneath the Ethiopian Plateau to around 25–31 km beneath the MER and 18–25 km beneath the Afar depression (Dugda et al., 2007; Hammond et al., 2011; Mahatsente et al., 1999; Maguire et al., 2006; Stuart et al., 2008). Correspondingly, the lithospheric mantle is thicker in a transect from the Ethiopian Plateau (down to 70–80 km depth) to the MER and to the Afar depression (down to around 50 km depth). The depth of the Moho beneath MER is around 38–40 km (Dugda et al., 2005).

There is a general agreement that in the MER magma is stored at the crust–mantle boundary before ascending to shallow levels and erupting. Within the crust both a complex dyke system and/or a large magma chamber have been proposed (e.g., Acocella & Korme, 2002; Gleeson et al., 2017; Maccaferri et al., 2014; Mahatsente et al., 1999; Peccerillo et al., 2007; Rooney et al., 2007, 2011).

Volcanological models for the MER have focused mainly on explaining the trend from subaluminous to peralkaline products and the nature of the Daly Gap. These models suggest that evolution processes (fractional crystallization (FC), assimilation and fractional crystallization (AFC), etc.) occurred within the crust, although different pressures have been suggested (Caricchi et al., 2006; Gleeson et al., 2017; Rooney et al., 2007). According to the model of Rooney et al. (2007), ponding proceeds at the base of the crust, following which the MER magmas ascend very quickly to a shallow crustal level in the Wonji Fault Belt. In contrast, a complex magmatic system with multiple storage levels within the crust has been proposed for the Debre Zeyit Fault Zone (Rooney et al., 2007).

Caricchi et al. (2006) investigated isobaric and polybaric FC processes, starting from a transitional basalt composition (sample AA1). Under hydrous conditions (4.0 wt% initial H_2O), clinopyroxene is a stable phase (with olivine) from the liquidus temperature down to 1000°C at 10 kbar and then (with plagioclase) down to 1100°C at 7 kbar. From these experiments, clinopyroxene is no longer a stable phase at around 950–970°C depending on pressure and composition. In agreement with Caricchi et al. (2006) our geobarometric data suggest that polybaric differentiation started from deep crustal levels (i.e., ~10 kbar, ~30 km) to the shallow crust. Additionally the MER clinopyroxene crystallization pressure revealed a significant storage level at ~5–6 kbar (i.e., 15–20 km).

Most of the tholeiitic basalts from the plateau are aphyric but the occasional clinopyroxene present could not provide geobarometric information. The almost continuous range of crystallization pressure indicated by the MER clinopyroxene (from the middle to the lower crust) can be explained by a plumbing system where magmas start differentiating olivine and clinopyroxene at depth (i.e., 10 kbar; Figure 6.9) reaching a main storage depth that can be placed at about 3–5 kbar (i.e., 15–20 km). The most probable temperature intervals at the storage depth range from 1125 to 1140°C (Figure 6.8) in agreement with the data reported by Caricchi et al. (2006). At a pressure of 5 kbar and a temperature of about 1130°C, the degree of crystallization of the system is still largely below 50 vol%. As a consequence, magmas are still eruptible, and they can easily reach the Earth's surface following either the structural path defined by the ongoing extensive tectonics, or by dikes. The melt

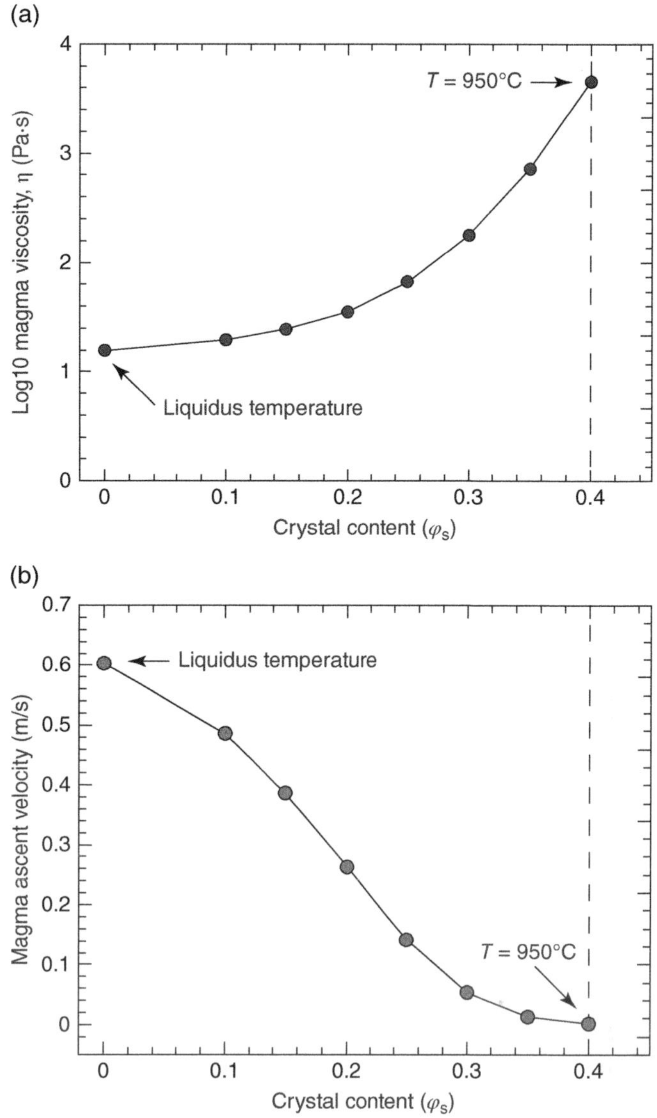

Figure 6.9 Calculated magma (a) viscosity and (b) ascent velocity versus crystal content. See text for discussion.

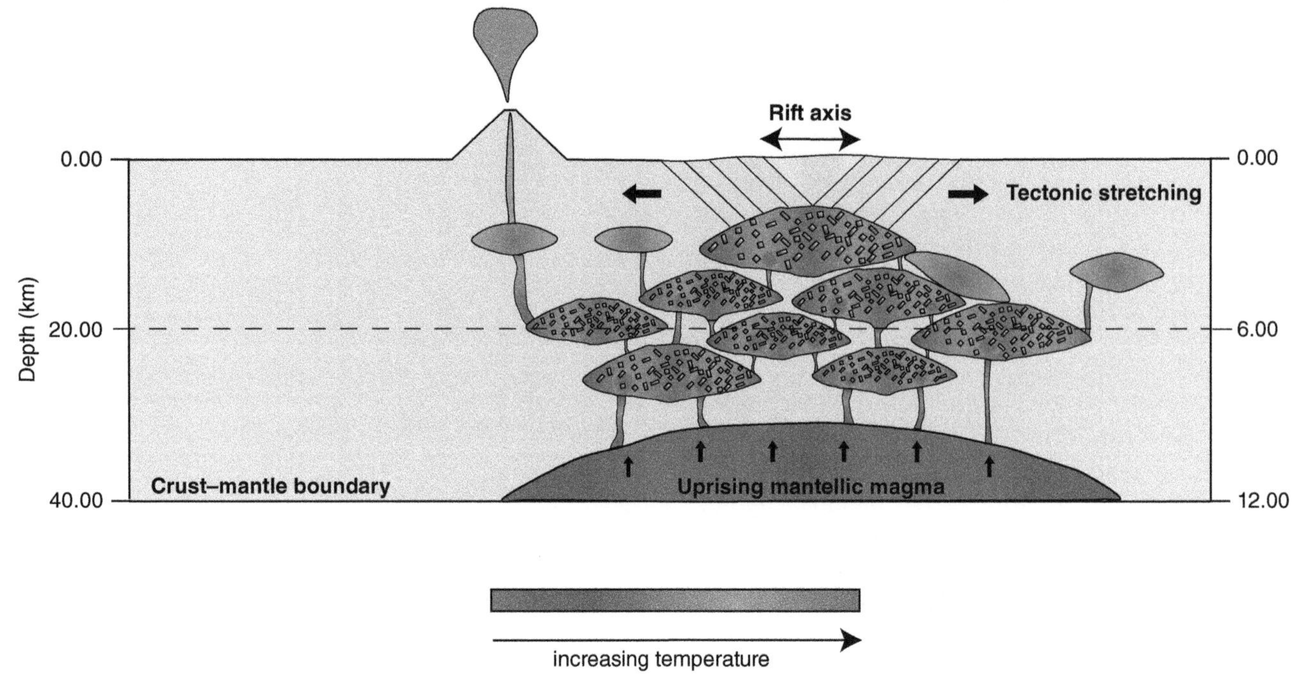

Figure 6.10 Schematic cross-section (not to scale) showing a possible model for distribution of magma reservoirs below the MER. The cartoon suggests that the magmatic plumbing system is composed of a complex of dikes promoting the ascent of magma towards the surface. See text for further details.

viscosity, η_{melt}, has been calculated according to the model of Giordano et al. (2008) in the temperature range 1125–1140°C considering an initial H_2O content varying from 2.0 to 4.0 wt%. The logaritmic values (η in Pa·s) obtained range from 1.18 to 1.09 and from 0.92 to 0.84 for 2.0 and 4.0 wt% H_2O, respectively. Also, we calculated the viscosity of the MER magmas at the estimated conditions of pressure and temperature of the main storage levels of mafic magmas (i.e., 1130°C and 5–6 kbar), taking into account the effects of crystal in agreement with Laumonier et al. (2014). We considered a melt with an initial H_2O content of 2.0 and 4.0 wt%, and a crystal content varying from 0 to 40 vol%, obtaining logarithmic values of 1.19 to 3.66 respectively (i.e., η from ~15 to ~4570 Pa·s; Figure 6.9) and from 0.93 to 3.4, respectively (i.e., η from ~8.5 to ~2512 Pa·s; Figure 6.9). Finally, considering the same range of crystal content variation, we estimated the potential ascending velocities (Rubin, 1993) to vary between 0.002 and 0.6 m s^{-1} (Figure 6.9).

All together our data suggest a complex architecture for the MER magmatic system, with multiple storage levels and an interconnected system of dikes and sills through which magma rises (e.g., Marsh, 2004; Figure 6.10). Within this framework, silicic magmas are formed by polybaric fractionation and assimilation and potentially (AFC; e.g., Peccerillo et al., 2007) by reactive flow (Jackson et al., 2018) at shallow crustal levels.

ACKNOWLEDGMENTS

Editor Francesco Vetere is gratefully acknowledged for handling the manuscript. We thank two anonymous referees for their constructive reviews. We gratefully acknowledge Angelo Peccerillo for providing the samples and for scientific discussions. We also thank Andrea Orlando and Eleonora Braschi (CNR-IGG, Section of Florence) for their help during the microprobe analysis. LC was supported by the European Research Council (ERC) under the European Union's Horizon 2020 research and innovation program (grant agreement No 677493- FEVER).

SUPPLEMENTARY MATERIAL

Details on the crystallographic and chemical data (single-crystal XRD, EMPA and LA-ICPMS) for clinopyroxene pheno- and megacrystals, as well as parameters used for calculation of magma ascent rate, are provided in Supplementary Material Tables SM1–SM3. Figure SM1 is also available as Supplementary Material at http://www.fisgeo.unipg.it/sabrina.nazzareni/Ch6_DynamicMagmaEvolution

REFERENCES

Acocella, V., & Korme, T. (2002). Holocene extension direction along the Main Ethiopian Rift, East Africa. *Terra Nova*, *14*, 191–197.

Alvaro, M., Domeneghetti, C. M., Fioretti, A. M., Cámara, F., & Marinangeli, L. (2015). A new calibration to determine the closure temperatures of Fe–Mg ordering in augite from nakhlites. *Meteoritics and Planetary Science*, *50*, 499–507.

Ayalew, D. (2000). Origin by fractional crystallization of transitional basalt for the Asela-Ziway pantellerites. *Chemical Geology*, *168*, 1–3.

Ayalew, D., Barbey, P., Marty, B., Reisberg, L., Yirgu, G., & Pik, R. (2002). Source, genesis, and timing of giant ignimbrite deposits associated with Ethiopian continental flood basalts. *Geochimica et Cosmochimica Acta*, *66*(8), 1429–1448.

Ayalew, D., Jungb, S., Romer, R. L., Kerstend, F., Pfänder, J. A., & Garbe-Schönberg, D. (2016). Petrogenesis and origin of modern Ethiopian rift basalts: Constraints from isotope and trace element geochemistry. *Lithos*, *258–259*, 1–14.

Baker, B. H., Goles, G. G., Leeman, W. P., & Lindstrom, M. M. (1977). Geochemistry and petrogenesis of a basalt-benmoreite-trachyte suite from the southern part of the Gregory Rift, Kenya. *Contribution to Mineralogy and Petrology*, *64*, 303–332.

Baker, J. A., Macpherson, C. G., Menzies, M. A., Thirlwall, M. F., Al-Kadasi, M., & Mattey, D. P. (2000). Resolving crustal and mantle contributions to continental flood volcanism, Yemen: Constraints from mineral oxygen isotope data. *Journal of Petrology*, *41*, 1805–1820.

Baker, J. A., Snee, L., & Menzies, M. A. (1996). A brief period of Oligocene flood volcanism in western Yemen: Implications for the duration and rate of continental flood volcanism at the Afro-Arabian triple junction. *Earth and Planetary Science Letters*, *138*, 39–56.

Barberio, M. R., Donati, C., Donato, P., Yirgu, G., Peccerillo, A., & Wu, T. W. (1999). Petrology and geochemistry of Quaternary magmatism in the northern sector of the Ethiopian Rift between Debre Zeit and Awash Park. *Acta Vulcanologica*, *11*, 69–81.

Boccaletti, M., Mazzuoli, R., Bonini, M., Trua, T., & Abebe, B. (1999). Plio- Quaternary volcanotectonic activity in the northern sector of the Main Ethiopian Rift: Relationships with oblique rifting. *Journal of African Earth Sciences*, *29*(4), 679– 698.

Brizi, E., Nazzareni, S., Princivalle, F., & Zanazzi, P. F. (2003) Clinopyroxenes from mantle-related xenocrysts in alkaline basalts from Hannuoba (China): augite–pigeonite exsolutions and their thermal significance. *Contributions to Mineralogy and Petrology*, *145*(5), 578–584.

Brotzu, P., Ganzerli-Valentini, M. T., Morbidelli, L., Piccirillo, E. M., Stella, R., & Traversa, G. (1981). Basaltic volcanism in the northern sector of the main Ethiopian Rift. *Journal of Volcanology and Geothermal Research*, *10*, 365–382.

Brotzu, P., Kazmin, U., Morbidelli, L., Piccirillo, E. M., Seife, M. B., & Traversa, G. (1980). Petrochemistry of the volcanics in the northern part of the main Ethiopian Rift. Proceedings, Geodynamic Evolution of the Afro-Arabic Rift System. *Accademia Nazionale dei Lincei, Rome*, *47*, 367–386.

Caricchi, L., Ulmer, P., & Peccerillo, A. (2006). A high-pressure experimental study on the evolution of the silicic magmatism of the Main Ethiopian Rift. *Lithos*, *91*(1–4), 46–58.

Dal Negro, A., Carbonin, S., Molin, G. M., Cundari, A., & Piccirillo, E. M. (1982). Intracrystalline cation distribution in natural clinopyroxenes of tholeiitic, transitional, and alkaline basaltic rocks. In Saxena, S. K. (Ed.), *Advances in physical geochemistry* (Vol 2, pp. 117–150). Berlin: Springer.

Dal Negro, A., Molin, G. M., Salviulo, G., Secco, L., Cundari, A., & Piccirillo, E. M. (1989). Crystal chemistry of clinopyroxene and its petrogenetic significance: a new approach. In A. Boriani, M. Bonafede, G. B. Piccardo, & G. B. Vai (Eds.), The lithosphere in Italy: advances in earth science research (Italian National Committee for the International Lithosphere Program). *Accademia Nazionale Lincei, Atti Convegni Lincei*, *80*, 271–295.ϕ_s

Dugda, M. T., Nyblade, A. A., & Julia, J. (2007). Thin lithosphere beneath the Ethiopian Plateau revealed by a joint inversion of Rayleigh wave group velocities and receiver functions. *Journal of Geophysical Research: Solid Earth*, *112*, B08305.

Dugda, M., Nyblade, A., Julia, J., & Ammon, C. (2005). Crustal structure in Ethiopia and Kenya from receiver function analysis; implications for rift development in eastern Africa. *Journal of Geophysical Research: Solid Earth*, *110*, B01303.

Furman, T. (2007). Geochemistry of East African Rift basalts: an overview. *Journal of African Earth Sciences*, *48*, 147–160.

George, R., Rogers, N., & Kelley, S. (1998). Earliest magmatism in Ethiopia: Evidence for two mantle plumes in one flood basalt province. *Geology*, *26*, 923–926.

Giordano, D., Russell, J. K., & Dingwell, D. B. (2008). Viscosity of magmatic liquids: a model. *Earth and Planetary Science Letters*, *271*, 123–134.

Gleeson, M. L. M., Stock, M. J., Pyle, D. M., Mather, T. A., Hutchinson, W., & Yirgu, G. (2017). Constraining magma storage conditions using phase equilibria models at a restless volcano in the Main Ethiopian Rift. *Journal of Volcanology and Geothermal Research*, *337*, 44–61.

Hammond, J. O. S., Kendall, J. M., Stuart, G. W., Keir, D., Ebinger, C., Ayele, A., & Belachew, M. (2011). The nature of the crust beneath the Afar triple junction: evidence from receiver functions. *Geochemistry, Geophysics, Geosystems*, *12*, Q12004.

Hofmann, C., Courtillot, V., Féraud, G., Rochette, P., Yirgu, G., Ketefo, E., & Pik, R. (1997). Timing of the Ethiopian flood basalts event and implications for plume birth and global change. *Nature*, *389*, 838–841.

Jackson, M. D., Blundy, J., & Sparks, R. S. J. (2018). Chemical differentiation, cold storage and remobilization of magma in the Earth's crust. *Nature*, *564*, 405–409.

Laumonier, M., Scaillet, B., Pichavant, M., Champallier, R., Andujar, J., & Arbaret, L. (2014). On the conditions of magma mixing and its bearing on andesite production in the crust. *Nature Communications*, *5*, 6607.

Maccaferri, F., Rivalta, E., Keir, D., & Acocella, V. (2014). Off-rift volcanism in rift zones determined by crustal unloading. *Nature Geosciences*, *7*, 297–300.

Macdonald, R., Davies, G. R., Upton, B. G. J., Dunkley, P. N., Smith, M., & Leat, P. T. (1995). Petrogenesis of Silali volcano, Gregory Rift, Kenya. *Journal of Geological Society of London*, *152*, 703–720.

Macdonald, R., Rogers, N. W., Fitton, J. G., Black, S., & Smith, M. (2001). Plume–lithosphere interactions in the generation of the basalts of the Kenya rift, East Africa. *Journal of Petrology*, *42*, 877–900.

MacDonough, W. F. & Sun, S. (1995). The composistion of the Earth. *Chemical Geologgy*, *120*, 223–253.

Maguire, P., Keller, G. R., Klemper, S. L., Mackenzie, G., Keranen, K., Harder, S., et al. (2006). Crustal structure of the northern Main Ethiopian Rift from the EAGLE controlled source survey; a snapshot of incipient lithospheric break-up. In G. Yirgu, C. Ebinger, P. Maguire (Eds.), *The Afar Volcanic Province within the East African Rift System* (Special Publication 259, pp. 269–292). London: Geological Society.

Mahatsente, R., Jentzsch, G., & Jahr, T. (1999). Crustal structure of the main Ethiopian rift gravity data: 3-dimensional modeling. *Tectonophysics*, *313*, 363–382.

Malgarotto, C., Molin, G., & Zanazzi, P. F. (1993) Cooling history of a dyke on Alicudi (Aeolian Islands) from intracrystalline Fe^{2+}–Mg exchange reaction in clinopyroxene. *European Journal of Mineralogy*, *5*, 755–762.

Marsh, B. D. (2004). A magmatic mush column Rosetta Stone: The McMurdo Dry Valleys of Antarctica. *Eos, Transactions of the American Geophysical Union*, *85*, 497–502.

Marty, B., Pik, R., & Yirgu, G. (1996). Helium isotopic variations in Ethiopian plume lavas: nature of magmatic sources and limit on lower mantle contribution. *Earth and Planetary Science Letters*, *144*, 223–237.

McKenzie, D. P. (1978). Some remarks on the development of sedimentary basins. *Earth and Planetary Science Letters*, *40*, 25–32.

Merla, G., Abatte, E., Azzaroli, A., Bruni, P., Fazzuoli, M., Sagri, M., & Tacconi, P. (1979). *Comments and a geological map of Ethiopia and Somalia* (scale 1:2,000,000). Florence: Consiglio Nazionale delle Ricerche, 89 pp.

Mohr, P. A. (1971). Ethiopian rift and plateaus: some volcanic and petrochemical differences. *Journal of Geophysical Research*, *76*, 1967–1983.

Mohr, P. A., & Zanettin, B. (1988). The Ethiopian flood basalt province. In: Macdougall, J. D. (Ed.), *Continental flood basalts* (pp. 63–110). Dordrecht: Kluwer Academic.

Murri, M., Camara, F., Adam, J., Domeneghetti, M.C., & Alvaro, M. (2018) Intracrystalline "geothermometry" assessed on clino and orthopyroxene bearing synthetic rocks. *Geochimica et Cosmochimica Acta*, *227*, 133–142.

Nazzareni, S., Busà, T., & Cristofolini, R. (2003). Magmatic crystallisation of Cr–Al diopside and Al–Fe^{3+} diopside from the ancient alkaline basalts (Mt. Etna, Sicily). *European Journal of Mineralogy*, *15*, 81–93.

Nazzareni, S., Molin, G., Peccerillo A., & Zanazzi P. F. (1998). Structural and chemical variations in clinopyroxenes from the Island of Alicudi (Aeolian arc) and their implications for conditions of crystallization. *European Journal of Mineralogy*, *10*(2), 291–300.

Nazzareni, S., Molin, G., Peccerillo, A., & Zanazzi, P. F. (2001). Volcanological implications of crystal-chemical variations in clinopyroxenes from the Aeolian Arc, Southern Tyrrhenian Sea (Italy). *Bulletin of Volcanology*, *63*, 73–82.

Nazzareni, S., Skogby, H., & Zanazzi P. F. (2011) Hydrogen content in clinopyroxene phenocrysts from Salina mafic lavas (Aeolian arc, Italy). *Contributions to Mineralogy and Petrology*, *162*(2), 275–288.

Neave, D. A., & Putirka, K. D. (2017). A new clinopyroxene-liquid barometer, and implications for magma storage pressures under Icelandic rift zones. *American Mineralogist*, *102*, 777–794.

Nimis, P. (1995). A clinopyroxene geobarometer for basaltic systems based on crystal-structure modeling. *Contributions to Mineralogy and Petrology*, *121*, 115–125.

Nimis, P., & Ulmer, P. (1998). Clinopyroxene geobarometry of magmatic rocks Part 1: An expanded structural geobarometer for anhydrous and hydrous, basic and ultrabasic systems. *Contributions to Mineralogy and Petrology*, *133*(1–2), 122–135.

Paton, C., Hellstrom, J., Paul, B., Woodhead, J., & Hergt, J. (2011). Iolite: Freeware for the visualisation and processing of mass spectrometric data. *Journal of Analytical Atomic Spectrometry*, *26*, 2508.

Peccerillo, A., Barberio, M. R., Yirgu, G., Ayalew, D., Barbieri, M., & Wu, T. W. (2003). Relationships between mafic and peralkaline acid magmatism in continental rift settings: a petrological, geochemical and isotopic study of the Gedemsa volcano, central Ethiopian rift. *Journal of Petrology*, *44*, 2003–2032.

Peccerillo, A., Donati, C., Santo, A.P., Orlando, A., Yirgu, G., & Ayalew, D. (2007). Petrogenesis of silicic peralkaline rocks in the Ethiopian rift: geochemical evidence and volcanological implications. *Journal of African Earth Sciences*, *48*, 161–173.

Perugini, D., Busà, T., Poli, G., & Nazzareni, S. (2003). The role of chaotic dynamics and flow fields in the development of disequilibrium textures in volcanic rocks. *Journal of Petrology*, *44*(4), 733–756.

Petrelli, M., Morgavi, D., Vetere, F., & Perugini, D. (2016). Elemental imaging and petro-volcanological applications of an improved laser ablation inductively coupled quadrupole plasma mass spectrometry. *Periodico di Mineralogia*, *85*, 1–20.

Petrelli, M., Perugini, D., Alagna, K. E., Poli, G., & Peccerillo, A. (2008). Spatially resolved and bulk trace element analysis by laser ablation–inductively coupled plasma–mass spectrometry (LA-ICP-MS). *Periodico di Mineralogia*, *77*, 3–21.

Piccirillo, E. M., Justin-Visentin, E., Zanettin, B., Joron, J. K., & Treuil, M. (1979). Geodynamic evolution from plateau to rift: major and trace element geochemistry of the central eastern Ethiopian Plateau volcanics. *Neues Jahrbuch für Geologie und Palaontologie*, *258*, 139–179.

Pik, R., Deniel, C., Coulon, C., Yirgu, G., Hofmann, C., & Ayalew, D. (1998). The northwestern Ethiopian Plateau flood basalts: classification and spatial distribution of magma types. *Journal of Volcanology and Geothermal Research*, *81*, 91–111.

Pik, R., Deniel, C., Coulon, C., Yirgu, G., & Marty, B. (1999). Isotopic and trace element signatures of Ethiopian flood basalts: Evidence for plume–lithosphere interactions. *Geochimica et Cosmochimica Acta*, *63*(15), 2263–2279.

Putirka, K. D. (1999). Clinopyroxene + liquid equilibria to 100 kbar and 2450 K. *Contributions to Mineralogy and Petrology*, *135*, 151–163.

Putirka, K. D. (2008). Thermometers and barometers for volcanic systems. *Reviews in Mineralogy and Geochemistry*, *69*, 61–120.

Rooney, T. O. (2010). Geochemical evidence of lithospheric thinning in the southern Main Ethiopian Rift. *Lithos*, *117*, 33–48.

Rooney, T. O., Bastow, I. D., & Keir, D. (2011). Insights into extensional processes during magma assisted rifting: Evidence from aligned scoria cones. *Journal of Volcanology and Geothermal Research*, *201*, 83–96.

Rooney, T. O., Furman, T., Bastow, I., Ayalew, D., & Yirgu, G. (2007). Lithospheric modification during crustal extension in the Main Ethiopian Rift, *Journal of Geophysical Research*, *112*, B10201.

Rooney, T., Furman, T., Yirgu, G., & Ayalew, D. (2005). Structure of the Ethiopian lithosphere: xenolith evidence in the Main Ethiopian Rift. *Geochimica et Cosmochimica Acta*, *69*(15), 3889–3910.

Rossi, G., Oberti, R., Dal Negro, A., Molin, G. M., & Mellini, M. (1987). Residual electron density at the M2 site in C2/c clinopyroxenes: relationships with bulk chemistry and subsolidus evolution. *Physics and Chemistry of Minerals*, *14*, 514–520.

Rubin, A. (1993). Tensile fracture of rock at high confining pressure: implications for dike propagation. *Journal of Geophysical Research*, *98*, 15919–15935.

Rubin, A. (1995). Propagation of magma-filled cracks. *Annual Review of Earth and Planetary Sciences*, *23*, 287–336.

Sheldrick, G. M. (2008). A short history of SHELX. *Acta Crystallographica*, *A64*, 112–122.

Stuart, G. W., Bastow, I. D., & Ebinger, C. J. (2008). Crustal structure of the northern Main Ethiopian Rift from receiver function studies. In G. Yirgu, C. Ebinger, P. Maguire (Eds.), *The Afar Volcanic Province within the East African Rift System* (Special Publication 259, pp. 253–267). London: Geological Society.

Trua, T., Deniel, C., & Mazzuoli, R. (1999). Crustal control in the genesis of Plio-quaternary bimodal magmatism of the main Ethiopian rift (MER): geochemical and isotopic (Sr Nd, Pb) evidence. *Chemical Geology*, *155*, 201–231.

Ubide, T., Mollo, S., Zhao, J., Nazzari, M., & Scarlato, P. (2019). Sector-zoned clinopyroxene as a recorder of magma history, eruption triggers, and ascent rates. *Geochimica et Cosmochimica Acta*, *251*, 265–283.

White, R. S., & McKenzie, D. (1989). Magmatism at rift zones: the generation of volcanic continental margins and flood basalts. *Journal of Geophysical Research*, *94*, 7685–7729.

Williams, F. M., Williams, M. A. J., & Aumento, F. (2004). Tensional fissures and crustal extension rates in the northern part of the Main Ethiopian Rift. *Journal of African Earth Sciences*, *38*, 183–197.

Yirgu, G. (1997). Magma–crust interaction during emplacement of Cenozoic volcanism in Ethiopia: geochemical evidence from Sheno-Megezez area, Central Ethiopia. *Ethiopian Journal of Science*, *20*, 49–72.

7

Rheological Behavior of Partly Crystallized Silicate Melts Under Variable Shear Rate

Francesco Vetere[1,2] and François Holtz[2]

ABSTRACT

High-temperature experiments were performed in order to shed new light on the shear-rate controls on magma viscosity. We compare viscosity data from three different compositions: an andesite from the Calbuco volcano, a basalt from the Etna 122 BCE Plinian eruption and a synthetic pyroxenite with a composition similar to the Canadian Theo's Flow. In addition to the determination of melt viscosity (at 1545–1715 K), we performed viscosity determinations at subliquidus conditions in partially crystallized materials, under controlled shear rates of 0.1 and 1 s^{-1} and at temperatures of 1483, 1493, and 1503 K for the Calbuco andesite, Etna basalt, and synthetic pyroxenite, respectively. The two different shear rates allow us to retrieve information about shear-rate influences on viscosity of partly crystallized systems. A decrease in viscosity is observed with increase in shear rate. This behavior is known in the literature as the "shear thinning effect." Our data show that changes of shear rates from 0.1 to 1 s^{-1} may cause a viscosity difference of 0.5 to one order of magnitude. This effect should be taken into account when considering magmatic processes occurring in volcanic conduits. The rheological properties of partly crystallized systems could drastically change depending on the dynamics of the magmatic system.

7.1. INTRODUCTION

One example illustrating the complexity of the viscosity behavior of suspensions is the role of the shape of particles (or crystals). Early studies estimating the viscosity of magmas applied the Einstein–Roscoe equation (Einstein, 1906; Roscoe, 1952), which is only valid if the shape of the particles can be approximated to that of a sphere. The maximum packing fraction, which corresponds to the crystal fraction at which a transition of the system to the rigid solid state occurs (according to Marsh, 1981), is often estimated to be 0.6 for magmatic systems, but this value can change significantly since it is dependent on the crystal aspect ratio. It has been demonstrated that this value is not always adequate for natural magmatic systems (Costa, 2005).

In addition, in fast crystallizing systems encountered in volcanic environments, kinetic processes related to crystal nucleation and growth play an extremely important role in (a) influencing magma rheology during ascent and (b) possibly deciphering residence time in magma conduits or reservoirs. The role of kinetic processes may also be important for industrial products such as glass and ceramics, to name only a couple of relevant research fields (e.g., Halder et al., 1999; Shelby, 2005; Tsuchiyama, 1983).

The kinetics of melt solidification processes has been investigated in the pioneering work of Carmichael (1974) and different cooling experiments have been conducted in aluminosilicate systems in order to understand this process (e.g., Cashman, 1989; Dowty, 1980; Hammer, 2008—for a general summary; Kirkpatrick, 1981; Kolzenburg

[1] *Department of Physics and Geology, University of Perugia, Perugia, Italy*
[2] *Institute of Mineralogy, Leibniz University Hannover, Hannover, Germany*

Dynamic Magma Evolution, Geophysical Monograph 254, First Edition. Edited by Francesco Vetere.

et al., 2016, 2017; Lofgren, 1983; Lasaga, 1997). In this work, Carmichael (1974) states that:

> The sharp break in grain size between phenocrysts and groundmass is correlated with some corresponding change in conditions prevailing during freezing of the magma. Such a break occurs where slow cooling of magma deep within the crust has given way to rapid cooling following uprise of the magma and extrusion at the surface or injection into cooler rocks of the upper crust.

The previous sentence holds the principal mechanisms controlling the evolution of a magmatic system that behaves as a crystal–liquid mush or suspension. It also implies that the transition from liquid to solid needs to be considered as a dynamic process that is dependent on cooling rates, on decompression rates (for volcanic systems), and on flow or shear rates. Crystallization kinetics plays a major role for volcanic systems since this parameter can be responsible for sudden changes in rheological behaviors of magmas, resulting in low or high energetic eruption. It is obvious that a better understanding of magma dynamics will improve our abilities to monitor and forecast eruptive styles and volcanic activity.

A considerable number of studies on the rheology of crystal-bearing silicate melts and analogue materials have explored the effect of variable solid fraction and particle shape on the rheology of suspensions (Caricchi et al., 2007; Chevrel et al., 2015; Cimarelli et al., 2011; Del Gaudio et al., 2013; Ishibashi & Sato, 2007, 2010; Klein et al., 2018; Kolzenburg et al., 2016; Lejeune & Richet, 1995; Mader et al., 2013; Moitra & Gonnermann, 2015; Mueller et al., 2010; Sato, 2005; Sehlke & Whittington, 2015; Soldati et al., 2016, 2017; Soule & Cashman, 2005; Thies & Deubener, 2002; Vetere et al., 2010, 2013, 2017; Vona et al., 2011, 2013). Despite this huge effort, a general law governing the effect of particle size, shape, and proportions on the rheology of magmatic and volcanic systems still evades us and further investigation is required. In particular, due to the great technological challenge of performing experiments in such dynamic conditions, there are only few data on the effect of crystals on viscosity in dynamic systems simulating magma ascent in conduits or convection processes. Nowadays, we are able to reproduce in one atmosphere, high-temperature furnaces different dynamic scenarios simulating natural geological conditions and to retrieve very important data that can be used for modeling the rheology of rising magmas and lava flows.

Magmas can display Newtonian or non-Newtonian behavior. A Newtonian behavior is characteristic for a system that does not show any change in viscosity as shear stress is changing. Newtonian viscosity can also be defined as the measure of the resistance of a liquid to shear deformation (i.e., a measure of the ratio between the applied shearing force (σ = shear stress) and the rate of flow of the liquid ($\dot{\gamma}$ = shear rate). Fluid regime (laminar or turbulent) is also an important consideration and can be described by the Reynolds number (*Re*), generally defined as:

$$Re = \rho \cdot u \cdot h / \eta \tag{7.1}$$

where ρ is the bulk lava density (kg m^{-3}), η viscosity (Pa·s), u (m s^{-1}) the flow velocity, and h the flow thickness. The Reynolds number is a dimensionless number that provides a first-order approximation about the dynamic regime of a fluid and, as such, it is reported in the literature in many different formulations depending on the rheological properties of the fluid and the geometrical constrains. The formulation that is generally used is that reported by Byrne et al. (2013), as well as that by Williams et al. (2001). In particular, Williams et al. (2001) report that the equation above is appropriate for describing a fluid in a lava flow with a width that is much larger than its depth. This implies that equation (7.1) is only valid if $Re << 2000$. In the case of high Re ($Re >> 2000$, implying high velocity of magma due to very low viscosity) it is preferable to apply the following equation, which differs by a factor of two from equation (7.1) (Byrne et al., 2013):

$$Re = 2 \cdot \rho \cdot u \cdot h / \eta \tag{7.2}$$

In systems containing melt and crystals, the following considerations and approaches in modeling viscosity data need to be taken into account. One of the most frequently used equations to evaluate the relative viscosity ($\eta_r = \eta_{eff}/\eta_m$, where η_{eff} is the effective viscosity of the suspension with a volume fraction of crystals and η_m is the viscosity of the melt) of a suspension is the Krieger–Dougherty (KD) equation (Krieger & Dougherty, 1959):

$$\eta_r = (1 - \Phi/\Phi_m)^{-\nu\Phi_m} \tag{7.3}$$

where Φ is a volume fraction of suspended particles, Φ_m is the maximum packing density, and ν is the so-called intrinsic viscosity (a measure of the crystal influence on the magmatic system viscosity). Assuming that $\nu\Phi_m = 2.5$, the effective viscosity of crystal + melts systems is often estimated with the Einstein–Roscoe (ER) equation (Einstein, 1906; Roscoe, 1952):

$$\eta_{eff} = \eta_m \left(1 - \Phi/\Phi_m\right)^{-2.5} \tag{7.4}$$

However, this equation is only valid if the shape of the particles can be approximated to that of a sphere. Moreover, Maron and Pierce (1956) found that better data reproduction could be obtained by changing the ER exponent from −2.5 to −2.0; this value was used by Pabst et al. (2006) and by Mueller et al. (2010). According to Marsh (1981) the Φ_m value can be estimated to be 0.6, which corresponds to the crystal fraction at which a transition of the system to the rigid solid state occurs. However, it has been demonstrated that this value is not always adequate for natural magmatic systems (Costa, 2005) and that it strongly depends on the aspect ratio of the crystals (Mader et al., 2013).

Recalling the above viscosity definition, a fluid is considered to be "purely viscous" if the shear stress (σ) is a function only of the shear rate ($\dot{\gamma}$). For non-Newtonian fluids (such as magmas composed of melts + crystals) in simple shear flow a viscosity function $\eta(\dot{\gamma})$ is:

$$\eta(\dot{\gamma}) = \sigma / \dot{\gamma} \tag{7.5}$$

The viscosity function is also called the apparent viscosity (this term is used to indicate that the result depends on the particular strain rate at which it was measured). Rheological parameters of two-phase suspensions are usually treated by using the classical approach for pseudoplastic (power law) flow. By using such an approach, it is possible to identify parameters such as the flow index n and consistency K by the following:

$$\sigma = K\dot{\gamma}^{n} \tag{7.6}$$

n defines the magnitude of shear-rate dependence. Smaller values of n indicate a greater dependence of apparent viscosity on shear rate. For a Newtonian liquid, $n = 1$. More details can be found in Moitra and Gonnermann (2014) for the Herschel–Bulkley model (Herschel & Bulkley, 1926). Then, in order to retrieve the flow index one can expressed it in ln or log terms for stress (equation 7.7) or for apparent viscosity (equation 7.8):

$$\ln\sigma = \ln k + n \ln\dot{\gamma} \tag{7.7}$$

and/or

$$\log\eta = \log k + (n-1) \log \dot{\gamma} \tag{7.8}$$

A few studies also aimed to directly monitor viscosity evolution using field measurements, by inserting a viscometer into the lava while it was flowing (Chevrel et al., 2018) and/or experimentally applying defined cooling rates and shear stress.

Studies on the viscosity of magmatic suspensions can be undertaken by performing experiments at temperatures between liquidus and solidus conditions. In such experiments, crystals will nucleate and grow from melts and the magma rheological evolution can be investigated during crystallization (Chevrel et al., 2015; Ishibashi, 2009; Ishibashi & Sato, 2007; Pinkerton & Norton, 1995; Sato, 2005; Vetere et al., 2013;Vona et al., 2011). This approach has the advantage of tracking the sequential variation of both the crystal texture and rheological properties during cooling of magmas, and is adopted here.

Literature and field observations reveal that the point at which a magma rheologically solidifies (preventing flow) is directly linked to dynamics. Considering a fast rising magma or lava flow the above-mentioned rheological threshold is enhanced due to the fast crystal growth process. On the other hand, slowing down the magma/lava velocity profile suppresses the crystallization process and rheological solidification will be delayed (Kolzenburg et al., 2018, and reference therein). Rheological investigations should consider shear-enhanced crystallization processes in order to model magmatic/volcanic processes in a properly. Thus, since viscosity of crystal-bearing systems is non-Newtonian, it is crucial to perform experiments in dynamic conditions simply, considering the tremendous evolution due to the crystallization process. However, a comparison of the models available with the few experimental data obtained so far show that there might be large deviations from predictions. In this chapter we present results describing the role of shear rate on viscosity of partly crystallized silicate melts and we discuss the link between different shear rates with viscosity of partly crystallized systems.

7.2. STARTING MATERIALS

Three compositions were used to test the shear-rate effect on viscosity and related crystal nucleation: a Chilean Andesite sampled after the recent eruption at Calbuco volcano (Romero et al., 2016), an Etna basalt from the 122 BCE Plinian eruption (Coltelli et al., 1988), and a synthetic pyroxenite with a composition derived from the so-called Theo's Flow (Lentz et al., 2011). The synthetic composition is based on the Theo's Flow pyroxenite unit located in the Abitibi Greenstone Belt. This area is well known as a komatiite location in Canada. For this composition, a glass was prepared by melting a mixture of oxides and carbonate in appropriate proportions to reproduce the natural samples. Melting was performed in a Nabertherm HT 04/17 $MoSi_2$-heated box furnace (Nabertherm GmbH, Lilienthal, Germany). The natural sample powder and the mixture of oxides and carbonate were treated in identical ways to produce glasses: the materials were crushed and melted at 1873 K for 4 h in a Pt crucible in air. Then, melts were poured on a brass plate to rapidly quench. The quenched glasses were recrushed and remelted again for 4 h in order to improve homogeneity. The samples were quenched rapidly again. This technique ensures compositional homogeneity of the glasses (Vetere et al., 2015, 2017).

7.3. EXPERIMENTAL AND ANALYTICAL METHODS

Viscosity measurements were performed in a Gero HTRV 70-250/18 high-temperature tube furnace with $MoSi_2$ heating elements (Gero GmbH, Neuhausen, Germany) operating up to 2073 K at room pressure. A rotational Anton Paar RheolabQC viscometer head installed at the Petro-Volcanology Research Group (PVRG) laboratory was used to measure viscosity. A sketch representing the experimental assemblage is shown in Figure 7.1. This instrument consists of a sample-filled crucible and a rotating measuring spindle that is immersed into the sample. Two Pt_{80}–Rh_{20} crucibles (inner diameter of 25 or 37 mm, outer diameter 28 or 40 mm and height of 50 or 70 mm for small and large crucible, respectively) were used to host the silicate melt and melt + crystals systems. The spindle is made of Al_2O_3 (with a diameter of 12.2 mm and length of 60 cm) and fixed to the head of the viscometer by a standard collet chuck. The lower end of the Al_2O_3 rod is sheathed by a tight-fitting 0.2 mm thick Pt_{80}–Rh_{20} foil in order to avoid any contamination of the silicate melts during the experimental runs. The rotational viscometer allows measurements under controlled shear rate ($\dot{\gamma}$). This allows us to investigate the possible shear effect on viscosity of melts and partly crystallized systems. Methods and procedures described by Dingwell (1986) and Ishibashi (2009) were applied in order to determine melt and melt + crystals viscosities. Briefly, by

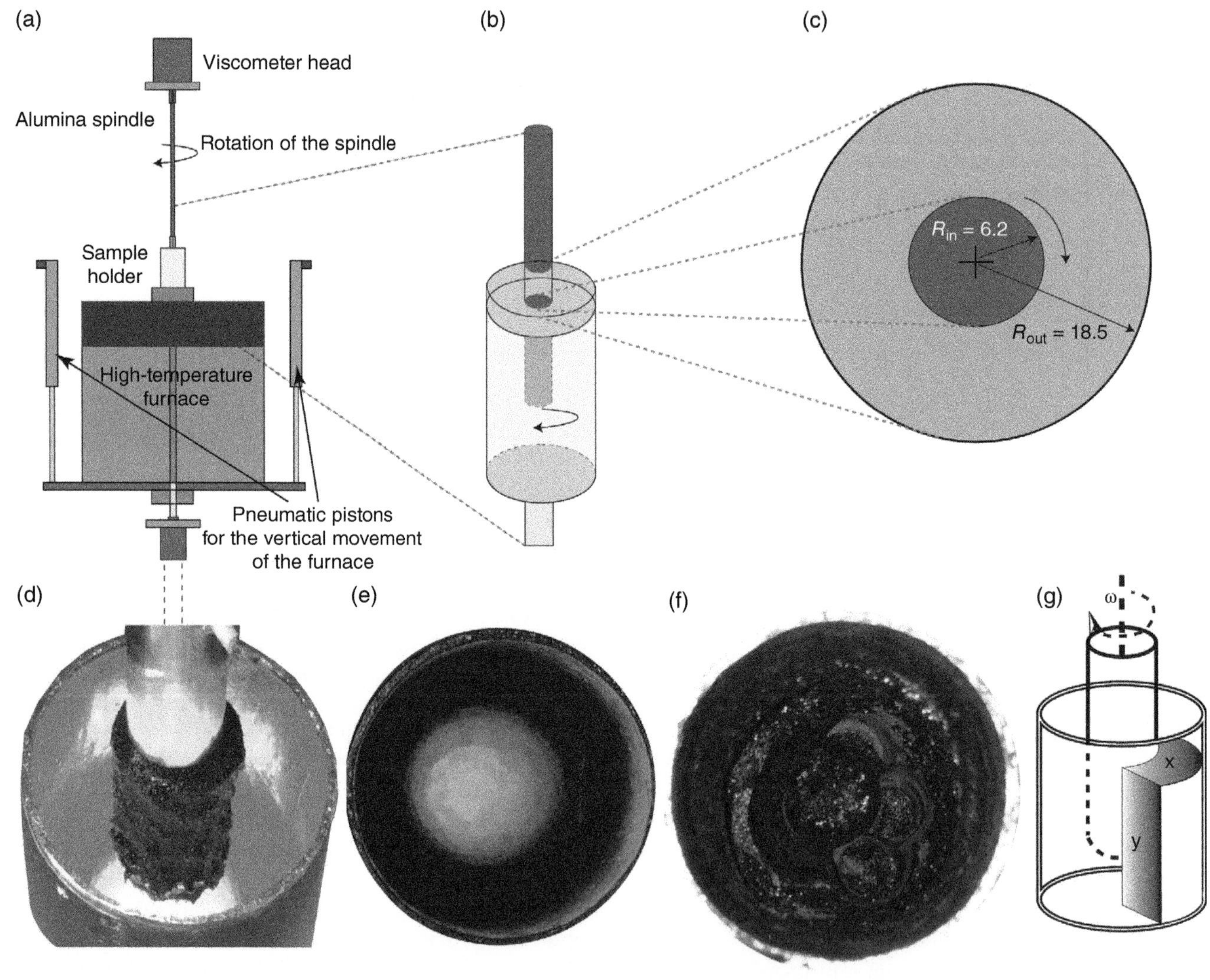

Figure 7.1 Details of the furnace, viscometer, and sample holder used in this work. Note the movable furnace (a), the Al spindle covered with Pt cup (b), the size of the spindle and the crucible (c), details of the sample soon after quench in the spindle (d) and crucible (e), the holes (f) after drilling procedure to collect samples, and the relative sections (g) used for analytical and textural analyses (see text; *x* and *y* are perpendicular and parallel sections of the sample relative to the spindle).

using this equipment, viscosity can be measured in a range from 0.1 to 10^5 Pa·s (Hess et al., 1996). The viscometer was calibrated against NIST 717a standard glass, for which the temperature–viscosity relationship is accurately known (https://www.nist.gov). Reproducibility of measurements on the standard glass is on the order of ±0.03 log units. Since melt viscosity was expected to be relatively low at high temperature, before running experiments we calibrated the viscometer using a Wacker silicone standard having viscosity of 10 Pa·s (Spina, Cimarelli, et al., 2016; Spina, Scheu, et al., 2016). One hundred measurements were performed and the results showed good reproducibility, with average values of 9.8 Pa·s (±0.3 standard deviation).

The furnace hosting the experimental charge is equipped with aluminum cooling heads. These are positioned on the top and the bottom openings of the furnace tube in order to prevent overheating of the viscometer head. Cooling is achieved by a continuous flux of cold water (313 K). The furnace can move vertically using two pneumatic cylinders (details are reported in Morgavi et al., 2015). This has two major advantages as it allows us to:

1. Carefully prepare the experimental geometry by precisely positioning the outer and inner cylinders outside the furnace.
2. Bring the furnace directly to the experimental temperature while the sample is still outside the furnace, preventing the sample from undergoing the entire thermal ramp.

Considering that our sample container has a capacity of ca. 70 cc, time required to thermally equilibrate the system is directly linked to the amount of sample we use for these experiments. In particular, 50 cc of sample that needs to be heated up to 1673 K requires about 11 min. This means a heating rate ca. 9000 K h^{-1}, much faster than the maximum heating rate of "normal furnaces," thus preventing any possible crystal nucleation and growth processes.

Two cross-mounted Thorlabs single-axis translation stages with a standard micrometer allow the correct positioning of the spindle and the crucible. An alumina rod (length = 600 mm; FRIATEC Aktiengesellschaft, Mannheim, Germany) is fixed to the lower part of the structure holding the outer cylinder. Temperature was monitored using an inhouse built S-type thermocouple ($Pt_{10}Rh_{90}$ vs. Pt) within an Al_2O_3 sheath, positioned at the bottom of the crucible. As the rotation of the viscometer prevents the use of thermocouples directly wired to a controller, OMEGA wireless thermocouple transmitters UWTC-Series were used (OMEGA Engineering, INC., Stamford, Connecticut, USA). Uncertainty on temperature measurements is on the order of 3 K.

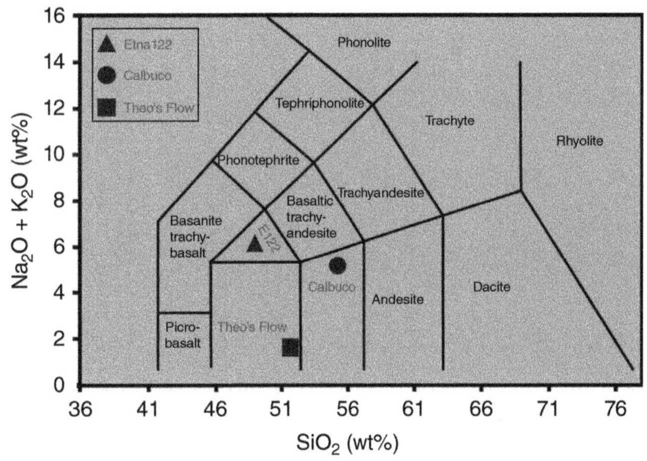

Figure 7.2 Total alkali versus silica diagram showing the three natural and synthetic compositions used in this study.

Prior to viscosity measurements, ca. 20 cm^3 (small crucible) to 50 cm^3 (large crucible) of melt was stirred at 1773 K for 2 h at strain rates of 1–10 s^{-1}. This allowed the complete removal of possible gas bubbles and the attainment of a compositionally homogeneous melt (Armienti, 2008; Davis & Ihinger, 1998; Dowty, 1980; Iezzi et al., 2008, 2011; Lofgren, 1983; Pupier et al., 2008; Vetere et al., 2015). Samples used in the subliquidus temperature experiments were first melted at superliquidus conditions. The temperature was then decreased continuously to the required subliquidus value at a rate of 5 K min^{-1}. At the end of the experiments the samples were quenched by moving them into the cooled head of the furnace. As the thermocouples are located inside the spindle (and thus inside the crucible containing the investigated materials) the quench rate was measured to be of the order of 100 K min^{-1}, which was sufficient to avoid the formation of quench crystals. After quenching, run products where drilled out from the crucible by using a diamond core drill (Figure 7.1) and were mounted in epoxy, ground flat and polished for textural and chemical analysis.

Chemical compositions of all three starting materials and of the products after viscosity measurements (solid phases and glasses) were determined by electron probe microanalysis (EPMA) installed at the Institute for Mineralogy of the Leibniz Universität Hannover. The analytical conditions for glasses are reported in Rossi et al. (2017) and Gonzalez-Garcia et al. (2017), while for solid phases we refer to Vetere et al. (2017). Figure 7.2 reports the three compositions in a total alkalis versus silica (TAS) diagram. Melt densities were calculated according to Klöß (2000) and Ochs and Lange (1999). The nonbridging oxygens per tetrahedron (NBO/T) were calculated following the methods described in Vetere et al. (2015). Data are summarized in Table 7.1.

Table 7.1 Starting and Residual Melt (RM) Compositions Representing an Average of 20 Measurements

Material	SiO_2	TiO_2	Al_2O_3	FeO	MgO	CaO	Na_2O	K_2O	ρ ($g cm^{-3}$)	NBO/T
Theo's Flow	52.01	0.73	5.65	10.16	14.03	15.78	1.08	0.04	2.793[a]	1.26
SD	0.52	0.02	0.08	0.09	0.17	0.15	0.2	0.02	2.695[b]	
Etna 122	49.41	1.72	18.81	9.53	3.58	9.23	4.15	1.69	2.678[a]	0.36
SD	0.5	0.03	0.08	0.1	0.13	0.12	0.17	0.04	2.579[b]	
Calbuco	55.11	0.95	19.12	8.03	2.6	7.99	4.06	0.83	2.605[a]	0.22
SD	0.45	0.02	0.09	0.1	0.11	0.12	0.15	0.07	2.526[b]	
Theo's Flow RM	51.23	1.23	8.29	13.71	11.21	12.31	1.78	0.2	2.791[a]	0.97
SD	0.47	0.02	0.09	0.08	0.17	0.13	0.23	0.05	2.721[b]	
Etna 122 RM	50.53	1.98	17.38	10.89	4.13	8.61	4.4	1.95	2.670[a]	0.41
SD	0.47	0.05	0.09	0.12	0.13	0.12	0.17	0.04	2.607[b]	
Calbuco RM	57.99	1.19	15.79	6.75	4.97	6.52	4.99	0.82	2.587[a]	0.34
SD	0.38	0.02	0.1	0.11	0.1	0.1	0.15	0.07	2.518[b]	

Note. Results are reported in wt%. SD, 1 standard deviation. NBO/T, nonbridging oxygens per tetrahedron (as reported in Mysen & Richet, 2005); calculated by considering 1:1 ratio between Fe^{2+} and Fe^{3+} from the total Fe as measured by EPMA.
[a] Glass density was estimated using the model of Klöß (2000) adding 0.02 wt% H_2O.
[b] Melt density was calculated using the Ochs and Lange (1999) model and adding water content equal to 0.02 wt%.

Back-scattered electron (BSE) images were collected by field emission scanning electron microscopy (FE-SEM; FEG LEO 1525, Zeiss Company, Germany) at the Department of Physics and Geology of the University of Perugia (Italy). Energy dispersive X-ray spectroscopy (EDS) analysis (Quantax 200 with Xflash 400 detector, Bruker Company, Germany), coupled with SEM, was used to detect phases. The textural features were analyzed with FE-SEM using BSE. To accurately characterize the textures and the distribution of phases, as well as the possible tendency to nucleate and crystallize preferentially on Pt walls, images were collected at magnifications between 150× and 3000×. As a whole, about 150 FE-SEM images with section areas ranging between 100 and ~105 μm^2 were collected; about 70% of them, mainly at magnifications of 150×, 400×, and 800×, were used for the determination of phase content by image analysis.

We used Image-ProPlus 6.0 software to determine phase abundance and distribution in each run product; the applied analytical protocol of image analysis is the same as reported in Iezzi et al. (2008, 2011) and Vetere et al. (2013, 2015). For each run product, the identification of phases was determined by linking their grey-level ranges with the compositions.

7.4. RESULTS

7.4.1. Calbuco, Etna, and Theo's Flow Melt Viscosities

Viscosity was measured at different shear rates and a Newtonian behavior was observed for all melts (Figure 7.3). Melt-viscosity experimental data of the Calbuco andesite and the Etna basalt are provided in

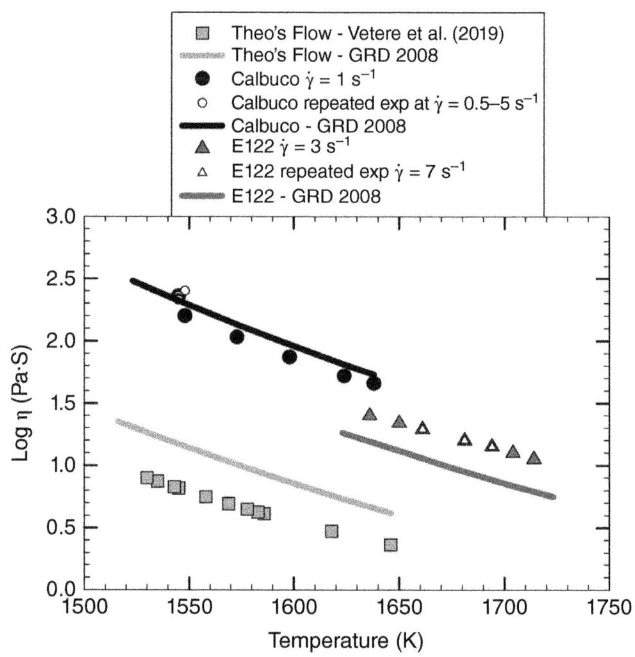

Figure 7.3 Viscosity data of melts at superliquidus conditions. Light gray, black, and dark gray solid lines (GRD 2008) represent the viscosity predicted using the general model of Giordano et al. (2008) for pyroxenite, Calbuco andesite, and Etna basalt, respectively. The light gray square represents the predictive model given in equation (1) by Vetere et al. (2019) for Theo's Flow. Black and white circles are Calbuco andesite viscosity data collected at $\dot{\gamma}$ = 0.5 to 5 s^{-1}. No viscosity differences are noted when using different shear rate for the three compositions investigated, suggesting a Newtonian behavior of such melts at the experimental high temperatures. Error bar is less that the symbol size (0.03 log unit; see text for details).

Table 7.2 Experimental Conditions and Results of Melts Viscosity Measurements (η) Using the Concentric Cylinder Apparatus

Material	T (K)	$\dot{\gamma}$ (s^{-1})	η (Pa·s)	Log η (Pa·s)
Calbuco andesite	1638	1	45.9	1.66
	1624	1	52.6	1.72
	1598	1	74.3	1.87
	1573	1	106.2	2.03
	1548	1	150.0	2.18
	1545	0.5	254.6	2.41
	1545	0.7	239.9	2.38
	1545	1	230.4	2.36
	1545	2	221.1	2.34
	1545	3	216.3	2.34
	1545	5	212.2	2.33
ETNA 122 basalt	1636	3	41.6	1.62
	1650	3	36.7	1.56
	1661	3 and 7	32.2	1.51
	1681	3 and 7	26.2	1.42
	1694	3 and 7	23.5	1.37
	1604	3	21.0	1.32
	1714	3	18.5	1.27
Pyroxenite[a]	A (Pa·s)	B (K)	T_0 (K)	
	−4.22	4982.0	556.1	

Note. $\dot{\gamma}$, the applied shear rate.
[a] Log $\eta = A + B/(T - T_0)$ from Vetere et al. (2019).

Table 7.2 and are well reproduced by the general model of Giordano et al. (2008). The data for the pyroxenite melt are given in Vetere et al. (2019). This composition is out of the calibration range of the model of Giordano et al. (2008) and can be modeled by an empirical equation proposed by Vetere et al., (2019) based on the Vogel–Fulcher–Tammann (VFT) equation (Vogel, 1921):

$$\log \eta (\mathrm{Pa \cdot s}) = A + B/(T - T_0) \tag{7.9}$$

In this equation T is the temperature in Kelvin and A, B, and T_0 are fitting parameters, representing the preexponential term, the pseudoactivation energy (related to the barrier of potential energy obstructing the structural rearrangement of the liquid), and the VFT temperature, respectively. The proposed values for the constants are: $A = -4.22$ (Pa·s), $B = 4982.0$ (K) and $T_0 = 556.1$ (K).

7.4.2. Experimental Products and Viscosity of Partially Crystallized Andesite, Basalt, and Pyroxenite

Viscosity experiments under controlled shear rates of 0.1 and 1 s^{-1} were conducted at temperature of 1483, 1493, and 1503 K for Calbuco andesite, Etna basalt, and pyroxenite, respectively. At these conditions, crystalline phases are present in the samples. Backscattered electron

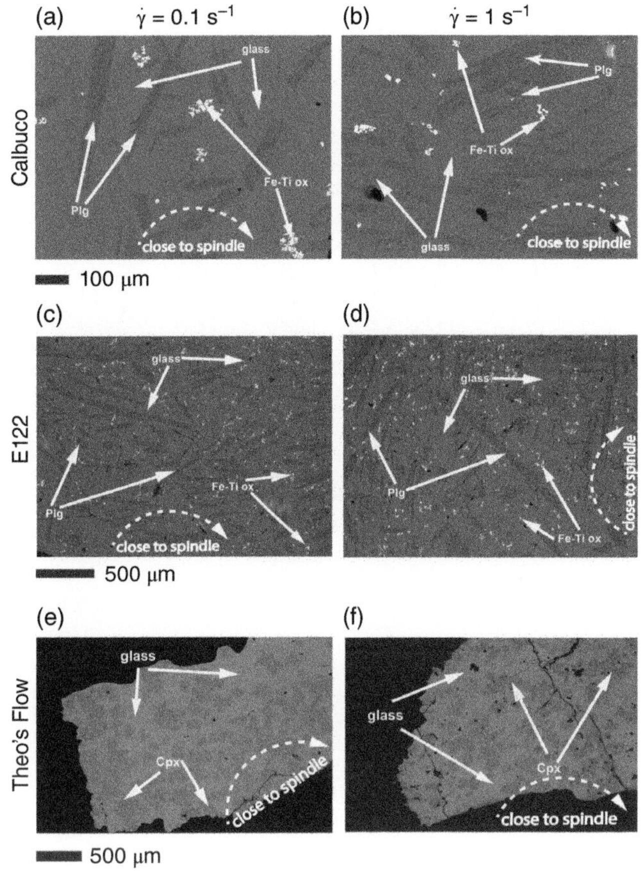

Figure 7.4 Selected BSE images obtained after experiments performed at shear rates of 0.1 and 1 s^{-1}: (a, b) Calbuco andesite compositions; (c, d) the E122 Etna basalt sample; (e, f) Theo's Flow pyroxenite quench products after viscosity measurement. Note that figures on the right-hand side are experiments at $\dot{\gamma}$=1 s^{-1} while those on the left-hand side refer to experiments performed at $\dot{\gamma}$=0.1 s^{-1}.

images of the investigated partially crystallized samples are presented in Figure 7.4a–f. Detected phases are glass, plagioclase, and iron titanium oxide for both Calbuco andesite and the Etna basalt, while the pyroxenite contains glass, pyroxene, and iron titanium oxide. The total area investigated in the samples obtained at shear rate $\dot{\gamma}$ = 0.1 and 1 s^{-1} is approximately between 25 and 30 mm^2, consisting of two surfaces cut along a parallel plane of the spindle and along a perpendicular plane. The amount of solid phases was calculated via image analyses and was found to be 12 ± 1, 21 ± 2 and 26 ± 5 vol% for the partially crystallized andesite, basalt, and pyroxenite, respectively. Electron microprobe analysis was used to check for the composition of solid phases. Results are shown in Figure 7.5a–c.

The Calbuco andesite and the Etna basalt have plagioclase as a main phase and the compositional variation is $An_{64\text{-}67}$ to $An_{53\text{-}70}$, respectively (Figure 7.5a and b). The

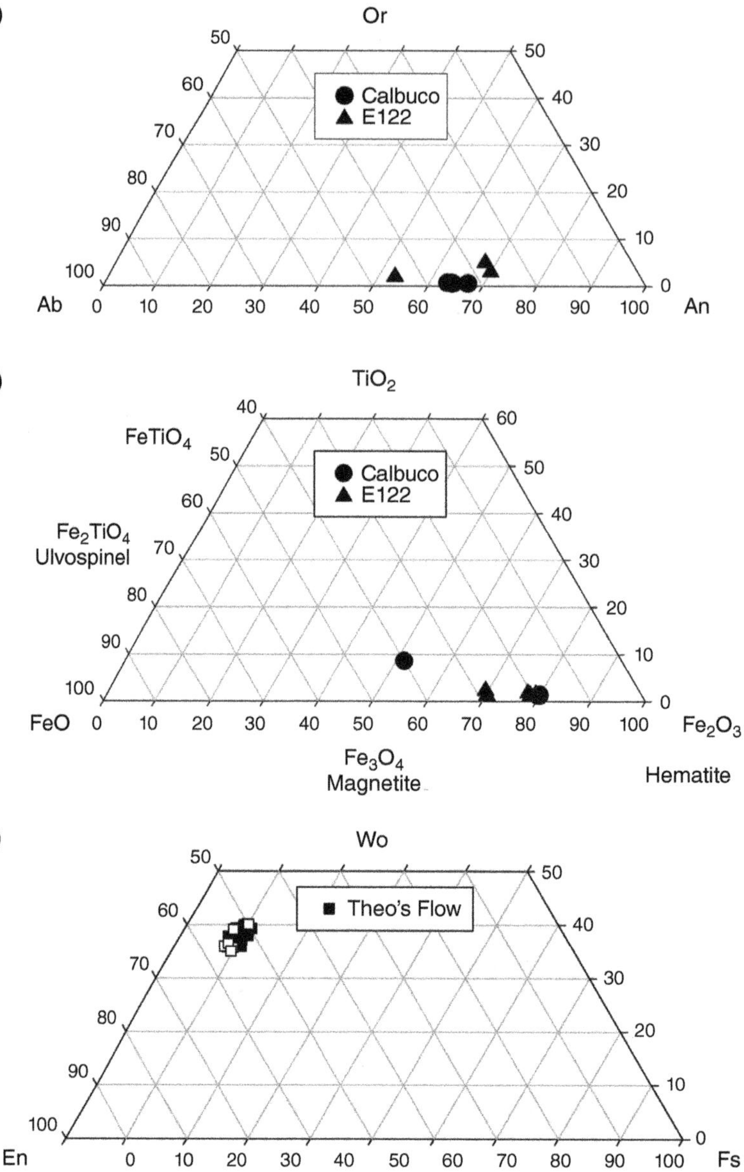

Figure 7.5 Solid phases chemistry: (a) plagioclase crystal chemistry for the andesite (black circles) and basalt (black triangles); (b) Fe–Ti oxides for andesite and basalt (symbols as for (a)); (c) pyroxenes chemistry for the pyroxenite.

Fe–Ti oxides, on the other hand, are close to magnetite–hematite composition. The pyroxenite contains clinopyroxene with composition ranging from to $Wo_{35}En_{55}$ to $Wo_{41}En_{49}$ (Figure 7.5c). The Calbuco andesite plagioclases are elongated with aspect ratio $4 < AS < 10$ (where AS = major axis/minor axis) and frequently show a skeletal habit, while Fe–Ti oxides are irregular and sometimes also showing a skeletal habit. As a whole, the plagioclase crystal fraction (Φ) varies between 0.11 and 0.13 for $\dot{\gamma} = 0.1\ s^{-1}$ and 0.10–0.14 for $\dot{\gamma} = 1\ s^{-1}$ and we observed a maximum deviation of ca. 3 vol% as $\dot{\gamma}$ varies from 0.1 to 1 s^{-1}. The volume of Fe–Ti oxides is < 1.5 vol%.

The overall crystal fraction (Φ) for the Etna basalt varies between 0.15 and 0.17 for $\dot{\gamma} = 0.1\ s^{-1}$ and between 0.15 and 0.18 for $\dot{\gamma} = 1\ s^{-1}$. No significant difference of Φ was found between the two experiments and the maximum deviation is ca. 3 vol% as also observed for the Calbuco andesite. Generally, experiments exhibit a relatively homogeneous distribution of both plagioclase and Fe-Ti oxides along the entire crucible. The Fe-Ti oxides fraction is very low and varies from 0.75 to 1 vol%. Plagioclase frequently shows a skeletal habit and is elongated in shape, pointing to an aspect ratio of 5<AS<11. Both higher and lower shear rates show a similar distribution

Table 7.3 Experimental Conditions and Results of the Apparent Viscosity Measurements (η), Using the Concentric Cylinder Apparatus

Material	(s^{-1})	Apparent viscosity (log Pa·s)	Relative viscosity (η_r)	Crystallinity (vol%)	Temperature (K)	AS (average)
Pyroxenite	0.1	3.61	135.2	26 ± 5	1503	1.5–3.1
	1	2.44	9.2			
Etna 122 basalt	0.1	3.81	53.7	21 ± 3	1493	3.9–10.2
	1	3.12	11.0			
Calbuco andesite	0.1	3.76	10.7	12 ± 2	1483	5.0–11.3

Note. $\dot{\gamma}$, the applied shear rate. AS, aspect ratio.

of particle aspect ratio. Differences rely, for the fastest rate, on a relatively higher alignment of the particle's major axis with the flow direction.

Clinopyroxene is the major phase crystallizing in the pyroxenite. Its crystals are much more rounded compared to the plagioclase. They show $2 < AS < 4$ and a slight compositional variability (Figure 7.5c and Table 7.3).

The viscosity evolution versus time is reported in Figure 7.6a-c. All three experiments show a classical S-shape curve. The resulting viscosity change is, from left to right, from melt relaxation, crystal nucleation, and crystal growth, as proposed by Vona et al. (2011). The plateau with constant viscosity observed in the right-hand part of Figure 7.6 indicates that near-equilibrium conditions are reached (constant melt and crystal proportions) for the chosen temperature (Figure 7.6a–c). All experiments show a clear viscosity dependence on the applied shear rate. In particular, the lower applied shear rate results in the highest measured viscosity. In detail, as shear rate decreases from 1 to 0.1 s^{-1} the apparent viscosity plateau increases from $10^{3.29}$ to $10^{3.76}$ Pa·s for the andesite, from $10^{3.12}$ to $10^{3.81}$ Pa·s for the basalt and from $10^{2.44}$ to $10^{3.61}$ Pa·s for the pyroxenite (Table 7.3).

7.5. DISCUSSION

7.5.1. General Observations

During recent decades, the effect of magma cooling and relative crystal content, size distributions, and crystal shapes on magma viscosity, has been studied by both theoretical and experimental approaches (Arbaret et al., 2007; Caricchi et al., 2007; Costa et al., 2009; Ishibashi & Sato, 2007; Lejeune & Richet, 1995; Mader et al., 2013; Murase & McBirney, 1973; Murase et al., 1985; Pinkerton & Norton, 1995; Pinkerton & Stevenson, 1992; Ryerson et al., 1988; Sato, 2005; Shaw, 1969; Sehlke & Whittington, 2015; Sehlke et al., 2014; Vetere et al., 2010, 2013, 2017; Vona et al., 2011, 2013).

In terms of relative viscosity, we can compare our experimental data with literature data as shown in Figure 7.7. All data presented derive from experiments performed at variable shear rates (0.01 to 5 s^{-1}) and all are pointing to a shear thinning behavior of magmas. In other words, the higher the shear rate, the lower is the apparent viscosity. Note that, for better visualization of literature data, Figure 7.7 has a vertical axis limited to 20 and as consequence two experimental runs at $\dot{\gamma} = 0.1$ s^{-1} for basalt and pyroxenite are not shown. This is because the relative viscosity determined for basalt and pyroxenite have values larger than 50. Only the values obtained at shear rates of both 0.1 and 1 s^{-1} are shown for andesite. Although there is a decrease of apparent viscosity with shear rate, the data show that this decrease is difficult to predict, even for a constant value of Φ. As expected the highest variations are observed for the highest crystal fractions, but the texture and the melt viscosity are also expected to play an important role. For example, the variation of apparent viscosity obtained by Vetere et al. (2017) for a Mercury Mg-rich low viscosity melt is much lower than that obtained for the pyroxenite, which is also highly depolymerized. At similar values of Φ the difference between the relative viscosity at 0.1 and 1 s^{-1} is much higher for the pyroxenite, in which mineral phases are roughly spherical in shape (diameter 100–200 μm, Figure 4e and f). In contrast, the minerals formed in the Mercury melt are elongated (AS up to 14) with a clear preferred orientation (see figure 7 in Vetere et al., 2017). Another general observation is that the differences between data obtained at different strain rate (same value of Φ) are lower if the melt viscosity is higher. Considering the three systems, andesite, basalt, and pyroxenite, the differences in apparent viscosity between shear rates of 0.1 and 1 s^{-1} are 0.47, 0.69, and 1.17 log units, respectively (see Table 7.3).

It is important to note that the evolution of melt viscosity with crystallization does not change significantly during the experiments. For example, the melt SiO_2 content from Calbuco andesite increases from 55 to 58 wt% and the Al_2O_3 content decreases from 19 to 15 wt% as a result of crystallization. Thus, this chemical shift (see Table 7.1 for details) leads to an increase in viscosity of the melt by only 0.06 log units at 1483 K (using the model

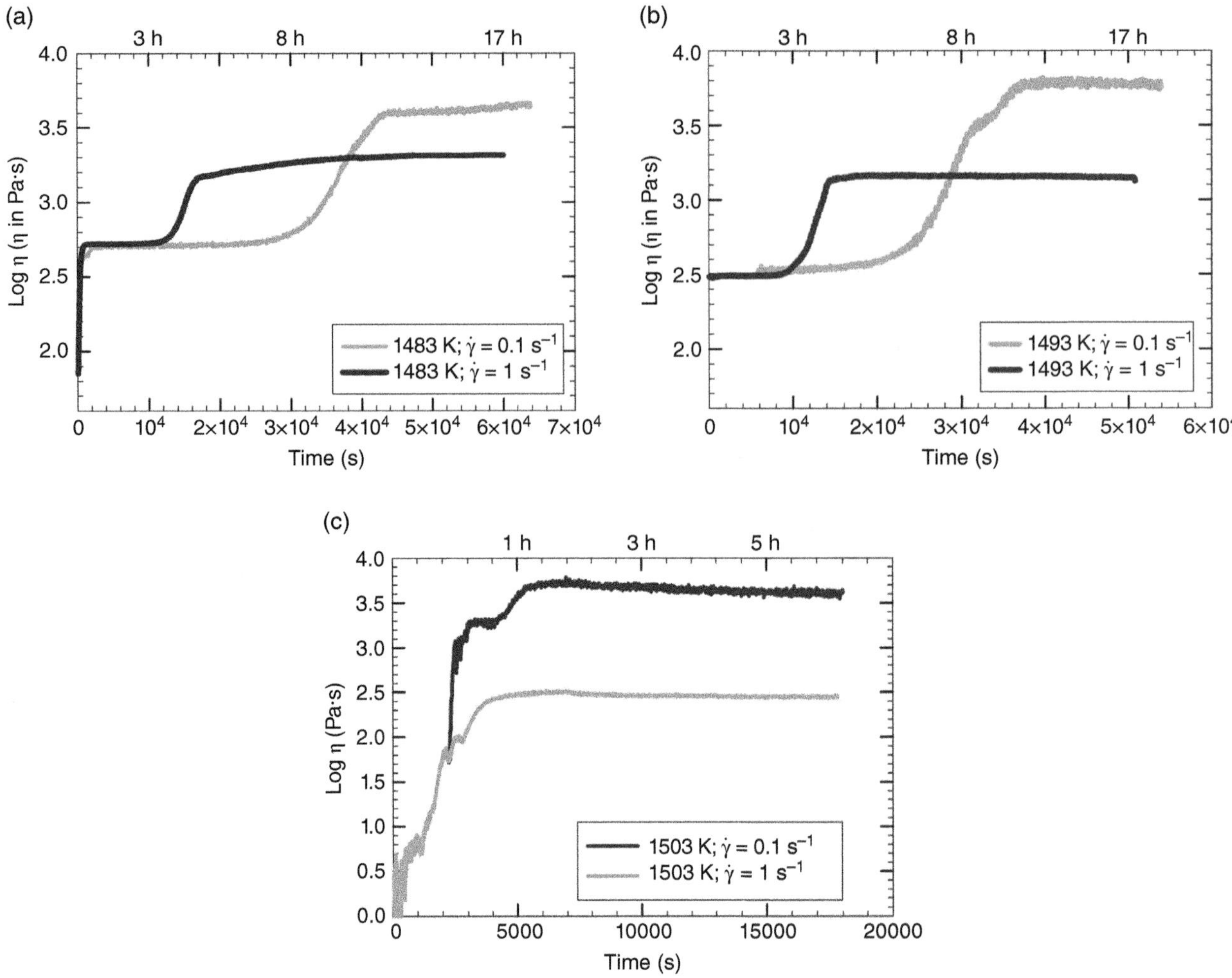

Figure 7.6 Effective viscosity-time paths for (a) Calbuco, (b) E122 and (c) Theo's Flow samples. Blue and red curves refer to effective viscosity obtained at 1 and 0.1 s^{-1}. All compositions show a shear-thinning behavior with nucleation and crystallization processes related to the applied shear rates (see text for details).

of Giordano et al., 2008). Similar observations are evident for the basalt and the pyroxenite at 1493 and 1503 K, where differences of 0.02 and 0.07 log units, respectively, are calculated. Thus, for the compositions investigated in this study, the melt viscosity variation is lower than 0.1 log units and can be considered to be negligible.

7.5.2. Comparison with Previous Data Sets

The change in viscosity as shear rate changes has a profound implication on magma ascent and emplacement as well as on the potential of a magmatic system to change its eruptive style, which basically is controlled by viscosity. It is obvious that the rheological properties of an ascending magma are changing if crystallization occurs. However, at a given crystal content, the intrinsic dynamics also play a role and viscosity decreases as the dynamics (or shear rates) of the system increases. Although the literature confirms our experimental results (e.g., Caricchi et al., 2007; Champallier et al., 2008; Cordonnier et al., 2009; Costa, 2005; Costa et al., 2009; Deubelbeiss et al., 2011; Ishibashi, 2009; Mueller et al., 2010; Petford, 2009; Pinkerton & Stevenson, 1992; Vetere et al., 2017; Vona et al., 2011), at the present stage of our knowledge it is difficult to establish a general law relating relative viscosity with shear rate, crystal fraction, and crystal shape.

The experimental data set obtained so far is compared with various models proposed in the literature in Figure 7.7. Clearly, the Einstein–Roscoe model (using parameters proposed by Marsh, 1981) predicts the lowest effect of crystal fraction on relative viscosity. The two models directly calibrated on natural compositions proposed by Sato (2005) and Vona et al. (2011) for basalts predict a strong increase of relative viscosity at lower Φ

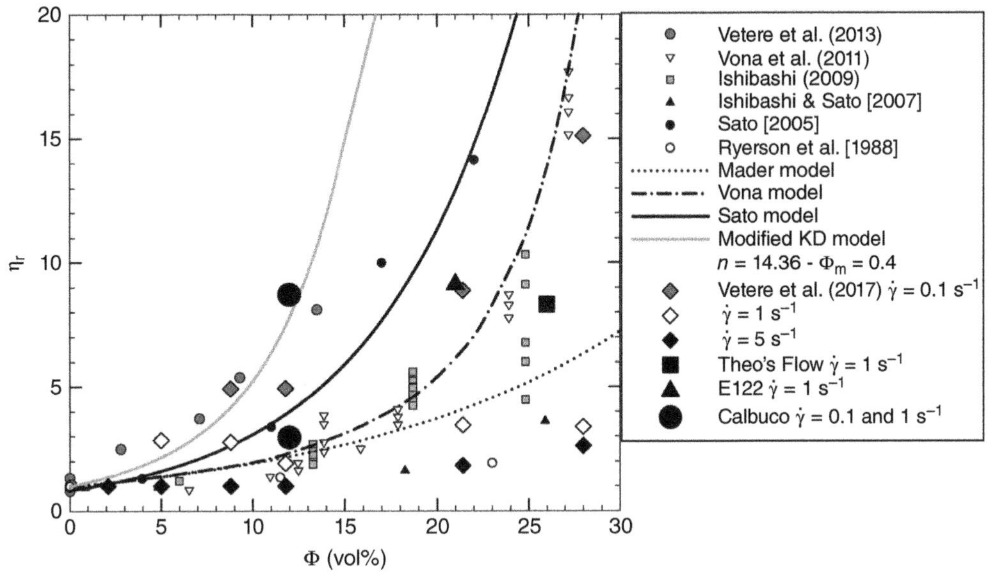

Figure 7.7 Relationship between relative viscosity (η_r) and crystal fraction (Φ; measured by image analysis, see text for details) for the systems used in this study. Literature data (Ishibashi, 2009; Ishibashi & Sato 2007; Ryerson et al., 1998; Sato, 2005; Sehlke & Whittington, 2015; Sehlke et al., 2014; Vetere et al., 2013, 2017; Vona et al., 2011) are also reported for comparison and the curves correspond to different models: ER, Einstein–Roscoe model (Einstein, 1906; Roscoe, 1952); KD, Krieger–Dougherty model (Krieger & Dougherty, 1959); Sato model (Sato, 2005); Mader model (Mader et al., 2013); Vona model (Vona et al., 2011).

values compared to the Einstein–Roscoe model. Strain rates are not clearly defined in the study of Sato (2005) but the model proposed by Vona et al. (2011) predicts only a small, almost negligible relative viscosity variation as a function of shear rate in the range 0.02–2.2 s^{-1}. In this study the relative viscosity variations between 0.1 and 1 s^{-1} are much more pronounced, independently of the compositions (andesite, basalt, pyroxenite). At this stage, there is no clear explanation for this difference. One possible explanation may be related to the configuration of the experimental assemblage and to the geometry of cylinder measuring systems. In the literature such a method is related to the so-called Seerle method, where the bob (spindle) is set in motion and the cup is fixed and stationary (details are reported in Mezner, 2014). When using the large-gap cylinder measurement system, the shear stress and shear rate are related to the bob surface and therefore principally to: torque (M), immersion of the spindle (L') into the filled crucible, to the radius (r) of the spindle with the torque M (Nm), and to the length L (m) of the cylindrical part of the bob, as reported in Mezner (2014).

In the study of Vona et al. (2011) the ratio between the radius of the spindle and the radius of the crucible is 0.13 (spindle and crucible radii = 1.60 and 12.80 mm, respectively). Ishibashi (2009) as well as Sato and Ishibashi (2007) used a configuration with a ratio spindle/crucible = 0.20 (spindle radius of 3 mm and crucible radius of 15 mm). Sehlke and Whittington (2015) used an assemblage with a spindle/crucible ratio of 0.25 (spindle and crucible radii = 3.75 and 15.08 mm, respectively). Finally, the geometry adopted in this study has a spindle/crucible ratio of 0.32 (spindle and crucible radii equal to 6.00 and 18.50 mm, respectively), which is larger than the previous configurations, and which allows us to use a larger sample volume. Qualitatively, different radii of spindle and samples (and related spindle/crucible ratios) could possibly influence particle alignment along the flow direction and the effective viscosity, which is directly linked to the interactions between elongated crystals and the spindle.

Figure 7.8 provides a comparison of apparent viscosities determined experimentally, including the available literature data, with the model proposed by Vona et al. (2011). As expected, the apparent viscosity data utilized for building the model itself are close to the 1:1 line, but the calculated apparent viscosity is mostly underestimated for other literature data and our new experimental data. This indicates that much more experimental effort is necessary to further increase our understanding of the evolution of viscosity in crystallizing magmatic systems, in which crystal shape, size, and volume changes may change. In particular, new high temperature, high pressure experiments at variable shear rates and also cooling rates are necessary to closely match the natural scenario.

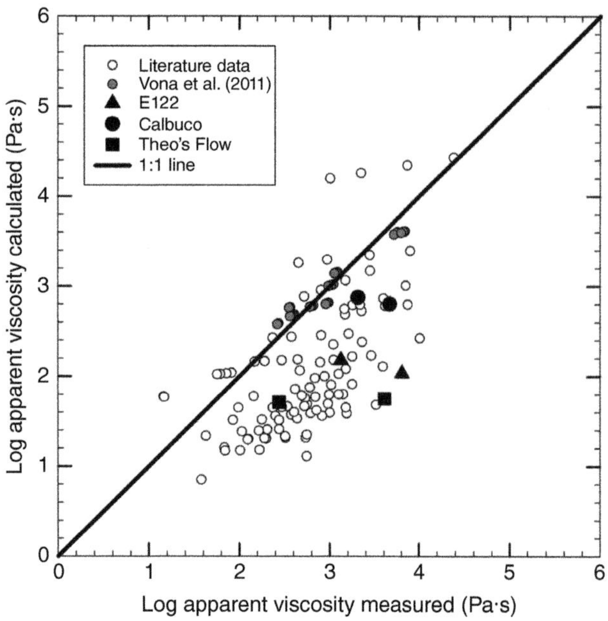

Figure 7.8 Experimental and literature apparent viscosity data (see references in caption of Fig. 7.7) compared with the Vona et al. (2011) model.

7.5.3. Viscosity–Time Evolution and Implications for Crystallization Rates

As stated above and also described by Vona et al. (2011) each viscosity–time path shows a classical S-shape curve that can be divided into four parts from left to right (Figure 7.6): (a) relaxation of melts, (b) crystal nucleation, (c) crystal growth, and (d) final plateau indicating the end of crystallization and the achievement of a constant effective viscosity (η_{eff}). Those four segment are linked to time. Thus, we can estimate the growth rates of crystal during the applied shear rates. Although there is 10 K difference between andesite and basalt experimental temperatures, Figure 7.6 shows that the time needed to reach a final plateau for these compositions is much longer than that of the pyroxenite (for a given shear rate). One explanation is the higher experimental temperature of the pyroxenite (1503 K), but a second reason may be related to the high reactivity of the pyroxenite in terms of nucleation and crystallization processes. At the experimental conditions pyroxenite needs only ca. 1.5 h for both shear rates ($\dot{\gamma}$ = 0.1 and 1 s^{-1}) in order to reach the final viscosity value and complete crystallization. In contrast, both basalt and andesite show a similar time evolution in viscosity at high shear rate and the final plateaus are reached after 1.7×10^4 s for the andesite and 1.5×10^4 s for the basalt. Considering a crystal average size of 200 μm (see Figure 7.4) a growth rate of the order of 1.2×10^{-7} and 1.3×10^{-7} cm s^{-1} can be calculated for andesite and basalt, respectively. Low-shear-rate runs required ca. 4.3×10^4 s and 3.7×10^4 s to reach a constant viscosity for the andesite and basalt, respectively (Figure 7.6). In this case growth rates of 4.65×10^{-7} and 5.4×10^{-7} cm s^{-1} for andesite and basalt, respectively, are obtained. In the pyroxenite run at high $\dot{\gamma}$ only ca. 4000 s are required in order to complete the crystallization process, resulting in a growth rate of 5.0×10^{-6} cm s^{-1}. At low $\dot{\gamma}$, the plateau is reached after 5600 s, resulting in a growth rate of 3.5×10^{-6} cm s^{-1}.

Our experimental viscosity data confirm that higher shear rates result in faster crystallization when compared with low shear rates, as reported by Kolzenburg et al. (2017). In general, these observations were also reported for isothermal experiments at subliquidus conditions by Kouchi et al. (1986), Vona and Romano (2013), Chevrel et al. (2015), and Vetere et al. (2017) and it is now widely accepted that the rheological departure (i.e., the change of η_r as shear rate varies) at elevated shear rates increases crystallization rate. However, the melt viscosity also plays a crucial role; although the crystallinity is higher in the pyroxenite (26 vol% crystals when reaching the viscosity plateau), the time difference is lower for the much more reactive pyroxenite melts, which crystallized almost in the same time interval by changing shear rate from 0.1 to 1 s^{-1}. The possible mechanisms that could be involved to explain this difference in dynamic crystallization process are related to the continuous "feed growth ingredients" (elements necessary to build a crystalline structure) on individual crystal surfaces that facilitate crystal growth at high shear rate (Petrelli et al., 2016). Another factor may be the nature of the phase that crystallizes; for example, at the same undercooling, the growth rate of plagioclase (main phase in andesite and basalt) is not expected to be similar to that of clinopyroxene (main phase in pyroxenite). Finally, the increased stirring velocity may also, statistically, more efficiently increase the crystal nuclei, as suggested by Cashman et al. (1999), Emerson (1926), and Vona et al. (2013).

7.6. CONCLUSION

In this study we performed a series of viscosity measurements applying different shear rates on compositions relevant for natural scenarios (basalt, andesite, and pyroxenite). The new data set confirms the shear-thinning effect on magma and lava as shear rates vary. Although the mineral/melt ratios do not remarkably change in our experiments as shear rates vary, the effective viscosity measurements show that as shear rate varies from 0.1 to 1 s^{-1}, effective viscosity decreases from 3.61 to 2.44, 3.81 to 3.12 and from 3.76 to 3.29 log units (in Pa·s) for pyroxenite, basalt, and andesite, respectively.

The analysis of available literature data shows that many parameters may affect the viscosity of magmas in a dynamic system and that a prediction model taking compositional variations and shear rates into account is difficult to establish. Although the evolution of viscosity in dynamic crystal-bearing systems as a function of shear rate is a long way from being established with a good precision, some general observations can be formulated.

• For a given composition, the viscosity evolution of samples deformed at different shear rates have approximately the same shape in a diagram of time versus apparent viscosity, which implies that they have the same nucleation rate but that the stirring velocity plays a role in controlling the feed growth ingredients.

• The configuration of the experimental assemblage, the sample volume, and, especially, the spindle/crucible ratio, may play a role on the viscosity values determined in dynamic crystal-bearing systems.

• In modeling magma ascent and emplacement of lava flows, rheological change imposed by slower/faster dynamics is an important parameter to be taken into account.

The data set presented in this study contributes to our understanding of the behavior of magmas in conduits during magma rise and lava flow on Earth's surface and on other planetary bodies. However, further experimental studies under variable shear and cooling rates at temperature and pressure conditions relevant for magmatic and volcanic systems are crucially needed in order to improve available general models.

ACKNOWLEDGMENTS

This research was funded by the European Research Council Consolidator Grant ERC-2013-COG No. 612776 (CHRONOS project) to D. Perugini and by the F.R.B. TESLA to F. Vetere. Alexander von Humboldt foundation senior research grant to F. Vetere is also acknowledged. Data of experiments are reported in Tables 7.1, 7.2 and 7.3 and in Figures 7.1–7.8

REFERENCES

Armienti, P. (2008). Decryption of igneous rock textures: Crystal size distribution tools. *Reviews in Mineralogy and Geochemistry*, *69*, 623–649. https://doi:10.2138/rmg.2008.69.16

Byrne, P. K., Klimczak, C., Williams, D. A., Hurwitz, D. M., Solomon, S. C., Head, J. W., Preusker, F., & Oberst, J. (2013). An assemblage of lava flow features on Mercury. *Journal of Geophysical Research E: Planets*, *118*(6), 1303–1322.

Carmichael, R. S. (1974). *Igneous petrology*. New York: McGraw-Hill, 793 pp.

Caricchi, L., Burlini, L., Ulmer, P., Gerya, T., Vassalli, M., & Papale, P. (2007). Non-Newtonian rheology of crystal-bearing magmas and implications for magma ascent dynamics. *Earth and Planetary Science Letters*, *264*, 402–419. https://doi:10.1016/j.epsl.2007.09.032

Cashman, K. V. (1989). Relationship between plagioclase crystallization and cooling rate in basaltic melts. *Contributions to Mineralogy and Petrology*, *113*, 126–142.

Cashman, K. V., Thornber, C., & Kauahikaua, J. P. (1999). Cooling and crystallization of lava in open channels, and the transition of Pahoehoe lava to ‘a‘a. *Bulletin of Volcanology*, *61*, 306–323.

Champallier, R., Bystricky, M., & Arbaret, L. (2008). Experimental investigation of magma rheology at 300 MPa: From pure hydrous melt to 76 vol.% of crystals. *Earth and Planetary Science Letters*, *267*(3–4), 571–583. https://doi:10.1016/j.epsl.2007.11.065

Chevrel, M. O., Cimarelli, C., deBiasi, L., Hanson, J. B., Lavallée, Y., Arzilli, F., & Dingwell, D. B. (2015). Viscosity measurements of crystallizing andesite from Tungurahua volcano (Ecuador). *Geochemistry, Geophysics, Geosystems*, *16*, 870–889. https://doi.org/10.1002/2014GC005661

Chevrel, M. O., Harris, A. J. L., James, M. R., Calabrò, L., Gurioli, L., & Pinkerton, H. (2018). The viscosity of pāhoehoe lava: In situ syn-eruptive measurements from Kilauea, Hawaii. *Earth and Planetary Science Letters*, *493*, 161–171.

Cimarelli, C., Costa, A., Mueller, S. P., & Mader, H. M. (2011). Rheology of magmas with bimodal crystal size and shape distributions: insights from analog experiments. *Geochemistry, Geophysics, Geosystems*, *12*. https://doi.org/10.1029/2011GC003606

Coltelli, M., Del Carlo, P., & Vezzoli, L. (1998). The discovery of a Plinian basaltic eruption of Roman age at Etna volcano, Italy. *Geology*, *26*, 1095–1098.

Cordonnier, B., Hess, K.-U., Lavallee, Y., & Dingwell, D. B. (2009). Rheological properties of dome lavas: Case study of Unzen volcano. *Earth and Planetary Science Letters*, *279*(3–4), 263–272. https://doi:10.1016/j.epsl.2009.01.014

Costa, A. (2005). Viscosity of high crystal content melts: Dependence on solid fraction. *Geophysical Research Letters*, *32*, 1–5. https://doi:10.1029/2005GL024303

Costa, A., Caricchi, L., & Bagdassarov, N. (2009). A model for the rheology of particle-bearing suspensions and partially molten rocks. *Geochemistry, Geophysics, Geosystems*, *10*, 1–13. https://doi:10.1029/2008GC002138

Davis, M. J., & Ihinger, P. D. (1998). Heterogenous crystal nucleation on bubbles in silicate melt. *American Mineralogist*, *83*, 1008–1015.

Deubelbeiss, Y., Kaus, B. J. P., Connolly, J. A. D., & Caricchi, L. (2011). Potential causes for the non-Newtonian rheology of crystal-bearing magmas. *Geochemistry, Geophysics, Geosystems*, *12*, Q05007. https://doi:10.1029/2010GC003485

Del Gaudio, P. D., Ventura, G., & Taddeucci, J. (2013). The effect of particle size on the rheology of liquid–solid mixtures with application to lava flows: Results from analogue experiments. *Geochemistry, Geophysics, Geosystems*, *14*, 2661–2669. https://doi:10.1002/ggge.20172.

Dingwell, D. B. (1986). Viscosity–temperature relationships in the system $Na_2Si_2O_5$–$Na_4Al_2O_5$. *Geochimica et Cosmochimica Acta*, *50*(6), 1261–1265. https://doi:10.1016/0016-7037(86)90409-6

Dowty, E. (1980). Crystal growth and nucleation theory and the numerical simulation of igneous crystallization, in R. B. Hargraves (Ed.), *Physics of magmatic processes* (pp. 419–485). Princeton, NJ: Princeton University Press.

Einstein, A. (1906). Eine neue Bestimmung der Molekul dimensionen. *Annalen der Physik, 19*, 289–306.

Emerson, O. H. (1926). The formation of aa and pahoehoe. *American Journal of Science, 68*, 109–114.

Giordano, D., Russell, J. K., & Dingwell, D. B. (2008). Viscosity of magmatic liquids: a model. *Earth and Planetary Science Letters, 271*, 123–134.

Gonzalez-Garcia, D., Behrens, H., Petrelli, M., Vetere, F., Morgavi, D., Zhang, C., & Perugini, D. (2017). Water-enhanced interdiffusion of major elements between natural shoshonite and high-K rhyolite melts at 1200°C. *Chemical Geology, 466*, 86–101. http://dx.doi.org/10.1016/j.chemgeo.2017.05.023

Halder, N. K., Chatterjee, B. K., & Roy, S. C. (1999). The change of viscosity with concentration of suspended particles and a new concept of gelation. *Journal of Physics: Condensed Matter, 9*, 8873–8878. https://doi:10.1088/0953-8984/9/42/004

Hammer, J. E. (2008). Experimental studies of the kinetics and energetics of magma crystallization. *Reviews in Mineralogy and Geochemistry, 69*, 9–59. https://doi:10.2138/rmg.2008.69.2

Herschel, W., & Bulkley, R. (1926). Measurement of consistency as applied to rubber–benzene solutions. *Kolloid—Zeitschrift und Zeitschrift für Polymere, 39*, 231.

Hess, K.-U., Dingwell, D. B., & Rössler, E. (1996). Parametrization of viscosity–temperature relations of aluminosilicate melts. *Chemical Geology, 128*(1–4), 155–163. https://doi:10.1016/0009-2541(95)00170-0

Iezzi, G., Mollo, S., Ventura, G., Cavallo, A., & Romano, C. (2008). Experimental solidification of anhydrous latitic and trachytic melts at different cooling rates: The role of nucleation kinetics. *Chemical Geology, 253*, 91–101.

Iezzi, G., Mollo, S., Torresi, G., Ventura, G., Cavallo, A., & Scarlato, P. (2011). Experimental solidification of an andesitic melt by cooling. *Chemical Geology, 283*, 261–273.

Ishibashi, H. (2009). Non-Newtonian behavior of plagioclase-bearing basaltic magma: Subliquidus viscosity measurement of the 1707 basalt of Fuji volcano, Japan. *Journal of Volcanology and Geothermal Research, 181*, 78–88. https://doi:10.1016/j.jvolgeores.2009.01.004

Ishibashi, H., Sato, H. (2007). Viscosity measurements of subliquidus magmas: Alkali olivine basalt from the Higashi-Matsuura district, Southwest Japan. *Journal of Volcanology and Geothermal Research, 160*, 223–238. https://doi:10.1016/j.jvolgeores.2006.10.001

Ishibashi, H., & Sato, H. (2010). Bingham fluid behavior of plagioclase-bearing basaltic magma: Reanalyses of laboratory viscosity measurements for Fuji 1707 basalt. *Journal of Mineralogical and Petrological Sciences, 105*, 334–339.

Kirkpatrick, R. J. (1981). Kinetics of crystallization of igneous rocks. *Reviews in Mineralogy, 8*, 321–397.

Klein, J., Mueller, S. P., Helo, C., Schweitzer, S., Gurioli, L., & Castro, J. M. (2018). An expanded model and application of the combined effect of crystal-size distribution and crystal shape on the relative viscosity of magmas. *Journal of Volcanology and Geothermal Research, 357*, 128–133.

Klöß, G. H. (2000). *Dichtefluktuationen natürlicher Gläser (Dissertation)*. University of Jena.

Kolzenburga, S., Giordano, D., Cimarelli, C., & Dingwell, D.B. (2016). In situ thermal characterization of cooling/crystallizing lavas during rheology measurements and implications for lava flow emplacement. *Geochimica et Cosmochimica Acta, 195*, 244–258.

Kolzenburg, S., Giodano, D., Hess, K. U., & Dingwell, D. B. (2018). Shear rate-dependent disequilibrium rheology and dynamics of basalt solidification. *Geophysical Research Letters, 45*, 6466–6475. doi:10.1029/2018GL077799

Kolzenburg, S., Giordano, D., Thordarson, T., Höskuldsson, A., & Dingwell, D. B. (2017). The rheological evolution of the 2014/2015 eruption at Holuhraun, central Iceland. *Bulletin of Volcanology, 79*, 45.

Kouchi, A., Tsuchiyama, A., & Sunagawa, I. (1986). Effect of stirring on crystallization kinetics of basalt: texture and element partitioning. *Contributions to Mineralogy and Petrology, 93*, 429–438. http://dx.doi.org/10.1007/BF00371713

Krieger, I., & Dougherty, T. (1959). A mechanism for non-Newtonian flow in suspensions of rigid spheres. *Transactions of the Society of Rheology, 3*, 137–152.

Lasaga, A. C. (1997). *Kinetic theory in the Earth sciences*. Princeton, NJ: Princeton University Press.

Lentz, R. C. F., McCoy, T. J., Collins, L. E., Corrigan, C. M., Benedix, G. K., Taylor G. J., & Harvey, R. P. (2011). Theo's Flow, Ontario, Canada: A terrestrial analog for the Martian nakhlite meteorites. *The Geological Society of America Special Paper, 483*, 263–277.

Lejeune, A., & Richet, P. (1995). Rheology of crystal-bearing silicate melts: An experimental study at high viscosity. *Journal of Geophysiacal Research, 100*, 4215–4229. https://doi:10.1029/94JB02985

Lofgren, G. E. (1983). Effect of heterogeneous nucleation on basaltic textures: A dynamic crystallization study. *Journal of Petrology, 24*(3), 229–255.

Mader, H. M., Llewellin, E. W., & Mueller, S. P. (2013). The rheology of two-phase magmas: A review and analysis. *Journal of Volcanology and Geothermal Research, 257*, 135–158. https://doi:10.1016/j.jvolgeores.2013.02.014

Maron, S. H., & Pierce, P. E. (1956). Application of Ree–Eyring generalized flow theory to suspensions of spherical particles. *Journal of Colloid Science, 11*, 80–95.

Marsh, B. D. (1981). On the crystallinity, probability of occurrence, and rheology of lava and magma. *Contributions to Mineralogy and Petrology, 78*, 85–98.

Mezner, T. G. (2014). *The rheology handbook—for users of rotational and oscillatory rheometers* (4th edn). Hanover: Vincentz Network.

Moitra, P., & Gonnermann, H. M. (2015). Effects of crystal shape- and size-modality on magma rheology, *Geochemistry, Geophysics, Geosystems, 16*, 1–26. https://doi:10.1002/2014GC005554

Morgavi, D., Petrelli, M., Vetere, F. P., González-García, D., & Perugini, D. (2015). High-temperature apparatus for chaotic mixing of natural silicate melts. *Review of Scientific Instruments, 86*(10), 105108. https://doi:10.1063/1.4932610

Mueller, S., Llewellin, E. W., & Mader, H. M. (2010). The rheology of suspensions of solid particles. *Proceedings of the Royal Society A, 466*, 1471–2946. https://doi:10.1098/rspa.2009.0445

Mysen, B., & Richet, P. (2005). *Silicate glasses and melts* (Vol. 10, Developments in Geochemistry, pp. 291–327). Amsterdam: Elsevier Science.

Ochs III, F. A., & Lange, R. A. (1999). The density of hydrous magmatic liquids. *Science, 283*, 1314–1317.

Pabst, W., Gregorova, E., & Berthold, C. (2006). Particle shape and suspension rheology of short-fiber systems. *Journal of the European Ceramics Society*, *26*, 149–160.

Petford, N. (2009). Which effective viscosity? *Mineralogical Magazine*, *73*(2), 167–191. https://doi:10.1180/minmag.2009.073.2.167

Petrelli, M., El Omari, K., Le Guer, Y., & Perugini, D. (2016). Effects of chaotic advection on the timescales of cooling and crystallization of magma bodies at mid crustal levels. *Geochemistry, Geophysics, Geosystems*, *17*(2), 425–441.

Pinkerton, H., & Stevenson, R. J. (1992). Methods of determining the rheological properties of magmas at subliquidus temperatures. *Journal of Volcanology and Geothermal Research*, *53*(1–4), 47–66. https://doi:10.1016/0377-0273(92)90073-M

Pupier, E., Duchene, S., & Toplis, M. J. (2008). Experimental quantification of plagioclase crystal size distribution during cooling of a basaltic liquid. *Contributions to Mineralogy and Petrology*, *155*(5), 555–570. https://doi:10.1007/s00410-007-0258-9

Romero, J. E., Morgavi, D., Arzilli, F., Daga, R., Caselli, A., Reckziegel, F., et al. (2016). Eruption dynamics of the 22–23 April 2015 Calbuco Volcano (Southern Chile): Analyses of tephra fall deposits. *Journal of Volcanology and Geothermal Research*, *317*, 15–29.

Roscoe, R. (1952). The viscosity of suspensions of rigid spheres. *Journal of Applied Physics*, *2*, 267–269.

Rossi, S., Petrelli, M., Morgavi, D., Gonzalez, D., Fischer, L., Vetere, F., & Perugini, D. (2017). Exponential decay of concentration variance during magma mixing: robustness of a volcanic chronometer and implication for the homogeneity in magmatic systems. *Lithos*, *286–287*, 396–407. https://doi:10.1016/j.lithos.2017.06.022

Ryerson, F., Weed, H., & Piwinskii, A. (1988). Rheology of subliquidus magmas: 1. *Picritic compositions. Journal of Geophysical Research: Solid Earth*, *93*,3421–3436

Sato, H. (2005). Viscosity measurement of subliquidus magmas: 1707 basalt of Fuji volcano. *Journal of Mineralogical and Petrological Science*, *100*, 133–142. https://doi:10.2465/jmps.100.133

Sehlke, A., & Whittington, A. G. (2015). Rheology of lava flows on Mercury: An analog experimental study. *Journal of Geophysiacal Research: Planets*, *120*, 1924–1955. https://doi:10.1002/2015JE004792.

Sehlke, A., Whittington, A., Robert, B., Harris, A., Gurioli, L., & Médard, E. (2014). Pahoehoe toaa transition of Hawaiian lavas: an experimental study. *Bulletin of Volcanology*, *76*(11), 876.

Shaw, H. R. (1969) Rheology of basalt in the melting range. *Journal of Petrology*, *10*, 510–535.

Shelby, J. E. (2005). *Introduction to glass science and technology* (2nd edn). *Padstow, UK*; Royal Society of Chemistry.

Soldati, A., Beem, J., Gomez, F., Huntley, J. W., Robertson, T., & Whittington, A. (2017). Emplacement dynamics and timescale of a Holocene flow from the Cima Volcanic Field (CA): Insights from rheology and morphology. *Journal of Volcanology and Geothermal Research*, *347*, 91–111.

Soldati, A., Sehlke, A., Chigna, G., & Whittington, A. G. (2016). Field and experimental constraints on the rheology of arc basaltic lavas: the January 2014 Eruption of Pacaya (Guatemala). *Bulletin of Volcanology*, *78*, 43.

Soule, S. A., & Cashman, K. V. (2005). Shear rate dependence of the pahoehoe to ‘a‘a transition: Analog experiments. *Geology*, *33*, 361–364.

Spina, L., Cimarelli, C., Scheu, B., Di Genova, D., & Dingwell, D. B. (2016). On the slow decompressive response of volatile- and crystal-bearing magmas: An analogue experimental investigation. *Earth and Planetary Science Letters*, *433*, 44–53.

Spina, L., Scheu, B., Cimarelli, C., Arciniega-Ceballos, A., & Dingwell, D. B. (2016). Time scales of foam stability in shallow conduits: Insights from analogue experiments. *Geochemistry, Geophysics, Geosystems*, *17*(10), 4179–4194.

Thies, M., & Deubener, J. (2002). Onset of non-Newtonian flow of foamed soda–lime–silica glasses. *European Journal of Glass Science and Technology Part A Glass Technology*, *43C*, 43–45.

Tsuchiyama, A. (1983). Crystallization kinetics in the system $CaMgSi_2O_6$–$CaAl_2Si_2O_8$: the delay in nucleation of diopside and anorthite. *American Mineralogist*, *68*, 687–698.

Vetere, F., Behrens, H., Holtz, F., Vilardo, G., & Ventura, G. (2010). Viscosity of crystal-bearing hydrous andesite and its implication for magma ascent. *Journal of Mineralogical and Petrological Sciences*, *105*(4), 151–163. https://doi:10.2465/jmps.090402

Vetere, F., Iezzi, G., Behrens, H., Holtz, F., Ventura, G., Misiti, V., et al. (2015). Glass forming ability and crystallisation behaviour of sub-alkaline silicate melts. *Earth-Science Reviews*, *150*, 25–44. https://doi:10.1016/j.earscirev.2015.07.001

Vetere, F., Murri, M., Alvaro, M., Domeneghetti, C.M., Rossi, S., Pisello, A., Perugini, D., & Holtz, F. (2019). Viscosity of pyroxenite melt and its evolution during cooling. *Journal of Geophysical Research: Planets*, *124*(5), 1451–1469.

Vetere, F., Rossi, S., Namur, O., Perugini, D., Morgavi, D., Misiti, V., et al. (2017). Experimental constraints on the rheology, eruption and emplacement dynamics of lavas from Mercury Northern Volcanic Plains. *Journal of Geophysical Research: Planets*, *122*, 1–17. https://doi:10.1002/2016JE005181

Vetere, F., Sato, H., Ishibashi, H., De Rosa, R., Donato, P., Ishebashi, H., De Rosa, R., & Donato, P. (2013). Viscosity changes during crystallization of a shoshonitic magma: new insights on lava flow emplacement. *Journal of Mineralogical and Petrological Sciences*, *108*, 144–160. https://doi:10.2465/jmps.120724

Vogel, H. (1921). The law of the relation between the viscosity of liquids and the temperature. *Physikalische Zeitschrift*, *22*, 645–646.

Vona, A., Romano, C., Dingwell, D. B., & Giordano, D. (2011) The rheology of crystal-bearing basaltic magmas from Stromboli and Etna. *Geochimica et Cosmochimica Acta*, *75*, 3214–3236. https://doi:10.1016/j.gca.2011.03.031

Vona, A., Romano, C., Giordano, D., & Russell, J. K. (2013). The multiphase rheology of magmas from Monte Nuovo (Campi Flegrei, Italy). *Chemical Geology*, *346*, 213–227. https://doi:10.1016/j.chemgeo.2012.10.005

Williams, D. A., Kerr, R. C., Lesher, C. M., & Barnes, S. J. (2001). Analytical/numerical modeling of komatiite lava emplacement and thermal erosion at Perseverance, Western Australia. *Journal of Volcanology and Geothermal Research*, *110*(1–2), 27–55.

8

Investigating the Crystallization Kinetics Via Time-Resolved Neutron Diffraction

Marco Zanatta[1], Caterina Petrillo[2,3], and Francesco Sacchetti[2]

ABSTRACT

We present the use of time-resolved neutron diffraction to investigate the crystallization kinetics at high temperature and high pressure with a time resolution of a few minutes. To highlight the potential of this approach, we present a study on the isothermal crystallization of the GeO_2 in the deep supercooled liquid region at 1100 K, well below the melting temperature T_m = 1388 K. The analysis of the diffraction patterns acquired over about 67 h shows a continuous reorganization of the amorphous structure towards the α-quartz phase. The relative fractions of crystal and amorphous material obtained from the experimental data provide a perfect test bench to develop and improve precise models for the crystallization kinetics in different thermodynamic conditions. In particular, we developed an empirical model based on a predator–prey-like mechanism between the crystal and the surrounding amorphous medium, where the density variation controls the process. The approach presented naturally can be extended towards more complex crystallizing fluids like volcanic magmas. These are silicate melts with crystals and volatiles and the interplay between phases is pivotal in understanding and predicting their rheological properties and thus macroscopic flow and eruptive behavior.

8.1. INTRODUCTION

The most common phase transition occurring inside and at the surface of the Earth is the transformation of silicate melts and magmas into solidified material. This process can give rise to two classes of products, namely crystals and glasses.

Crystals are the stable phases of a system below its melting temperature T_m and are characterized by long-range order in the atomic structure. Crystallization is not an inevitable result, as liquids can be supercooled to form a glass below the glass transition temperature T_g. Supercooled liquids are characterized by a dramatic increase in their viscosity and, below T_g, the structure becomes frozen on any experimentally accessible timescale in a disordered liquid-like arrangement (e.g., Cavagna, 2009; Denbenedetti & Stillinger, 2001). However, despite their long lifetimes and apparent stability even on geological timescales, supercooled liquids and glasses are intrinsically metastable states, thermodynamically unstable towards crystals. Crystallization in supercooled liquids and glasses can thus occur at any time, depending on the thermodynamics and characteristics of systems (Turnbull, 1969).

The classic physical picture for crystallization is based on two processes: *nucleation* and *growth*. During nucleation, crystalline nuclei are formed inside an amorphous system by spontaneous fluctuations (homogeneous nucleation) or assisted by the presence of impurities and boundaries (heterogeneous nucleation). Under certain conditions, these small crystallites can grow, leading to complete crystallization of the material. In nature, the interplay between crystallization and vitrification is clearly visible in volcanic rocks, where about 90 vol% is

[1] *Department of Informatics, University of Verona, Verona, Italy*

[2] *Department of Physics and Geology, University of Perugia, Perugia, Italy*

[3] *National Institute for Nuclear Physics (INFN), Perugia Section, Perugia, Italy*

Dynamic Magma Evolution, Geophysical Monograph 254, First Edition. Edited by Francesco Vetere.

composed of a mixture of crystals and glasses. For a system with a defined chemical composition x, the two major parameters controlling the timescale for this process are temperature T and pressure P.

Despite the very general interest in studying the kinetics of the crystallization process, understanding its fundamental features still represents an unresolved issue in many fields (e.g., Kelton & Greer, 2010). Theoretical studies typically approach the problem by investigating each single-phase separately from the others, e.g., the classical nucleation theory (CNT) for nucleation (e.g., Kelton, 1991; Oxtoby, 1988), or the overall process, like the Johnson–Mehl–Avrami–Kolmogorov model (JMAK; Avrami, 1939, 1940, 1941; Johnson & Mehl, 1938; Kolmogorov, 1937). However, whereas the general scheme for crystallization seems well accepted, even simple model systems show the limitations in the description of this phenomenon (e.g., Russo & Tanaka 2012; Sanz et al., 2014; Zanatta et al., 2017).

From an experimental point of view, one of the main problems is to quantitatively describe a complex process that can have relevant timescales spanning from nanoseconds (e.g., Greer, 2015) to geological times (e.g., Zhao et al., 2013). Moreover, in nature, intrinsically dynamic environments with a large variation of T–P conditions can add further degrees of complexity.

In this chapter we will focus on the study of isothermal crystallization, presenting the use of time-resolved neutron diffraction as a probe for its kinetics. To this purpose, we will analyse the case of the strong glass-former GeO_2, whose crystallization kinetics in deeply supercooled liquid has been investigated recently (Zanatta et al., 2017). The data obtained allowed us to carry out a quantitative analysis of the process. A simple empirical model to account for the observed behavior is also presented.

8.2. NEUTRON DIFFRACTION

Diffraction techniques using X-ray (XRD) and neutrons (ND) have been pivotal for studying the atomic arrangement of crystals, liquids, and glasses. A detailed description of XRD or ND is beyond the purpose of this chapter and can be found elsewhere (for XRD Ewald, 1962; Stout & Jensen, 1968; Warren, 1990; for ND Bacon, 1975; Furrer et al., 2009; Lovesey, 1984). Here we recount only the basic principles of ND and we provide a schematic comparison with XRD.

In a typical neutron diffraction experiment, a beam of neutrons with initial momentum $\boldsymbol{k}_i$ impinges on a sample. Then, neutrons are scattered with a final momentum $\boldsymbol{k}_f$ and revealed, without any energy discrimination, on a detector placed at a given scattering angle 2θ and covering a solid angle $\Delta\Omega$. The geometry of the scattering

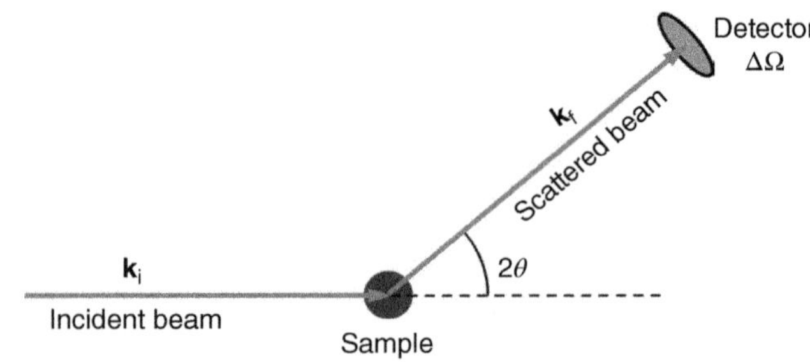

Figure 8.1 Schematic diagram of the scattering process.

process is schematically depicted in Figure 8.1. Assuming that $k_i \approx k_f$, i.e., the static approximation, the exchanged momentum $\boldsymbol{Q} = \boldsymbol{k}_i - \boldsymbol{k}_f$ is given by

$$Q = 2k_i \sin(\theta) \tag{8.1}$$

where $k_i = 2\pi/\lambda_i$, with λ_i the wavelength of the incident neutrons.

Neutrons interact with matter mainly by nuclear and magnetic interactions. Considering only nuclear scattering, the neutron–nuclei interaction can be described by a single parameter, the scattering length b, which is sensitive to the isotope and spin state.

The experimental observable in a diffraction experiment is the intensity I as a function of the scattering angle 2θ, which is proportional to the differential cross-section

$$\frac{d\sigma}{d\Omega} = \frac{d\sigma}{d\Omega}\Big|_{coh} + \frac{d\sigma}{d\Omega}\Big|_{inc}$$

The first term represents the coherent scattering and contains information about the structure of the sample. The second term accounts for incoherent scattering, which contributes to the diffracted intensity with a constant background.

For isotropic samples, like glasses and supercooled liquids, the coherent differential cross-section is proportional to the static structure factor $S(Q)$, which is the Fourier space-transform of the pair distribution function $g(r)$. For a monoatomic sample, $g(r)$ is proportional to the probability of finding an atom at a position r relative to a reference atom taken to be at the origin. Knowledge of the pair distribution function allows the atomic arrangement in liquids and glasses to be characterized, at least on the first neighbour's scale, i.e. within the "short-range order" (Elliott, 1991).

While XRD and ND basically produce the same quantity, each of the two probes has several peculiarities. A schematic summary is reported in Table 8.1 and shows how the purpose of our experiment should drive the choice of the most suitable technique. In a time-resolved study, we aim at recording a "movie" based on diffraction patterns acquired over a short time that catch the struc-

Table 8.1 Summary of the Main Properties of X-Rays and Neutrons as a Probe for Diffraction Experiments

Property	X-rays	Neutrons
Interaction	Electromagnetic interaction	Nuclear interaction (neutron–nuclei); magnetic interaction via neutron dipolar moment
Scattering amplitude	Proportional to Z and on $\sin\theta/\lambda$ and determined by calculation of limited accuracy (>2%)	Dependence on isotope and spin state, no dependence on Z and $\sin\theta/\lambda$
Scattering type	Only coherent scattering (incoherent contribution by Compton scattering)	Coherent and incoherent scattering
Absorption	High absorption. Dependence on λ^2 and Z^4 (far from resonances)	Generally weak absorption except for Li, He, B, Cd, In, Sm, Gd, Hf, Hg, actinides. Dependence on $1/\lambda$ (far from resonances) NB: many isotopes are activated even becoming radioactive for low absorption
Sources	X-ray tubes (weak flux) and synchrotrons sources (high flux)	Reactors and spallation sources, weak flux

tural evolution of the process. Consequently, the minimum acquisition time to obtain a meaningful pattern produces the time resolution Δt of the experiment. XRD generally offers high incident flux, in particular when synchrotron sources are exploited, thus allowing fast acquisitions and very fine time resolution. Conversely, neutrons have generally weak sources but their strength is a penetration length larger than that of X-rays, thus allowing a safe use of bulky samples with a complex and heavy sample environment. This property largely simplifies the experimental setup and, as long as a time resolution of the order of minutes is sufficient, neutrons provide the perfect probe for time-resolved studies at high T or high P. Finally, it is worth mentioning that the coherent and incoherent cross-sections can be very different for isotopes of the same atoms. A striking example is hydrogen (see Dianoux & Lander, 2003), where ^{1}H has a huge incoherent cross-section σ_{inc} with a small coherent cross-section σ_{coh} ($\sigma_{coh}(^1\mathrm{H})$ = 1.7583 b; $\sigma_{inc}(^1\mathrm{H})$ = 80.27 b; the cross-section is measured in barn b, 1 b = 10^{-24} cm^2) whereas ^{2}H (deuterium) has prevalently coherent cross-section ($\sigma_{coh}(^2\mathrm{H})$ = 5.592 b; $\sigma_{inc}(^2\mathrm{H})$ = 2.05 b). Based on this property, isotopic substitution can be used to highlight specific atoms, thus providing a key to understanding complex systems (e.g., Fischer et al., 2006).

8.2.1. Instruments and Data Analysis

Several neutron diffraction instruments were designed specifically for neutron diffraction studies in liquid and glasses. In particular, we mention D4 (Fischer et al., 2002) and D20 (Hansen et al., 2008) at the Institut Laue-Langevin (ILL; Grenoble, France), 7C2 (Cuello et al., 2016) at the Laboratoire Leon Brillouin (LLB; CEA Saclay, France). The typical layout of these instruments is shown in Figure 8.2. Both D4 and 7C2 are installed on hot sources, thus allowing the use of neutrons with a wavelength below 1 Å, ideally down to 0.5 Å. This increases the accessible Q range, thus reducing truncation errors while performing the Fourier transform to calculate the $g(r)$. Moreover, it minimizes corrections due to inelastic effects (see Bacon, 1975; Lovesey, 1984). Conversely, D20 is installed on a thermal source and originally designed as a powder diffractometer, so it can provide a higher Q resolution, which can ease the identification of multiple phases through their Bragg peaks.

In a neutron diffraction experiment, the intensity I collected by a detector positioned at a scattering angle 2θ and covering a solid angle $\Delta\Omega$ with an efficiency η can be written as

$$I = \phi N \frac{d\sigma}{d\Omega} \eta \Delta\Omega T_s, \tag{8.2}$$

where ϕ is the incident neutron flux, N is the number of scattering centres, $d\sigma/d\Omega$ the differential scattering cross-section, and T_s the transmission of the sample in the scattering direction. This quantity is the sum of all the possible scattering processes. In a typical experiment, the sample is contained in a sample holder and surrounded by a sample environment. Consequently, the measured signal also includes their contributions as well as multiple events involving only the sample or both the sample and the sample holder. The $S(Q)$ is proportional to the sample single scattering intensity (coherent), hence a careful experimental strategy and data analysis are mandatory. In addition to the measurement of the diffraction pattern from the sample, a set of ancillary measurements is needed. These include measurements of (a) empty sample holder, (b) absorber (e.g., cadmium) with the same dimensions as the sample holder, (c) an empty beam, and (d) an incoherent standard (vanadium) with the same dimensions as the sample. The absorber and the empty beam allow determination of the instrument background, i.e., neutrons that are revealed by the

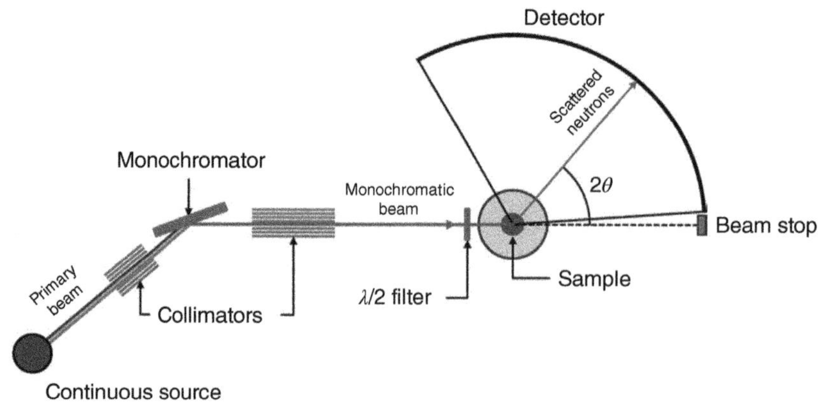

Figure 8.2 Schematic diagram of a typical diffractometer for liquids and amorphous systems installed at continuous neutron sources. A monochromatic beam is extracted from the primary white beam using a crystal monochromator at the appropriate Bragg angle. High-order contaminations are removed using a filter. The beam so obtained impinges on the sample and is scattered onto the detector, where neutrons are collected as a function of the scattering angle 2θ. No energy analysis is carried out by the detector.

detector without being scattered by the sample. The incoherent sample, typically vanadium, is needed to normalize data by removing beam and instrument-related effects. Indeed, vanadium is an almost fully incoherent sample, so it should have an almost flat diffraction pattern with only weak Bragg peaks. Dedicated measurements or calculations are needed to estimate the proper transmission coefficients. Finally, multiple scattering contributions must be estimated and subtracted. A detailed account for the analysis of ND data for the complex environment can be found in Petrillo and Sacchetti (1990, 1992).

8.3. A CASE STUDY: ISOTHERMAL CRYSTALLIZATION IN SUPERCOOLED GeO_2

In this section, we present an application of the above-mentioned method by studying the isothermal crystallization of GeO_2 in the deeply supercooled liquid, i.e., for $T << T_m$ (Zanatta et al., 2017).

GeO_2 is a covalent oxide glass and a prototype of the strong network forming systems, with a $T_g \approx 818$ K and $T_m = 1388$ K (Böhmer et al., 1993). From a structural point of view, a vitrous (v) GeO_2 glass network is based on $Ge(O_{1/2})_4$ tetrahedra bound together in a corner-sharing network, which also is preserved in the supercooled liquid (Micolaut et al., 2006). Crystalline GeO_2 has two stable polymorphs at room pressure and temperature: a rutile-like tetragonal structure ($P4_2/mnm$; Baur & Khan, 1971), and an α-quartz-like structure ($P3_221$) (Smith & Isaacs, 1964), which is the stable phase for $T \geq 1281$ K.

The sample was prepared by melt-quenching, starting from Aldrich germanium (IV) oxide crystalline powder (purity higher than 99.998%). The powder was molten in an alumina crucible at about 1900 K and then quenched in air. Cylindrical specimens with a diameter of 8 mm were core-drilled from the bulk glass and piled up to obtain a sample 50 mm high. The sample was loaded in a vanadium cylindrical cell with an outer diameter of 10 mm and walls 0.5 mm thick. The cell was closed with a steel screw cap, which was carefully shielded with boron nitride masks to minimize its scattering contribution. High-temperature measurements were carried out using a vanadium oven under vacuum ($\sim 10^{-6}$ mbar). The temperature was monitored by two thermocouples fixed on the body of the sample holder.

Starting from the glass, we approached the supercooled liquid by heating the system up to T_{exp} 1100 K, where GeO_2 is still in a substantially arrested state with dynamical and structural properties very similar to those of the glass (Caponi et al., 2009; Zanatta et al., 2011). The static structure factor $S(2\theta)$ of v-GeO_2 was first measured at room temperature to exclude any appreciable crystallization of the original glass. The temperature was then slowly raised up to 975 K, monitoring the structure to detect any trace of crystallization. Finally, the sample was quickly heated to $T = 1100$ K with a slope of 20 K min^{-1}, and then the 67 h long isothermal measurement was initiated. In order to properly trace the time evolution of the crystallization, we chose two different acquisition times. During the first period of 16 h, the acquisition time was set to $\Delta t = 5$ min. This is the minimum time to obtain good statistics, and short enough to provide a detailed view of the beginning of the process. Once the crystallized fraction was clearly visible, the acquisition time was lengthened to $\Delta t = 30$ min.

The time evolution of the $S(2\theta)$ during the isothermal measurement is reported in Figure 8.3a. With increasing time, the initially amorphous $S(2\theta)$ shows growth of a crystalline phase through the appearance of Bragg peaks, clearly visible in the low 2θ part of the diffraction pattern,

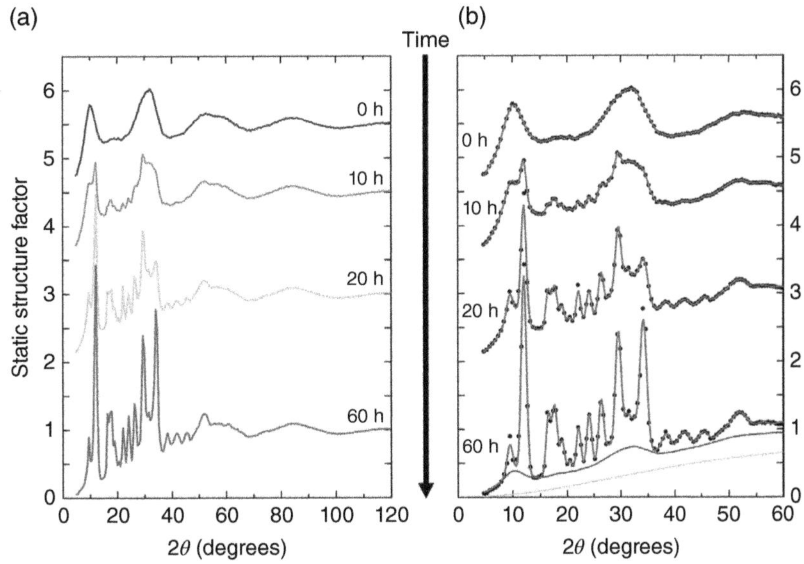

Figure 8.3 (a) Time evolution of the static structure factor of GeO_2 at T_{exp} = 1100 K. Time increases from top to bottom as shown by the black arrow. The static structure factors measured at different times are shifted along the vertical axis to appreciate the emergence of the crystalline features. Each $S(2\theta)$ is normalized to unity at high scattering angle. (b) Best fit of the static structure factors shown in (a) using the procedure outlined in the text. Black dots represent experimental data whereas the red line is the best fit according to equation (8.1). The different components of the fit are shown for the most crystalized sample (t = 60 h). The solid green line is $S_T(2\theta)$, and the solid blue line shows the sum of the amorphous and TDS components. [Data from Zanatta et al. (2017).]

below about 60°. The peak intensities increase and eventually saturate, but the pattern of the peaks remains the same, without intermediate phases. At high 2θ the Debye–Waller factor reduces the intensity of the Bragg peaks, and the $S(2\theta)$ appears substantially unchanged, still keeping glassy-like smooth features.

The position of the Bragg peaks visible in Figure 8.3 is compatible with that of the α-quartz GeO_2 structure, and no traces of rutile-like structure are visible. This also implies that the crystallization process preserves the chemical composition without any appreciable phase separation.

8.3.1. Data Analysis and Determination of Crystalline and Amorphous Fractions

Assuming that no contributions arise from the crystalline–amorphous interfaces, we can write the measured $S(2\theta)$ as the sum of two terms, accounting for the amorphous and the crystalline fraction, namely

$$S(2\theta) = S_A(2\theta) + S_C(2\theta) \tag{8.3}$$

The amorphous term $S_A(2\theta)$ can be written as

$$S_A(2\theta) = A_a S_g(2\theta) \tag{8.4}$$

where A_a is a parameter and S_g is the static structure factor of the fully amorphous system, obtained by considering the first scans at 1100 K, where no trace of crystallization is visible. Since the measurement of the static structure factor implies an integral over the energy, the crystalline term $S_C(2\theta)$ can be written as the sum of an elastic contribution $S_B(2\theta)$, accounting for the Bragg peaks, and an inelastic part identified as the thermal diffuse scattering (TDS). Consequently,

$$S_C(2\theta) = S_B(2\theta) + A_c S_T(2\theta) \tag{8.5}$$

where A_c is a parameter. Following Warren (1990), we use a very simple approximation for the TDS contribution $S_C(2\theta)$, i.e.,

$$S_T(2\theta) = 1 - e^{-2W} \tag{8.6}$$

where $\exp(-2W) = \exp[-2B(\sin\theta/2)^2]$ is the Debye–Waller factor. The parameter B for v-GeO_2 at 1100 K was calculated in harmonic approximation using the vibrational density of states from Fabiani et al. (2008). Finally, $S_B(2\theta)$ is modeled describing each Bragg peak with a Gaussian, whose position $2\theta_i$ is given by the α-quartz structure using the appropriate lattice parameters a and c. A preliminary analysis of the Bragg diffraction pattern did not show any appreciable time-evolution of the peak width, so we assumed that the peak full width at half maximum (FWHM) is given by the instrument resolution that can be measured using a polycrystalline sample (e.g., Ni or KBr). The Bragg contribution turns out to be:

$$S_B(2\theta) = \sum_i \frac{A_i}{\sigma_i\sqrt{2\pi}} \exp\left[-\frac{1}{2}\left(\frac{2\theta - 2\theta_i}{\sigma_i}\right)^2\right] \quad (8.7)$$

where A_i is the integrated intensity of the ith reflection, which is fitted independently for each peak, and $2\sigma_i\sqrt{2\log 2}$ the FWHM with σ_i derived from the resolution (Caglioti et al., 1958).

The lattice parameters a and c for the α-quartz type GeO_2 at T_{exp} were determined by fitting the most crystallized data with equation (8.3). This leads to $a(T_{exp})$ = 5.053 ± 0.002 Å and $c(T_{exp})$ = 5.66 ± 0.04 Å, which were then fixed to fit the whole time evolution. Figure 8.3b shows the fit to data obtained using equation (8.3) during the crystallization process.

To extract the fraction of atoms in the crystalline and amorphous phase we can resort to the coefficients A_c and A_a of equations (8.4) and (8.5). As a matter of fact, the scattering intensity at high scattering angle is proportional to the number of atoms, and the properly normalized $S(2\theta)$ is equal to 1. In this limit, Bragg peaks are suppressed by the Debye–Waller factor and smeared out by the instrument resolution, whereas the TDS and the amorphous static structure factor go to unity. Equation 8.3 reduces to $S(2\theta) = A_a + A_c \approx 1$, hence A_c represents the fraction of atoms in the crystalline phase, whereas A_a represents that in the amorphous one. The time evolution of these quantities is reported in Figure 8.4 and provides an insight into the kinetics of the crystallization process, as well as into the corresponding decrease of the amorphous matrix. In particular, a qualitative analysis of their shapes confirms that crystallization becomes appreciable after 4 h, and then it rapidly develops by subtracting material from the amorphous phase. After about 30 h, the crystallization rate slows down, leading to a final material where 77% of the atoms is organized in the α-quartz structure, whereas the remaining 23% still shows amorphous features.

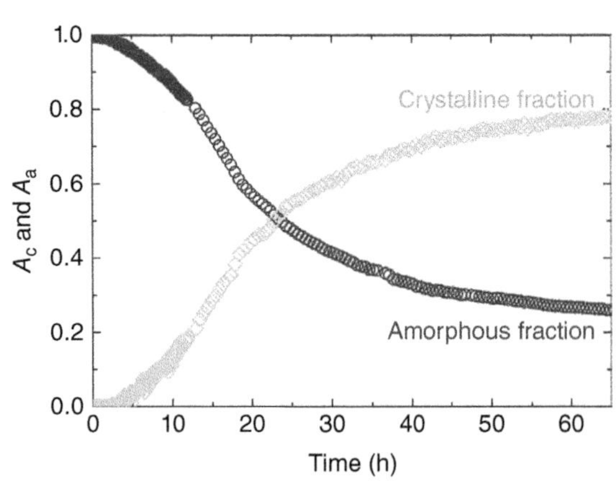

Figure 8.4 Time evolution of the crystalline and amorphous fractions A_c and A_a, green open diamonds and blue open circles, respectively. [Data from Zanatta et al. (2017).]

8.4. POSSIBLE FRAMEWORKS FOR DATA INTERPRETATION

The above-mentioned qualitative analysis of time-resolved $S(2\theta)$ can already provide many insights into the process. However, the precise determination of A_c and A_a suggests that a more quantitative approach to determine the crystallization kinetics is indeed possible. In this section we will present the results obtained with the standard analysis framework (the JMAK approach) as well as those by applying a predator–prey model, specifically developed for isothermal crystallization of highly viscous supercooled liquids (Zanatta et al., 2017).

8.4.1. Johnson–Mehl–Avrami–Kolmogorov Model

As mentioned above, the standard framework to describe the time evolution of the fraction of transformed material during isothermal crystallization is the JMAK model (Avrami, 1939, 1940, 1941; Johnson & Mehl 1939; Kolmogorov, 1937). This approach is based on nucleation and growth processes, and assumes that the nucleation occurs randomly with a large number of spherical transforming regions. Growth is the same for all these regions and stops at points of impingement, continuing elsewhere. According to the JMAK model, the time evolution of the fraction of transformed volume X can be written as:

$$X = 1 - e^{-kt\hat{}n} \quad (8.8)$$

where k is an effective rate constant, depending on the nucleation and growth rates. The exponent n is termed the Avrami exponent. It is expected to assume integer or half an integer values and depends on the characteristics of the process. As an example, for continuous nucleation and three-dimensional spherical growth, the Avrami exponent is n = 4 (Weinberg et al., 1997). However, equation (8.8) implies a complete crystallization of the amorphous medium, which appears in contrast with the observed time evolution of the crystallized fraction A_c. Figure 8.5 shows a fit of A_c with equation (8.8) using k and n as free parameters. The result is represented with a red dashed line and it reproduces neither the shape nor the long-time behavior. Conversely, considering only the first part of the process, e.g., the first 20 h, the JMAK provides a very good description of the experimental data, reproducing their behavior. A possible explanation is that, at the beginning of the crystallization process, the hypotheses of random nucleation and similar growth of spherical regions hold.

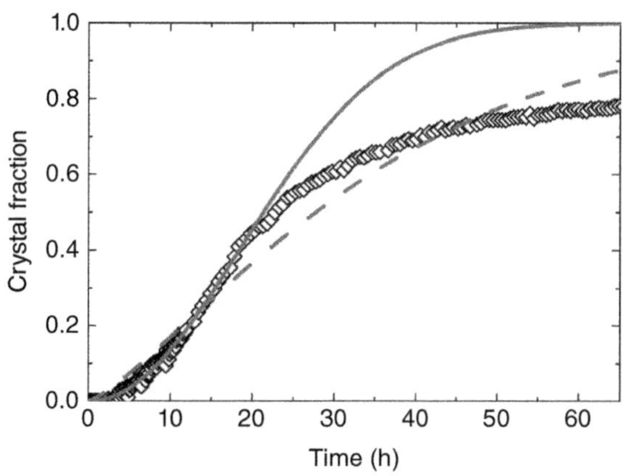

Figure 8.5 Time evolution of the crystalline fraction A_c (black open diamonds). The dashed red line is the fit with the JMAK equation (equation (8.8) using the whole time window. Conversely, the solid red line is the fit considering only the first 20 h of the process. [Data are from Zanatta et al. (2017).]

However, increasing the transformed fraction in a nondiffusive environment (i.e., $T << T_m$) introduces a slowing down mechanism that influences both the nucleation and the growth, thus producing the observed incomplete crystallization and the failure of the JMAK model.

8.4.2. A Predator–Prey Model

The qualitative analysis and the results shown in Figure 8.4 point towards a model with a *negative feedback* that slows down and eventually stops the crystallization. On this basis, we developed a predator–prey model for the crystallization at a temperature $T << T_m$, i.e., where diffusion can be safely neglected (Zanatta et al., 2017). The crystal growth is governed mainly by its surface, where structural rearrangements occur.

As a first guess, we assume that the observed process develops through a nucleation and growth mechanism. The creation of a nucleus is assumed to be a stochastic process that can be triggered by thermal fluctuations in the material and probably eased by even intrinsic heterogeneities (Russo & Tanaka, 2012; Tanaka, 2012). Conversely, since diffusion is practically arrested, the growth of each nucleus proceeds only through structural rearrangements involving the interface between the ordered and disordered regions in a self-limiting process. Indeed, each crystalline domain nucleates and grows at the expense of the surrounding amorphous region. However, since the crystal has a higher density than the supercooled liquid, this mechanism creates high-density fully ordered regions that become surrounded by depleted interfaces. In the absence of diffusion, this process slows down and stops the growth of the crystalline nuclei. A similar mechanism applies also to nucleation, which becomes less probable in overcrowded and depleted environments.

To be quantitative, we consider a system with N_{tot} atoms and $N_{tot} = N_c(t) + N_a(t)$, where $N_c(t)$ and $N_a(t)$ are the number of atoms in the crystalline or amorphous phase at a given time t, respectively. This relation holds at any time. Considering only the crystal, data suggest the presence of many different crystalline domains evolving in time. According to this, N_c can be written as

$$N_c(t) = \sum_{i=0}^{N(t)} n_i(t - t_i) \tag{8.9}$$

where $N(t)$ is the number of crystalline nuclei at a given time t, whereas n_i indicates the number of atoms in the ith domain created at t_i. Considering that the process is almost continuous and both $N(t) >> 1$ and $n_i(t) >> 1$, we can write the sum of equation (8.9) as a time integral, i.e.

$$N_c(t) = \int_0^t \frac{dN(t')}{dt'} n(t - t')dt' \tag{8.10}$$

However, in order to solve equation (8.10), we need to write the relations for the time evolution of both N and n.

According to the previous considerations, the equation for the evolution of the number of nuclei $N(t)$ is assumed to be related to a constant rate and can be written as:

$$\frac{dN}{dt} = \frac{1}{\tau_n}(N_m - N) \tag{8.11}$$

where $1/\tau_n$ is the rate of the nucleation process and N_m is the maximum number of nuclei.

For the n, we can assume that growth occurs on the surface with a rate α up to n_m, the maximum number of atoms in each crystalline domain. For the sake of simplicity, n_m is assumed to be constant throughout the sample and independent of time. The time evolution of n can be thus written as

$$\frac{dn}{dt} = \alpha n^{\mu}(n_m - n) \tag{8.12}$$

The exponent μ accounts for the fraction of atoms involved in the process, hence in general $0 \le \mu < 1$. Actually, μ conveys information on the geometry of the nucleus and on the dimensionality d of the process. In the case of spherical nuclei, $d = 3$ and μ turns out to be 2/3.

By changing the variables to $n^* = n/n_m$ and $t^* = t/\tau_d$, with $\tau_d = (\alpha n_m^{\mu})^{-1}$, equation (8.12) can be rewritten in the universal form

$$\frac{dn^*}{dt^*} = n^{*\mu}(1 - n^*), \tag{8.13}$$

which can be integrated numerically.

Furthermore, due to the previous thermal history of the sample, we have to consider the possibility that

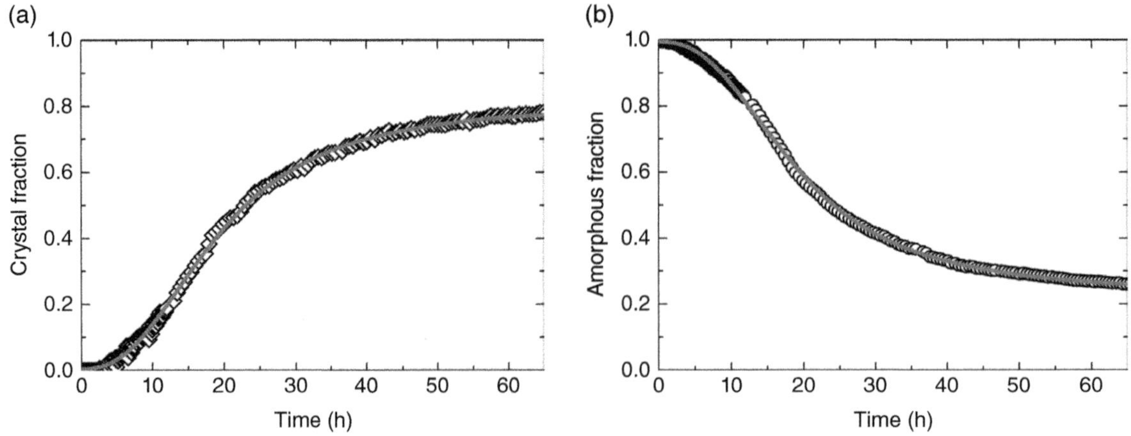

Figure 8.6 (a) Best fit of the time evolution of the crystalline fraction A_c (black open diamonds) using the predator–prey model (red line). (b) Corresponding best fit (red line) of the amorphous fraction (black open circles). Please note that the fit is obtained using equation (8.13) and the condition $N_{tot} = N_c(t) + N_a(t)$ considering both the datasets within an overall optimization procedure. [Data are from Zanatta et al. (2017).]

crystallization has already started at $t = 0$. This can be easily incorporated into the model by integrating equation (8.10) with the condition $N(t = 0) = N_0$. Within these assumptions, equation (8.10) can be rewritten as

$$N_c(t) = N_0 n_0 + \frac{N_m - N_0}{\tau_n} \int_0^t \exp\left(\frac{t'}{\tau_n}\right) n_m \frac{n^*(t - t' + t_0)}{\tau_d} dt', \tag{8.14}$$

where $N_c^0 \equiv N_0 n_0$ is the initial number of crystallized atoms, being $n(t = 0) = n_0$.

Once divided by N_{tot}, equation (8.14) allows the fit of the crystalline fraction A_c. Moreover, thanks to $N_{tot} = N_a(t) + N_c(t)$, the amorphous fraction A_a also can be obtained. The model has thus five fitting parameters: the initial and the final fraction of atoms in the crystalline phase, $A_c^{(i)}$ and $A_c^{(f)}$, the timescales of nucleation and growth, τ_n and τ_d, and the fictive starting time t_0. The parameters can be fitted to A_c and A_a using an overall procedure. The result is reported in Figure 8.6. The model provides an accurate and coherent description of both the crystal growth and the amorphous reduction. At the beginning of the process, the system is almost fully amorphous and the crystalline fraction is about $A_c^{(i)} = 0.005 \pm 0.001$, while $A_c^{(f)} = 0.77 \pm 0.01$. However, the presence of a nonzero t_0 indicates that the crystallization process is already active, since it starts with a nonzero derivative. The resulting timescales for the nucleation and growth are $\tau_n = 16.9 \pm 0.2$ h and $\tau_d = 31.3 \pm 0.6$ h.

8.5. CONCLUSIONS

We have shown the potential of time-resolved neutron diffraction as a tool for the study of crystallization kinetics in glasses and supercooled liquids. A possible data analysis procedure has been thoroughly described in the case of the isothermal crystallization of the deeply supercooled v-GeO_2. Although this case study appears very simple, especially with respect to the volcanic context, it highlights the wealth of information that an accurate analysis of ND data can produce. It is worth noting that the model can be extended to a more complex environment. Indeed, in the case of multiple crystallizing phases, the analysis of the Bragg peaks can provide an unambiguous identification of the structures with proper lattice parameters and an idea of the crystallization timescales for each phase. Finally, the isotopic sensitivity of neutrons can be exploited to highlight a single species with respect to the surrounding medium, thus easing data interpretation for complex systems such as volcanic magmas.

Neutron-based techniques are particularly suitable to measure bulk samples, controlling environmental parameters such as T and P. This can be pivotal for geological studies aiming at measurements in situ during time-dependent processes such as crystallization in magmas. In addition, a recent development in neutron scattering instrumentation concerns the coupling with other techniques. This is presently carried out or is under implementation using a variety of beamlines, such as X-ray, and light-based techniques (Raman or IR), as well as calorimetry and dielectric conductivity.

Finally, we presented an empirical predator–prey model for the interpretation of the crystallization kinetics in supercooled liquids with negligible diffusion. The model provides a remarkably good fit to experimental data identifying the density variation as the negative feedback mechanism that slows down and stops the process. The model can be further extended to also account for the effects of diffusion. As a matter of fact, even if its contribution seems negligible in the observed timescale, diffusion could give rise to long-time contributions that are expected to become dominant as T_m is approached.

ACKNOWLEDGMENTS

The authors are grateful to L. Cormier, L. Hennet, J. Darpentigny, and J. Dupont for their help during the experiment, as well as to R. Dal Maschio for the sample preparation. We thank also F. P. Vetere, D. Morgavi, M. Petrelli and D. Perugini for the stimulating discussions on the role of crystallization in volcanic processes. Finally, the Laboratoire Leon Brillouin is gratefully acknowledged for the beamtime granted at the 7C2 diffractometer.

REFERENCES

Avrami, M. (1939). Kinetics of phase change. I General theory. *Journal of Chemistry and Physics*, *7*, 1103.

Avrami, M. (1940). Kinetics of phase change. II Transformation–time relations for random distribution of nuclei. *Journal of Chemistry and Physics*, *8*, 212.

Avrami, M. (1941). Granulation, phase change, and microstructure kinetics of phase change. III. *Journal of Chemistry and Physics*, 9, 177.

Bacon, G. E. (1975). *Neutron diffraction* (3rd edn). Oxford, UK: Clarendon.

Baur, W. H., & Khan, A. A. (1971). Rutile-type compounds. IV. SiO_2, GeO_2 and a comparison with other rutile-type structures. *Acta Crystallographica B*, *27*, 2133–2139.

Böhmer, R., Ngai, K. L., Angell, C. A., & Plazek, J. (1993). Nonexponential relaxations in strong and fragile glass formers. *Journal of Chemistry and Physics*, *99*, 4201–4209.

Caglioti, G., Paoletti, A., & Ricci, F. P. (1958). Choice of collimators for a crystal spectrometer for neutron diffraction. *Nuclear Instruments and Methods*, *3*, 223–228.

Caponi, S., Zanatta, M., Fontana, A., Bove, L. E., Orsingher, L., Natali, F., Petrillo, C., & Sacchetti, F. (2009). Ergodicity breaking in strong and network-forming glassy systems. *Physical Reviews B*, *79*, 172201.

Cavagna, A. (2009). Supercooled liquids for pedestrians. *Physics Reports*, *476*, 51–124.

Cuello, G. J., Darpentigny, J., Hennet, L., Cormier, L., Dupont, J., Homatter, B., & Beuneu, B. (2016). 7C2, the new neutron diffractometer for liquids and disordered materials at LLB. *Journal of Physics: Conference Series*, *746*, 012020.

Debenedetti, P. G., & Stillinger F. H. (2001). Supercooled liquids and the glass transition. *Nature*, 410, 259–267.

Dianoux, J. & Lander, G. (Eds.) (2003). *Neutron data booklet*. Philadelphia, PA: Old City Publishing.

Elliott, S. R. (1991). Medium-range structural order in covalent amorphous solids. *Nature*, *354*, 445–452.

Ewald, P. P. (Ed.) (1962). *Fifty years of X-ray diffraction*. Boston, MA: Springer.

Fabiani, E., Fontana, A., & Buchenau, U. (2008). Neutron scattering study of the vibrations in vitreous silica and germania. *Journal of Chemistry and Physics*, *128*, 244507.

Fischer, H. E., Barnes A. C., & Salmon, P. S. (2006). Neutron and X-ray diffraction studies of liquids and glasses. *Reports on Progress in Physics*, *69* , 233–299

Fischer, H. E., Cuello, G. J., Palleau, P., Feltin, D., Barnes, A. C., Badyal, Y. S., & Simonson, J. M. (2002). D4c: A very high precision diffractometer for disordered materials. *Applied Physics A*, *74*, S160–S162.

Furrer, A., Mesot, J., & Strassle, T. (2009). *Neutron scattering in condensed matter physics*. World Scientific.

Greer, A. L. (2015). New horizons for glass formation and stability. *Natural Materials*, *14*, 542–546.

Hansen, T. C., Henry, P. F., Fischer, H. E., Torregrossa, J., & Convert, P. (2008). The D20 instrument at the ILL: a versatile high-intensity two-axis neutron diffractometer. *Measurement Science and Technology*, *19*, 034001.

Johnson, W. A., & Mehl, R. (1939). Reaction kinetics in processes of nucleation and growth. *Transactions of the American Institute of Mechanical Engineers*, *135*, 416.

Kelton, K. F. (1991) Crystal nucleation in liquids and glasses. *Solid State Physics*, *45*, 75.

Kelton, K. F., & Greer, A. L. (2010). *Nucleation in condensed matter: Applications in materials and biology* (1st edn). Oxford, UK: Pergamon.

Kolmogorov, A. N. (1937). On the statistical theory of the crystallization of metals. *Izvestiya Akademii Nauk SSR, Seriya Fizicheskaya*, *3*, 355.

Lovesey, S. W. (1984). *Theory of thermal neutron scattering from condensed matter* (2nd edn, Vol. 1). Oxford, UK: Clarendon Press.

Micoulaut, M., Cormier, L., & Henderson, G. S. (2006). The structure of amorphous, crystalline and liquid GeO_2. *Journal of Physics: Condensed Matter*, *18*, R753–R784.

Oxtoby, D. W. (1988). Nucleation of crystals from the melt. *Advances in Chemistry and Physics*, *70*, 263.

Petrillo, C., & Sacchetti, F. (1990). Analysis of neutron diffraction data in the case of high-scattering cells. *Acta Crystallographica Section A: Foundations and Advances*, *46*, 440.

Petrillo, C., & Sacchetti, F. (1992). Analysis of neutron diffraction data in the case of high-scattering cells. II. Complex cylindrical cells. *Acta Crystallographica Section A: Foundations and Advances*, *48*, 508.

Russo, J., & Tanaka, H. (2012). The microscopic pathway to crystallization in supercooled liquids. *Nature: Scientific Reports*. *2*, 505.

Sanz, E., Valeriani, C., Zaccarelli, E., Poon, W. C. K., Cates, M. E., & Pusey, P. N. (2014). Avalanches mediate crystallization in a hard-sphere glass. *Proceedings of the National Academy of Sciences of the United States of America*, *111*, 75–80.

Smith, G. S. & Isaacs, P. B. (1964). The crystal structure of quartz-like GeO_2. *Acta Crystallographica*, 17 842–6.

Stout, G. H. & Jensen, L. H. (1968). *X-ray structure determination: a practical guide*. New York: Macmillan.

Tanaka, H. (2012). Bond orientational order in liquids: Towards a unified description of water-like anomalies, liquid-liquid transition, glass transition, and crystallization. *European Physical Journal E*, *35*, 113.

Turnbull, D. (1969). Under what conditions can a glass be formed? *Contemporary Physics*, *10*, 473–488.

Warren, B. E. (1990). *X-Ray diffraction* (2nd edn). New York, NY: Dover Publications.

Weinberg, M. C., Birnie D. P., & Shneidman V. A. (1997). Crystallization kinetics and the JMAK equation. *Journal of Non-Crystalline Solids*, *219*, 89–99.

Zanatta, M., Baldi, G., Caponi, S., Fontana, A., Petrillo, C., Rossi, F., & Sacchetti, F. (2011). Debye to non-Debye scaling of the Boson peak dynamics: Critical behaviour and local disorder in vitreous germania. *Journal of Chemistry and Physics*, *135*, 174506.

Zanatta, M., Cormier, L., Hennet, L., Petrillo, C., & Sacchetti, F. (2017). Real-time observation of the isothermal crystallization kinetics in a deeply supercooled liquid. *Nature: Scientific Reports*, *7*, 43671.

Zhao, J., Simon, S. L. & McKenna, G. B. (2013). Using 20-million-year-old amber to test the super-Arrhenius behaviour of glass-forming systems. *Nature Communications*, *4*, 1783.

9

Axial Melt-Lens Dynamics at Fast Spreading Midocean Ridges

Jürgen Koepke[1] and Chao Zhang[2]

ABSTRACT

Multichannel seismic studies performed at fast spreading midocean ridges revealed the presence of a thin (tens to hundreds of meters high), narrow (< 1–2 km wide) axial melt lens (AML) in the midcrust, which is underlain by crystal–melt mush that is in turn laterally surrounded by a transition zone of mostly solidified material. In this chapter we shed light on the complexity of magmatic and metamorphic processes ongoing within and at the roof of axial melt lenses, with a focus on the petrological and geochemical record provided by fossilized AMLs. Of particular significance is Hole 1256D in the equatorial Pacific drilled by the International Ocean Discovery Program, where for the first time the transition between sheeted dikes and gabbros in intact fast spreading crust was penetrated, providing a drill core with a more or less continuous record of the upper part of an AML. This can be regarded as a Rosetta Stone to answer long-standing questions on the complex magmatic evolution within an AML, as well as on metamorphic and anatectic processes ongoing at the roof of a dynamic AML, rising upward in the midcrust as a consequence of a replenishment event.

9.1. INTRODUCTION

The construction and evolution of oceanic lithosphere at midocean ridge systems is one of the dominant magmatic processes in the chemical differentiation and physical evolution of our planet. The approximately 60,000 km long mountain chain represented by midocean ridges is associated with 90% of Earth's volcanism and ~25% of the heat released from deep Earth. In contrast to the oceanic crust formed at slow spreading ridges, crust created at fast and intermediate spreading ridges shows a relatively uniform seismic stratigraphy and is usually regarded as layered and relatively homogeneous (e.g., Canales et al., 2003). Here, multichannel seismic studies revealed the presence of a thin (tens to hundreds of meters high), narrow (< 1–2 km wide) magma lens of partial to pure melt in the midcrust—the axial melt lens (AML)—which is underlain by crystal–melt mush that is in turn laterally surrounded by a transition zone of mostly solidified material grading into a zone completely frozen, which correspond to gabbros (e.g., Detrick et al., 1987; Sinton & Detrick, 1992; Vera et al., 1990) (Figure 9.1).

The role of the AML during magmatic accretion of the crust is still not clear: either, the AML is the source for the complete lower, gabbroic crust, formed by a suspension of crystals formed in the AML (the "gabbro glacier model"; e.g., Henstock et al., 1993; Phipps Morgan & Chen, 1993; Quick & Denlinger, 1993), or the AML is the source for a distinct portion of the lower crust, while other, significant parts, originate from in situ crystallization in the deep crust (the "sheeted sill model"; e.g., Bedard et al., 1988; Kelemen et al., 1997). These are both end-member models. However, most scientists believe that the magmatic processes responsible for crustal accretion at fast spreading ridges is too complex

[1] *Institute of Mineralogy Leibniz University Hannover, Hannover, Germany*

[2] *Institute of Mineralogy Leibniz University Hannover, Hannover, Germany and State Key Laboratory of Continental Dynamics, Northwest University, Xi'an, China*

Dynamic Magma Evolution, Geophysical Monograph 254, First Edition. Edited by Francesco Vetere.

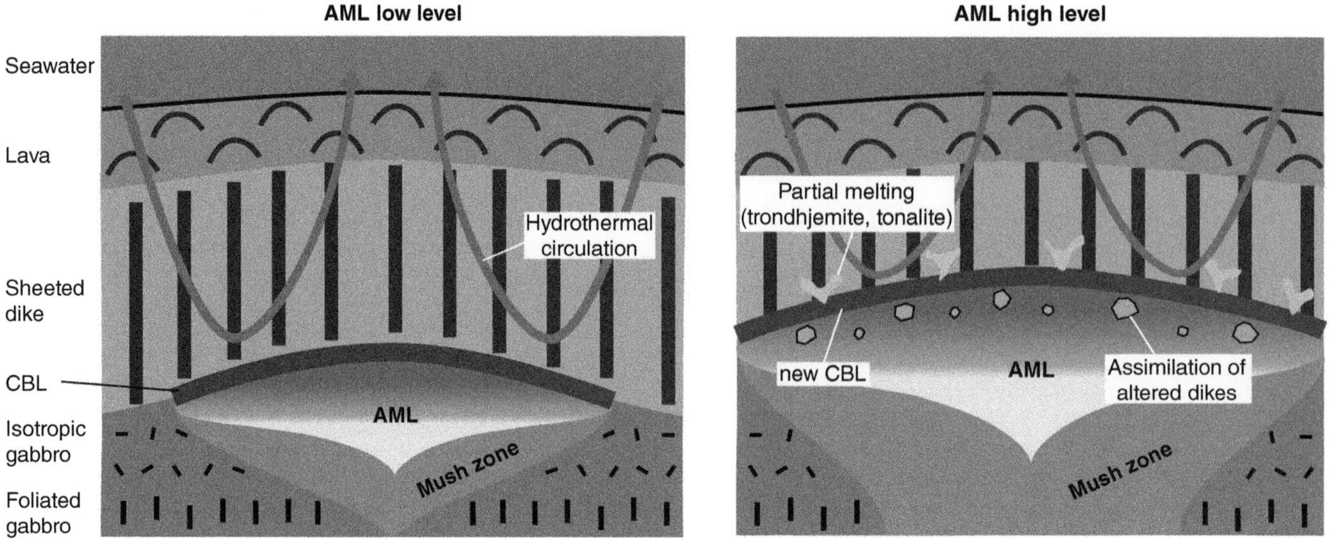

Figure 9.1 Schematic model depicting the axial melt-lens (AML) system present at the dike–gabbro transition of fast spreading systems at two different levels. (Left) The AML is in a low position. A deep conductive boundary layer (CBL) separates the 1200°C hot melt of the AML and the sheeted dikes above, which are hydrothermally cooled by seawater-derived fluids at maximum temperatures of 400–500°. (Right) After a replenishment event, the AML moves into a higher position, burning through the previously altered sheeted dikes. At the roof of the AML, a new CBL is established, under conditions of the two-pyroxene hornfels facies. Locally temperatures are so high that the solidus of altered dikes is approached, resulting in the formation of anatectic melts of trondhjemitic to tonalitic compositions. Parts of the older sheeted dike system are assimilated by the upward moving AML melt.

to be explained with such simple models, and that the truth probably lies somewhere in between these end members (e.g., Boudier et al., 1996; Natland & Dick, 2009). Such a hybrid model is also supported from the estimation of crystallization depths of MORB within the oceanic crust via the measurement of volatile concentrations in melt inclusions, revealing that <25% of the melt inclusions show crystallization pressures consistent with formation in the lower oceanic crust (Wanless & Shaw, 2012). A hybrid model is also indicated by recent multichannel seismic studies, verifying the presence of deep melt sills under recent midoceanic ridges from the Juan de Fuca Ridge (Canales et al., 2009; Nedimovic et al., 2005) and from the East Pacific Rise (EPR; Marjanovic et al., 2014).

The detection of the AML by seismic methods has raised the question of how it is related to the overlying sheeted dyke complex. Theoretical models predict that heat is exchanged across a thin, impermeable conductive boundary layer (CBL), which is sandwiched between the AML and the sheeted dyke complex (Figure 9.1; e.g., Gillis, 2008; Lowell & Burnell, 1991). The CBL separates two convective systems: the AML filled with a basaltic melt at ~1200°C below, and the seawater-derived hydrothermal circulation systems operating at maximum temperatures of 400–500°C above. Seismic, structural, petrological, and geochemical studies from fast spreading ridges imply that the AML and the overlying CBL are transient phenomena, moving up and down, probably as a consequence of the magmatic activity beneath the spreading segments (e.g., Coogan et al., 2003; France et al., 2009; Gillis & Coogan, 2002; Hooft et al., 1997; Koepke et al., 2008; Lagabrielle & Cormier, 1999; Zhang et al., 2014). During a phase of magmatic quiescence, the AML is in a relative deeply position. At that stage, a CBL several decameters thick may develop overlying the AML, separating the AML and the sheeted dike complex, which is cooled by circulating seawater-derived fluids. During a subsequent stage of magmatic activity, probably related to a replenishment process, the AML migrates upward into a higher position, developing a new CBL within the previously hydrothermally altered dikes. The heat of the upward moving magma is responsible for contact metamorphism at the roof of the AML, resulting in the formation of granoblastic hornfelses, and culminating in anatectic processes expressed by trondhjemitic and tonalitic veins crosscutting the hornfelses (e.g., Alt et al., 2010; Erdmann et al., 2015; France et al., 2014; Gillis, 2002; Gillis & Roberts, 1999; Koepke et al., 2008; Zhang, Koepke, France, et al., 2017).

The scope of this chapter is to shed light on the complexity of the magmatic processes ongoing in an AML, with emphasis on the petrological and geochemical record provided by fossilized AMLs located in the recent crust of the Pacific or in ophiolites. First we present corresponding results and conclusions from those few locations on our planet where frozen AMLs has been discovered and investigated. We then present new petrographic and analytical data of the contact between sheeted dikes and gabbro recovered during the International Ocean Discovery Program (IODP) legacy drilling campaign "Superfast Spreading Crust" at IODP Site 1256 in the equatorial Pacific, where for the first time the dike–gabbro transition (DGT) of intact oceanic crust was penetrated. Based on natural and experimental data, and supported by petrological modeling, we then present an overview of the assumed differentiation processes within the AML, considering both the main- (cocrystallization of olivine, plagioclase, and clinopyroxene) and late-stage crystallization (late precipitation of FeTi-oxides, orthopyroxene, apatite, and amphibole). Subsequent sections deal with all aspects on felsic melt generation related to the AML, including the anatectic processes at the roof of an AML, and with the subsolidus stage, where the distinction between magmatic and metamorphic processes within the AML is blurred. Finally, we summarize the constraints on timescales for the vertical fluctuations of the AML under fast spreading ridges.

9.2. FOSSILIZED AML FROM FAST SPREADING MIDOCEAN RIDGES: INSIGHTS INTO MAGMATIC AND HYDROTHERMAL PROCESSES

Frozen AML horizons from fast spreading ridges enabling the application of well-proven structural, petrological, and geochemical tools are best suited to shed light on the complexity of those magmatic and metamorphic processes ongoing in axial magma chambers. Unfortunately, there are only a few localities on our planet where fossilized AMLs within fast spreading ocean crust have been described: IODP Hole 1256D and Hess Deep, representing EPR crust from the equatorial eastern Pacific, and ophiolites, especially from the Oman.

Of particular significance is IODP Hole 1256D in the equatorial Pacific, where the transition between sheeted dikes and gabbros in intact fast spreading crust was penetrated, providing a drill core with a more or less continuous record of such a frozen AML horizon. Here, a fossil CBL was penetrated by drilling, and subsequent research revealed that this horizon consists of metamorphosed, granoblastic hornfelses, representing former hydrothermally altered sheeted dykes, sandwiched between gabbros and sheeted dykes (Alt et al., 2010; Koepke et al., 2008; Teagle et al., 2012). The availability of this core provides the potential to obtain fundamental information on the magmatic processes within the DGT zone and the related AML dynamics, with a focus on the vertical fluctuations, which are regarded as key for triggering related magmatic and metamorphic processes. Gabbros from the drilled DGT at IODP Site 1256, interpreted as the interior of a frozen AML, are isotropic and very inhomogeneous in texture and composition. Petrological and geochemical studies reveal that in situ crystallization of a fairly evolved crystal mush played a dominant role. Within the mush there are also rare relics of products of primitive midocean ridge basalt (MORB) melt crystallization, represented by cores of plagioclase very rich in An content (Coogan et al., 2002; Koepke et al., 2011). These patchy gabbros, often called "varitextured" due to the strong heterogeneities in texture, are known only from fast spreading ridges. This term was introduced for typical gabbros found in the Oman ophiolite between the gabbro sequence and the base of the sheeted dike, also known as "high-level," "isotropic," or "upper" gabbro (for details and photographs of typical varitextured gabbros see MacLeod & Yaouancq (2000)). By using a diffusion-modelling approach based on CaAl–NaSi interdiffusion in zoned plagioclases from the granoblastic hornfelses from Hole 1256D, Zhang et al. (2014) made the first estimation of cooling rates above the AML (see section 9.8). Recent experimental studies using starting material from Hole 1256D constrained the conditions and compositions of felsic melt generation due to partial melting of granoblastic hornfelses of the AML roof by an upward moving AML (Erdmann et al., 2015, 2017; Fischer et al., 2016; see section 9.6).

From the other locality at the EPR, the Hess Deep Rift, mostly gabbronorite samples from a high crustal level have been collected, which were regarded as belonging to a DGT (Natland & Dick, 1996, 2009). Three samples of granoblastic hornfelses from the same location were also collected—key rocks of the DGT of IODP Site 1256 and the Oman ophiolites (Gillis, 2008) —so it is obvious that typical DGTs with contact metamorphic overprinted sheeted dikes also exist at Hess Deep, but their exact context within the stratigraphy of the DGT, i.e., their role in the CBL, is not known.

Typical gabbros from the DGT in the Oman ophiolite consisting mostly of heterogeneous, often evolved, so-called "varitextured" gabbros (see above), have been interpreted by MacLeod & Yaouancq (2000) and Coogan et al. (2002) as a rock suite corresponding to a fossilized AML. Based on their observations, these authors suggested that the varitextured gabbro horizon could be regarded as the highly differentiated end product of MORB liquid crystallization at the margins of an AML. MacLeod & Yaouancq (2000) suggested that these rocks are produced by an ascending, highly differentiated melt flux. Other authors (Coogan et al., 2002; Mueller et al., 2017) observed within relatively evolved varitextured gabbros from the Oman ophiolite mineral relics derived from primitive MORB, which implies that the AML was filled during an earlier stage with primitive MORB, similar to that recorded in the modern EPR. Wanless and Shaw (2012) and France et al. (2009) compared the gabbros drilled at IODP Site 1256 and the overlying granoblastic hornfelses to corresponding lithologies from the DGT of the Oman ophiolite, and found an excellent correspondence between these two localities. These authors concluded that the AML in fast spreading systems is a very dynamic horizon characterized by upward and downward migration as a consequence of eruption and replenishment events, recrystallization in the overlying sheeted dykes, and assimilation of the roof rocks. For the Oman ophiolite, France et al. (2010, 2014) constrained the compositions and conditions for generating felsic melts and granoblastic residues by partial melting of basalts from the sheeted dikes. For a fossilized AML in the Wadi Gideah of the Oman ophiolite, Mueller et al. (2017) highlighted the complexity of magmatic evolution involving both extreme fractional crystallization (in situ crystallization) of primitive MORB, as well as partial melting of altered dikes triggered by the heat of an ascending AML.

9.3. IODP LEGACY DRILLING AT SITE 1256: FIRST PENETRATION OF AN INTACT DIKE/GABBRO BOUNDARY

In this chapter we present new petrographical and petrological data of the dike/gabbro contact recovered within the achievements in the IODP drilling campaign "Superfast Spreading Crust" at IODP Site 1256 in the equatorial Pacific. Here, the DGT of intact oceanic crust was penetrated by drilling what, to date, is the only section of a DGT from actual fast-spread oceanic crust. Because of the uniqueness and exemplary character of this legacy core, we present here, for the first time, detailed petrographic descriptions and phase analyses from this important contact, separating the eruptive and the plutonic regime in the sheeted dike-rooting zone. IODP Site 1256 was chosen to confirm the inverse relationship between spreading rate and the depth of the AML, inferred from seismic experiments (Wilson et al., 2006). During IODP Expedition 312, gabbroic rocks, interpreted as a frozen AML, were first encountered at 1406.6 m below sea floor, with a depth predicted by the extrapolation of the seismic low-velocity-zone depth versus spreading rate correlation for actual midocean ridges to the spreading rate from IODP Site 1256 implied by magnetic survey data (Wilson et al., 2006).

The dike–gabbro contact itself is an intrusive contact, with gabbroic rocks interpreted as the frozen interior of an AML intruding granoblastic hornfels, interpreted as part of the sheeted dikes complex, now contact metamorphic overprinted due to the heat of an upward moving AML. Two major bodies of gabbro have been drilled beneath this contact: a 52-m-thick upper and a 24-m-thick lower gabbro intrusion separated by a 24 m section of granoblastic hornfelses. The drilled gabbroic rocks are relatively evolved and range in their composition from gabbro, to disseminated oxide gabbro, oxide gabbro, orthopyroxene-bearing gabbro, quartz diorite, and quartz-rich oxide diorite, including a few late trondhjemite dikelets. Primitive olivine gabbros, typical for the lower part of the gabbroic sequence of fast-spread oceanic crust, have not been recovered (Teagle et al., 2006; Wilson et al., 2006).

In the following we focus on the recovered dike–gabbro transition, from which we analyzed two samples. The dike–gabbro contact itself is located in piece 13 of section 1 of core 213 (Teagle et al., 2006), as depicted in Figure 9.2, and is recorded in thin-section hay213-52 (expedition 312, core 213, section 1, 52–54 cm). This section records intrusion of a quartz-bearing oxide diorite interpreted as a frozen part of an AML into the granoblastic hornfels. While this gabbro marks the top of the 52-m-thick upper gabbroic intrusion, there is another, much smaller plutonic intrusion 4 cm above this contact, represented by a ~7-mm-thick vein of quartz oxide diorite, with an orientation similar to that of the dike–gabbro contact (thin-section hay213-45: expedition 312, core 213, section 1, 45–50 cm; Figure 9.2). A 1-mm-thick off shoot of the quartz oxide diorite into the hornfels clearly demonstrates the intrusive nature of the gabbroic rocks. Further arguments for the intrusive nature are apophyses of dioritic melts within the hornfels, as well as plagioclase crystals arranged in comb structures perpendicular to the contact with the hornfels (see thin-section scan and BSE image in Figure 9.3).

Electron probe microanalysis (EPMA) has been carried out on both sections and the results are presented in Table 9.1. The EPMA was performed at the Institute for Mineralogy of Leibniz Universität Hannover using a Cameca SX100 electron microprobe with five spectrometers, "Peak sight" software, Kα emission for all elements, a focused beam, 15 kV acceleration potential, beam current of 15 nA, 10 s analysis time per element, and the "PAP" matrix correction. For amphiboles, F and Cl were analyzed with an acquisition time of 30 s and with higher beam current. The energy dispersive X-ray (EDX) element maps were obtained using a field emission SEM Jeol JSM-7610F operated with two Bruker Quantax 400 6/30 EDX detectors.

9.3.1. Dioritic Intrusions

Both plutonic rocks named here as dioritic intrusions are medium grained, granular and isotropic in texture, with primary minerals plagioclase, clinopyroxene, amphibole, and Fe–Ti oxides. The evolved nature of these rocks is expressed by relatively high amounts of apatite, which is

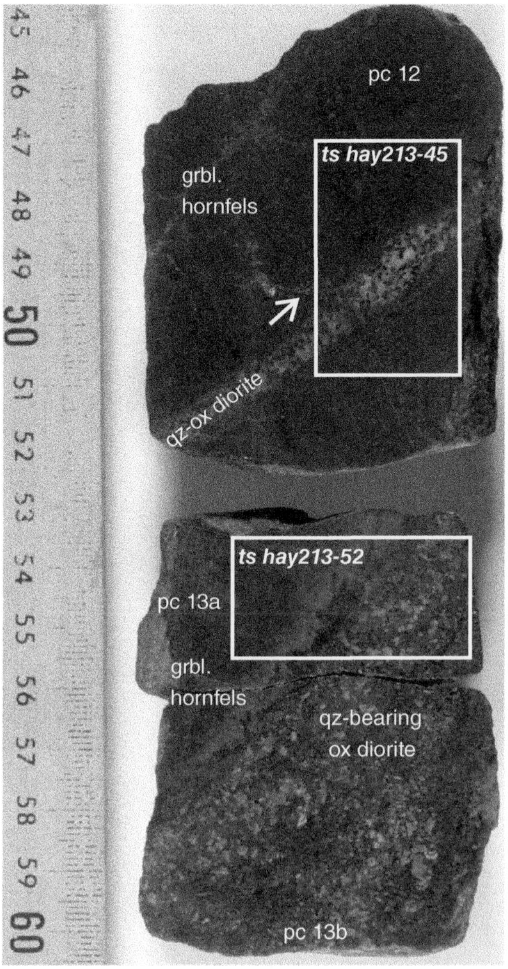

Figure 9.2 Core 312, section 1, from 45 to 60 cm, showing the dike–gabbro contact drilled by IODP Expedition 312 in piece 13. Approximately 5 cm above, in piece 12, a vein of oxide quartz diorite several millimeters thick cut the granoblastic hornfels, forming the host rock in which the gabbros were intruded. The arrow marks an off-shoot 1 mm thick of the quartz oxide diorite into the hornfels, demonstrating the intrusive nature of this rock. The locations of the two thin-sections investigated in this study are marked.

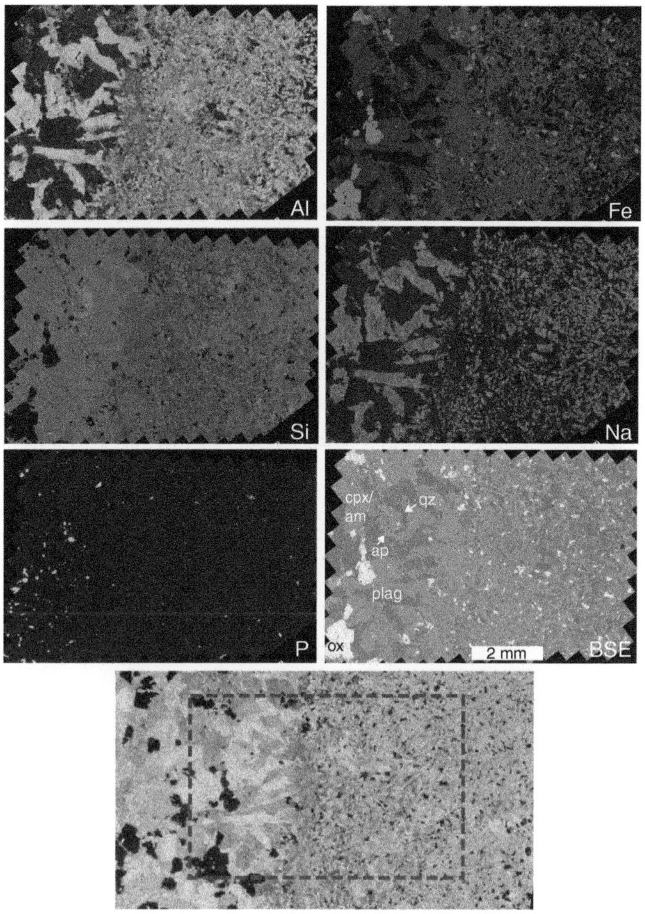

Figure 9.3 Energy dispersive X-ray (EDX) mosaic maps for Al, Fe, Si, Na, and P, as well as a backscattered electron (BSE) image and a scanned thin-section from the (right) dike and (left) gabbro contact exposed in sample hay213-52. In the scanned thin-section (bottom), the position of the area where the EDX maps and the BSE image were obtained is marked. Abbreviations in the BSE image: ox, Fe–Ti oxide; plag, plagioclase; cpx/am, former clinopyroxene crystals now totally overgrown by amphibole; ap, apatite; qz, quartz.

easily visible in the element map for P from the contact in thin-section hay213-52 (Figure 9.3). The primary features are strongly overprinted by a static greenschist-facies hydrothermal overprint (see thin-section scan in Figure 9.3 and Supplementary Material (SM) Figures SM1–SM3), which is quite common within the rocks from the DGT of the drilled core (Alt et al., 2010; Koepke et al., 2008; Teagle et al., 2006). Plagioclase shows a strong primary core–rim zoning, varying from An_{56} to An_{45} for the inner core and from An_{45} to An_9 for the outer core, while the outermost rims are often converted to albite (Figure 9.4 and Figure SM3), as a result of the hydrothermal overprint. Primary clinopyroxenes and amphiboles are nearly completely overprinted by secondary clinopyroxene and amphibole. Only a few relics of primary clinopyroxenes are present in the quartz oxide diorite of sample 213-45, with relatively high Al_2O_3 and TiO_2 contents, corresponding to those values from other gabbros (see compilation in Koepke et al., 2011) or from lavas and dikes from IODP Site 1256 (Figure 9.5). In contrast, the vast majority of clinopyroxene analyses of this sample correspond, with their low Al_2O_3 and TiO_2 contents, to those typical of a secondary hydrothermal overprint (Alt et al., 2010; Koepke et al., 2008), as shown in Figure 9.5. In thin-section hay213-52, no clinopyroxene was found in the dioritic domain; here all clinopyroxene was completely overgrown by secondary amphibole.

Table 9.1 Mineral Compositions of Two Samples from the Dike–Gabbro Contact at IODP Site 1256

Point	Sample	Lithology[a]	Phase[b]	Location analyzed[c]	Phase qualifier[d]	Number of samples[e]	SiO_2	TiO_2	Al_2O_3	Cr_2O_3	FeO	MnO	MgO	CaO	Na_2O	K_2O	Cl	Total	Mg#[f]	An[g]
	Hay213-52	grnbl	cpx	pa	hi-Al	4	51.52	0.44	1.04	0.01	11.01	0.38	13.65	21.39	0.30	0.00		99.74	68.86	
							0.35	*0.12*	*0.04*	*0.01*	*0.35*	*0.05*	*0.05*	*0.51*	*0.03*	*0.00*			*0.70*	
		grnbl	cpx	pa/ri	sec	15	52.40	0.06	0.39	0.01	9.62	0.35	13.42	23.23	0.35	0.00		99.82	71.27	
							0.42	*0.02*	*0.14*	*0.02*	*0.55*	*0.05*	*0.70*	*0.46*	*0.06*	*0.00*			*2.14*	
		grnbl	cpx	co		54	52.19	0.16	0.62	0.01	10.49	0.36	13.83	21.81	0.26	0.00		99.73	70.15	
							0.32	*0.06*	*0.10*	*0.02*	*0.47*	*0.04*	*0.26*	*0.51*	*0.06*	*0.00*			*1.20*	
13/6		grnbl	pl	co			51.78		29.07		0.63		0.02	11.98	4.69	0.07		98.25		58.51
12/8		grnbl	pl	co			54.17		27.69		0.59		0.02	10.61	5.85	0.04		98.97		50.07
15/13		grnbl	pl	co			58.79		25.07		0.44		0.00	7.64	7.53	0.13		99.59		35.93
9/1		grnbl	pl	ou			60.42		23.90		0.52		0.01	5.86	8.62	0.11		99.44		27.31
12/ 9		grnbl	pl	ri			63.52		22.29		0.15		0.00	3.35	10.30	0.05		99.66		15.23
11/7		grnbl	pl	ri			66.21		20.59		0.15		0.00	1.40	11.43	0.06		99.84		6.32
		grnbl	mt			62	0.04	2.21	0.48	0.23	89.26	0.16	0.02	0.04	0.01	0.02		92.46		
							0.03	*1.09*	*0.08*	*0.07*	*1.23*	*0.12*	*0.02*	*0.04*	*0.02*	*0.02*				
		grnbl	il			51	0.01	45.64	0.05	0.12	49.80	2.46	0.02	0.03	0.01	0.02		98.17		
							0.03	*1.07*	*0.03*	*0.07*	*1.31*	*0.81*	*0.03*	*0.03*	*0.02*	*0.01*				
29/10		grnbl	am		mg-hbl		50.78	0.78	3.74	0.07	13.14	0.28	15.73	11.46	1.07	0.05	0.23	97.32	68.09	
30/7		grnbl	am		act		52.68	0.48	2.62	0.00	12.15	0.33	17.01	10.74	0.86	0.03	0.15	97.04	71.39	
		ox-dio	pl	co		56	54.86		27.27		0.78		0.08	10.66	5.80	0.07		99.52		50.38
							0.78		*0.46*		*0.07*		*0.03*	*0.61*	*0.33*	*0.04*				*2.77*
2/23		ox-dio	pl	pa			56.65		26.31		0.14		0.00	8.78	6.89	0.07		98.83	57.00	41.32
3/29		ox-dio	pl	pa			66.13		21.22		0.16		0.00	1.95	11.00	0.04		100.50	71.00	8.91
		ox-dio	mt			11	0.05	1.51	0.52	0.05	90.58	0.09	0.04	0.03	0.01	0.00		92.86		
							0.02	*0.34*	*0.07*	*0.04*	*0.42*	*0.04*	*0.02*	*0.03*	*0.01*	*0.01*				
		ox-dio	il			30	0.02	49.01	0.02	0.01	47.75	1.50	0.26	0.04	0.02	0.00		98.63		
							0.04	*0.90*	*0.02*	*0.02*	*1.08*	*0.48*	*0.04*	*0.04*	*0.02*	*0.01*				
9/18		ox-dio	am		ede		44.71	2.28	7.85	0.10	13.40	0.27	14.30	11.02	2.36	0.31	0.32	96.92	65.56	
7/10		ox-dio	am		mg-hbl		50.28	0.51	3.97	0.01	15.68	0.31	13.49	12.05	0.87	0.03	0.07	97.26	60.53	
8/16		ox-dio	am		act		52.83	0.46	2.26	0.00	11.53	0.27	17.42	11.47	0.77	0.03	0.09	97.14	72.93	
7/13		ox-dio	am		fe-act		51.74	0.13	1.40	0.00	21.47	0.19	10.01	10.89	0.47	0.03	0.05	96.40	45.39	
	Hay213-45	grnbl	cpx	co		13	51.28	0.58	1.41	0.02	11.92	0.36	14.44	19.54	0.30	0.00		99.84	68.37	
							0.35	*0.09*	*0.23*	*0.02*	*0.73*	*0.04*	*0.37*	*0.91*	*0.03*	*0.00*			*1.33*	
		grnbl	opx	co		3	51.75	0.39	0.71	0.00	24.01	0.64	20.56	1.63	0.02	0.00		99.71	60.42	
							0.18	*0.05*	*0.10*	*0.00*	*0.07*	*0.03*	*0.19*	*0.10*	*0.01*	*0.00*			*0.17*	
		grnbl	pl	co		13	53.74		28.43		0.67		0.01	11.28	5.26	0.09		99.49		54.26
							0.76		*0.52*		*0.22*	*0.04*	*0.55*	*0.30*	*0.03*					*2.51*
		grnbl	pl	ri		12	60.47		24.25		0.45		0.00	5.89	8.33	0.08		99.47		26.14

						1.07		*0.52*		*0.08*		*0.00*	*0.65*	*0.31*	*0.01*				*7.39*
	grnbl	mt			10	0.08	2.23	0.55	0.21	90.00	0.13	0.04	0.03	0.01	0.02		93.31		
						0.03	*0.97*	*0.11*	*0.03*	*1.17*	*0.05*	*0.03*	*0.02*	*0.01*	*0.00*				
	grnbl	il			9	0.11	45.83	0.10	0.08	51.04	0.84	0.10	0.08	0.00	0.01		98.20		
						0.04	*1.32*	*0.04*	*0.03*	*1.34*	*0.07*	*0.05*	*0.04*	*0.01*	*0.00*				
	grnbl	am		act	5	52.38	0.22	2.71	0.00	13.01	0.22	16.39	11.05	0.76	0.05		96.79	69.18	
						1.06	*0.15*	*0.61*	*0.00*	*0.60*	*0.03*	*0.59*	*0.14*	*0.23*	*0.03*			*1.70*	
7/37	ox-dio	cpx		pri rel		50.85	0.45	3.33	0.00	14.62	0.35	13.29	15.42	0.69	0.02		99.03	61.84	
7/38	ox-dio	cpx		pri rel		50.46	0.25	2.19	0.08	13.34	0.38	12.34	18.85	0.58	0.03		98.50	62.25	
8/38	ox-dio	cpx		pri rel		51.69	0.37	2.28	0.05	13.96	0.39	12.13	18.35	0.51	0.02		99.76	60.77	
	ox-dio	cpx	co	sec	49	53.19	0.03	0.42	0.04	8.08	0.35	14.28	24.03	0.31	0.01		100.72	75.91	
						0.50	*0.03*	*0.32*	*0.03*	*0.84*	*0.05*	*0.55*	*1.10*	*0.10*	*0.01*			*2.44*	
	ox-dio	pl	co	pop1	103	54.99		27.58		0.76		0.08	10.67	5.62	0.12		99.82		51.20
						0.39		*0.26*		*0.11*		*0.04*	*0.23*	*0.14*	*0.04*				*1.05*
	ox-dio	pl	co	pop2	21	59.25		25.31		0.45		0.03	7.48	7.39	0.18		100.08		35.85
						0.84		*0.69*		*0.13*		*0.02*	*0.67*	*0.32*	*0.07*				*2.95*
	ox-dio	pl	co	pop3	16	54.39		27.82		0.76		0.08	11.16	5.44	0.10		99.75		53.12
						0.55		*0.26*		*0.07*		*0.04*	*0.19*	*0.10*	*0.02*				*0.80*
3/35	ox-dio	pl	ou	pop1		56.98		26.73		0.52		0.02	9.36	6.31	0.12		100.04		45.04
3/38	ox-dio	pl	ou	pop1		59.15		25.37		0.44		0.02	7.54	7.62	0.12		100.25	10.00	35.38
1/30.	ox-dio	pl	ou	pop1		61.24		23.84		0.31		0.02	5.37	8.46	0.06		99.29	17.00	25.95
2/1	ox-dio	pl	ri	pop1		67.28		20.58		0.15		0.00	1.69	10.54	0.17		100.40	21.00	8.15
	ox-dio	mt			16	0.08	2.30	0.29	0.09	90.50	0.10	0.03	0.02	0.00	0.00		93.41		
						0.04	*1.00*	*0.16*	*0.04*	*1.27*	*0.05*	*0.03*	*0.02*	*0.00*	*0.00*				
	ox-dio	il			27	0.05	49.50	0.01	0.01	47.59	1.98	0.12	0.03	0.00	0.00		99.29		
						0.03	*1.74*	*0.03*	*0.02*	*1.63*	*0.77*	*0.07*	*0.02*	*0.00*	*0.00*				
7/39	ox-dio	am		mg-hbl		49.63	0.53	4.72	0.04	16.46	0.26	12.82	12.39	1.01	0.04		97.91	58.13	
8/24.	ox-dio	am		mg-hbl		48.88	1.43	6.50	0.00	14.47	0.38	14.33	11.19	1.59	0.20		98.98	63.84	
	ox-dio	am		act	84	52.95	0.30	1.83	0.04	14.81	0.25	14.88	12.10	0.47	0.04	0.11	97.77	64.09	
						1.20	*0.18*	*0.74*	*0.03*	*1.93*	*0.05*	*1.46*	*0.48*	*0.19*	*0.03*	*0.05*		*5.28*	

Note. FeO = FeO^{tot}. [a] grnbl, granoblastic hornfels; ox-dio, oxide diorite. [b] am, amphibole; cpx, clinopyroxene; il, ilmenite; mt, magnetite; opx, orthopyroxene; pl, plagioclase. [c] co, core; ou, outer core; ri, rim; pa, patch; otherwise central part of the crystal. [d] act, actinolite; ede, edenite; fe-act, ferro-actinolite; mg-hbl, magnesiohornblende; hi-Al, high Al content; pop1,2,3, populations 1,2, and 3; pri rel, relic of a primary magmatic phase; sec, secondary. [e] Average values are determined for the number of samples analyzed, with one standard deviation values in italics; otherwise representative values are given; missing values indicate not analyzed. [f] $MgO/(MgO + FeO^{tot}) \times 100$, molar. [g] An content of the plagioclase, mol%.

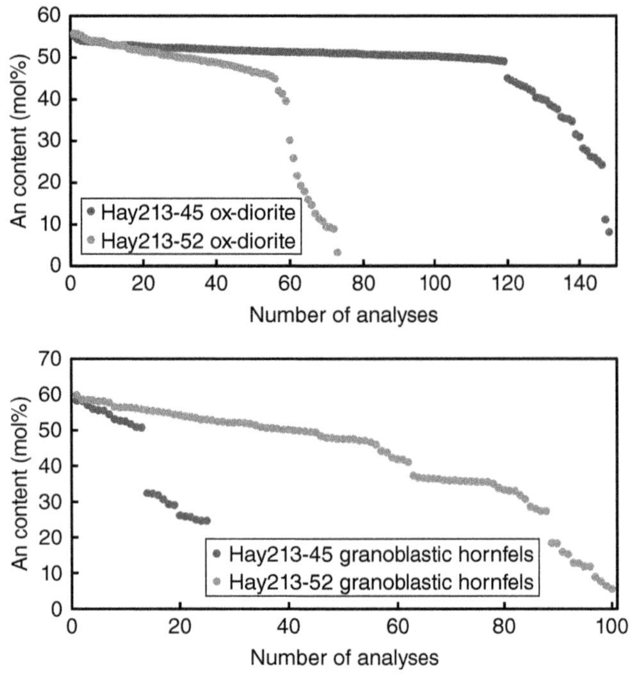

Figure 9.4 Anorthite content from plagioclase analyses acquired via profiles for samples hay213-45 and hay213-52 from the dike–gabbro transition observed at IODP Site 1256, for both the granoblastic hornfels and the oxide diorite domains. For details see text.

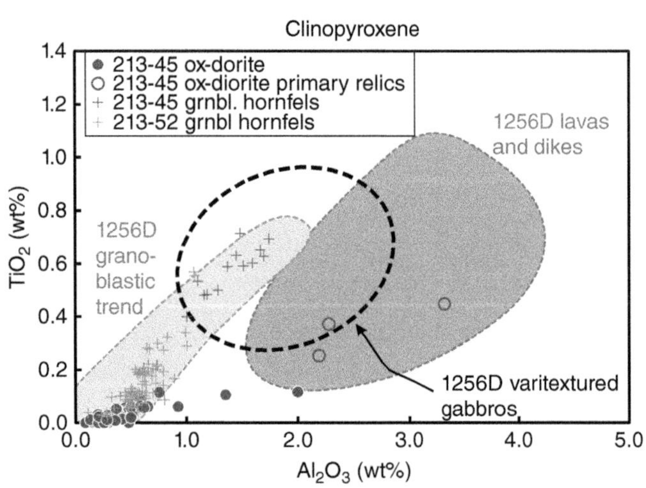

Figure 9.5 TiO_2 versus Al_2O_3 diagram for clinopyroxenes of two thin sections (hay213-45 and hay213-52) from the dike–gabbro transition at IODP Site 1256, for both the granoblastic hornfels and the oxide diorite domains. No clinopyroxene was found in the dioritic domain in thin section hay213-52; in which all clinopyroxene was completely overgrown by secondary amphibole. The fields for the IODP Hole 1256D lavas and dikes, gabbros, and granoblastic trend are from Koepke et al. (2011). For details see text.

Analyzed amphiboles in the dioritic domains of both samples are mostly actinolites, ferroactinolites, and magnesiohornblendes, corresponding to formation of the hydrothermal overprint (Figure 9.6). Ti-in-amphibole temperatures (Ernst & Liu, 1998) for this overprint vary continuously from < 500 to 670°C, thus spanning temperatures typical for greenschist to amphibolite facies (Figure 9.6). Due to the omnipresence of Fe–Ti oxides in both samples, the conditions for Ti saturation in the system are given, so that the geothermometer of Ernst & Liu (1998) can be applied successfully. Only in the oxide diorite of sample hay213-52 have a few true relics of primary magmatic amphibole been detected (<< 1% in the mode), visible as tiny brownish domains strongly overgrown by greenish flakes of actinolite and magnesiohornblende. These are edenitic in compositions, high in Al_2O_3, TiO_2, and Na_2O, with Ti-in-amphibole temperatures varying between 700 and 870°C, approaching typical temperatures for late crystallization processes in evolved, Fe-enriched MORB-type melts at ~800°C for water-saturated conditions (Currin et al., 2018a; Koepke et al., 2018). Amphiboles of such compositions have not been found in the coexisting domain from the granoblastic hornfels regarded as former sheeted dikes in the same sample (hay214-52). This is a strong argument for the presence of amphibole relics from the magmatic stage, since it is obvious that any hydrothermal overprint in a given rock would result in similar amphibole compositions in coexisting lithological domains.

Fe–Ti oxides in the dioritic domains of both samples are magnetites very low in ulvospinel (6.7 and 4.4 mol%) and high in ilmenite (94.1 and 93.4 mol%) components; calculated according to Sauerzapf et al. (2008; see Table 9.1). The magnetites show exsolutions of ilmenites, visible also in the element map for Fe in Figure 9.3.

9.3.2. Granoblastic Hornfelses

The host rock in which the gabbros are intruded at the DGT are characterized by microgranular mosaics of clinopyroxene, plagioclase, amphibole, magnetite, and ilmenite with a granoblastic, hornfelsic texture, and crypto- to microcrystalline grain sizes. A general characteristic of the granoblastic overprint is that the primary magmatic precursor pattern, the intersertal texture of typical sheeted dikes of Hole 1256D, can still be recognized (Figure 9.3, and Figure SM3). With these features, the hornfelses in the samples investigated correspond well to granoblastic dikes described in a horizon 60 m thick above the dike–gabbro contact of the drilled core, interpreted as a CBL separating the AML from the hydrothermally cooled sheeted dikes (e.g., Alt et al., 2010; France et al., 2009; Koepke et al., 2008; Teagle et al., 2006). Orthopyroxene, which is also a typical mineral of

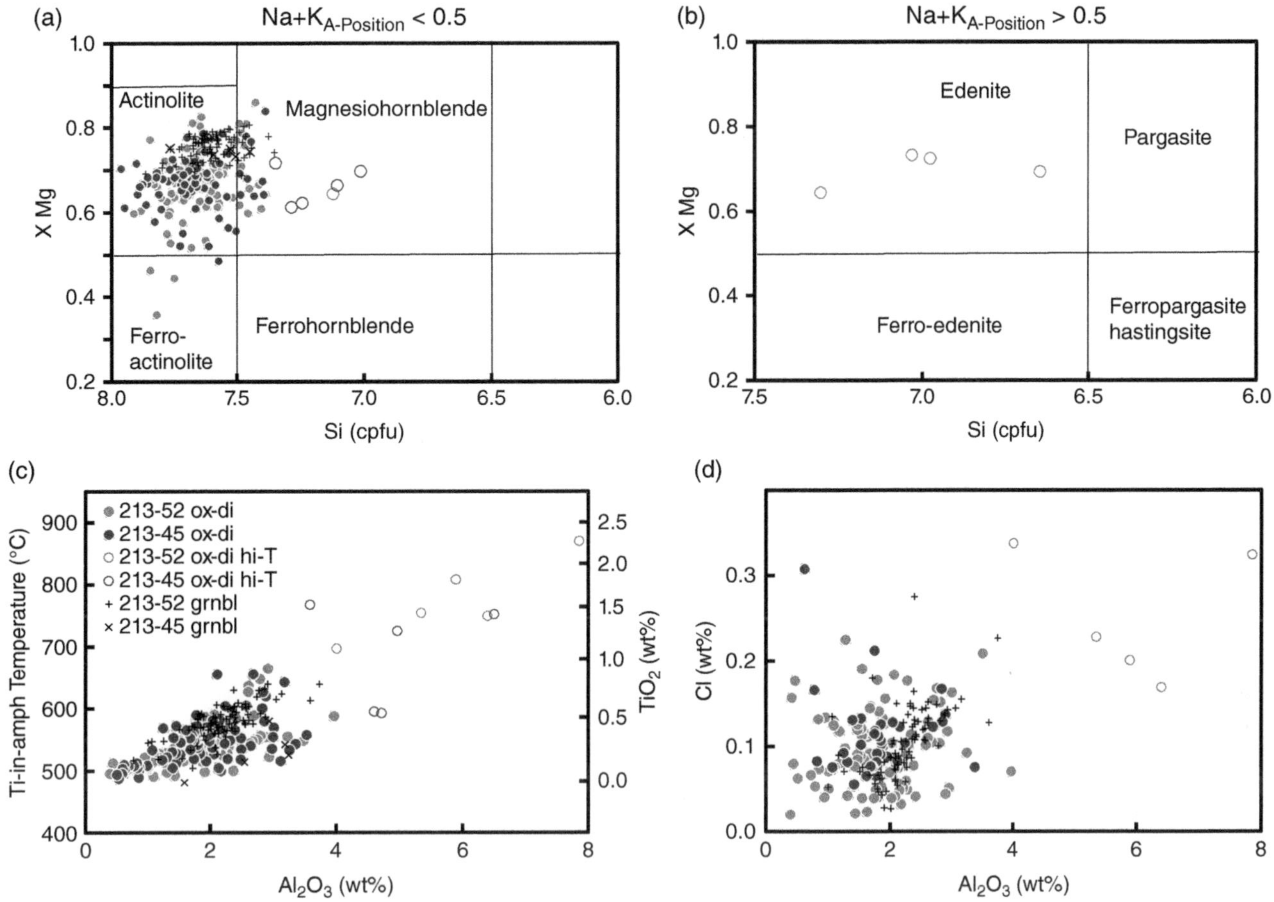

Figure 9.6 Composition diagrams for amphiboles of two thin sections (hay213-45 and hay213-52) from the dike–gabbro transition at IODP Site 1256, for both the granoblastic hornfels and the oxide diorite domain. (a, b) Composition diagrams and nomenclature according to Leake et al. (1997); (c) Ti-in-amphibole geothermometer according to Ernst and Liu (1998) versus Al_2O_3; (d) Cl versus Al_2O_3. For details see text.

the granoblastic assemblage from the granoblastic dikes of the 1256D drill core, was only rarely observed, as an inclusion in the clinopyroxene of sample hay213-45. We interpret this as a relic from a granoblastic equilibration at higher temperatures in the two-pyroxene hornfels facies, later overprinted under the conditions of the pyroxene hornfels facies. As with the dioritic domains in the same sample, the granoblastic domain is characterized by a strong overprint under greenschist facies conditions, as a consequence of subsequent hydrothermal cooling. Magmatic relics of plagioclase and clinopyroxene from the sheeted dike crystallization, which are often visible in the hornfelses from IODP Site 1256 (e.g., Koepke et al., 2008), have not been observed.

Plagioclase grains show an apparent zoning varying from An_{60} to An_{41} for the inner core area, and from An_{37} to An_{25} for the outer core area, while the outermost rims and some interstitial areas are filled with albite (Figure 9.4), regarded as a product from the hydrothermal overprint.

Clinopyroxenes of the granoblastic domains in both samples follow the typical granoblastic trend forming a narrow field in the TiO_2 versus Al_2O_3 diagram in Figure 9.5. With very low Al_2O_3 and TiO_2 contents, many of the analyzed clinopyroxenes correspond to secondary formations, probably equilibrated during the greenschist facies hydrothermal overprint, while others with moderate Al_2O_3 and TiO_2 contents correspond more to formations at peak conditions of the granoblastic overprint (Figure 9.5). Since orthopyroxene does not form an equilibrium paragenesis with clinopyroxene, reliable crystallization temperatures of the high-temperature stage cannot be evaluated. The Al-in-clinopyroxene geothermometer of France et al. (2010) revealed formation temperatures from 830 to 900°C, corresponding to the conditions of the clinopyroxene hornfels facies.

The vast majority of analyzed amphiboles in the granoblastic domains of both samples correspond to actinolites, and a few to magnesiohornblendes. Ti-in-amphibole temperatures (Ernst & Liu, 1998) vary continuously from < 500 to 670°C, thus spanning temperatures typical for greenschist to amphibolite facies (Figure 9.6). No relics of brownish amphiboles formed under the higher thermal conditions of the hornblende hornfels facies, as observed from time to time in other parts of the granoblastic hornfelses from the core drilled at IODP Site 1256 (e.g., Koepke et al., 2008), have been observed, implying that the original conditions of the metamorphic imprint of the granoblastic overprint proceeded under dry conditions in the pyroxene hornfels facies in the absence of amphibole stability.

Fe–Ti oxides in the granoblastic domains of both samples are magnetites very low in ulvospinel (6.6 mol% for both samples) and high in ilmenite (88.1 and 87.2 mol%) components; calculated according to Sauerzapf et al. (2008; see Table 9.1).

9.3.3. Petrogenesis

In spite of the strong, pervasive hydrothermal overprint recorded in both samples, relics of high-temperature stages under both magmatic and metamorphic conditions allow a reconstruction of the magmatic and metamorphic processes taking place at the dike–gabbro contact drilled at IODP Site 1256. The host rock was once a part of the sheeted dike complex that was overprinted by contact metamorphiism due to the heat of an upward moving AML, probably triggered by a replenishment process in the deeper part of the axial magma chamber. This metamorphic overprint proceeded in a dry environment (absence of high-temperature amphibole in the granoblastic domains) mainly under conditions of the pyroxene hornfels facies, with a peak metamorphism within the two-pyroxene hornfels facies (presence of orthopyroxene relics). This evaluation is fully in accord with the findings of Koepke et al. (2008), who investigated in detail the whole ~60-m-thick granoblastic horizon above the dike–gabbro contact at IODP Site 1256, combined with experimental simulations (as demonstrated in section 9.6.2). The resulting granoblastic hornfelses still show relics of the primary magmatic intersertal texture (Figure 9.3 and Figure SM3). Prior to this metamorphism, the sheeted dikes were affected by hydrothermal cooling, as shown in detail by Koepke et al. (2008). As a result of replenishment, evolved melt of the underlying AML intruded into the granoblastic dikes, forming the upper gabbroic intrusion (sample hay213-52), as well as thin dioritic veins some centimeters above the contact (sample hay213-45). Strong core–rim zoning in the plagioclase (as shown in Figure 9.S3) clearly points to the process of in situ crystallization, a process that has been identified as being significant during the freezing of an AML (e.g., Koepke et al., 2001; Mueller et al., 2017; Zhang, Koepke, France, et al., 2017); for details of the crystallization process in the fossilized AML from IODP Site 1256 see section 9.4. Plagioclase An contents correspond to those compositions known from the more evolved, granular domains in typical varitextured gabbros from frozen AMLs at both IODP Site 1256 (Koepke et al., 2011) and the Oman ophiolite (Mueller et al., 2017). In the drill core from Hole1256D, the variation of An content in plagioclase throughout the varitextured gabbro is from An_{82} (relics in the cores of the subophitic domains, Koepke et al., 2011) down to albite (in the diorites; Zhang, Koepke, France, et al., 2017). In the varitextured gabbro from the frozen AML in the Oman ophiolite in the Wadi Gideah plagioclase varies from An_{70} to albite (Mueller et al., 2017). Due to an intense subsequent hydrothermal overprint at mostly greenschist conditions, only a few magmatic relics of clinopyroxene (Figure 9.5) and amphibole (Figure 9.6) survived in the dioritic domains of the DGT samples.

The subsequent hydrothermal event affected both the dioritic intrusive rocks as well as the granoblastic host rocks with similar grade in both lithologies. The most visible record of this is the overgrowth of low-grade clinopyroxene (extremely low in Al and Ti) and amphibole (mostly actinolite) in both lithologies of both samples. This late hydrothermal event also affected the Fe–Ti oxides, in which the original compositions have been reset to greenschist facies conditions. According to the titanomagnetite–ilmenite thermo-oxybarometer (Sauerzapf et al., 2008) the calculated equilibrium temperatures for coexisting magnetite and ilmenite from both the dioritic and the granoblastic domains of both samples vary between 500 and 580°C, which is below the temperature range recommended for the application of this tool. The corresponding oxygen fugacities calculated for the same Fe–Ti oxide pairs reveal ΔNNO values of 2.8–3.1, thus too oxidizing for typical crystallization processes within MORBs, but in accord with the highly oxidizing conditions of a hydrothermal overprint. The temperature and oxygen fugacity estimations in both lithologies of both samples are very close to each other, which is another argument that the hydrothermal overprint affected both lithologies simultaneously.

In the Cl versus Al_2O_3 diagram for amphiboles (Figure 9.6) it is obvious that the vast majority of secondary amphiboles in all lithologies are significantly enriched in Cl, indicating formation in the presence of a high-saline fluid, such as those typically formed in seawater-derived fluids during phase separation under pressure and temperatures (Currin et al., 2018a, 2018b). Thus, it is obvious that these secondary amphiboles are the product of seawater-derived hydrothermal circulation. However, the highest Cl concentrations (> 0.3 wt% Cl, see Table 9.1) have been measured in those magmatic relics in the oxide diorite from sample hay213-52, implying that high-saline fluids were also present during the magmatic formation of the diorite. Such Cl concentrations are typical for experimental amphibole formed under magmatic conditions in the presence of highly saline fluids (NaCl–H_2O, with 6, 20, and 50 wt% NaCl; Currin et al., 2018a), revealing Cl contents varying between ~0.1 and 0.5 wt% Cl, well in accord with the Cl contents of the analyzed amphiboles. Moreover, high Cl activities in the melts that formed the diorites in the varitextured gabbros of the Hole 1256D core are also indicated by the presence highly Cl-rich cores (> 5 wt% Cl) within the magmatic apatites of these rocks (Zhang, Koepke, Albrecht, et al., 2017). The process of Cl enrichment was probably via the assimilation of previously hydrothermally altered sheeted dike of the AML roof, a mechanism for Cl incorporation into AML magmas suggested by several authors (e.g., Coogan et al., 2003; Erdmann et al., 2015; Fischer et al., 2016; Wanless et al., 2010; Zhang, Koepke, Albrecht, et al., 2017; Zhang, Wang, et al., 2017; see also section 9.6.3). The same process is also discussed for increasing water activity in the melts of the AML, necessary for crystallizing magmatic amphibole (e.g., Johnson et al., 1994; Pichavant & Macdonald, 2007). Direct evidence for this model is provided by careful petrographic investigations of varitextured gabbros, revealing domains with relics of typical textures from sheeted dikes (intersertal textures, fan-like arrangements of plagioclase laths; Teagle et al., 2006, Koepke et al., 2011). Several arguments indicating that the former sheeted dikes have been hydrothermally altered before incorporation into the axial melt lens are provided in Koepke et al. (2011).

One fundamental question is about the timescale of the magmatic–metamorphic processes at DGT's from fast spreading ridges. The preceding petrographic–petrological investigation within this section allows only the establishment of a chronological order for the processes observed. An evaluation of the absolute timescales for the replenishment cycles at fast spreading ridges is much more challenging, as presented in section 9.8.

9.4. MAGMATIC EVOLUTION WITHIN AML

Primitive and evolved intrusive rocks from the dike–gabbro transition zone of an intact oceanic crust have been sampled for the first time at IODP Hole 1256D (Wilson et al., 2006; Koepke et al., 2011), which provides a unique opportunity for studying magma evolution processes in the AML. The formation of dioritic rocks can be interpreted using the MELTS software package and taking the most primitive dike drilled in the IODP Hole 1256D as a starting system, with an initial H_2O content of 0–0.2 wt% and fO_2 similar to quartz–fayalite–magnetite (QFM) buffer as shown in Zhang, Koepke,

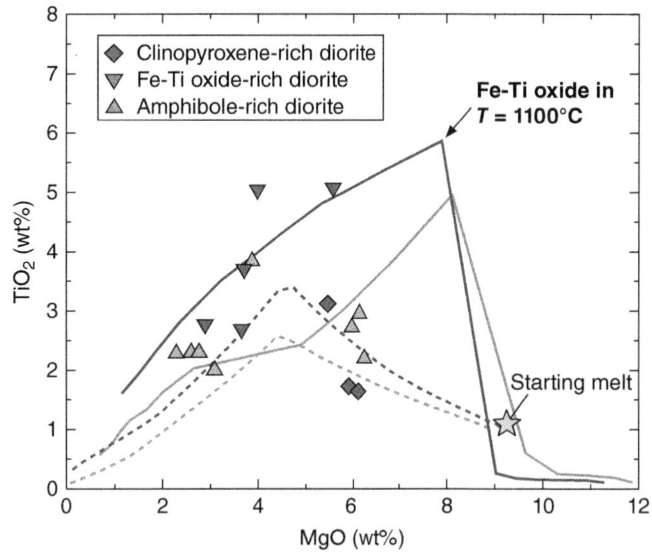

Fractionated melt	fO_2	Initial H_2O
- - - - - -	QFM	0 wt%
- - - - - -	QFM	0.2 wt%

Instantaneous cumulate	fO_2	Initial H_2O
———	QFM	0 wt%
———	QFM	0.2 wt%

Figure 9.7 Fractional crystallization modeling using the MELTS software package for dioritic intrusive rocks from IODP Hole 1256D. The composition of starting melt is from a typical MORB lava from the same site (sample 309-1256D-161R-1, piece 9; Teagle et al., 2006).

France, et al. (2017). The assumed initial H_2O content and prevailing fO_2 are consistent with analyses on fresh MORB glasses (Cottrell & Kelley, 2011; Wanless & Shaw, 2012). Some oxide-rich diorites from the Hole 1256D core have very high TiO_2 and $Fe_2O_3^{tot}$ contents up to 5 wt% and 27 wt% respectively, clearly indicating a major contribution of instantaneous cumulates in these intrusive rocks (Figure 9.7). These rocks fall on the path of the instantaneous cumulates in Figure 9.8, far away from the liquid path, implying a cumulate origin of these rocks. These rocks bear several percent of Fe–Ti oxides (Teagle et al., 2006), which drive the TiO_2 and $Fe_2O_3^{tot}$ to such high values. On the other hand, some of the clinopyroxene-rich diorites of the Hole 1256D core follow the liquidus lines of descent derived by the MELTS modeling in Figure 9.7, implying that these rocks correspond more to frozen liquids. Thus, the MELTS modeling indicates that the formation of diorites of the AML at IODP Hole 1256D is based on mixing between residual melt and instantaneous cumulates generated by fractional crystallization of a primitive MORB. More evidence for this based on major and trace element mineral compositions can be found in Zhang, Koepke, France, et al. (2017).

These evolved dioritic magmas in the AML appear to be uneruptible, as no analogous lavas has been recovered from the same drilling site, which is also implied when

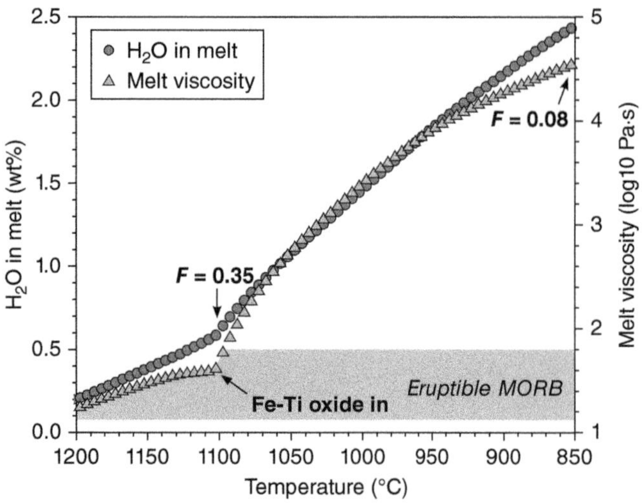

Figure 9.8 Water content and melt viscosity model as a result of fractional crystallization. Modeling is based on initial H_2O content of 0.2 wt% and fO_2 equal to the QFM buffer. The value of *F* indicates the proportion of residual melt relative to parental melt. The starting system for the calculation is the most primitive dike drilled from IODP Hole 1256D, which is similar in major element composition to the average composition of the East Pacific Rise derived from melt inclusions and glasses and reported by Wanless (sample 309–1256D-161R-1, piece 9; Mg# 62.3; Teagle et al., 2006).

considering the viscosities of the related melts (see below and Figure 9.8). As a result of fractional crystallization, both SiO_2 content and H_2O contents in the residual melts increase, which has compensating effects on the viscosity of the evolving liquid. We calculated the variations of H_2O and melt viscosity using the MELTS program for a starting melt containing 0.2 wt% H_2O at fO_2 equal to the QFM buffer, and the results show that viscosity of the residual melt increases continuously by three orders of magnitude from 1200 to 900°C. The increase in viscosity is sharply enhanced at the saturation of Fe–Ti oxide with a melt fraction of $F = 0.35$ (Figure 9.8), reaching values where melts are no longer eruptable (see discussion in Koepke et al., 2018), implying that the evolved dioritic melts stay in the AML rather than erupting. The strong increase in viscosity is due to the increase of silica after saturation of Fe–Ti oxides (see Koepke et al., 2018, and references herein), and the enrichment of H_2O in the residual melt with differentiation, which lowers the viscosity, is not able to compensate this effect (Figure 9.8). Another argument indicating that these melts are unable to erupt could be the high density of theses melts, although high-density Fe–Ti basalts can be found at several places at midocean ridges (for details see Koepke et al., 2018).

The large amount of dioritic components recovered from the fossilized AML at IODP Hole 1256D suggests that high-temperature fractional crystallization is the major process taking place in the upper part of the AML. This is intrinsically different from the lower crust where rare evolved melts occupy only the interstitial porous space (e.g., Koepke et al., 2018; Natland & Dick, 1996).

The gabbroic and dioritic rocks formed in a solidifying AML are variable not only in composition but also in texture, where granular and subophitic domains are spatially closely associated, leading to a spotty appearance of the these so-called varitextured gabbros. Plagioclase and clinopyroxene compositions indicate that the subophitic domains have crystallized at higher temperatures than the granular domains by the process of in situ crystallization (Koepke et al., 2011). The drilling results at IODP Hole 1256D show that the boundary between sheeted dike and AML is not smooth but characterized by intersections and assimilations (Teagle et al., 2012), which might result in very highly heterogeneous temperature distribution near the margin of the AML. Therefore, the cooling rate and resultant crystallization behaviors of minerals at even adjacent locations may be significantly different. Furthermore, because the thermal status of AML is controlled by both periodical melt replenishment and midocean-ridge spreading dynamics (Colman et al., 2012; Perfit & Chadwick, 1998; Singh et al., 1998; Zhang et al., 2014), magma crystallization at the margin of the AML must be dynamic and multistage magma differentiation products may coexist, as indicated by the different texture domains of gabbro contacts with each other due to complex intrusive relationships (Koepke et al., 2011). A simplified model of the magma evolution within the AML at a stage of peak magma replenishment is shown in the left panel of Figure 9.9, which depicts the order of mineral formation as a result of fractional crystallization and associated crustal assimilation at the marginal zone.

9.5. FELSIC MELT GENERATION WITHIN AML

9.5.1. Formation of Felsic Melt Within the Oceanic Lithosphere

Felsic magmatic rocks within oceanic lithosphere, also generally known as oceanic plagiogranites (for definition see Koepke et al., 2007), are commonly in the compositional range of quartz diorite, tonalite, trondhjemite, and rarely granodiorite (Coleman & Donato, 1979; Coleman & Peterman, 1975) and can be generated in both subduction-related and non-subduction settings (Furnes & Dilek, 2017). Of critical importance is that oceanic felsic magmatic rocks may be an analogue for early Earth crustal rocks (e.g., Martin, 1999; Rollinson, 2008). Here the oceanic felsic rocks, or plagiogranites, can be erupted lavas, enclosed glass inclusions in minerals, or intrusive bodies

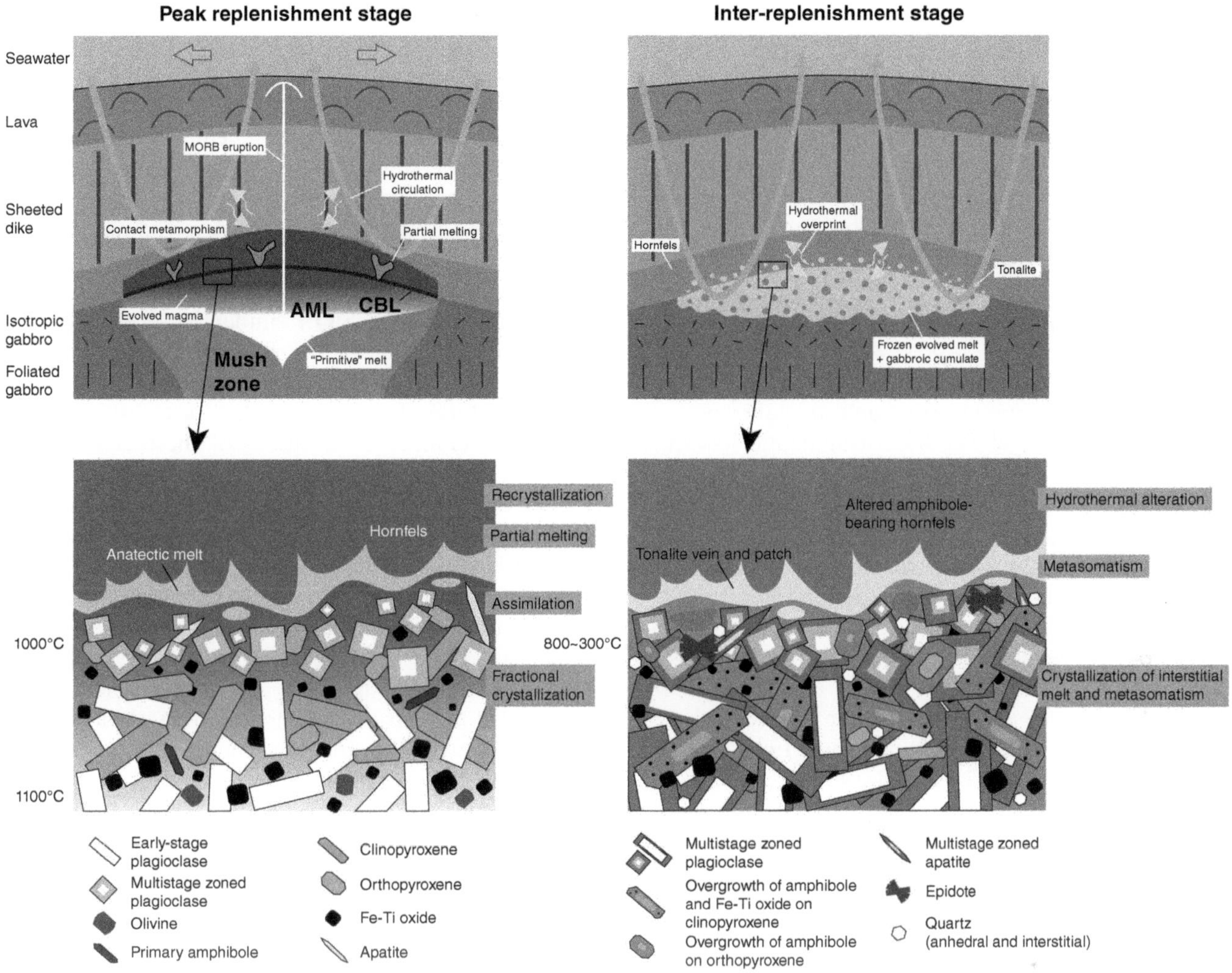

Figure 9.9 Schematic cross-axis view for axial melt lenses at fast spreading midocean ridges. (Left) Peak replenishment stage. Primary magma evolution from interior toward the margin of the AML is controlled by fractional crystallization. Peak contact metamorphism and partial melting in the sheeted dikes overlying the AML occurs at this stage, as well as crustal assimilation at the AML magma marginal zone. Hydrothermal fluids are restricted above the zone of contact metamorphism. Temperature at the top of the AML is about 1000°C. (Right) Interreplenishment stage. Further cooling, crystallization, and metasomatism induced by hydrothermal fluids. Crystallization of interstitial melts forms zoned plagioclase. Hydrothermal metasomatism induces formation of quartz, epidote, and zoned apatite, as well as overgrowth of amphibole on pyroxene. Temperature at the top of the fossilized AML can be variable within 800–300°C. Minerals in hornfels and anatectic melt/tonalitic vein and patch are not depicted. [Modified after Zhang, Koepke, France, et al. (2017).]

or dikes, and any given rock may not unambiguously correspond to one formation mechanism (see below).

In intraoceanic subduction-related settings, such felsic magmatic rocks are interpreted to have formed via partial melting of a subducted slab (Kay, 1978). In a midocean ridge setting, in which AML may exist, similar felsic magmatic rocks are rare and have been observed occurring as erupted andesitic–dacitic lavas (e.g., Cotsonika, 2006; Wanless et al., 2010) or as intrusive bodies or patches within sheeted dikes and gabbros (e.g., Haase et al., 2016; Zhang, Koepke, France, et al., 2017). Additionally, felsic intrusive rocks have been reported from mantle sections of some ophiolites, but their formation is mostly likely related to an intraoceanic subduction rather than to a spreading ridge (e.g., Rollinson, 2014; Haase et al., 2015). The slab-derived felsic melts in a subduction setting generally show higher MgO and other mantle-rich trace elements (e.g., Ni and Cr) than those from spreading ridges, because the former melts have a strong potential to be reacted with the mantle wedge during their ascent (Kelemen, 1995; Wood & Turner, 2009; Yogodzinski et al., 2001).

In a spreading-ridge setting in the absence of influence by intraoceanic subduction, the origin of felsic melts is not axiomatic but may be related to the following processes (see reviews in Koepke et al., 2007): (a) partial melting of gabbro at the presence of hydrothermal fluid; (b) partial melting of hydrothermally altered sheeted dikes at the roof of AMLs; (c) liquid immiscibility of MORB melts into separated high-Si and high-Fe liquids; (d) extensive differentiation of MORB melts within AMLs.

9.5.2. The Model of Fractional Crystallization

Since the pioneering work of Coleman and Peterman (1975), the model of extensive fractional crystallization of primitive MORB melts at shallow crustal depths has long been proposed for the formation of oceanic felsic magmas at spreading ridges (e.g., Byerly, 1980; Floyd et al., 1998; Haase et al., 2016; Jafri et al., 1995; Niu et al., 2002; Pallister & Knight, 1981; Perfit & Fornari, 1983; Stern, 1979). There are abundant experiments in the literature, on equilibrium crystallization and resultant liquidus line of descent for MORB-like mafic systems (e.g., Berndt et al., 2005; Botcharnikov et al., 2008; Feig et al., 2006; Husen et al., 2016; Juster et al., 1989; Kinzler & Grove, 1992; Koepke et al., 2018; Spulber & Rutherford, 1983; Tormey et al., 1987) and for andesitic–dacitic systems (e.g., Almeev et al., 2013; Erdmann & Koepke, 2016; Gaetani et al., 1994; Holtz et al., 2005), but much less attention has been paid to the simulation of "true" fractional crystallization processes (Grove et al., 1992; Villiger et al., 2004; see below). In fact, "real" experimental simulations on the formation of silicic melts via fractional crystallization of a MORB-like parental magma under the pressure condition of oceanic crust formation are still lacking.

Grove et al. (1992) performed several equilibrium crystallization experiments using MORB-like starting compositions at pressures 1 atm to 1 GPa, and discussed the evolution of melt composition as a result of fractional crystallization based on estimation of the compositions and proportions of solid phases that crystallize as a function of pressure and magma composition. The fractionation degree (i.e., weight proportion of fractionated solid phases relative to parental magma) involved in modeling in this study is below ~32 wt%. The resultant evolved melts always have MORB-like compositions but with slightly higher FeO^{tot}, TiO_2, Na_2O, and lower MgO contents compared to their parental magmas, but the variation in SiO_2 content (< 52 wt% in all cases) is very limited and negligible. Therefore, these experimental data cannot be applied directly for assessing the formation of oceanic felsic melts from a parental MORB melt, but it can be inferred that the fractionation degree of the potential differentiation process from MORB to plagiogranite should be much higher than ~30%. Instead of performing equilibrium crystallization experiments, Villiger et al. (2004, 2007) performed fractional crystallization experiments on a primitive low-K tholeiitic melt (49.10 wt% SiO_2, 13.10 wt% MgO, 0.08 wt% K_2O) at mantle pressures (1.0 and 0.7 GPa). In their experimental setup, near-perfect fractional crystallization is approached in a stepwise manner using equidistant temperature steps and starting compositions corresponding to the liquid composition of the previous, higher temperature glass composition. These authors observed that residual melts could achieve a felsic composition (~67 wt% SiO_2) similar to oceanic plagiogranites when fractionation degree is above 90 wt%. However, the large discrepancy in pressure of these experiments and those of the shallow AMLs at fast spreading ridges prevents direct applications of these experimental data to the formation of the felsic melts within an AML. Recently, Erdmann and Koepke (2016) conducted equilibrium crystallization experiments at a shallow pressure (200 MPa) using a dacitic starting glass to investigate the formation of silicic melts at varied a H_2O and fO_2. Their results indicate that natural oceanic felsic rocks can be formed only at low a_{H2O} if fractional crystallization of mafic parental magmas is the responsible process, because high a_{H2O} strongly depresses the stability of plagioclase and would result in unrealistically high Al_2O_3 content in the residual melt. The results also demonstrate that the prevailing fO_2 of the evolving system should be similar to or less oxidized than the QFM buffer, because higher fO_2 effectively promotes abundant crystallization of Fe–Ti oxides and results in over depletion of Fe in the residual melt; the inferred fO_2 is consistent with that determined from the Fe oxidation state of natural MORB glasses (Bézos & Humler, 2005; Cottrell & Kelley, 2011). Recently, Koepke et al. (2018) performed crystallization experiments within a typical MORB late-stage system strongly enriched in FeO and TiO_2, in order to understand the late magmatic processes that occur in the deep oceanic crust, leading to the formation of oxide gabbros. The results show that Fe–Ti oxides are the liquidus phases followed by clinopyroxene, apatite, and plagioclase, leading to residual melts at lower temperatures with typical compositions of oceanic plagiogranites. The experimental results of this study imply a relatively simple two-step differentiation model for reaching felsic residual melts. Initially, primitive MORBs differentiate along the 1 atm cotectic trend to typical ferrobasaltic compositions by fractionation of olivine, plagioclase, clinopyroxene. This trend continues as long as Fe–Ti oxides are not saturated, which is strongly dependent on the prevailing oxygen fugacity. Once such ferrobasaltic magmas are cut off from replenishment by fresh MORB, the possibility for subsequent differentiation to highly evolved melts is provided by fractionation of oxides, clinopyroxene, plagioclase, apatite, and finally amphibole and apatite.

Despite the available constraints from previous experimental studies, a full picture for the whole route from MORB to silicic melt via fractionation is vague. Besides the experimental tool, numerical simulation using dedicated phase-equilibria modeling programs (such as MELTS and Petrolog) has become an important technique for investigating magma differentiation (e.g., MacLeod et al., 2013; Niu et al., 2002; Toplis & Carroll, 1996) and also for exploring the formation mechanism and condition of the felsic magmas within oceanic crust (Freund et al., 2014; Mueller et al., 2017; Wanless et al., 2010; Zhang, Koepke, France, et al., 2017).

9.5.3. The Model of Partial Melting

Two modes of hydrous partial melting of mafic source rocks within the oceanic crust can be observed: partial melting of deep gabbro with the presence of hydrothermal fluid (Koepke et al., 2004), and partial melting of hydrothermally altered sheeted dikes at the roof of AMLs. Both models can explain the formation of felsic dikes and patches as partial melts in sheeted dikes and gabbros, due to heating and partial melting involving breakdown of hydrous phases or input of fluid (France et al., 2010, 2013; Gillis & Coogan, 2002; Koepke et al., 2014). The robustness of the two models has been confirmed by partial melting experiments using appropriate protoliths at corresponding conditions (Erdmann et al., 2015; Fischer et al., 2016; France et al., 2010; Koepke et al., 2004, 2007; Wolff et al., 2013). Partial melting of crustal rocks can be seen as a consequence and boundary condition in the investigation of the magmatic evolution within the AML, and assimilation of partially melted roof materials by the AML is considered as an important process for inducing significant variations in trace-element abundances and some isotopes (such as Cl concentration and O isotopes; Fischer et al., 2016; France et al., 2014; Freund et al., 2013; Wanless et al., 2011). For details of this topic see also section 9.6.3. In principle, the question of whether melts produced by anatexis of hydrothermal altered source rocks have been involved or not, could be answered by applying specific isotope systems like Os and B (e.g., Gannoun et al., 2017), which to date has not been done for felsic rocks from the oceanic crust. A potential difficulty, however, would be to distinguish between the primary signal and that from the later hydrothermal alteration, which is known to be very strong in all rocks related to the AML.

9.5.4. The Model of Liquid Immiscibility

The model of formation of felsic melts due to liquid immiscibility of primitive and/or evolved MORB melt and within AMLs is still under debate. Although liquid immiscibility has been observed experimentally for a dry primitive MORB system at 1 atm (Dixon & Rutherford, 1979; Philpotts, 1979), there is no evidence of liquid immiscibility from abundant experiments with similar compositions performed at conditions with elevated pressure, water activity (a_{H2O}), and/or oxygen fugacity (fO_2) (e.g., Berndt et al., 2005; Botcharnikov et al., 2008; Feig et al., 2006; Grove & Baker, 1984; Koepke et al., 2018). Some other experimental studies on MORB liquid immiscibility also revealed that a two-liquid field is strongly depressed by increasing pressure, water content, and fO_2 (Charlier & Grove, 2012; Lester et al., 2013; Toplis & Carroll, 1995). For the magmatic system within AMLs, the water content in the evolved melts should be much higher than primary MORB melts (Figure 9.8), which are mostly 0.1–0.3 wt% H_2O (Dixon et al., 1988; Wanless & Shaw, 2012). In addition, a slightly more oxidized condition than that of primary MORB melts, which are mostly 0.4–0.0 log units below the QFM buffer (Bézos & Humler, 2005; Cottrell & Kelley, 2011), is expected for the evolved melts within AMLs. Furthermore, evolved mafic and felsic lithologies sampled at the DGT usually show continuous variations in SiO_2 and FeO^{tot} contents (e.g., Freund et al., 2014; Zhang, Koepke, France, et al., 2017), which contradicts with the significant compositional gap that would be expected from the model of liquid immiscibility. Thus, we argue that the formation of felsic melts within AML cannot be solely or primarily controlled by liquid immiscibility of primitive or evolved MORB melts within AMLs (see also discussion on this in Koepke et al., 2018).

9.5.5. Composition of Oceanic Plagiogranites

We provide a selective up to date compilation of studies reporting oceanic plagiogranites from current oceans and ophiolites in Table 9.2. The published data on plagiogranites are classified and compared based on calculated CIPW (Cross–Iddings–Pirsson–Washington) norms from different settings (Figure 9.10). Although it is difficult to reconcile all data with the origins proposed by the respective authors, several distinctive fields in the Ab–Or–An (albite–orthoclase–anorthite) diagram can be inferred for the specific origins of plagiogranites, in combination with experimental and modeling results (see the lowest panels in Figure 9.10), and comparison with natural plagiogranites is briefly discussed below.

For the felsic melts formed via experimental partial melting of crustal rocks at shallow pressures, low-degree partial melts are usually trondhjemitic for low-K protoliths (≤0.1 wt% K_2O) but can be granitic if protoliths contain slightly higher K_2O contents (0.2–0.3 wt%). This is consistent with the wide range of normative Or proportions in the natural plagiogranites from the current

Table 9.2 Compilation of Studies Reporting Oceanic Plagiogranite from Current Ocean and Ophiolite

Background	Setting[a]	Location[b]	Rock type	Lithology[c]	Origin[d]	Reference[e]
Current ocean	MOR	Galapagos, EPR	Lava	And, Rhy	FC	Byerly et al., 1976
Current ocean	MOR	Galapagos, EPR	Lava	And, Dac, Rhy	FC	Byerly, 1980
Current ocean	MOR	Galapagos, EPR	Lava	Bas-and, And, Dac	FC	Perfit & Fornari, 1983
Current ocean	MOR	Galapagos, EPR	Lava	Bas-and, And, Dac	AFC	Perfit et al., 1992
Current ocean	MOR	9°N, EPR	Lava	Dac	AFC	Wanless et al., 2010
Current ocean	MOR	9°N, EPR	Lava	Bas-and, And, Dac	AFC	Wanless et al., 2011
Current ocean	MOR	IODP Hole 1256D, EPR	Intrusion	Ton	PMC	Zhang et al., 2017
Current ocean	MOR	IODP Hole 1256D, EPR	Intrusion	Dio	AFC	Zhang et al., 2017
Current ocean	MOR	45°N, MAR	Intrusion	Dio, Ton	FC	Aumento, 1969
Current ocean	MOR	5°S, MAR	Intrusion	Gbn, Ton	PMC	Silantyev et al., 2014
Current ocean	MOR	13°N, MAR	Melt inclusion	Dac, Rhy	PMC	Aranovich et al., 2015
Current ocean	MOR	37-40°S, PAR	Lava	Bas-and, And, Dac	AFC	Freund et al., 2013
Current ocean	MOR	ODP Hole 735B, SWIR	Intrusion	Dio, Ton, Tro	FC	Niu et al., 2002
Current ocean	MOR	IODP Hole 1473A, SWIR	Intrusion	Dio, Tro, Ton	FC?	MacLeod et al., 2017
Ophiolite	MOR	Corsica, France	Intrusion	Ton	FC	Beccaluva et al., 1977
Ophiolite	MOR	Newfoundland, Canada	Intrusion	Ton, Tro	PMC	Malpas, 1979
Ophiolite	MOR	Newfoundland, Canada	Intrusion	Ton, Tro	FC	Malpas, 1979
Ophiolite	MOR/SSZ	Oman	Intrusion	Dio, Tro	FC	Alabaster et al., 1982
Ophiolite	MOR/SSZ	Oman	Intrusion	Tro	PMC	Rollinson, 2009
Ophiolite	MOR/SSZ	Oman	Intrusion	Dio, Ton, Tro	FC	Rollinson, 2009
Ophiolite	MOR/SSZ	Oman	Intrusion	Grd, Ton	PMM	Rollinson, 2014
Ophiolite	MOR/SSZ	Oman	Intrusion	Dio, Ton, Tro	FC	Haase et al., 2016
Ophiolite	MOR/SSZ	Oman	Intrusion	Ton	FC	Mueller et al., 2017
Ophiolite	MOR/SSZ	Oman	Intrusion	Dio, Tro	PMC	Mueller et al., 2017
Ophiolite	SSZ	Troodos, Cyprus	Intrusion	Grd, Ton	FC	Coleman & Peterman, 1975
Ophiolite	SSZ	Troodos, Cyprus	Intrusion	Ton, Tro	FC	Freund et al., 2014
Ophiolite	SSZ	Troodos, Cyprus	Lava	And, Dac	FC	Moores & Vine, 1971
Ophiolite	SSZ	Troodos, Cyprus	Intrusion	Dio, Ton, Tro	FC	Moores & Vine, 1971
Ophiolite	SSZ	Troodos, Cyprus	Intrusion	Dio, Ton, Tro	FC	Huggett, 2015
Ophiolite	SSZ	Karmoy, Norway	Intrusion	Grd, Ton, Tro	FC	Pedersen & Malpas, 1984
Ophiolite	SSZ	Karmoy, Norway	Intrusion	Ton, Tro	PMC	Pedersen & Malpas, 1984
Ophiolite	SSZ	Canyon, Oregen	Intrusion	Ton, Tro	PMC	Gerlach et al., 1981
Ophiolite	SSZ	Fidalgo, Washington	Intrusion	Dio, Tro	PMC?	Brown et al., 1979
Ophiolite	SSZ	Sarmento, Chile	Intrusion	Ton	FC	Saunders et al., 1979
Ophiolite	SSZ	Anatoly, Turkey	Lava	Rhy	FC	Floyd et al., 1998
Ophiolite	SSZ	Anatoly, Turkey	Intrusion	Tro	FC	Floyd et al., 1998
Ophiolite	SSZ	Zagros, Iran	Intrusion	Gra, Ton, Tro	PMC	Sepidbar & Mirnejad, 2016
Ophiolite	SSZ	Mingora, Pakistan	Intrusion	Ton, Tro	FC	Barbieri et al., 1994
Ophiolite	SSZ	Muslim Bagh, Pakistan	Intrusion	Gra, Ton, Tro	PMC	Cox et al., 2018
Ophiolite	SSZ	Nagaland–Manipur, India	Intrusion	Dio, Tro	PMM	A. K. Singh et al., 2016
Ophiolite	SSZ	Andaman, India	Intrusion	Ton, Tro	FC	Jafri et al., 1995
Ophiolite	SSZ	Altyn-Tagh, China	Intrusion	Ton	PMS	Gao et al., 2011
Ophiolite	SSZ	Altyn-Tagh, China	Intrusion	Ton	PMS	Gai et al., 2015
Ophiolite	SSZ	Myitkyina, Myanmar	Intrusion	Ton, Tro	PMS	Xu et al., 2017
Ophiolite	BAB	Makran, Iran	Intrusion	Dio, Tro	FC	Ghazi et al., 2004
Ophiolite	BAB	Cuomoqu, China	Intrusion	Tro	PMC	Yin et al., 2015

Note. [a] MOR, midocean ridge; SSZ, suprasubduction zone; BAB, back-arc basin. [b] EPR, East Pacific Rise; MAR, Mid-Atlantic Ridge; PAR, Pacific-Antarctic Rise; SWIR, Southwest Indian Ridge. [c] Bas-and, basaltic andesite; And, andesite; Dac, dacite; Rhy, rhyolite; Dio, diorite; Gbn, gabbronorite; Grd, granodiorite; Gra, granite; Ton, tonalite; Tro, trondhjemite. [d] FC, fractional crystallization of MORB magma; AFC, assimilation of crustal material and fractional crystallization of MORB magma; PMC, partial melting of crustal rocks adjacent to axial magma chamber; PMS, partial melting of subducted oceanic slab; PMM, partial melting of mantle. [e] Provided separately in Table SM1 of the Supplementary Material.

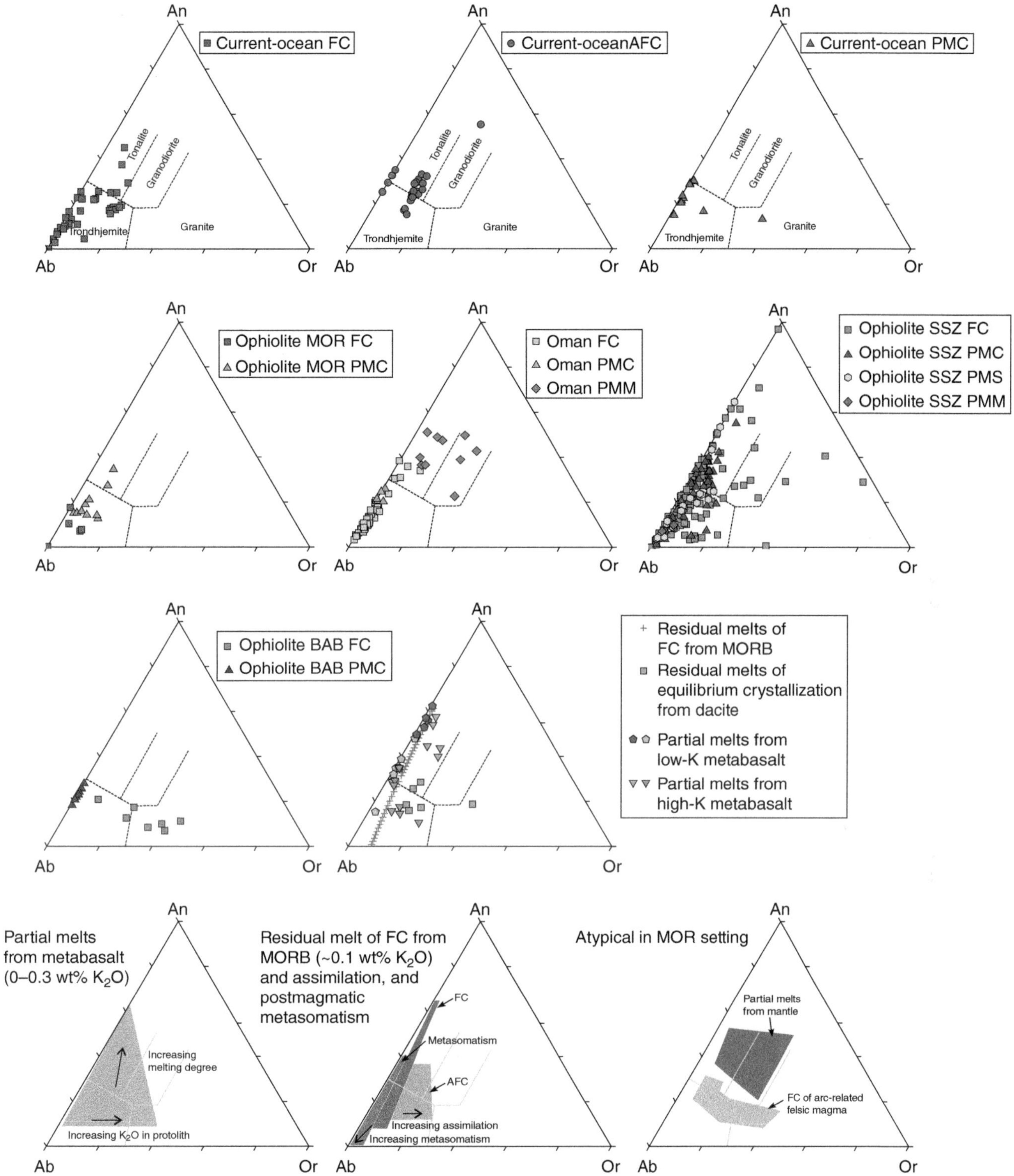

Figure 9.10 Classification of oceanic felsic magmatic rocks ($SiO_2 > 63$ wt%) based on CIPW norms (after Barker, 1979). Residual melts of fractional crystallization (FC) from MORB are modeled with initial K_2O content of 0.13 wt% and H_2O content of 0.2 wt% with oxygen fugacity (fO_2) along the quartz–fayalite–magnetite (QFM) buffer (after Zhang, Koepke, France, et al., 2017). Residual melts of equilibrium crystallization from dacite are from Erdmann and Koepke (2016). Partial melts from low-K metabasalt (0.07 and 0.10 wt% K_2O) are from Koepke et al. (2004) and Erdmann et al. (2015). Partial melts from high-K metabasalt (0.31 and 0.20 wt% K_2O) are from Koepke et al. (2004) and France et al. (2010). MOR, midocean ridge; SSZ, suprasubduction zone; BAB, back-arc basin; FC, fractional crystallization of MORB magma; AFC, assimilation of crustal material and fractional crystallization of MORB magma; PMC, partial melting of crustal rocks adjacent to axial magma chamber; PMS, partial melting of subducted oceanic slab; PMM, partial melting of mantle. See Table 9.2 for data sources.

ocean floor that are believed to have formed from partial melting of crustal rocks (i.e., hydrous altered basalt or gabbro). Tonalitic melts are generally formed at a higher melt fraction compared to the formation of trondhjemitic melts, but the degree of melting is still generally low (< 30%, Koepke et al., 2004). Modeling of fractional crystallization of MORB melt shows that sole residual melts have a low and strictly small range of normative Or proportions, which are inconsistent with many natural plagiogranites from the current ocean floor that are believed have formed from fractional crystallization of parental MORB magmas. On the one hand, the wide occurrence of high-Or plagiogranites suggests that assimilation of crustal-derived melts into the evolving magma system within AMLs may be very common, which is also supported by evidence from trace elements (Fischer et al., 2016; France et al., 2014). On the other hand, some plagiogranites are extremely depleted in normative Or and rich in albite, which may have formed by post-magmatic metasomatism with Na-rich hydrothermal fluids, such as those albitites recovered from IODP Hole 1256D of the EPR (Zhang, Koepke, France, et al., 2017). These rocks are composed exclusively of low-temperature alteration minerals, as indicated by abundant albite, epidote, and quartz. Bulk geochemistry reveals strong enrichments of Na and Al, and low Fe and Mg contents, suggesting metasomatism of the original rocks by hydrothermal, sodium-rich fluids, which is also supported by the trace-element geochemistry. The most immobile trace-element contents such as Zr, Hf, and Ti imply a gabbro as the original rock material (for details see Zhang, Koepke, France, et al., 2017; see also section 9.7).

The plagiogranites from ophiolites, due to potentially complicated geological settings related to their formation, show more complex compositional features. The plagiogranites from the MOR and/or SSZ (suprasubduction zone) settings believed to have formed via fractional crystallization (FC) of MORB parental magmas generally show a parallel compositional trend relative to the modeled FC trend, with influences from crustal assimilation and postmagmatic metasomatism. However, the FC-related plagiogranites from the back-arc basin (BAB) related ophiolite (Makran, Iran) show a low-An parental magma and strong enrichment in K with increasing SiO_2 content, implying that the primary magma should be characterized by high Na and intermediate K. Erdmann and Koepke (2016) experimentally modeled a similar crystallization-controlled differentiation at oxidized and water-rich conditions for a dacitic parental magma, and the residual melts show a comparative K-enrichment trend. Those plagiogranites believed to have formed via partial melting of crustal rocks are generally interpreted from experimental partial melting results, with variable protolith K_2O contents and postmagmatic metasomatism. Some plagiogranites from SSZ-related ophiolites are proposed to have derived from subducted slab, which generally overlaps the range of shallow crustal partial melts in the normative Ab–Or–An diagram. Finally, the mantle-derived plagiogranites in the Oman ophiolite have distinctively higher Or and lower Ab, as well as much higher MgO and other mantle-rich trace elements (Rollinson, 2014). However, the plagiogranites from the Nagaland–Manipur ophiolite (India), which are also proposed by A. K. Singh et al. (2016) to have been derived from a depleted mantle source, have similar major and trace elemental features as that of crustal plagiogranites.

9.6. MAGMATIC PROCESSES AT THE AML ROOF

In our model of vertically fluctuating AMLs at fast spreading ridges, one critical step is the rising of an AML after a replenishment event, which must somehow burn its way through solid rocks, which are most likely hydrothermally altered sheeted dikes, which may or may not be granoblastically overprinted. Two different magmatic processes can be assumed then: (a) assimilation of the rocks above the top of an upmoving AML and (b) partial melting of the AML roof rocks. According to our limited understanding, we assume for the assimilation process that (a) the upward moving of the AML must be relatively fast, whereas for (b) the partial melting event we expect lower velocities for ascent, because here we assume that a metamorphic gradient must be developed with peak conditions approaching the solidus of the system. For both events we found evidence in the DGT drilled by the IODP at IODP Site 1256, which are highlighted in the following.

9.6.1. Assimilation of the AML Roof Rocks

In many of the gabbros of the IODP Hole 1256D core relics of former sheeted dikes have been identified, both macroscopically as well as microscopically. During the macroscopic characterization of gabbro sections, fragments up to centimeter size of former basalts from the sheeted dike sequence could be observed, which were typically transformed into fine grained two-pyroxene hornfelses (e.g., France et al., 2009; Teagle et al., 2006; Wilson et al., 2006). More evidence for dike relics was reported in thin-sections from gabbros of IODP Site 1256, where both textural and mineralogical relics from former dike basalts were recorded (Koepke et al., 2011; Teagle et al., 2006). Similar observations have been reported from varitextured gabbros of the Oman ophiolite, interpreted as part of a frozen melt lens system (Mueller et al., 2017). Thus, there is plenty of evidence that assimilation occurs at the roof of AMLs with the potential to contaminate MORB, expressed by enrichment in Cl and other volatile and fluid-mobile elements, provided that the assimilated material was previously hydrothermally altered. We focus on this in section 9.6.3.

9.6.2. Anatectic Processes at the AML Roof

Tonalitic to trondhjemitic patches and veins cutting granoblastic hornfelses are well known both from the DGT horizons drilled at IODP Site 1256 (e.g., France et al., 2009, 2014; Teagle et al., 2006; Zhang, Koepke, France, et al., 2017) and from the Troodos (Gillis & Coogan, 2002) and Oman (France et al., 2014; Haase et al., 2016; Mueller et al., 2017) ophiolites. Gillis & Coogan (2002) were the first to relate these felsic rocks to a partial melting event at the roof of an AML, a model that was later confirmed by geochemical studies (France et al., 2014; Haase et al., 2016; Zhang, Koepke, France, et al., 2017). Zhang, Koepke, France, et al. (2017) showed for the DGT from IODP Site 1256, using trace element modeling in combination with MELTS simulations, that thin tonalitic dikes cutting granoblastic hornfelses ~2 m above the dike–gabbro contact in core 212 (Teagle et al., 2006) have an anatectic origin by hydrous partial melting of basalts from the sheeted dike sequence. This is in contrast to dioritic and gabbroic rocks from the plutonic section of IODP Site 1256, which could be derived from fractional crystallization of MORB. Similar results have been reported from the Oman ophiolite by Mueller et al. (2017), who studied an outcrop in the mound of Wadi Gideah in the southernmost ophiolite massif, showing very complex intrusion relationships, which was interpreted as part of a fossilized AML. The plagiogranitic rocks of mainly dioritic composition could be derived, according to trace-element modeling in combination with MELTS simulations, from fractional in situ crystallization of typical MORB, while late, trondhjemite dikes cutting the dioritic rocks could be interpreted as anatectic melts derived from partial melting of sheeted dikes. A similar result was found by Haase et al. (2016) from a bulk geochemical study on Oman plagiogranites, i.e., that most of the plagiogranites were formed by fractional crystallization of MORB, except some small-scale tonalitic to trondhjemitic intrusions in the sheeted dike rooting zone, which were interpreted as melts formed by anatexis.

Very strong evidence that trondhjemite and tonalite melts occurring in the DGT could be derived from partial melts of former sheeted dikes was provided by experimental studies (Erdmann et al., 2015, 2017; Fischer et al., 2016). These studies experimentally simulated partial melting of AML roof rocks, using basalts and hornfelses from the IODP Hole 1256D core as starting materials. The experimental products—melts and residual crystals—generated at low melt fractions at temperatures between 910°C and 970°C match the natural lithologies from IODP Site 1256 in terms of major and trace element compositions. The compositions of the partial melts correspond to the compositions of the tonalitic and trondhjemitic veins cutting the granoblastic hornfelses, while the residual minerals match the compositions of the two-pyroxene hornfels (Figure 9.11), indicating that hydrous partial melting of AML roof rocks is an essential magmatic process in the GDT of fast spreading midoceanic ridges. For details on the trace-element distribution between products and residues of the partial melting event see the detailed experimental studies of France et al. (2014) and Erdmann et al. (2017).

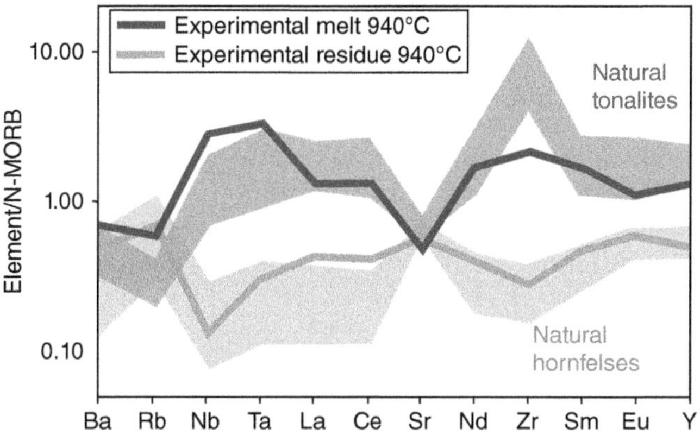

Figure 9.11 Complementary normal MORB (Gale et al., 2013) trace-element patterns for experimental melts and residuals produced by hydrous partial melting of basalts from the sheeted dike complex of IODP Site 1256 at a temperature of 940°C and a pressure of 100 MPa. The experimental products match the corresponding natural lithologies, the tonalitic veins and the two-pyroxene hornfelses, respectively, from the dike-gabbro transition at IODP Site 1256, implying that the tonalites can be regarded as anatectic melts, and the hornfelses as a residual phase after a partial melting event. [Diagram modified from Fischer et al. (2016).]

One benefit of the experimental studies is that they may help to constrain details of the physical conditions of anatectic processes at the roof of AMLs from fast spreading systems. Erdmann et al. (2015, 2017) performed partial melting experiments using six different protoliths from the IODP Site 1256 sheeted dike complex, spanning a lithological range from poorly to strongly altered basalts to partially or fully recrystallized granoblastic hornfels. Results show that extensively altered starting material lacking primary magmatic minerals cannot reproduce the chemistry of the natural tonalites and trondhjemite dikes of the drilled DGT at IODP Site 1256, especially in terms of volatile and fluid-mobile elements (e.g., K, Cl). Geochemical trends of the natural rocks from the IODP Hole 1256D core are reproduced through partial melting of moderately altered basalts. Two-pyroxene hornfels, the assumed residual association after partial melting of sheeted dike basalts, were reproduced only at low melt fractions (<0.2). In none of the experiments by Erdmann et al. (2015) was amphibole present, which confirms the observation in the natural hornfelses that amphibole was not produced during peak metamorphism, leading to the hypothesis that the presence of residual amphibole is not required for anatectic processes involving the DGT, even though residual amphibole is often used as an important phase for explaining trace element characteristics in relevant SiO_2-enriched lithologies formed at fast to medium spreading midocean ridge systems when modeling anatectic processes (e.g., Haase et al., 2005; Wanless et al., 2010). The comparison between the experimental products of Erdmann et al. (2015) and the natural equivalents revealed that water activity was significantly reduced during anatectic processes (based on a relatively low Al_2O_3 content in melt and relatively low An content of residual plagioclase).

9.6.3. Contamination of MORB by Magmatic Processes at the AML Roof

The MORBs from fast spreading ridges are in general enriched in Cl, thus regarded as "contaminated," when using Cl as a proxy for contamination (e.g., Michael & Chase, 1987; Michael & Cornell, 1998). Studies using other proxies such as Os isotopes came to the same conclusion (e.g., Gannoun et al., 2007).

Three end-member models are discussed to explain the mechanism of contamination: (a) assimilation of AML roof rocks that are overprinted by seawater-derived hydrothermal circulation and thus enriched in Cl; (b) contamination by felsic melts produced by partial melting of previously hydrothermally altered AML roof-rocks (for details see Coogan et al., 2003); and (c) assimilation of seawater-derived Cl-rich brines (e.g., Bischoff & Rosenbauer, 1989; Kendrick et al., 2013; Zhang, Koepke, Albrecht, et al., 2017). With the help of trace-element compositions of experimental melts produced by hydrous partial melting of dikes from IODP Site 1256, Fischer et al. (2016) were able to perform calculations of mixing and of assimilation and fractional crystallization (AFC) using the experimental partial melts as contaminant/assimilant in order to clarify which process is more important. In accord with theoretical evaluations of Coogan et al. (2003), these authors showed that anatectic melts can only be a minor contributor to the contamination process, revealing that the assimilation of AML roof rocks is obviously the main source of MORB contamination.

9.7. THE TRANSITION FROM MAGMATIC TO METAMORPHIC PROCESSES: THE ROLE OF HYDROTHERMAL FLUIDS

Cycling of hydrothermal fluids plays a key role in the formation of lower oceanic crust by extracting the heat influx from magmatic input at MORs (Maclennan et al., 2005; Phipps Morgan & Chen, 1993; Theissen-Krah et al., 2016). At the stage of peak magma replenishment, the overlying sheeted dikes of the AML undergo contact metamorphism and anatexis to form tonalitic partial melts, and in the meantime the activity of hydrothermal fluids is restricted above the contact metamorphism zone. Consequently, this hornfelsic layer serves as a conductive boundary layer between the shallow hydrothermal system and the AML (Gillis, 2008; Koepke et al., 2008). Although abundant low-density fluid can be excluded as a component involved in the magma crystallization (Erdmann & Koepke, 2016; Zhang, Koepke, France, et al., 2017) and crustal anatexis (*Erdmann et al.*, 2015), a small amount of high-salinity brine that resulted from hydrothermal boiling of seawater-derived fluids may sink and assimilate AML interstitial melts, as suggested by high-Cl magmatic apatite (Zhang, Koepke, Albrecht, et al., 2017).

At the interreplenishment stage, magma supply from the mantle shrinks and finally stops, which allows cooling and enhanced crystallization of interstitial melts in the AML. Simultaneously, the overlying hydrothermal fluids are able to penetrate the cooled hornfelsic layer and tonalitic rocks, and beyond to the gabbroic and dioritic rocks within the fossilized AML (see right panel of Figure 9.9). Strontium isotopes indicate that hydrothermal fluids are preferentially channeling along the boundary between the intrusive body and hornfelsic rocks (Harris et al., 2015), implying a structural dependence of fluid pathways. Characteristic hydrothermal minerals, such as epidote, chlorite, actinolite, quartz, and overgrowth of apatite have been uncovered at the dike–gabbro transition zone (Alt et al., 2010; Zhang, Koepke, Albrecht, et al., 2017, Zhang, Koepke, France, et al., 2017; Zhang, Wang, et al., 2017). Estimated temperatures of hydrothermal fluids based on fluid inclusions in quartz can be as low as 300°C (Alt et al., 2010). Fe–Ti oxides have also been recrystallized at subsolidus and more oxidized conditions (fO_2 higher than the QFM buffer by 1–3 log units; Zhang, Koepke, France, et al., 2017). The anatectic tonalites and some dioritic intrusive rocks have also been strongly influenced by hydrothermal metasomatism, as indicated by multistage overgrowth of apatite and quartz, which resulted in remarkable depletions in mobile elements (e.g., Th, U, REE) and relative enrichment of immobile elements (e.g., Zr, Hf; Zhang, Koepke, France, et al., 2017). There are also albititic rocks that have been strongly metasomatized by Na-rich fluids, which shows characteristic positive Eu anomalies and extremely variable trace-element abundances that are highly dependent on the local occurrence of accessory minerals (Zhang, Koepke, France, et al., 2017). Most intrusive and hornfelsic rocks in this zone have been influenced by seawater-derived hydrothermal fluids, because they have higher bulk Cl concentrations than MORB lavas (Zhang, Wang, et al., 2017); if no extra Cl was introduced by hydrothermal fluids, these rocks should be relatively depleted in Cl since magma-originated Cl is expected to be highly partitioned into the fluid phase that forms near solidus conditions (Botcharnikov et al., 2015; Webster & Holloway, 1990). In addition, the high-SiO_2 tonalitic and albititic rocks show abnormally high Br/Cl ratios relative to other less altered dioritic and gabbroic rocks (Zhang, Wang, et al., 2017), indicating strong Br/Cl fractionation relative to the mantle budget and that these rocks have been to the greatest extent metasomatized by seawater-derived fluids.

As the period of an interreplenishment stage (normally tens of years (Zhang et al., 2014); for details see section 9.8) is shorter than the time it takes for the fossilized AML intrusive rocks to be pushed the distance to an off-axis position (Colman et al., 2012; Zhang et al., 2014), these hornfelsic and intrusive gabbroic–dioritic–tonalitic rocks, which have been hydrothermally altered to variable extents, must be country rocks of the new AML formed by the next peak magma replenishment (Alt et al., 2010). When heated by fresh AML magmas, these rocks would tend to be recrystallized and/or partially melted, with dehydration of hydrous minerals but without a free fluid phase ($a_{H2O} < 1$; see Erdmann et al., 2015). Nevertheless, hydrothermal alteration is the last formative force that shaped the final appearance of those rocks encountered at a mature oceanic crust, which has introduced remarkable variations in mineralogy, trace elements, and isotopes that greatly deviate from their initial igneous characteristics. These alteration effects will also finally influence the process and outcome of oceanic lithosphere subduction.

9.8. CONSTRAINTS ON TIMESCALES FOR THE VERTICAL FLUCTUATIONS OF THE AML

The periodic activity of the AML at fast spreading ridges is characterized by cyclical processes of waxing and waning due to discontinuous magma supply, which is a key parameter that dominates the dynamics of crustal accretion. The alternating rate of magma replenishment and volcanic eruption largely determines the timescale for the vertical fluctuation of AML, with the highest position of its upper boundary corresponding to the peak stage of magma replenishment (Coogan et al., 2003; France et al., 2009; Gillis, 2002, 2008; Hooft et al., 1997; Koepke et al., 2008; Zhang et al., 2014). The timescales of the periodic activity of the AML have been inferred by geophysical, geochemical, petrological, and mineralogical approaches, and the estimated results in the literature from different locations and/or approaches differ greatly from tens of years to hundreds of thousands of years.

9.8.1. Evidence for a Long Timescale

Based on the changes in axial depth, morphology, and basalt geochemistry from different segments along the southern EPR, Hooft et al. (1997) proposed that the variation in magma supply is on a timescale of ~100,000 years. Similarly, based on the differences in ridge propagation at 16–19°S of the southern EPR that have occurred prior to 1 Ma and since 1 Ma, and the assumption that magma supply is the triggering force for such a change, Cormier (1997) proposed that the timescale of transition from a starved to a robust magma supply is about 100,000 years. Such a long duration has also been proposed by Reynolds et al. (1992) to interpret the spatial and temporal variability in the composition of basalts from 12°N of the northern EPR. Similarly, based on the spatial variations in basalt compositions, Sinton et al. (2002) and Pollock et al. (2009) proposed that a roughly constant magmatic temperature might have persisted within a timescale of tens of thousands of years. Colman et al. (2012) studied eruptive units in two areas along the Galápagos Spreading Center, and observed significant differences in eruptive volume, lava temperature, morphology, and inferred eruption rates between these two areas that represent a time span of hundreds to thousands of years.

9.8.2. Evidence for a Short Timescale

Using a geochemical approach on basalt glasses, Perfit et al. (1994) observed that the timescale of replenishment of an AML should be much shorter than 5000 years. Mutter et al. (2008) compared the results of multi-streamer reflection imaging at the EPR at 9°50'N obtained in 1985 and in 2008, and they observed a significant variation in the depth of the AML reflector, which was much shallower and less complexly structured in 1985. The deeper top boundary of the AML observed in 2008 was likely due to the known volcanic eruptions in 1991 and 2005–2006 in the ridge segment. Therefore, a very short timescale of ~20 years for the AML from waxing to waning can be inferred from the observations of Mutter et al. (2008). Based on the ^{210}Pb–^{226}Ra–^{230}Th radioactive disequilibria analyzed in recently erupted lavas, Rubin et al. (2005) revealed that melt transport, accumulation, and eruption can be accomplished within a short timescale of a few decades, which implies that the vertical fluctuation of the AML can also occur within such a short timescale. Using a mathematical approach to model the magma supply and geochemical variations in the erupted lavas from 17–19°S of the southern EPR, Rannou et al. (2006) determined that the timescale of a cycle between two replenishment events, i.e., refilling of a Mg-rich MORB magma (MgO = 9.5 wt%) into the AML, is about 750 years, in which a magma residence time of about 300 years is included prior to volcanic eruption.

Based on studies of partial melting in the roof of the AML recorded in the migmatized sheeted dyke complex of the Troodos ophiolite, Gillis and Coogan et al. (2002) proposed that the duration of melting events should be tens to hundreds of years long, suggesting that the timescale of vertical fluctuation of the AML, which is the cause of partial melting in the roof, also should be in the order of tens to hundreds of years. Gillis (2008) further estimated a heating timescale based on the diffusion of Mg in plagioclase from a hornfels sampled at Hess Deep at the EPR applying a mean temperature of 990°C, and the results indicate that a single contact metamorphism event may have a duration as short as 50 years. Similarly, Koepke et al. (2008) estimated the metamorphism timescale for hornfelsic rocks from IODP Hole 1256D, based on NaSi–CaAl interdiffusion in plagioclase and an assumed mean temperature of 800°C. They obtained a duration of 23,000 to 5400 years for the contact metamorphism event, which is significantly longer than that estimated by Gillis (2008). The large discrepancy between the estimated timescales of Gillis (2008) and Koepke et al. (2008) is primarily due to the contrasting mean temperature values assumed in their calculations (990°C vs. 800°C). Zhang et al. (2014) further investigated the timescale of heating and cooling associated with the contact metamorphism of the hornfelsic rocks from IODP Hole 1256D. Their more detailed modeling of the NaSi–CaAl interdiffusion and Mg trace diffusion in plagioclase with a cooling process from peak metamorphism temperature (~1000°C) to the boundary condition at the margin of a spreading center (~600°C), indicated that the timescale of a cycle of thermal overprint in the roof of the AML is about 20–60 years. Zhang et al. (2014) also conducted modeling on heat balance, and the results show that, for a fast spreading ridge with a half spreading rate of 110 mm $year^{-1}$, the hydrothermal cooling above the AML must be rapid in order to extract the latent heat from the solidifying magma, and the inferred cooling rate is consistent with the timescale derived from the modeling of element diffusion.

9.8.3. Different Timescales for Different Situations

The long timescales mentioned above seem more relevant to a fundamental change in the nature of magma source rather than to a fluctuation in the depth or geometry of the AML. In contrast, the short timescale appears to be more closely linked to spreading events that reflect the vertical fluctuation of the AML as a result of alternating magma replenishment and volcanic eruption. We may consider a simple scenario for the roof rock overlying the AML at its highest level, during a period of on-ridge presence (Fig. 9.12). Such a rock can be regarded as initially being a sheeted dike generated at the ridge center and

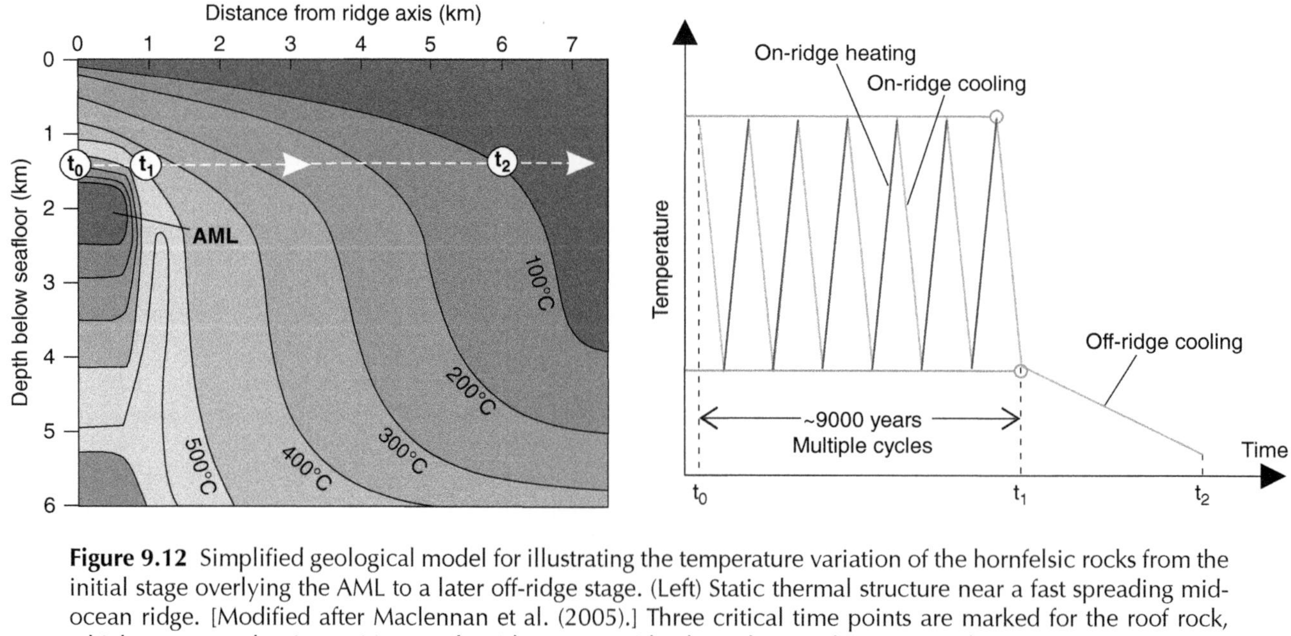

Figure 9.12 Simplified geological model for illustrating the temperature variation of the hornfelsic rocks from the initial stage overlying the AML to a later off-ridge stage. (Left) Static thermal structure near a fast spreading mid-ocean ridge. [Modified after Maclennan et al. (2005).] Three critical time points are marked for the roof rock, which correspond to its positions at the ridge center, ridge boundary, and 100°C. (Right) Temperature variation with time for the hornfelsic rock. The on-ridge duration (from t0 to t1) of ~9000 years is calculated assuming a width of 2 km for the AML and a half spreading rate of 110 mm $year^{-1}$.

laterally heated by the AML to form hornfels at the stage of peak replenishment. For a fast spreading ridge like the EPR 19–12 Ma, when IODP Site 1256 was formed under superfast conditions with a half spreading rate of ~110 mm $year^{-1}$, assuming an AML width of 2 km, ~9000 years are needed for the considered roof rock to complete the whole period of being on-ridge (Figure 9.12). Therefore, if the cycling of vertical fluctuation of the AML is of a short timescale, for instance 100 years, the roof rock should have experienced 90 cycles of heating and cooling. The dynamic nature of the AML is probably the most important factor controlling short-lived hydrothermal venting systems that are common at fast spreading ridges (Lowell & Germanovich, 1994).

ACKNOWLEDGMENTS

This research used samples and/or data provided by the International Ocean Drilling Program (IODP). The IODP is sponsored by the United States National Science Foundation (NSF) and participating countries under management of the Consortium for Ocean Leadership (COL). Funding for this research was provided by grants from the Deutsche Forschungsgemeinschaft (DFG) in the framework of the SPP "International Ocean Discovery Program (IODP)" (SPP 527).

SUPPLEMENTARY MATERIAL

Figures SM1–SM3 and Table SM1 are available as supplementary material at https://doi.org/10.17605/osf.io/cxfp7

REFERENCES

Alabaster, T., Pearce, J. A., & Malpas, J. (1982). The volcanic stratigraphy and petrogenesis of the Oman ophiolite complex. *Contributions to Mineralogy and Petrology*, *81*(3), 168–183.

Almeev, R., Holtz, F., Ariskin, A., & Kimura, J.-I. (2013). Storage conditions of Bezymianny Volcano parental magmas: results of phase equilibria experiments at 100 and 700 MPa. *Contributions to Mineralogy and Petrology*, *166*(5), 1389–1414.

Alt, J. C., Laverne, C., Coggon, R. M., Teagle, D. A. H., Banerjee, N. R., Morgan, S., et al. (2010). Subsurface structure of a submarine hydrothermal system in ocean crust formed at the East Pacific Rise, ODP/IODP Site 1256. *Geochemistry Geophysics Geosystems*, *11*, Q10010. doi: 10010.11029/12010gc003144

Aranovich, L. Y., Prokofiev, V. Y., Pertsev, A. N., Bortnikov, N. S., Ageeva, O. A., Bel'tenev, V. E., et al. (2015). Composition and origin of a K_2O-rich granite melt in the Mid-Atlantic Ridge, 13°34′ N: Evidence from the analysis of melt inclusions and minerals of the gabbro-plagiogranite association. *Doklady Earth Sciences*, *460* (2), 174–178.

Aumento, F. (1969). Diorites from the mid-Atlantic ridge at 45°N. *Science*, *165*(3898), 1112–1113.

Barbieri, M., Caggianelli, A., Di Florio, M. R., & Lorenzoni, S. (1994). Plagiogranites and gabbroic rocks from the Mingora ophiolitic melange, Swat Valley, NW Frontier Province, Pakistan. *Mineralogical Magazine*, *58*(4), 553–566.

Barker, F. (1979). *Trondhjemites, dacites and related rocks*. Amsterdam: Elsevier.

Beccaluva, L., Ohnenstetter, D., Ohnenstetter, M., & Venturelli, G. (1977). The trace element geochemistry of Corsican ophiolites. *Contributions to Mineralogy and Petrology*, *64*(1), 11–31.

Bedard, J. H., Sparks, R. S. J., Renner, R., Cheadle, M. J., & Hallworth, M. A. (1988). Peridotite sills and metasomatic gabbros in the Eastern Layered Series of the Rhum Complex. *Journal of the Geological Society*, *145*, 207–224.

Berndt, J., Koepke, J., & Holtz, F. (2005). An experimental investigation of the influence of water and oxygen fugacity on differentiation of MORB at 200 MPa. *Journal of Petrology*, *46*(1), 135–167.

Bézos, A., & Humler, E. (2005). The Fe^{3+}/Fe ratios of MORB glasses and their implications for mantle melting. *Geochimica et Cosmochimica Acta*, *69*, 711–725.

Bischoff, J. L., & Rosenbauer, R. J. (1989). Salinity variations in submarine hydrothermal systems by layered double-diffusive convection. *Journal of Geology*, *97*(5), 613–623.

Botcharnikov, R. E., Almeev, R., Koepke, J., & Holtz, F. (2008). Phase relations and liquid lines of descent in hydrous ferrobasalt—implications for the Skaergaard Intrusion and Columbia River Flood Basalts. *Journal of Petrology*, *49*, 1687–1727.

Botcharnikov, R. E., Holtz, F., & Behrens, H. (2015). Solubility and fluid–melt partitioning of H_2O and Cl in andesitic magmas as a function of pressure between 50 and 500 MPa. *Chemical Geology*, *418*, 117–131.

Boudier, F., Nicolas, A., & Ildefonse, B. (1996). Magma chambers in the Oman ophiolite: Fed from the top and the bottom. *Earth and Planetary Science Letters*, *144*(1–2), 239–250.

Brown, E., Bradshaw, J., & Mustoe, G. (1979). Plagiogranite and keratophyre in ophiolite on Fidalgo Island,Washington. *Geological Society of America Bulletin*, *90*(5), 493–507.

Byerly, G. (1980). The nature of differentiation trends in some volcanic rocks from the Galapagos Spreading Center. *Journal of Geophysical Research: Solid Earth*, *85*(B7), 3797–3810.

Byerly, G. R., Melson, W. G., & Vogt, P. R. (1976). Rhyodacites, andesites, ferro-basalts and ocean tholeiites from the Galapagos spreading center. *Earth and Planetary Science Letters*, *30*(2), 215–221.

Canales, J. P., Detrick, R. S., Toomey, D. R., & Wilcock, W. S. D. (2003). Segment-scale variations in the crustal structure of 150–300 kyr old fast spreading oceanic crust (East Pacific Rise, 8 degrees 15' N–10 degrees 5' N) from wide-angle seismic refraction profiles. *Geophysical Journal International*, *152*, 766–794.

Canales, J. P., Nedimovic, M. R., Kent, G. M., Carbotte, S. M., & Detrick, R. S. (2009). Seismic reflection images of a near-axis melt sill within the lower crust at the Juan de Fuca ridge. *Nature*, *460*(7251), 89–100.

Charlier, B., & Grove, T. L. (2012). Experiments on liquid immiscibility along tholeiitic liquid lines of descent. *Contributions to Mineralogy and Petrology*, *164*(1), 27–44.

Coleman, R. G., & Donato, M. M. (1979). Oceanic plagiogranite revisited. In F. Barker (Ed.), *Trondhjemites, Dacites, and Related Rocks* (pp. 149–167). Amsterdam: Elsevier.

Coleman, R. G., & Peterman, Z. E. (1975). Oceanic plagiogranite. *Journal of Geophysical Research*, *80*, 1099–1108.

Colman, A., Sinton, J. M., White, S. M., McClinton, J. T., Bowles, J. A., Rubin, K. H., et al. (2012). Effects of variable magma supply on mid-ocean ridge eruptions: Constraints from mapped lava flow fields along the Galápagos Spreading Center. *Geochemistry, Geophysic,s Geosystems*, *13*(8), Q08014.

Coogan, L. A., Mitchell, N. C., & O'Hara, M. J. (2003). Roof assimilation at fast spreading ridges: An investigation combining geophysical, geochemical, and field evidence. *Journal of Geophysical Research-Solid Earth*, *108*(B1), doi:10.1029/2001JB001171

Coogan, L. A., Thompson, G., & MacLeod, C. J. (2002). A textural and geochemical investigation of high level gabbros from the Oman ophiolite: Implications for the role of the axial magma chamber at fast-spreading ridges. *Lithos*, *63*(1–2), 67–82.

Cormier, M. H. (1997). The ultrafast East Pacific Rise: instability of the plate boundary and implications for accretionary processes. *Philosophical Transactions of the Royal Society of London A: Mathematical, Physical and Engineering Sciences*, *355*(1723), 341–367.

Cotsonika, L. A. (2006). *Petrogenesis of andesites and dacites from the Southern Juan de Fuca Ridge*. MA thesis, University of Florida,

Cottrell, E., & Kelley, K. A. (2011). The oxidation state of Fe in MORB glasses and the oxygen fugacity of the upper mantle. *Earth and Planetary Science Letters*, *305*(3), 270–282.

Cox, D., Kerr, A. C., Hastie, A. R., & Kakar, M. I. (2018). Petrogenesis of plagiogranites in the Muslim Bagh Ophiolite, Pakistan: implications for the generation of Archaean continental crust. *Geological Magazine*, *156*(5), 874–888.

Currin, A., Koepke, J., Almeev, R., & Beermann, O. (2018a). Interaction of highly saline fluid and olivine gabbro: experimental simulation of deep hydrothermal processes involving amphibole at the base of the oceanic crust. *Lithos*. https://doi.org/10.1016/j.lithos.2018.09.017

Currin, A., Koepke, J., Almeev, R., & Beermann, O. (2018b). Chlorine-rich amphibole in deep layered gabbros as evidence for brine/rock interaction in the lower oceanic crust: A case study from the Wadi Wariyah, Samail Ophiolite, Sultanate of Oman. *Lithos*. https://doi.org/10.1016/j.lithos.2018.09.015

Detrick, R. S., Buhl, P., Vera, E., Mutter, J., Orcutt, J., Madsen, J., & Brocher, T. (1987).Multichannel seismic imaging of a crustal magma chamber along the East Pacific Rise. *Nature*, *326*, 35–41.

Dixon, S., & Rutherford, M. J. (1979). Plagiogranites as late-stage immiscible liquids in ophiolite and mid-oceanic ridge suites: an experimental study. *Earth and Planetary Sciences Letters*, *45*, 45–60.

Dixon, J. E., Stolper, E., & Delaney, J. R. (1988). Infrared spectroscopic measurements of CO_2 and H_2O in Juan de Fuca Ridge basaltic glasses. *Earth and Planetary Sciences Letters*, *90*, 87–104.

Erdmann, M., Fischer, L. A., France, L., Zhang, C., Godard, M., & Koepke, J. (2015). Anatexis at the roof of an oceanic magma chamber at IODP Site 1256 (equatorial Pacific): an experimental study. *Contributions to Mineralogy and Petrology*, *169*(4).

Erdmann, M., France, L., Fischer, L. A., Deloule, E., & Koepke, J. (2017). Trace elements in anatectic products at the roof of mid-ocean ridge magma chambers: An experimental study. *Chemical Geology*, *456*, 43–57.

Erdmann, M., & Koepke, J. (2016). Silica-rich lavas in the oceanic crust: experimental evidence for fractional crystallization under low water activity. *Contributions to Mineralogy and Petrology*, *171*(10).

Ernst, W. G., & Liu, J. (1998). Experimental phase-equilibrium study of Al- and Ti-contents of calcic amphibole in MORB—a semiquantitative thermobarometer. *American Mineralogist*, *83*, 952–969.

Feig, S., Koepke, J., & Snow, J. (2006). Effect of water on tholeiitic basalt phase equilibria: An experimental study under oxidizing conditions. *Contributions to Mineralogy and Petrology*, *152*, 611–638.

Fischer, L. A., Erdmann, M., France, L., Wolff, P. E., Deloule, E., Zhang, C., et al. (2016). Trace element evidence for anatexis at oceanic magma chamber roofs and the role of partial melts for contamination of fresh MORB. *Lithos*, *260*, 1–8.

Floyd, P. A., Yaliniz, M. K., & Goncuoglu, M. C. (1998). Geochemistry and petrogenesis of intrusive and extrusive ophiolitic plagiogranites, Central Anatolian Crystalline Complex, Turkey. *Lithos*, *42*, 225–241.

France, L., Ildefonse, B., & Koepke, J. (2009). Interactions between magma and the hydrothermal system in the Oman ophiolite and in IODP hole 1256D: fossilisation of a dynamic melt lens at fast spreading ridges. *Geochemistry, Geophysics, Geosystems*, *10*. doi:10.1029/2009GC002652

France, L., Ildefonse, B., & Koepke, J. (2013). Hydrous magmatism triggered by assimilation of hydrothermally altered rocks in fossil oceanic crust (northern Oman ophiolite). *Geochemistry, Geophysics, Geosystems*, *14*(8), 2598–2614.

France, L., Koepke, J., Ildefonse, B., Cichy, S., & Deschamps, F. (2010). Hydrous partial melting in the sheeted dike complex at fast spreading ridges: experimental and natural observations. *Contributions to Mineralogy and Petrology*, *159*. doi: 10.1007/s00410-00010-00502-00416

France, L., Koepke, J., MacLeod, C. J., Ildefonse, B., Godard, M., & Deloule, E. (2014). Contamination of MORB by anatexis of magma chamber roof rocks: Constraints from a geochemical study of experimental melts and associated residues. *Lithos*, *202*, 120–137.

Freund, S., Beier, C., Krumm, S., & Haase, K. M. (2013). Oxygen isotope evidence for the formation of andesitic–dacitic magmas from the fast-spreading Pacific–Antarctic Rise by assimilation–fractional crystallisation. *Chemical Geology*, *347*(0), 271–283.

Freund, S., Haase, K. M., Keith, M., Beier, C., & Garbe-Schönberg, D. (2014). Constraints on the formation of geochemically variable plagiogranite intrusions in the Troodos Ophiolite, Cyprus. *Contributions to Mineralogy and Petrology*, *167*(2), 978.

Furnes, H., & Dilek, Y. (2017), Geochemical characterization and petrogenesis of intermediate to silicic rocks in ophiolites: A global synthesis. *Earth-Science Reviews*, *166*, 1–37.

Gaetani, G. A., Grove, T. L., & Bryan, W. B. (1994). Experimental phase relations of basaltic andesite from hole 839B under hydrous and anhydrous conditions. *Proceedings of the ODP Scientific Results*, *135*, 557–563.

Gai, Y. S., Liu, L., Kang, L., Yang, W. Q., Liao, X. Y., & Wang, Y. W. (2015). The origin and geological significance of plagiogranite in ophiolite belt at North Altyn Tagh. *Acta Petrologica Sinica*, *31*(9), 2549–2565.

Gale, A., Dalton, C.A., Langmuir, C.H., Su, Y., & Schiling, J. (2013), The mean composition of ocean ridge basalts. *Geochemistry, Geophysics, Geosystems*, *14*, doi 10.1029/2012GC004334.

Gannoun, A., Burton, K.W., Parkinson, I.J., Alard, O., Schiano, P., & Thomas, L.E. (2007), The scale and origin of the osmium isotope variations in mid-ocean ridge basalts. *Earth and Planetary Sciences Letters*, *259*, 541–556.

Gao, X., Xiao, P., Guo, L., Dong, Z., & Xi, R. (2011). Opening of an early Paleozoic limited oceanic basin in the northern Altyn area: Constraints from plagiogranites in the Hongliugou-Lapeiquan ophiolitic mélange. *Science China Earth Sciences*, *54*(12), 1871–1879.

Gerlach, D. C., Leeman, W. P., & Lallemant, H. G. A. (1981). Petrology and geochemistry of plagiogranite in the Canyon Mountain Ophiolite, Oregon. *Contributions to Mineralogy and Petrology*, *77*(1), 82–92.

Ghazi, A. M., Hassanipak, A. A., Mahoney, J. J., & Duncan, R. A. (2004). Geochemical characteristics, ^{40}Ar–^{39}Ar ages and original tectonic setting of the Band-e-Zeyarat/Dar Anar ophiolite, Makran accretionary prism, S.E. Iran. *Tectonophysics*, *393*(1–4), 175–196.

Gillis, K. M. (2002). The rootzone of an ancient hydrothermal system exposed in the Troodos ophiolite, Cyprus. *Journal of Geology*, *110*(1), 57–74.

Gillis, K. M. (2008). The roof of an axial magma chamber: A hornfelsic heat exchanger. *Geology*, *36*(4), 299–302.

Gillis, K. M., & Coogan, L. A. (2002), Anatectic migmatites from the roof of an ocean ridge magma chamber. *Journal of Petrology*, *43*(11), 2075–2095.

Gillis, K. M., & Roberts, M. D. (1999), Cracking at the magma–hydrothermal transition: Evidence from the Troodos ophiolite, Cyprus. *Earth and Planetary Sciences Letters*, *169*, 227–244.

Grove, T. L., & Baker, M. B. (1984). Phase equilibrium controls on the tholeiitic versus calc-alkaline differentiation trends. *Journal of Geophysical Research*, *89*(B5), 3253–3274.

Grove, T. L., Kinzler, R. J., & Bryan, W. B. (1992). Fractionation of Midoceanic Ridge Basalt (MORB). In J. P. Morgan, D. K. Blackman, J. M. Sinton (Eds.), *Mantle flow and melt generation at mid-ocean ridges* (Geophysical Monograph Series, Vol. *71*, pp. 281–310). Washington, DC: American Geophysical Union.

Haase, K. M., Freund, S., Beier, C., Koepke, J., Erdmann, M., & Hauff, F. (2016). Constraints on the magmatic evolution of the oceanic crust from plagiogranite intrusions in the Oman ophiolite. *Contributions to Mineralogy and Petrology*, *171*(5).

Haase, K. M., Freund, S., Koepke, J., Hauff, F., & Erdmann, M. (2015). Melts of sediments in the mantle wedge of the Oman ophiolite. *Geology*, *43*(4), 275–278.

Haase, K. M., Stroncik, N. A., Hekinian, R., & Stoffers, P. (2005), Nb-depleted andesites from the Pacific–Antarctic Rise as analogs for early continental crust. *Geology*, *33*(12), 921–924.

Harris, M., Coggon, R. M., Smith-Duque, C. E., Cooper, M. J., Milton, J. A., & Teagle, D. A. H. (2015). Channelling of hydrothermal fluids during the accretion and evolution of the upper oceanic crust: Sr isotope evidence from ODP Hole 1256D. *Earth and Planetary Science Letters*, *416*(0), 56–66.

Henstock, T. J., Woods, A. W., & White, R. S. (1993), The accretion of oceanic-crust by episodic sill intrusion. *Journal of Geophysical Research-Solid Earth*, *98*(B3), 4143–4161.

Holtz, F., Sato, H., Lewis, J., Behrens, H., & Nakada, S. (2005). Experimental petrology of the 1991–1995 Unzen dacite, Japan. Part I: Phase relations, phase composition and pre-eruptive conditions. *Journal of Petrology*, *46*, 319–337.

Hooft, E. E. E., Detrick, R. S., & Kent, G. M. (1997). Seismic structure and indicators of magma budget along the southern East Pacific Rise. *Journal of Geophysical Research—Solid Earth*, *102*(B12), 27319–27340.

Huggett, N. L. (2015). *Making oceanic plagiogranite bodies by thermal migration: evidence from the Agros transect, Troodos ophiolite, Cyprus*. MS thesis, University of Illinois.

Husen, A., Almeev, R. R., & Holtz, F. (2016). The effect of H_2O and pressure on multiple saturation and liquid lines of descent in basalt from the Shatsky Rise. *Journal of Petrology*, *57*(2), 309–344.

Jafri, S., Charan, S., & Govil, P. (1995). Plagiogranite from the Andaman ophiolite belt, Bay of Bengal, India. *Journal of the Geological Society*, *152*(4), 681–687.

Johnson, M. C., Anderson, A. T., & Rutherford, M. J. (1994). Pre-eruptive volatile contents of magmas. In M. R. Carroll, J. R. Holloway (Eds.), *Volatiles in magmas* (Vol. *30*, pp. 281–330). Washington, DC: Mineralogical Society of America,

Juster, T. C., Grove, T. L., & Perfit, M. R. (1989). Experimental constraints on the generation of Fe–Ti basalts, andesites, and rhyodacites at the Galapagos spreading centre, 85°W and 95°W. *Journal of Geophysical Research*, *94*, 9251–9274.

Kay, R. W. (1978). Aleutian magnesian andesites—melts from subducted Pacific Ocean crust. *Journal of Volcanology and Geothermal Research*, *4*(1–2), 117–132.

Kelemen, P. B. (1995). Genesis of high Mg# andesites and the continental crust. *Contributions to Mineralogy and Petrology*, *120*(1), 1–19.

Kelemen, P. B., Hirth, G., Shimizu, N., Spiegelman, M., & Dick, H. J. B. (1997). A review of melt migration processes in the adiabatically upwelling mantle beneath oceanic spreading ridges. *Philosophical Transactions of the Royal Society A—Mathematical Physical and Engineering Sciences*, *355*(1723), 283–318.

Kendrick, M. A., Arculus, R., Burnard, P., & Honda, M. (2013). Quantifying brine assimilation by submarine magmas: Examples from the Galápagos spreading centre and Lau Basin. *Geochimica et Cosmochimica Acta*, *123*, 150–165.

Kinzler, R. J., & Grove, T. L. (1992). Primary magmas of midocean ridge basalts .1. Experiments and methods. *Journal of Geophysical Research-Solid Earth*, *97*(B5), 6885–6906.

Koepke, J., Berndt, J., Feig, S. T., & Holtz, F. (2007). The formation of SiO_2-rich melts within the deep oceanic crust by hydrous partial melting of gabbros. *Contributions to Mineralogy and Petrology*, *153*, 67–84.

Koepke, J., Berndt, J., & Holtz, F. (2001). Experimental generation of SiO_2-rich melts at shallow pressures (200 MPa): Differentiation of water-bearing MORB and partial melting of gabbro. In J. C. Andersen (Ed.), *Processes in dynamic magmatic systems* (pp. 24–27). Exeter: Camborne School of Mines.

Koepke, J., Berndt, J., Horn, I., Fahle, J., & Wolff, P. E. (2014). Partial melting of oceanic gabbro triggered by propagating water-rich fluids: a prime example from the Oman ophiolite. In H. R. Rollinson, M. P. Searle, I. A. Abbasi, A. Al-Lazki, M. H. Al-Kindi (Eds.), *Tectonic evolution of the Oman Mountains* (Special Publication 392, pp. 187–204). London: Geological Society of London.

Koepke, J., Botcharnikov, R., & Natland, J. H. (2018). Crystallization of late-stage MORB under varying water activities and redox conditions: Implications for the formation of highly evolved lavas and oxide gabbro in the ocean crust. *Lithos*. https://doi.org/10.1016/j.lithos.2018.10.001

Koepke, J., Christie, D. M., Dziony, W., Holtz, F., Lattard, D., Maclennan, J., et al. (2008). Petrography of the dike/gabbro transition at IODP Site 1256D (Equatorial Pacific): The evolution of the granoblastic dikes. *Geochemistry, Geophysics, Geosystems*, *9*. doi:10.1029/2008GC001939

Koepke, J., Feig, S. T., Snow, J., & Freise, M. (2004). Petrogenesis of oceanic plagiogranites by partial melting of gabbros: An experimental study. *Contributions to Mineralogy and Petrology*, *146*, 414–432.

Koepke, J., France, L., Müller, T., Faure, F., Goetze, N., Dziony, W., & Ildefonse, B. (2011). Gabbros from IODP Site 1256 (Equatorial Pacific): Insight into axial magma chamber processes at fast-spreading ocean ridges *Geochemistry, Geophysics, Geosystems*, *12*. doi:10.1029/2011GC003655

Lagabrielle, Y., & Cormier, M. H. (1999). Formation of large summit troughs along the East Pacific Rise as collapse calderas: An evolutionary model. *Journal of Geophysical Research-Solid Earth*, *104*(B6), 12971–12988.

Leake, B. E., Woolley, A. R., Birch, W. D., Gilbert, M. C., Grice, J. D., Hawthorne, F. C., et al. (1997). Nomenclature of amphiboles—report of the Subcommittee on Amphiboles of the International Mineralogical Association Commission on New Minerals and Mineral Names. *European Journal of Mineralogy*, *9*, 623–651.

Lester, G., Clark, A., Kyser, T., & Naslund, H. (2013). Experiments on liquid immiscibility in silicate melts with H_2O, P, S, F and Cl: Implications for natural magmas. *Contributions to Mineralogy and Petrology*, *166*(1), 329–349.

Lowell, R. P., & Burnell, D. K. (1991). Mathematical-modeling of conductive heat-transfer from a freezing, convecting magma chamber to a single-pass hydrothermal system—implications for sea-floor black smokers. *Earth and Planetary Science Letters*, *104*(1), 59–69.

Lowell, R. P., & Germanovich, L. N. (1994). On the temporal evolution of high-temperature hydrothermal systems at ocean ridge crests. *Journal of Geophysical Research*, *99*(B1), 565–575.

Maclennan, J., Hulme, T., & Singh, S. C. (2005). Cooling of the lower oceanic crust. *Geology*, *33*, 357–360.

MacLeod, C. J., Dick, H. J. B., Blum, P., & and_the_Expedition_360_Scientists. (2017). Southwest Indian Ridge Lower Crust and Moho. *Proceedings of the Integrated Ocean Drilling Program*, 360. http://dx.doi.org/10.14379/iodp.proc.14360.14103.12017

MacLeod, C. J., Lissenberg, C. J., & Bibby, L. E. (2013). "Moist MORB" axial magmatism in the Oman ophiolite: The evidence against a mid-ocean ridge origin. *Geology*, *41*(4), 459–462.

MacLeod, C. J., & Yaouancq, G. (2000). A fossil melt lens in the Oman ophiolite: Implications for magma chamber processes at fast spreading ridges. *Earth and Planetary Sciences Letters*, *176*, 357–373.

Malpas, J. (1979). Two contrasting trondhjemite associations from transported ophiolites in Western Newfoundland: initial report. *Developments in Petrology* (Vol. *6*, pp. 465–487). Amsterdam: Elsevier.

Marjanovic, M., Carbotte, S. M., Carton, H., Nedimovic, M. R., Mutter, J. C., & Canales, J. P. (2014). A multi-sill magma plumbing system beneath the axis of the East Pacific Rise. *Nature Geoscience*, *7*(11), 825–829.

Martin, H. (1999). Adakitic magmas: modern analogues of Archaean granitoids. *Lithos*, *46*(3), 411–429.

Michael, P. J., & Chase, R. L. (1987). The Influence of primary magma composition, H_2O and pressure on Midocean Ridge basalt differentiation. *Contributions to Mineralogy and Petrology*, *96*, 245–263.

Michael, P. J., & Cornell, W. C. (1998). Influence of spreading rate and magma supply on crystallization and assimilation beneath mid-ocean ridges: Evidence from chlorine and major element chemistry of mid-ocean ridge basalts. *Journal of Geophysical Research-Solid Earth*, *103*(B8), 18325–18356.

Moores, E., & Vine, F. J. (1971). The Troodos Massif, Cyprus and other ophiolites as oceanic crust: evaluation and implications. *Philosophical Transactions of the Royal Society A—Mathematical Physical and Engineering Sciences*, *268*(1192), 443–467.

Mueller, T., Koepke, J., Garbe-Schonberg, C. D., Dietrich, M., Bauer, U., & Wolff, P. B. (2017). Anatomy of a frozen axial melt lens from a fast-spreading paleo-ridge (Wadi Gideah, Oman ophiolite). *Lithos*, *272*, 31–45.

Mutter, J., Carton, H., Carbotte, S., Canales, J., Nedimovic, M., Newman, K., et al. (2008). Searching for changes in AMC characteristics on the EPR using comparisons of reflection images obtained in 1985 and 2008. Paper presented at the AGU Fall Meeting.

Natland, J. H., & Dick, H. J. B. (1996). Melt migration through high-level gabbroic cumulates of the East Pacific Rise at Hess Deep: the origin of magma lenses and the deep crustal structure of fast-spreading ridges. *Proceedings of the ODP Scientific Results*, *147*, 21–58.

Natland, J. H., & Dick, H. J. B. (2009). Paired melt lenses at the East Pacific Rise and the pattern of melt flow through the gabbroic layer at a fast-spreading ridge. *Lithos*, *112*, 73–86.

Nedimovic, M. R., Carbotte, S. M., Harding, A. J., Detrick, R. S., Canales, J. P., Diebold, J. B., et al. (2005). Frozen magma lenses below the oceanic crust. *Nature*, *436*, 1149–1152.

Niu, Y., Gilmore, T., Mackie, S., Greig, A., & Bach, W. (2002). Mineral chemistry, whole-rock compositions, and petrogenesis of Leg 176 gabbros: data and discussion. *Proceedings of the ODP Scientific Results*, *176*, 1–60.

Pallister, J. S., & Knight, R. J. (1981). Rare earth element geochemistry of the Samail Ophiolite near Ibra, Oman. *Journal of Geophysical Research*, *86*, 2673–2697.

Pedersen, R. B., & Malpas, J. (1984). The origin of oceanic plagiogranites from the Karmoy ophiolite, western Norway. *Contributions to Mineralogy and Petrology*, *88*(1–2), 36–52.

Perfit, M. R., & Chadwick, W. W. (1998). Magmatism at mid-ocean ridges: Constraints from volcanological and geochemical investigations. In *Faulting and magmatism at mid-ocean ridges* (Geophysical Monograph Series, Vol. *106*, pp. 59–115). Washington, DC: American Geophysical Union.

Perfit, M. R., & Fornari, D. J. (1983). Geochemical studies of abyssal lavas recovered by DSRV Alvin from Eastern Galapagos Rift, Inca Transform, and Ecuador Rift: 2. Phase Chemistry and Crystallization History. *Journal of Geophysical Research*, *88*, B12, 10530–10550.

Perfit, M. R., Fornari, D. J., Smith, M. C., Bender, J. F., Langmuir, C. H., & Haymon, R. M. (1994). Small-scale spatial and temporal variations in mid-ocean ridge crest magmatic processes. *Geology*, *22*(4), 375–379.

Philpotts, A. R. (1979), Silicate liquid immiscibility in tholeiitic basalts. *Journal of Petrology*, *20*(1), 99–118.

Phipps Morgan, J. P., & Chen, Y. J. (1993), The genesis of oceanic-crust—magma injection, hydrothermal circulation, and crustal rlow. *Journal of Geophysical Research-Solid Earth*, *98*(B4), 6283–6297.

Pichavant, M., & Macdonald, R. (2007). Crystallization of primitive basaltic magmas at crustal pressures and genesis of the calc-alkaline igneous suite: experimental evidence from St Vincent, Lesser Antilles arc. *Contributions to Mineralogy and Petrology*, *154*, 553–558.

Pollock, M. A., Klein, E. M., Karson, J. A., & Coleman, D. S. (2009). Compositions of dikes and lavas from the Pito Deep Rift: Implications for crustal accretion at superfast spreading centers. *Journal of Geophysical Research*, *114*(B3), B03207.

Quick, J. E., & Denlinger, R. P. (1993). Ductile deformation and the origin of layered gabbro in ophiolites. *Journal of Geophysical Research-Solid Earth*, *98*, 14015–14027.

Rannou, E., Caroff, M., & Cordier, C. (2006). A geochemical approach to model periodically replenished magma chambers: Does oscillatory supply account for the magmatic evolution of EPR 17–19°S? *Geochimica et Cosmochimica Acta*, *70*(18), 4783–4796.

Reynolds, J. R., Langmuir, C. H., Bender, J. F., Kastens, K. A., & Ryan, W. B. F. (1992). Spatial and temporal variability in the geochemistry of basalts from the East Pacific Rise. *Nature*, *359*(6395), 493–499.

Rollinson, H. (2008). Ophiolitic trondhjemites: a possible analogue for Hadean felsic "crust." *Terra Nova*, *20*(5), 364–369.

Rollinson, H. (2009). New models for the genesis of plagiogranites in the Oman ophiolite. *Lithos*, *112*(3–4), 603–614.

Rollinson, H. (2014). Plagiogranites from the mantle section of the Oman Ophiolite: models for early crustal evolution. In H. R. Rollinson, M. P. Searle, I. A. Abbasi, A. Al-Lazki, M. H. Al-Kindi (Eds.), *Tectonic evolution of the Oman Mountains* (Special Publication 392, pp. 247–261). London: Geological Society of London.

Rubin, K. H., van der Zander, I., Smith, M. C., & Bergmanis, E. C. (2005). Minimum speed limit for ocean ridge magmatism from Pb-210–Ra-226–Th-230 disequilibria. *Nature*, *437*(7058), 534–538.

Saunders, A. D., Tarney, J., Stern, C. R., & Dalziel, I. W. (1979). Geochemistry of Mesozoic marginal basin floor igneous rocks from southern Chile. *Geological Society of America Bulletin*, *90*(3), 237–258.

Sauerzapf, U., Lattard, D., Burchard, M., & Engelmann, R. (2008). New experimental data and a simple version of the titanomagnetite–ilmenite thermo-oxybarometer for high temperature and reduced to moderately oxidised conditions. *Journal of Petrology*, *49*. doi:10.1093/petrology/egn1021

Sepidbar, F., & Mirnejad, H. (2016). Mineralogy, geochemistry and geotectonic of plagiogranites from Shahre-Babak ophiolite, Zagros zone, Iran. *Journal of Earth Science*, *27*(3), 507–518.

Silantyev, S. A., Koepke, J., Ariskin, A. A., Anosova, M. O., Krasnova, E. A., Dubinina, E. O., & Suhr, G. (2014). Geochemical nature and age of the plagiogranite-gabbronorite association of the oceanic core complex of the Mid-Atlantic ridge at 5°10′S. *Petrology*, *22*(2), 109–127.

Singh, A. K., Chung, S.-L., Bikramaditya, R. K., & Lee, H. Y. (2016). New U–Pb zircon ages of plagiogranites from the Nagaland–Manipur Ophiolites, Indo-Myanmar Orogenic Belt, NE India. *Journal of the Geological Society*, *174*, 170–179.

Singh, S. C., Kent, G. M., Collier, J. S., Harding, A. J., & Orcutt, J. A. (1998). Melt to mush variations in crustal magma properties along the ridge crest at the southern East Pacific Rise. *Nature*, *394*(6696), 874–878.

Sinton, J., Bergmanis, E., Rubin, K., Batiza, R., Gregg, T. K. P., Grönvold, K., et al. (2002). Volcanic eruptions on mid-ocean ridges: New evidence from the superfast spreading East Pacific Rise, 17°–19°S. *Journal of Geophysical Research*, *107*(B6), 2115.

Sinton, J. M., & Detrick, R. S. (1992). Midocean ridge magma chambers. *Journal of Geophysical Research-Solid Earth*, *97*(B1), 197–216.

Spulber, S. D., & Rutherford, M. J. (1983). The origin of rhyolite and plagiogranite in oceanic crust: An experimental study. *Journal of Petrology*, *24*(1), 1–25.

Stern, C. R. (1979). Open and closed system igneous fractionation within two Chilean ophiolites and the tectonic implications. *Contributions to Mineralogy and Petrology*, *68*, 243–259.

Teagle, D. A. H., Alt, J. C., Umino, S., Miyashita, S., Banerjee, N. R., Wilson, D. S., & and the Expedition 309/312 Scientists (2006). *Proceedings Integrated Ocean Drilling Program, 309/312*. Washington, DC. doi:10.2204/iodp.proc.309312.302006

Teagle, D. A. H., Ildefonse, B., Blum, P., & and_the_Expedition_335_Scientists. (2012). *Proceedings of the Integrated Ocean Drilling Program*, *335*. Washington, DC. doi:10.2204/iodp.proc.2335.2102.2012

Theissen-Krah, S., Rupke, L. H., & Hasenclever, J. (2016). Modes of crustal accretion and their implications for hydrothermal circulation. *Geophysical Research Letters*, *43*(3), 1124–1131.

Toplis, M. J., & Carroll, M. R. (1995). An experimental study of the influence of oxygen fugacity on Fe–Ti oxide stability, phase relations, and mineral–melt equilibra in ferro-basaltic systems. *Journal of Petrology*, *36*(5), 1137–1170.

Toplis, M. J., & Carroll, M. R. (1996), Differentiation of ferro-basaltic magmas under conditions open and closed to oxygen: Implications for the Skaergaard intrusion and other natural systems. *Journal of Petrology*, *37*(4), 837–858.

Tormey, D. R., Grove, T. L., & Bryan, W. B. (1987). Experimental petrology of normal MORB near the Kane Fracture Zone: 22°–25° N, mid-Atlantic ridge. *Contributions to Mineralogy and Petrology*, *96*(2), 121–139.

Vera, E. E., Mutter, J. C., Buhl, P., Orcutt, J. A., Harding, A. J., Kappus, M. E., et al. (1990). The structure of 0-my to 0.2-my old oceanic crust at 9° N on the East Pacific Rise from expanded spread profiles. *Journal of Geophysical Research—Solid Earth and Planets*, *95*(B10), 15529–15556.

Villiger, S., Ulmer, P., & Muntener, O. (2007). Equilibrium and fractional crystallization experiments at 0.7 GPa; the effect of pressure on phase relations and liquid compositions of tholeiitic magmas. *Journal of Petrology*, *48*(1), 159–184.

Villiger, S., Ulmer, P., Muntener, O., & Thompson, A. B. (2004). The liquid line of descent of anhydrous, mantle-derived, tholeiitic liquids by fractional and equilibrium crystallization—an experimental study at 1 center dot 0 GPa. *Journal of Petrology*, *45*(12), 2369–2388.

Wanless, V. D., Perfit, M. R., Ridley, W. I., & Klein, E. (2010). Dacite petrogenesis on mid-ocean ridges: Evidence for oceanic crustal melting and assimilation. *Journal of Petrology*, *51*(12), 2377–2410.

Wanless, V. D., Perfit, M. R., Ridley, W. I., Wallace, P. J., Grimes, C. B., & Klein, E. M. (2011). Volatile abundances and oxygen isotopes in basaltic to dacitic lavas on mid-ocean ridges: The role of assimilation at spreading centers. *Chemical Geology*, *287*(1–2), 54–65.

Wanless, V. D., & Shaw, A. M. (2012). Lower crustal crystallization and melt evolution at mid-ocean ridges. *Nature Geoscience*, *5*(9), 651–655.

Webster, J. D., & Holloway, J. R. (1990). Partitioning of F and Cl between magmatic hydrothermal fluids and highly evolved granitic magmas. *Geological Society of America Special Papers*, *246*, 21–34.

Wilson, D. S., Teagle, D. A. H., Alt, J. C., Banerjee, N. R., Umino, S., Miyashita, S., et al. (2006). Drilling to gabbro in intact ocean crust. *Science*, *312*, 1016–1020.

Wolff, P. E., Koepke, J., & Feig, S. T. (2013). The reaction mechanism of fluid-induced partial melting of gabbro in the oceanic crust. *European Journal of Mineralogy*, *25*(3), 279–298.

Wood, B. J., & Turner, S. P. (2009). Origin of primitive high-Mg andesite: Constraints from natural examples and experiments. *Earth and Planetary Science Letters*, *283*(1–4), 59–66.

Xu, Y., Liu, C.-Z., Chen, Y., Guo, S., Wang, J.-G., & Sein, K. (2017). Petrogenesis and tectonic implications of gabbro and plagiogranite intrusions in mantle peridotites of the Myitkyina ophiolite, Myanmar. *Lithos*, *284–285*, 180–193.

Yin, Z., Yuan, Y., Lü, B., CAI, Z., Zheng, H., Huang, Q., et al. (2015). Zircon U–Pb geochronology and Hf isotopic constraints on petrogenesis of plagiogranite from the Cuomuqu Ophiolite, Bangong Lake Area, North Tibet. *Acta Geologica Sinica*, *89*(2), 418–440.

Yogodzinski, G., Lees, J., Churikova, T., Dorendorf, F., Werner, G., & Volynets, O. (2001). Geochemical evidence for the melting of subducting oceanic lithosphere at plate edges. *Nature*, *409*(6819), 500–504.

Zhang, C., Koepke, J., Albrecht, M., Horn, I., & Holtz, F. (2017). Apatite in the dike–gabbro transition zone of mid-ocean ridge: Evidence for brine assimilation by axial melt lens. *American Mineralogist*, *102*(3), 558–570.

Zhang, C., Koepke, J., France, L., & Godard, M. (2017). Felsic plutonic rocks from IODP Hole 1256D, Eastern Pacific: Implications for the nature of the axial melt lens at fast-spreading mid-ocean ridges. *Journal of Petrology*, *58*, 1535–1565.

Zhang, C., Koepke, J., Kirchner, C., Goetze, N., & Behrens, H. (2014). Rapid hydrothermal cooling above the axial melt lens at fast-spreading mid-ocean ridge. *Scientific Reports*, *4*, 6342.

Zhang, C., Wang, L. X., Marks, M. A. W., France, L., & Koepke, J. (2017). Volatiles (CO_2, S, F, Cl, Br) in the dike–gabbro transition zone at IODP Hole 1256D: Magmatic imprint versus hydrothermal influence at fast-spreading mid-ocean ridge. *Chemical Geology*, *459*, 43–60.

INDEX

Page numbers in *italics* refers to figures and those in bold refer to tables

Dynamic Magma Evolution, Geophysical Monograph 254, First Edition. Edited by Francesco Vetere.